Methods of Analysis of Food Components and Additives

Second Edition

Chemical and Functional Properties of Food Components Series

SERIES EDITOR

Zdzisław E. Sikorski

Methods of Analysis of Food Components and Additives, Second Edition
Edited by Semih Ötleş

Food Flavors: Chemical, Sensory and Technological Properties
Edited by Henryk Jelen

Environmental Effects on Seafood Availability, Safety, and Quality
Edited by E. Grażyna Daczkowska-Kozon and Bonnie Sun Pan

Chemical and Biological Properties of Food Allergens
Edited by Lucjan Jędrychowski and Harry J. Wichers

Food Colorants: Chemical and Functional Properties
Edited by Carmen Socaciu

Mineral Components in Foods
Edited by Piotr Szefer and Jerome O. Nriagu

Chemical and Functional Properties of Food Components, Third Edition
Edited by Zdzisław E. Sikorski

Carcinogenic and Anticarcinogenic Food Components
Edited by Wanda Baer-Dubowska, Agnieszka Bartoszek and Danuta Malejka-Giganti

Toxins in Food
Edited by Waldemar M. Dąbrowski and Zdzisław E. Sikorski

Chemical and Functional Properties of Food Saccharides
Edited by Piotr Tomasik

Chemical and Functional Properties of Food Lipids
Edited by Zdzisław E. Sikorski and Anna Kolakowska

Chemical and Functional Properties of Food Proteins
Edited by Zdzisław E. Sikorski

Methods of Analysis of Food Components and Additives

Second Edition

EDITED BY

Semih Ötleş

CRC Press is an imprint of the
Taylor & Francis Group, an **informa** business

CRC Press
Taylor & Francis Group
6000 Broken Sound Parkway NW, Suite 300
Boca Raton, FL 33487-2742

First issued in paperback 2016

Version Date: 20110801

ISBN 13: 978-1-138-19914-9 (pbk)
ISBN 13: 978-1-4398-1552-6 (hbk)

Library of Congress Cataloging-in-Publication Data

Methods of analysis of food components and additives / editor, Semih Otles. -- 2nd ed.
    p. cm. -- (Chemical & functional properties of food components)
    Includes bibliographical references and index.
    ISBN 978-1-4398-1552-6 (alk. paper)
    1. Food additives--Analysis--Methodology. 2. Food additives--Analysis--Standards. 3. Food--Analysis--Methodology. I. Ötles, Semih.

TP455.M48 2012
664'.06--dc23   2011027105

Visit the Taylor & Francis Web site at
http://www.taylorandfrancis.com

and the CRC Press Web site at
http://www.crcpress.com

Contents

Preface

The first edition of this book *Methods of Analysis of Food Components and Additives* was published by CRC Press/Taylor & Francis Group in 2005. The book quickly became a highly successful and popular educational and scientific tool among academicians, students, and staff working in private companies and governmental institutions worldwide. The readers valued this book as good teaching material and a useful reference. Thus, it is of no surprise that the book enjoyed a high ranking on the CRC Press' book list. It has turned out that, as methodological and instrumental progress occurs, new applications materialize, which, in turn, provide a driving force for further methodological and instrumental developments. Advances in instrumentation and applied instrumental analysis methods have allowed scientists concerned with food and beverage quality, labeling, compliance, and safety to meet ever-increasing analytical demands.

The second edition of the book is an extended and updated version of the original. The new edition follows the same format as the first edition and covers topics such as selection of techniques, statistical assessments, rapid microbiological techniques, and demonstration of the applications of chemical, physical, microbiological, sensorial, and instrumental novel analysis to food components and additives such as proteins, peptides and amino acids, lipids, trace elements, vitamins, carotenoids, chlorophylls, polyphenols, drinking water, food allergens, genetically modified components, pesticide residues, pollutants, chemical preservatives, and radioactive components in foods. New additions include three chapters on analytical quality assurance, carbohydrates, and natural toxins. The second edition includes important developments in analytical quality assurance and the analysis of carbohydrates and natural toxins in foods. The chapter on the analysis of natural toxins is an especially welcome addition, a welcomed addition as it covers various methods of analysis of natural toxins, which is very important for food quality control related to toxicology. Information provided in the chapter "Analysis of Carbohydrates in Foods" would be equally valuable for scientists, students, and staff working in the field of food quality control. The text in the remaining chapters has been updated and is supported by numerous examples.

As the editor of this book, I would like to express my sincere thanks to all the authors for their excellent contributions to this book.

Semih Ötleş

Acknowledgments

I would like to thank the contributors for all the hard work they have put in and the valuable contributions they have made to the various chapters of the book. I want to thank the Taylor & Francis Group/CRC Press team for their help in production of this book and to express my sincere gratitude to Stephen M. Zollo, Taylor & Francis Group, for his help in publishing the book. Finally, special thanks to my wife, Sema Ötleş, for her patience during the preparation and publication steps of this book.

Semih Ötleş

Editor

A native of Izmir, Turkey, Professor Ötleş obtained his B.Sc. degree from the Department of Food Engineering (Ege University) in 1980. During his assistantship at Ege University, in 1985, he received an M.S. in Food Chemistry, and in 1989, after completing his research thesis on the instrumental analysis and chemistry of vitamins in foods, he received a Ph.D. in Food Chemistry from Ege University. In 1991–1992, he completed postdoctoral training including OECD—Postdoctoral Fellowship—in the Research Center Melle at Gent University, Belgium. Later, he joined the Department of Food Engineering at Ege University as a scientist of Food Chemistry and was promoted as Associate Professor in 1993 and as Professor in 2000. From 1996–1998, he was Deputy Director at the Ege Vocational School of Higher Studies. He was also Vice Dean at Engineering Faculty from 2003–2009 and head of the Department of Nutrition and Dietetics from 2008–2011.

The research activities of Professor Ötleş have been focused on instrumental analysis of foods. Ötleş began a series of projects on separation and analysis using HPLC techniques, first for analysis of vitamins in foods, then proteins, carbohydrates, and, most recently, carotenoids. Other activities span the fields of GC, GC/MS analysis, soy chemistry, aromatics, medical and functional foods, and nutraceutical chemistry. This includes multiresidue analysis of various foods, n-3 fatty acids in fish oils, and medical and functional foods.

Professor Ötleş is the author or coauthor of more than 250 publications (technical papers, book chapters, and books) and has presented some seminars in these areas. He is a member of several scientific societies, associations, and organizations, including the Asian Pacific Organization for Cancer Prevention and the International Society of Food Physicists. He is a member of the steering committee of APOCP local scientific bureau and a Turkish representative of ISFP and has organized international congresses on diet/cancer and food physics.

Professor Ötleş is a member of the Editorial Advisory Board of *Asian Pacific Journal of Cancer Prevention* (APJCP), *Food Science and Technology Abstracts* (FSTA) of International Food Information Service (IFIS), *Current Topics in Nutraceutical Research, Electronic Journal of Environmental, Agricultural and Food Chemistry* (EJEAFChe), *Newsline* (IUFoST, Corr.), *Journal of Oil, Soap, Cosmetics, Tr. World Food, Tr. Food Science and Technology, Pakistan Journal of Nutrition, Journal of Food Technology, Academic Food, Australian Journal of Science and Technology, Journal of Oil Research Institute, Journal of Food Safety, Key Food Magazine, Advances in Food Sciences* and is a referee/reviewer for *AOAC International, Journal of Experimental Marine Biology and Ecology, Journal of Medical Foods, die Nahrung, Journal of Alternative and Complementary Medicine, the Analyst, Journal of Scientific and Industrial Research, the Science of Total Environment, Current Microbiology, Journal of Essential Oil Bearing Plants*, and *Journal of Agricultural and Food Chemistry*.

Contributors

Marek Biziuk
Gdansk University of Technology
Gdansk, Poland

Richard G. Brereton
University of Bristol
Bristol, United Kingdom

Stephen G. Capar
U.S. Food and Drug Administration
College Park, Maryland

Alejandro Cifuentes
Institute of Food Science Research
    (CIAL-CSIC)
Madrid, Spain

Maria Dolores del Castillo
Institute of Food Science Research
    (CIAL-CSIC)
Madrid, Spain

Francisco Diez-Gonzalez
University of Minnesota
St. Paul, Minnesota

Douglas G. Hayward
U.S. Food and Drug Administration
College Park, Maryland

Miguel Herrero
Institute of Food Science Research
    (CIAL-CSIC)
Madrid, Spain

Elena Ibáñez
Institute of Food Science Research
    (CIAL-CSIC)
Madrid, Spain

Yildiz Karaibrahimoglu
U.S. Department of Agriculture
Wyndmoor, Pennsylvania

Edward Kolakowski
Agriculture University of Szczecin
Szczecin, Poland

Jae Hwan Lee
Sungkyunkwan University
Suwon, Korea

Keith A. Lampel
U.S. Food and Drug Administration
Laurel, Maryland

Steven J. Lehotay
U.S. Department of Agriculture
Wyndmoor, Pennsylvania

Dan Levy
U.S. Food and Drug Administration
Laurel, Maryland

Kannapon Lopetcharat
Unilever Corporation
Edgewater, New Jersey

Katerina Mastovska
Covance
Greenfield, Indiana

Mina McDaniel
Oregon State University
Corvallis, Oregon

Malgorazata Michalska
Gdansk University of Technology
Gdansk, Poland

Robert A. Moreau
U.S. Department of Agriculture
Wyndmoor, Pennsylvania

Marian Naczk
St. Francis Xavier University
Nova Scotia, Canada

Palmer A. Orlandi
U.S. Food and Drug Administration
Laurel, Maryland

Semih Ötleş
Ege University
Izmir, Turkey

Beata Plutowska
European Commission, IRMM
Geel, Belgium

Tõnu Püssa
Estonian University of Life Sciences
Tartu, Estonia

Marta Dabrio Ramos
European Commission, IRMM
Geel, Belgium

Richard B. Raybourne
U.S. Food and Drug Administration
Laurel, Maryland

Adriaan Ruiter
IRAS, Utrecht University
Utrecht, the Netherlands

Peter Scherpenisse
IRAS, Utrecht University
Utrecht, the Netherlands

Steven J. Schwartz
Ohio State University
Columbus, Ohio

Fereidoon Shahidi
Memorial University of Newfoundland
Newfoundland, Canada

Andras S. Szabo
Budapest Corvinus University
Budapest, Hungary

Piotr Szefer
Medical University of Gdansk
Gdansk, Poland

Sandor Tarjan
National Food Control Institute
Budapest, Hungary

Mary W. Trucksess
U.S. Food and Drug Administration
College Park, Maryland

Michael H. Tunick
U.S. Department of Agriculture
Wyndmoor, Pennsylvania

Carmen D. Westphal
U.S. Food and Drug Administration
Laurel, Maryland

Kristina M. Williams
U.S. Food and Drug Administration
Laurel, Maryland

Jill K. Winkler-Moser
National Center for Agricultural
    Utilization Research
Peoria, Illinois

1 Choosing Techniques for Analysis of Food

Michael H. Tunick

CONTENTS

1.1 CLASSIFICATION OF TECHNIQUES

Any technique selected for food analysis depends on what the scientist is looking for, and there is a host of food properties from which to choose. A lengthy list of the physical properties of food was assembled by Jowitt.[1] Using that list as a starting point, a classification of food properties was created by Rahman[2] and revised by Rahman and McCarthy.[3] In reorganized form, it is as follows:

- Mechanical properties (mass–volume–area related, morphometric and structural, rheological, and surface)
- Other physical and physicochemical properties (electromagnetic, mass transfer, physicochemical constants, thermal, and thermodynamic)
- Kinetic properties (quality kinetic constants and microbial growth–decline–death kinetic constants)
- Sensory properties (color and appearance, odor, sound, tactile, taste, and texture)
- Health properties (functional properties, medical properties, nutritional composition, toxicity, and unbalanced intake)

Health properties are beyond the scope of this book, and sensory analysis will be addressed in Chapter 12. Compositional analyses, which were not included among physical properties in the classifications above, are covered in many of the remaining chapters.

1.2 CONSIDERATIONS

Scientists decide which characteristics are important for a particular food and which techniques should be used to examine them. The principles for method selection include the following criteria:

- Ability to conduct analysis (sample size, reagents, instruments, cost, and possible destruction of sample)
- Fundamental characteristics (precision, accuracy, sensitivity, specificity, detection limit, and reproducibility)
- Personnel concerns (safety, simplicity, and speed)
- Method status (official or in-house method)

Several techniques may fit these criteria. The remainder of this chapter will deal with different properties of food and the choices commonly available for analysis.

1.3 MECHANICAL PROPERTIES

The mechanical properties of a food deal with changes in its structure and behavior when force is applied. The Rahman and McCarthy classification[3] separated morphometric and structural properties, but these will be combined here since the same techniques are used for analyzing both.

1.3.1 ACOUSTIC

Acoustic measurements are used to evaluate crispness, which has been claimed to be the most important food attribute affecting acceptability by consumers.[4] According to a review by Duizer,[5] the two acoustical measurements used for examining crispness and crunchiness are measurement of the perception of sounds by a trained panel and recording of the sounds produced during application of a force to a food. Perception is evaluated by playing prerecorded biting and chewing sounds to panelists or by having the panelists masticate the food themselves. Recording of sounds produces an amplitude–time plot in which the force required to displace molecules is measured with a microphone. Combinations of number, height, and duration of peaks, sound pressure and energy, and other factors (Table 1.1) are used to formulate regression equations that characterize crispness and crunchiness.[5] When used in combination with fracture, compression, and puncture tests, acoustic measurements can accurately describe food sounds.

1.3.2 MASS–VOLUME–AREA RELATED

This class of properties includes density, shrinkage, and porosity and is important for material handling and process design and characterization. Shrinkage is the volume change upon thermal or other treatment, and porosity is the void volume per total volume. True density is defined as mass per unit volume, and other types of density used in food analysis are apparent density (includes all pores in material), particle density (includes closed pores but not externally connected ones), material or substance density (when no closed pores exist), and bulk density (when packed or stacked, as in a shipment). Rahman reviewed the techniques for mass–volume–area determinations,[6] which are summarized in Table 1.2.

1.3.3 MORPHOLOGY AND STRUCTURE

In food analysis, morphology is the study of the size and shape of particles and their relation with material properties. Morphometric and structural characteristics affect the mechanical and sensory properties of food, which are responsible for overall quality. Microscopy, x-ray microtomography, magnetic resonance imaging, and

---

TABLE 1.1
Acoustic Measurement Parameters

Parameter	Definition
Amplitude	Height of sound wave
Mean height × number of sound bursts	Mean height of peaks multiplied by number of peaks in amplitude–time plot
Acoustic intensity	Energy per second over 1-m² area
Sound pressure	Force per unit area of sound wave proportional to square root of acoustic intensity
Equivalent sound level	Average of energy of sound produced
Fast Fourier analysis	Determines predominant sound frequencies

Source: Duizer, L., Trends Food Sci. Technol., 12, 17–24, 2001.

TABLE 1.2

Mass–Volume–Area Measurement Techniques

Technique	Description	Application
	Density and Shrinkage	
Bulk density	Measures mass and volume of container	Bulk density of stacked or packed materials
Buoyant force	Weighs in air and in water	Apparent density of irregularly shaped coated solids
Gas pycnometer	Volume of gas displaced by sample	Apparent shrinkage, apparent density, and particle density of irregularly shaped solids, material density of ground samples
Geometric dimension	Measures height, width, depth	Apparent shrinkage and apparent density of regularly shaped solids
Liquid displacement	Volume of nonaqueous liquid displaced by sample	Apparent shrinkage and apparent material and particle density of irregularly shaped solids
Mercury porosimetry	Pressure at which mercury intrudes into pores	Pore volume and distribution
Solid displacement	Volume of solid (sand and glass beads) displaced by sample	Apparent shrinkage and apparent density of irregularly shaped solids
	Porosity	
Direct	Compares bulk and compacted volumes	Porosity of soft materials
Optical microscopy	Views random section under microscope	Porosity of solids
X-ray microtomography	Attenuation of x-rays provides three-dimensional image	Visualize pores in solids, estimates void volume

Source: Rahman, M.S., Mass-volume-area-related properties of foods, Pages 1–39 in *Engineering Properties of Foods*, 3rd edn., M.A. Rao, S.S.H. Rizvi, A.K. Datta, eds, CRC Press, Boca Raton, FL, 2005.

ultrasound measurements are available for examining structure and morphology. Falcone et al.[7] compared these techniques, which are shown in Table 1.3.

1.3.4 RHEOLOGY

Food rheology may be defined as the study of flow and deformation of ingredients, intermediate products, and final products of food processing. The three classes of rheological instruments are empirical, imitative, and fundamental. The first two correlate well with sensory methods but are arbitrary and cannot be converted to physical equivalents. Fundamental tests give well-defined results that can be treated mathematically but require continuous, homogeneous, and isotropic

TABLE 1.3
Morphology and Structure Analysis Techniques

Technique	Description	Application
Light microscopy	Visible light illuminates stained sample	View components such as fat droplets and protein
Scanning electron microscopy	Electrons illuminate sample surface	Visualize surface
Transmission electron microscopy	Electrons pass through thin sample layer	Two-dimensional picture of internal structure
Confocal laser scanning microscopy	Laser spot induces fluorescence	Three-dimensional topographic map
Atomic force microscopy	Deflection of cantilevered probe by repulsive forces on sample surface	Surface roughness
X-ray microtomography	Attenuation of x-rays provides three-dimensional image	Three-dimensional picture of internal structure
Magnetic resonance imaging	Detects proton spin alignment	Spatial distribution of water, salt, fat, etc.
Ultrasound	Attenuation of ultrasonic pulses	Two- and three-dimensional images of structure and physical state

Source: Falcone, P.M., et al., *Adv. Food Nutr. Res.*, 51, 205–263, 2006.

(same physical properties in all directions) samples and do not correlate well with sensory methods.

1.3.4.1 Empirical and Imitative Tests

There are dozens of empirical instruments for measuring rheological characteristics of food, and these are required for evaluating samples that are not homogeneous or isotropic. Imitative tests include texture profile analysis (TPA), which mimics chewing, and instruments that simulate handling of bread dough. The quantities obtained in TPA include hardness (maximum force during first compression), springiness (height specimen recovers between end of first compression and start of second), cohesiveness (ratio of area of second force–distance peak to first peak), and chewiness (product of these three parameters).

1.3.4.2 Fundamental Tests

The two major categories of fundamental tests for liquid and semisolid samples are rotational type and tube type; for solid samples, compression, dynamic mechanical analysis, tension, and torsion instruments are used. These types have been discussed by Steffe[8] and Bourne[9] and are shown in Table 1.4. Dynamic mechanical analysis is often performed using oscillatory shear analyses in the linear viscoelastic range. All foods exhibit viscoelastic properties, and the equation governing this behavior is

$$G^* = G' + iG'' \tag{1.1}$$

TABLE 1.4

Fundamental Rheological Measurement Techniques

Technique	Description	Comments
Solid Samples		
Dynamic mechanical analysis	Response of material to oscillating strain or stress	Can be applied to liquids, measures viscous and elastic properties
Tension	Sample stretches and fractures in plane perpendicular to plane of applied tension	Measures adhesion to surface
Torsion	Response to twisting	Measures shear stress of highly deformable foods
Uniaxial compression, nondestructive	Sample compressed in one direction and unrestrained in other directions without damage	Force required to squeeze sample by hand
Uniaxial compression, destructive	Sample compressed with enough force to break sample	Force required to bite down on sample
Viscometry of Liquid Samples		
Capillary	Time required to pass through capillary tube	Simple and rapid
Tube	Pressure drop through horizontal pipe	Needs constant temperature and large quantity of sample
Cone and plate	Torque caused by drag of fluid on small-angle cone that touches flat surface	Rotational method, can measure shear stress and shear rate of liquids
Parallel plate	Torque caused by drag of fluid between flat surfaces	Rotational method, can measure shear stress and shear rate of liquids and suspensions

Source: Steffe, J.F., *Rheological Methods in Food Process Engineering*, Freeman Press, E. Lansing, MI, 1996; Bourne, M., *Food Texture and Viscosity: Concept and Measurement*, Academic Press, London, 2002.

where $G*$ is complex modulus (measure of energy required to deform a sample), G' is elastic modulus (measure of energy stored per deformation cycle), G'' is viscous modulus (measure of energy lost as heat per deformation cycle), and $i = \sqrt{-1}$.

1.3.5   SURFACE PROPERTIES

The surface properties of primary concern in food are surface tension, interfacial tension, and stickiness or adhesion.

1.3.5.1   Surface Tension and Interfacial Tension

Surface tension is the energy per unit area of a liquid surrounded by a gas (such as oil and air), and interfacial tension is the energy per unit area between two immiscible liquids (such as oil and water). A drop placed on a solid forms a contact angle with the surface, and the angle is measured with a goniometer. Contact angle calculations are used to determine the degree to which a solid will wet. Surfactants are added to reduce surface tension, which is important when rate of wetting and dissolution are a consideration. Surface and interfacial tension are measured by techniques as described by Lee et al.[10] (Table 1.5).

1.3.5.2 Stickiness

Attraction between particles (cohesion) causes lumping and caking, and particles adhering to another surface (adhesion) affect processing and handling. The temperature at which particles exhibit the highest tendency to stick to each other or other surfaces is the sticky-point temperature. The collapse temperature of freeze-dried material is reached when its stickiness can no longer support its structure against gravity. Adhikari and Bhandari[11] and Boonyai et al.[12] reviewed methods for evaluating stickiness, sticky-point temperature, and collapse temperature, which are summarized in Table 1.6. Stickiness and adhesiveness of nonparticulate food is often measured by tensile tests.

1.4 PHYSICAL AND PHYSICOCHEMICAL PROPERTIES

1.4.1 ELECTROMAGNETIC

According to Ryynänen,[13] one of the most important variables affecting the heating of food is the dielectric properties, which describe how materials interact with

---

TABLE 1.5
Surface and Interfacial Tension Measurement Techniques

Technique	Description	Comments
Capillary rise	Measures height liquid rises in immersed capillary tube	For strongly adhesive liquids
Drop weight	Weighs drops from tip of known radius	Simple and inexpensive
Du Noüy ring (or needle)	Force required to lift ring (or needle) from liquid surface	Ring is the traditional method for surface tension measurements; needle is used for small sample volumes
Pendant drop	Measures contact angle of drop suspended from end of tube	Can calculate surface and interfacial tension
Sessile drop	Places drop on surface of solid sample and measure contact angle	For solid samples
Spinning drop	Optically measures diameter of drop within heavier phase when both are rotated	For low interfacial tension measurements
Wilhelmy plate	Force exerted on plate perpendicular to liquid	Useful for films

Source: Lee, B.-B., et al., *Colloid Surface A*, 332, 112–120, 2009.

electromagnetic radiation. The ability to absorb, transmit, and reflect electromagnetic energy is measured by dielectric permittivity, ε. As in Equation 1.1, there is a real component (ε') and an imaginary component (ε''):

$$\varepsilon = \varepsilon' - j\varepsilon'' \qquad (1.2)$$

where ε' is the dielectric constant (ability to store electrical energy), ε'' is the dielectric loss factor (related to energy dissipation), and j is the complex constant. Measurement techniques were reviewed by İçier and Baysal[14] and by Venkatesh and Raghavan[15] and are summarized in Table 1.7.

1.4.2 MASS TRANSFER

Mass transfer is related to the transport of components in food. Water molecules move in food during concentration, drying, freeze-drying, evaporation, and heating,

TABLE 1.6
Direct Methods for Measuring Stickiness of Particles

Technique	Description	Advantages	Disadvantages
	Pneumatic Methods		
Blow test	Creates hole of specified diameter and depth by blowing air into sample	Easy to make and operate	Long (>1 day) preparation time
Cyclone	Swirls sample in chamber and varies temperature and humidity until particles fail to swirl	Interior of particles do not have to come to equilibrium	Particles are dried and therefore may not match true industrial situation
Fluidization	Humidified air blows through beds of sample	Can obtain sticky-point temperature as function of humidity	Disadvantages are same as cyclone test, but apparatus is more complex
	Other Methods		
Ampule	Visually inspects ampules filled with freeze-dried sample after heating at different temperatures	Can obtain collapse temperature	Subjective
Impeller driven	Measures increase in stirring force as sample is heated	Can obtain sticky-point temperature	Does not measure particle-to-equipment adhesion and thus not applicable to stickiness during drying
Optical probe	Measures changes in surface reflectance	Online monitoring of sticky-point temperature	Novel and not yet tested on transparent particles

Source: Adhikari, B.P. and Bhandari, B.R. Sticky and collapse temperature: Measurement, data, and predictions, Pages 347–379 in *Food Properties Handbook*, 2nd edn., M.S. Rahman, ed., CRC Press, Boca Raton, FL, 2009; Boonyai, P., et al., *Powder Technol.*, 145, 34–46, 2004.

and control of this movement is vital for product quality and economy. Mass transfer is also involved in salting, sugaring, and introduction and removal of air. Cayot et al.[16] described several methods for evaluating mass transfer of small molecules in food, which are summarized in Table 1.8.

1.4.3 PHYSICOCHEMICAL CONSTANTS

Many physicochemical constants are encountered in food analysis, including some covered elsewhere in this chapter. An important area that makes use of several physicochemical constants is the study of food emulsions. Table 1.9 shows how these constants are obtained, as described by McClements.[17]

TABLE 1.7
Measurement of Dielectric Properties of Food

Technique	Description	Advantages	Disadvantages
Resistivity cell	Applies test voltage to material between two plates	Simple calculations, inexpensive	Sample must be a flat sheet, DC measurements only
Parallel plate	Measures capacitance between two electrodes	Simple calculations, inexpensive, high accuracy	Sample must be a flat sheet, frequency range of <100 MHz
Coaxial probe	Probe on coaxial line touches sample and reflected signal is measured	Easy, no sample preparation, nondestructive, large frequency range	Sample must have flat surface and be >1 cm thick, accuracy ±5%
Transmission line	Sample completely fills cross-section of transmission line causing changes in impedance and propagation	Accurate and sensitive, can be used for solids and liquids	Sample preparation is difficult because a precise shape is required
Cavity resonator (cavity perturbation)	Cavity in waveguide filled with sample causes changes in resonant frequency and absorption	Easy sample preparation, high or low temperatures can be used, accurate	Complex analysis, no broadband data
Free space	Microwave beam directed at sample	No contact with sample, nondestructive	Large and thin sample with parallel faces required, needs special calibration
Time-domain spectroscopy (reflectometry)	Measures reflection of signal pulse	Rapid measurement, high accuracy, large frequency range	Sample must be very small and homogeneous, expensive instrument

Source: İçier, F. and Baysal, T., *Crit. Rev. Food Sci. Nutr.*, 44, 473–478, 2004; Venkatesh, M.S. and Raghavan, G.S.V., *Can. Biosyst. Eng.*, 47(7), 15–30, 2005.

TABLE 1.8

Measurement of Mass Transfer in Food

Technique	Description	Advantages	Disadvantages
Concentration profiles	Two cylinders containing different concentrations of the compound are brought together, and measurements of thin layers are taken	Adaptable to many small molecules	Destructive
Gravimetric sorption/ desorption	Measures gain or loss of volatile compounds by weighing sample	Rapid, accurate	Possible boundary layer may resist mass transfer
Permeation	Gradient produced by placing sample between two compartments with different permeate partial pressures	Good for solid and dense interfaces	Results greatly affected by conditions
Magnetic resonance imaging	Detects proton spin alignment	Nondestructive or invasive	Not a routine technique

Source: Cayot, N., et al. *Food Res. Int.*, 41, 349–362, 2008.

1.4.4 THERMAL AND THERMODYNAMIC

Thermodynamics deal with the conversion of energy into work and heat, and thermal properties quantify the response of a material to heat. In food analysis, thermodynamic and thermal properties include phase transitions, enthalpy, specific heat, thermal conductivity, thermal diffusivity, surface heat transfer coefficients, and water activity. These values are important for predicting the behavior of food and ingredients when they undergo thermal processing.

1.4.4.1   Thermal Profile

Differential scanning calorimetry (DSC) is commonly used for measuring phase transitions of food, including freezing point, melting point, glass transition temperature, and crystallization. DSC measures the difference in the amount of heat required to maintain the same temperature in a sample and a reference while they are being heated or cooled linearly with time. The resulting curve represents the thermal profile of the sample. DSC is also used to measure enthalpy (the amount of heat that must be absorbed or evolved to create a phase transition) and specific heat C_p (the heat required to increase the temperature of 1 g of sample by 1 K).

1.4.4.2   Diffusivity, Conductivity, and Heat Transfer

Thermal diffusivity α is the speed at which heat propagates through a sample, and thermal conductivity k is the ability of a material to conduct heat. They are related by

$$\alpha = \frac{k}{\rho C_p}$$

(1.3)

where ρ is density. Surface heat transfer coefficient is used to measure the rate at which heat passes between a fluid in motion and a solid. Thermal diffusivity is affected by sample composition, porosity, temperature, and water content and is calculated from density, specific heat, and thermal conductivity data.[18] Thermal conductivity and surface heat transfer coefficients are measured by steady-state (temperature does not change with time) and transient (temperature varies with

TABLE 1.9

Measurement of Physicochemical Constants of Droplets in Food Emulsions

Technique	Description	Examples	Advantages	Disadvantages
Size				
Dynamic light scattering	Droplets scatter light beam depending on their motion	Photon correlation, diffusing-wave spectroscopy	Useful for small droplets and protein aggregates	Cannot be used for emulsions containing thickening agents
Static light scattering (laser diffraction)	Droplets scatter light beam depending on their size	Angular scattering, turbidometric	Rapid (several minutes)	Cannot be used for opaque samples
Electrical pulse counting	Impedance changes when emulsion is drawn through small aperture	Coulter counter	Most common method	Must be diluted in electrolyte, cannot measure droplets <0.4 μm in diameter
Microscopy	See Table 1.2	See Table 1.2	Can view image of sample	Difficult to prepare samples without altering structure
Ultrasonic	Measures velocity and attenuation of ultrasonic waves	Pulsed wave, continuous wave	Useful for opaque and concentrated samples	Much data needed to interpret results, air bubbles interfere
Concentration (Disperse Phase Volume Fraction)				
Electrical conductivity	Conductivity depends on electrolyte concentration	Conductivity cell	No sample preparation	Sensitive to physical state and ionic strength
Proximate analysis	Various wet chemical methods	Solvent extraction, evaporation	Common equipment	Labor intensive and time consuming
Charge (ξ-potential)				
Electroacoustic	Applies electric signal and records resulting acoustic signal, or vice versa	Electrosonic amplitude, colloid vibration potential	Can perform on undiluted samples	Much data needed to interpret results
Particle electrophoresis	Droplets migrate toward oppositely charged electrode	Electrophoretic mobility cell	Provides image of moving droplets	Possible water uptake along walls of cell

Source: McClements, D.J. *Food Emulsions: Principles, Practice, and Techniques*, 2nd edn., CRC Press, Boca Raton, FL, 2005.

time) techniques. A common steady-state method for measuring thermal conductivity uses guarded hot plate, where the sample is on each side of a hot plate sandwiched between two cold plates. The temperature gradient is then measured.[18] The most common transient method in thermal conductivity measurements uses hot disk, where a disk is heated between two sample planes and temperature change is measured as a function of time.[18] Surface heat transfer coefficients of food are calculated from thermal conductivity or specific heat values using one of the several equations.

1.4.4.3   Water Activity

Water activity is a thermodynamic property that measures the status of water in a system. Water activity is defined as the ratio of the vapor pressure of water in a sample to the vapor pressure of pure water at the same temperature. The optimum shelf life of a food is related to its water activity; lowering it reduces or eliminates bacterial growth. Dew point hygrometers measure water activity by cooling a mirror in a sealed sample chamber and optically detecting the dew point, which is related to vapor pressure. Capacitance hygrometers measure the capacity of a polymer membrane to hold a charge, which increases as the membrane absorbs water from the sample. Dew point hygrometers can be affected by volatiles but are more accurate than capacitance hygrometers, which require regular calibration.

1.5   KINETIC PROPERTIES

The quality of food is affected by the growth of the microorganisms it contains, enzymatic activity, and reactions that take place because of heat, moisture, oxygen, or other factors. The rates of reactions under various conditions are obtained by kinetic modeling.

1.5.1   CHANGES IN QUALITY

Kinetic modeling of chemical and physical changes involves the estimation of rate constant, activation energy, and frequency or preexponential factor. Two-step regression is normally used to obtain kinetic parameters from experimental results. A quality property is plotted versus time at different temperatures T to determine a value for the rate constant k at each temperature. Activation energy E_a and frequency factor A are then determined from the Arrhenius equation.

$$k = A \exp\left(\frac{-E_a}{RT}\right) \tag{1.4}$$

where R is the gas constant. Integral or differential methods are used to estimate the dependence of reaction rate on reactant concentration. Van Boekel[19] reviewed statistical aspects of kinetic modeling in food, and Hindra and Baik[20] reviewed a specific situation the kinetics of quality changes during frying of food.

1.5.2 RATES OF MICROBIAL GROWTH AND DECLINE

Kinetic modeling of microbial growth, survival, and inactivation is called "predictive microbiology." Bacteria in food may undergo lag, growth, maximum population density (stationary), and death phases, and kinetic data may be obtained from all of these. Lag phase duration (the time required for a bacterial population to adjust to its environment before cells can start dividing), growth rate, and maximum population density (the highest bacterial concentration achievable) are then calculated using mathematical models.[21] Death rates from lethal agents such as heat may be obtained by linear and nonlinear models.[22]

1.6 SUMMARY

Food has many physical properties, and there are many ways to measure them. Once the analyst decides on the attributes to be investigated, he or she must consider the strengths and limitations of appropriate techniques before deciding which one to use. It is hoped that this chapter and the subsequent ones in this book will aid in technique selection.

REFERENCES

1. Jowitt, R. 1974. Classification of foodstuffs and physical properties. *Lebensm. Wiss. Technol.* 7: 358–371.
2. Rahman, M.S. 1998. Editorial. *Int. J. Food Prop.* 1(1): v–vi.
3. Rahman, M.S., McCarthy, O.J. 1999. Classification of food properties. *Int. J. Food Prop.* 2: 1–6.
4. Szczesniak, A.S. 1990. Texture: Is it still an overlooked food attribute? *Food Technol.* 44(9): 86–95.
5. Duizer, L. 2001. A review of acoustic research for studying the sensory perception of crisp, crunchy and crackly textures. *Trends Food Sci. Technol.* 12: 17–24.
6. Rahman, M.S. 2005. Mass-volume-area-related properties of foods. Pages 1–39 in *Engineering Properties of Foods.* 3rd edn., M.A. Rao, S.S.H. Rizvi, A.K. Datta, eds. CRC Press: Boca Raton, FL.
7. Falcone, P.M., Baiano, A., Conte, A., Mancini, L., Tromba, G., Zanini, F., and Del Nobile, M.A. 2006. Imaging techniques for the study of food microstructure: A review. *Adv. Food Nutr. Res.* 51: 205–263.
8. Steffe, J.F. 1996. *Rheological Methods in Food Process Engineering.* 2nd edn. Freeman Press: E. Lansing, MI.
9. Bourne, M. 2002. *Food Texture and Viscosity: Concept and Measurement.* 2nd edn. Academic Press: London.
10. Lee, B.-B., Ravindra, P., Chan, E.-S. 2009. New drop weight analysis for surface tension determination of liquids. *Colloid Surface A* 332: 112–120.
11. Adhikari, B.P., Bhandari, B.R. 2009. Sticky and collapse temperature: Measurement, data, and predictions. Pages 347–379 in *Food Properties Handbook.* 2nd edn., M.S. Rahman, ed. CRC Press: Boca Raton, FL.
12. Boonyai, P., Bhandari, B., Howes, T. 2004. Stickiness measurement techniques for food powders: A review. *Powder Technol.* 145: 34–46.
13. Ryynänen, S. 1995. The electromagnetic properties of food materials: A review of the basic principles. *J. Food Eng.* 26: 409–429.

14. İçier, F., Baysal, T. 2004. Dielectrical properties of food materials—2: Measurement techniques. *Crit. Rev. Food Sci. Nutr.* 44: 473–478.
15. Venkatesh, M.S., Raghavan, G.S.V. 2005. An overview of dielectric properties measuring techniques. *Can. Biosyst. Eng.* 47(7): 15–30.
16. Cayot, N., Dury-Brun, C., Karbowiak, T., Savary, G., Voilley, A. 2008. Measurement of transport phenomena of volatile compounds: A review. *Food Res. Int.* 41: 349–362.
17. McClements, D.J. 2005 . Pages 461–514 in *Food Emulsions: Principles, Practice, and Techniques.* 2nd edn. CRC Press: Boca Raton, FL.
18. Sweat, V.E. 1995. Pages 99–138 in *Engineering Properties of Foods.* 2nd edn., M.A. Rao, S.S.H. Rizvi, eds. Marcel Dekker: New York.
19. Van Boekel, M.A.J.S. 1996. Statistical aspects of kinetic modeling for food science problems. *J. Food Sci.* 61: 477–485, 489.
20. Hindra, F., Baik, O.-D. 2006. Kinetics of quality changes during food frying. *Crit. Rev. Food Sci. Nutr.* 46: 239–258.
21. Tamplin, M.L. 2006. Predicting the growth of microbial pathogens in food. Pages 205–218 in *Advances in Microbial Food Safety.* V.K. Juneja, J.P. Cherry, M.H. Tunick, eds. American Chemical Society: Washington, DC.
22. Peleg, M., Penchina, C.M. 2000. Modeling microbial survival during exposure to a lethal agent with varying intensity. *Crit. Rev. Food Sci. Nutr.* 40: 159–172.

2 Statistical Assessment of Results of Food Analysis

Richard G. Brereton

CONTENTS

2.1   INTRODUCTION

Most quantitative analytical techniques aim to obtain a measurement such as from chromatography or spectroscopy and relate this to the concentration of a compound in a material such as food. There are two principal needs for statistical methods. The first is determining how well the concentration of a single sample can be estimated in a laboratory. This may, for example, be a reference sample using a standard method of analysis, and it may be important to compare this against published data or with other laboratories. A second need is during calibration, when establishing a new method, using a series of standards of different concentrations to develop an analytical technique that will be employed to estimate the concentration of an unknown. There are more comprehensive guides [1–4] to the statistical methods described in this chapter to which the reader who wishes to delve further is referred.

2.2 UNCERTAINTY AND PRECISION

2.2.1 CONCEPT

A key concept is that of uncertainty of measurements [5,6]. It is often desirable to be able to cite a range between which we are confident that the true value of a concentration lies. For example, after performing a measurement, we may be 99% confident that the true concentration of a compound in a food is between 21.52 and 23.65 mg/kg. The broader this range, the larger the uncertainty. It is assumed that there is an underlying population uncertainty for any measurement process, which we want to assess.

2.2.2 ORIGIN OF UNCERTAINTY

Uncertainty is influenced by two main factors. The first class of factors relates to sampling. If we want to determine the amount of additive in a food, each sample may be slightly different because the additive is not evenly distributed. In addition, the production process will not always result in products that are identical in composition. This is especially important if the source of material varies, for example, according to time of year, cultivation, geography, genetic makeup, and even what time of day the plant was harvested.

The second class of factors is caused by analytical error. An important source involves measurement error. Most measurements consist of several steps such as extraction, weighing, dilution, and recording an instrumental response. Each step involves uncertainty. If a 100-mL volumetric flask is used, the amount of solvent is not always exactly 100 mL. There is a range of flasks dependent on manufacturing process; one flask may have a volume of 99.93 mL, the next 100.21 mL, and so on. In addition, the volumes will depend on temperature and also on the skill and consistency of the analyst. Thus, the amount of liquid measured will form a distribution; the wider the distribution, the greater the uncertainty. Another source is a consequence of calibration uncertainty. This may arise, for example, from bias in the calibration model. Although replicate measurements on a sample may be very similar, and the sampling may be performed well, there may be problems with the original calibration, adding an additional source of error.

2.2.3 ERRORS

Analytical or measurement error will always be present. The relative importance of sampling error depends on how broad is the question we want to answer. In some cases, the main objective is to determine the level of a component in a specific batch or sample from a discrete origin. In other cases, we might want to pose a broader question, for example, the amount of tannin in a commercial brand on tea. The more generic the question, the greater the uncertainty.

The experimentally estimated value of a concentration, c, relates to the intrinsic value in absence of measurement error, $\tilde{c}$, by

$$c = \tilde{c} + e_{\text{measurement}} + e_{\text{sampling}} + e_{\text{calibration}}$$

where the e terms are called "error terms." Note that the semantic meaning of measurement error is different from that of measurement uncertainty. For example, assume that the underlying concentration of a chemical in a food is 1.73 mg/L. We may take a sample of this food and find the concentration to be 1.68 mg/L, resulting in an error of 0.05 mg/L. However, the uncertainty in measurement may be 0.08 mg/L. It just happens that in one of our measurements, we have determined a sample that is 0.05 mg/L away from the bulk mean, but the next time we might, for example, be 0.13 mg/L away from the bulk mean, due to inhomogeneity of the samples.

Each of the measurement and sampling error terms in themselves forms a distribution and originates from several different sources. In practice, the error terms are sums of several distinct distributions, but an important theorem is that the sums of different symmetric distributions normally approximate to what is described as a normal distribution [7], characterized by mean and standard deviation. The larger the standard deviation, the larger the uncertainty.

2.2.4   DETERMINATION OF UNCERTAINTY

To determine uncertainty, it is normal to first decide the factors that are to be studied. For example, do we wish to include variability between samples, or are we studying a single sample? Are the measurements restricted to using a single set of apparatus or analyst? For an official method of analysis, it is essential to define these factors in advance, and organizations such as International Union of Pure and Applied Chemistry, Eurachem, and National Institute of Standards and Technology provide suitable guidelines [8–10].

Once the limits to the range of factors have been determined, it is usual then to perform replicate analyses and measure the spread of the results over these replicate analyses. Normally, some form of experimental design is required to ensure that there is no bias in the replication. For example, if four analysts and two instruments are to be used in the replication study, and we make 24 measurements, then, ideally, each analyst should perform six measurements, three on each of the two instruments. The more the factors to be studied, the more complex the design.

Sufficient replicates should be performed to provide a good estimate of the standard deviation of the measurements. If, for example, we measure 100 replicates and the underlying population standard deviation is 0.136 mg/kg, it is quite likely that we will obtain a good estimate of this value. If only five replicates are measured, the value of the standard deviation may be seriously in error. It is important to realize that many more replicates are required to determine uncertainties than means, to a given level of confidence. There are methods for overcoming this problem, but they are a bit complex, and it is easiest to use a large number of replicates.

2.2.5 SAMPLING UNCERTAINTY

Sampling, although an important source of uncertainty, is often treated separately from analytical uncertainty: this historic division often relates to organizational divisions, for example, one section might be responsible for determining the concentration of an additive using instrumental methods and another section for collecting the samples in the field.

Over the past few years, there has been special interest in the sampling theory [11–13]. Most sources of uncertainty can be well modeled by normal distributions, but this is not necessarily so for primary sampling. A recent paper discusses approaches for estimating the amount of aflatoxins in pistachio nuts [12]. Gy's sampling theory or Theory of Sampling [14] is frequently employed in this context: it identifies seven types of error. The problem is to define a "representative sample," as the distribution of a compound is usually inhomogeneous. In the study on pistachio nuts, two different approaches yield very different values of uncertainty differing sixfold (136% of the mean compared with 22.5% of the mean).

2.2.6 CALCULATION OF UNCERTAINTY

Often a procedure can be broken down into a series of steps, j. It is usual to calculate the standard deviation of measurements obtained by replicating step j, defined by

$$s = \sqrt{\sum_{i=1}^{I} (c_i - \overline{c})^2 / (I - 1)}$$

where c_i is the ith value of the measured concentration and $\overline{c}$ is the mean. This is sometimes called the "precision." This step may be at any stage of the procedure, for example, it may involve weighing, dilution, recording spectra, or sampling. This value is the experimental estimate of the precision, u_j, of each step.

The overall precision can be calculated by

$$u = \sqrt{\sum_{j=1}^{J} u_j^2}$$

For some steps, such as calibration, it is not always possible to measure the value of u_j by replication, but alternatives such as mean square calibration errors can be used instead.

Normally, the individual uncertainties are expressed in percentage terms: a volumetric flask may have uncertainty of 2%; a balance, 1%; an extraction procedure, 5%; and so on. Hence, if there are five factors with uncertainties of 8%, 1%, 3%, 11%, and 4%,

$$u = \sqrt{(0.08^2 + 0.01^2 + 0.03^2 + 0.11^2 + 0.04^2)} = 0.145 \quad \text{or} \quad 14.5\%$$

If the intrinsic or true concentration of a compound in a food is 32.61 mg/kg, then a 14.5% uncertainty corresponds to a precision of 4.74 mg/kg. Note that if we neglect the two smallest sources of uncertainty, the overall uncertainty changes only by 0.3% to 14.2%, despite the fact that their levels are 1% and 3%, respectively. This means that it is fairly safe to neglect these sources, and so we will get satisfactory answers by replicating just three out of the five factors.

Sometimes, it is not necessary to determine all the uncertainties experimentally as these are often provided either as a standard reference or by the manufacturer. For example, it may be specified that 95% of 100-mL volumetric flasks from a given manufacturer are certified within 0.6 mL. This means that the volume of 95% of the flasks is between 99.4 and 100.6 mL. To convert from this to an uncertainty, it is usual to use the normal distribution, in which it is expected that 95% of all measurements are within 1.96 standard deviations of the mean, so that 0.4 mL is equivalent to 1.96 times the uncertainty, meaning that $u = 0.6/1.96 = 0.306$ mL, representing 1 standard deviation.

Sometimes, manufacturers quote a range, for example, a 5-mL pipette may have a minimum volume of 4.92 mL and a maximum of 5.08 mL. It is usual to divide this range by 3 to provide an estimate of the standard deviation, which then allows uncertainties to be calculated in the usual manner.

Note that for these calculations to have meaning, the method of analysis must be similar in all cases, for example, using balance, volumetric flask, measuring cylinder, and instrumental conditions that are identical. If the levels of concentrations of a compound in a sample vary substantially, this is not always possible, so it may be necessary to concentrate or dilute samples prior to analysis to obtain comparable results. Alternatively, one could calculate different uncertainties according to the concentration level of a compound because a different analytical procedure may be employed to measure concentrations centering on 1, 10, or 100 mg/kg.

If it is known that measurement error is heteroscedastic, and it depends on the concentration of analyte or intensity of spectroscopic peaks, there are methods for adjusting the basic equations for uncertainty. In some cases, certain apparatus in the analytical procedure remain the same; therefore, uncertainty is independent of analyte, and only one step is influenced by concentration. The equation for uncertainty can be modified under such circumstances with each of the individual uncertainties given a weight c_j as follows:

$$u = \sqrt{\sum_{j=1}^{J} c_j u_j^2}$$

2.2.7 CONFIDENCE

The concept of confidence is closely related to that of uncertainty (see Miller and Miller [15] and chapter 2 of Brereton [16]). If we assume that the residuals are normally distributed, we can compute this from the second column of Table 2.1, whether the number of degrees of freedom equals the number of samples minus 1. If the number of samples used to calculate the uncertainty is large, then this implies that 50% of samples, or one in two samples, are within 0.674 standard deviations either side of the mean, and 95%, or 19 out of 20, samples are within 1.960 standard deviations of the mean. Since the uncertainty is usually represented by the standard deviation, if, for example, the uncertainty of an analysis is 3.2%, and the experimental measurement of a concentration is 56.3 mg/kg, then the uncertainty is 1.80 mg/kg. Hence, we are 95% confident that the true value is between 52.77 and 59.83 mg/kg.

Sometimes, the measurement standard deviation is measured on a small number of samples. The t-distribution corrects for this. For the normal distribution, it is assumed that there are a large number of degrees of freedom for the determination of the standard deviation. If there are less samples, then the number of degrees of freedom equals one less than the number of samples, so if we use 11 samples for determination of the uncertainty, there will be 10 degrees of freedom. The resultant measurements do not exactly form a normal distribution, and it is usual to use a t-distribution. The right-hand columns of Table 2.1 represent the equivalent number of standard deviations away from the mean when sample sizes are restricted. So, if only 11 samples are used to determine uncertainty, 95% of samples will lie within 2.228 rather than 1.960 standard deviations of the experimentally determined mean. In many practical situations, it is acceptable to use a smaller sample size, for example, if the main objective is to see whether the concentration of a compound is likely to exceed a given limit, rather than to provide an exact measurement of concentration.

It is important to realize that these uncertainty calculations depend on the experimental errors falling roughly on a normal distribution. There are several tests for normality available if in doubt [4]. However, unless the number of samples is very large, tests for normality usually can only adequately model the center of the distribution and not the tails. Due to the central limit theorem, many distributions do indeed exhibit approximate normality in the center, but may not necessarily follow this in the tails. However, to adequately determine whether a 1% tail (representing 99% confidence limits) follows closely a normal distribution, it is necessary to

TABLE 2.1

Number of Standard Deviations Either Side of the Mean Required to Obtain a Given Confidence Level Assuming a Normal Distribution

Probability	Degrees of Freedom				
	Large	100	50	10	5
50%	0.674	0.677	0.679	0.700	0.727
90%	1.645	1.660	1.676	1.812	2.015
95%	1.960	1.984	2.009	2.228	2.571
99%	2.576	2.626	2.678	3.169	4.032

obtain many more samples than the percentile (at least 10 times so 1000 samples are required to have an approximate ability to model 1% tails). In practice, this is not usually possible, so very accurate assessments of 99% confidence limits are not usually possible. It is, however, usually possible to obtain a good estimate of the 50% percentile and to determine whether a sample is an extreme outlier. Confidence limits such as 99% or 99.9% should be taken with a pinch of salt and not interpreted too literally, but provide guidance whether something unusual may be happening.

2.2.8 REPORTING UNCERTAINTY

Uncertainty can be reported in various ways. The simplest is to state that the uncertainty of a batch of 5-mL volumetric flasks is 0.32 mL. this implies that the standard deviation of their volumes is 0.32 mL.

More usually, uncertainty is reported by using a range, for example, the estimate of the concentration of compound is reported as 86.69 ± 3.27 mg/kg. The number 3.27 is recommended to be twice the uncertainty or standard deviation of the errors. It is said that the "coverage factor" is 2, and so we report the concentration as being $c \pm 2u$. This corresponds to approximately 95% confidence in the analysis if the samples are normally distributed, so 19 in 20 measurements should be within this range.

2.3 ACCURACY AND BIAS

2.3.1 DEFINITIONS

Accuracy relates to how close a result is to the underlying true result (see chapter 1 of Miller and Miller [15] and chapter 2 of Caulcutt and Boddy [17]). For example, if an underlying mean concentration is 103.24 mg/kg and the mean measured concentration is 105.61 mg/kg, there is a 2.37 mg/kg inaccuracy in the measurement process. The closer the measured concentration (which is usually the average of several individual measurements) is to the true concentration, the more accurate it is. Accuracy is different from precision or uncertainty in that the latter measures the spread of results, whereas the former measures how well the result agrees with the underlying true value.

Sometimes, a measurement process can be precise but not accurate, for example, a balance might be poorly calibrated; although the results of replicate analyses might appear quite similar, in fact they are all in error by a given amount. This type of error is often called "bias."

2.3.2 DETERMINATION OF ACCURACY

It is often harder to determine accuracy than precision or uncertainty because this cannot easily be performed using replicate analysis.

There are two principal approaches. The first is to use a certified reference material, whose properties are well established. Major national and international organizations are in the business of producing such standards. The second involves interlaboratory tests, in which one material is sent to several different laboratories (or several analysts in one laboratory or to several instruments, if it is suspected that there is an in-house bias), and the results are compared.

In all cases, it is recommended that several replicated measurements are performed on the material, so in each case, mean and standard deviation of the measured quantity, such as concentration, are obtained.

2.3.3 SIGNIFICANCE OF DIFFERENCE IN MEANS

The main statistical technique is to ask whether a mean measurement from one source (e.g., laboratory A – $\overline{c}_A$) is significantly different from measurement from another source (e.g., laboratory B – $\overline{c}_B$). In order to determine this, we compare the difference between the means from both sources with their standard deviations. If the mean measurements are many standard deviations apart, then the difference between means is significant, and we can state that there is bias in one of the procedures relative to the other.

To do this, the first step is to calculate the pooled standard of the two procedures, defined by

$$s = \sqrt{\frac{(I_A - 1)s_A^2 + (I_B - 1)s_B^2}{(I_A + I_B - 2)}}$$

where s_A is the standard deviation obtained from procedure or laboratory A, and I_A is the corresponding number of samples. This averages the standard deviation for each of the procedures.

The next step is to compare the means and calculate the t-statistic, defined by

$$t = \frac{\overline{c}_A - \overline{c}_B}{s\sqrt{1/I_A + 1/I_B}}$$

which is normally presented as a positive number; so, procedure A is the one that results in a higher average concentration estimate. The larger this number, the more likely it is that the two procedures differ. Note that there are several definitions of t in the literature, the one above being relevant in this situation.

Consider the following example. For reference procedure A, we record 50 measurements with a mean of 12.1 and a standard deviation of 1.8. For our laboratory procedure B, we record 10 measurements with a mean of 10.4 and a standard deviation of 2.1. The t-statistic, calculated using the formula above, is 2.65.

The final step is to convert this into a significance value. In order to do this, it is necessary to know the number of degrees of freedom, which in this case equals 49 + 9 or 58, and look at the probability value of the t-statistic (see chapter 3 of Miller and Miller [15], appendix A.3 of Brereton [16], and chapter 4 of Caulcutt and Boddy [17]). The higher the t-statistic, the lower the chance that the difference of means could occur by sampling the same population of measurements, and so the more likely it is that there is a real difference between means. Many traditional statistical texts present critical values of this statistic in tabular form, so we can determine whether the likelihood this occurs by chance is less than 10%, 5%, or 1%. When using these tables, be sure to use the two-tailed distribution, if the question being asked is whether procedure B differs significantly from procedure A, because a mean that is significantly higher or lower than the reference is equally significant; the number of degrees of freedom equals the number of measurements minus 2; note that when the number of measurements becomes large, the t-distribution becomes the same as the normal distribution. However, it is not always necessary to use these tables, and values can also be obtained using common packages such as Excel. In order to do this, for the example above, type the function TDIST(2.65,58,2), where the first number represents the t-value, the second the total number of degrees of freedom, and the third that the test is two-tailed. We find, in this case, the probability is .010, so we can be 99% sure that there is a significant difference between the means; thus, the two procedures are quite different. Sometimes, this is presented as a p value for accepting the "null hypothesis" that there is no difference between the means: $p = .010$ suggests that the null hypothesis is not a very good one and can be rejected with 99% certainty.

2.4 CALIBRATION

Calibration is an important procedure in analytical chemistry [18,19]. It involves forming a mathematical relationship, or model, between concentration and a measured variable such as a chromatographic peak area or a spectroscopic absorbance. This model is first developed using a series of standards of known concentration, and then, in a separate step, used to predict the concentration of unknowns.

2.4.1 CLASSICAL CALIBRATION

There are several ways in which a calibration model can be formed between concentration (c) and an analytical response such as a peak height in chromatography (x). Note that we use the x/c rather than x/y notation in this chapter because in multivariate statistics, often "x" and "y" are swapped around compared with traditional analytical chemistry, resulting in confusion when moving from univariate to multivariate models.

The traditional approach is to employ the so-called classical calibration (see chapter 5 of Brereton [16] and chapter 3 of Martens and Naes [20]), where we obtain a model of the form

$$\hat{x} = a_1 c$$

or

$$\hat{x} = a_0 + a_1 c$$

where the latter includes the intercept and may be useful, for example, when there are baseline problems. The "^" over the x means that this is a predicted rather than an experimentally observed value. The equation for determining the value of a_1 without the intercept is quite simple and given by

$$a_1 = \frac{\sum_{i=1}^{I} x_i c_i}{\sum_{i=1}^{I} c_i^2}$$

where each measurement is denoted by i.

In order to include an intercept term, we can center the data to give

$$a_1 = \frac{\sum_{i=1}^{I} (x_i - \bar{x})(c_i - \bar{c})}{\sum_{i=1}^{I} (c_i - \bar{c})^2}$$

and

$$a_0 = \bar{c} - a_1 \bar{x}$$

where the "–" over the numbers denotes the mean.

For a sample dataset presented in Table 2.2, the best-fit equations are $\hat{x} = 2.576c$ or $\hat{x} = 0.611 + 2.436c$.

2.4.2 INVERSE CALIBRATION

Although most textbooks introduce classical calibration, in fact, inverse calibration is more relevant to analytical chemistry. In classical calibration, it is assumed that all the errors are in x or the response, and none in the c or concentration values. If the calibration is from a well-established reference standard, this may be true, but if the calibration standards are prepared in a laboratory, there are often sample preparation errors that in some cases can be more significant than the instrumental measurement

---

TABLE 2.2

Sample Data for Univariate Calibration

Concentration (mM)	Response
1	3.803
1	3.276
2	5.181
3	6.948
3	8.762
4	10.672
4	8.266
5	13.032
6	15.021
6	16.426

---

errors. This is especially true for modern instrumentation, which is quite reproducible compared with instrumentation several years ago. In addition, the aim of calibration, once a model has been established, is to predict c (or concentration) from x (e.g., a signal intensity). In such cases, an inverse regression equation of the form

$$\hat{c} = b_1 x$$

or

$$\hat{c} = b_0 + b_1 x$$

is fitted to the data. For the dataset of Table 2.2, the best-fit equation is $\hat{c} = 0.3847x$ or $\hat{c} = -0.0851 + 0.3923x$ according to whether the intercept is included or not. Note that the values of the coefficients for the classical and inverse models are only approximately related, for example, the inverse of the coefficient for the single-parameter classical model is 0.3882 compared with 0.3847 for the inverse model. This is because each model involves different assumptions about errors.

A good rule of thumb is to determine both models and see whether they give comparable predictions. If they do not, there are probably samples that are outliers, perhaps samples prepared in error or not typical of the main distribution, which have undue influence on the calibration and should be eliminated before repeating the calculations again.

Below, we will illustrate calculations using classical calibration models.

2.4.3 Error Analysis

It is important to have an idea of how precise an instrumental calibration model is and how well we can predict concentrations from the analytical method.

In addition, it is important to determine how well a calibration model is obeyed: whether there is really a linear relationship between analyte concentration and response. Sometimes, the relationship is nonlinear. The most common problems are that the concentration is either too high, when the detector is overloaded, or too low, when the signal is dominated by noise. Normally, there are concentration ranges within which the relationship is expected to be linear. In order to achieve these, it is sometimes necessary to dilute or concentrate samples, or to change instrumental conditions such as chromatographic injection volumes to reach a range of linearity. When using optical spectroscopy, typically, spectra are linear and obey the Beer–Lambert law between about 0.1 and 1.5 AU: this range is only slightly more than 10-fold, and if compounds have strongly different chromophores, the range may differ considerably between compounds.

There are a large number of approaches for studying the goodness of fit to calibration models, ranging from the graphical to the statistical. In most cases, the first step is to predict the value of x from c using the best-fit model. The residuals are calculated by

$$(x - \hat{x})$$

One of the simplest approaches is to represent these graphically. For the data in Table 2.2 and a two-parameter classical model, the residuals are represented in Figure 2.1. Such graphs can be used to spot if there are obvious difficulties, for example, an outlier that may have a very large residual, or heteroscedasticity, in which case the residuals may change in magnitude as the concentration increases.

If it appears that there is no significant trend in the residuals, the next step is normally to calculate a root mean square error, given by

$$E = \sqrt{\sum_{i=1}^{I}(x_i - \hat{x}_i)^2 / (I - P)}$$

where there are P parameters in the model, and I samples, so that in our case, $I - P = 10 - 2$ or 8 and the error = 1.0229 for the two-parameter classical model. This error is often reported as a percentage of the mean (11.19%) of the data and can be used as an indication of the average uncertainty of the measurements and therefore to indicate whether the technique is acceptable or not.

A second aim is to determine how well the underlying model is obeyed (see chapter 2 of Brereton [16]). In order to do this, it is usual to compare the replicate or experimental error to the lack of fit to the linear model. Provided there are sufficient replicates in the calibration, it is possible to obtain this information quite easily. In our example of Table 2.2, we performed four replicates, at concentrations of 1, 3, 4, and 6 mM. Note that having only four replicates will give us only an approximate idea of the experimental uncertainty and should be used for guidance only; if it is important to achieve a more accurate estimate, for example, for regulatory purposes, many more replicates are required. It is important though to realize that such rough and ready approaches will probably allow us to obtain an approximate estimate of

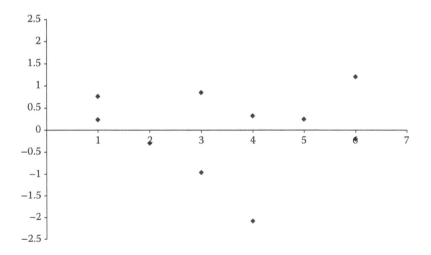

FIGURE 2.1   Residual plot of the data of Table 2.2, after fitting to a two-parameter classical model.

uncertainty, and very accurate estimates of errors often can only be interpreted in terms of probabilities and confidence if we are very sure that the residuals fall into a normal distribution, which is not necessarily the case; hence, it is not always necessary to measure replicate errors to a great deal of precision.

In order to perform this calculation, we need to calculate the average response at each concentration level, $\bar{x}_i$. At 1 mM, this is 3.540, the average of the two values 3.803 and 3.276. At 2 mM, there is only one measured response, so the average is simply the value at that concentration. The total sum-of-square replicate error is defined by

$$S_{rep} = \sum_{i=1}^{I} (\bar{x}_i - x_i)^2$$

equaling 5.665 in this example.

However, the total sum-of-square residual error is given by

$$S_{resid} = \sum_{i=1}^{I} (x_i - \hat{x}_i)^2$$

equaling 8.371 for the two-parameter classical model (the root mean square residual error is given by $\sqrt{8.371/8}$ or 1.0229 as above).

A third sum-of-square error, called the "lack of fit," can be calculated by subtracting the replicate error from the total residual error:

$$C = S_{resid} - S_{rep} = 8.371 - 5.665 = 2.706$$

This latter error indicates how well an underlying linear model is obeyed and helps answer whether we are justified in assuming the relationship is linear. It is normal to compare the size of this error to that of the replicate error. To do this, one first divides each error by the associated number of degrees of freedom to create an average sum-of-square error. For the replicate error, the number of degrees of freedom, R, equals the number of replicates, or 4 in this case. For the lack of fit, the number of degrees of freedom equals the number of experiments minus the number of parameters in the model minus the number of replicates, or

$$L = N - P - R$$

equalling $10 - 2 - 4 = 4$ in the case of the two-parameter classical model. We then calculate a statistic called the F-ratio

$$F = \frac{(S_{lof}/L)}{(S_{rep}/R)}$$

or 0.478. Since the average lack of fit is less than the replicate error, it is not significant; this means that there is no evidence to say that the model is incorrect, and we can assume that the method is good enough to provide a linear relationship. If the average lack of fit is larger than the replicate error, it can be assessed using an F-test (see appendix A.3 of Brereton [16]), and the confidence that there is a linear relationship is often expressed as a probability. However, using the F-test to convert this ratio to probabilities depends on assumptions of underlying normality.

2.4.4 CONFIDENCE INTERVALS

It is sometimes useful to determine the confidence limits for the calibration model [3,21]. In Section 2.2, we were primarily concerned with confidence in prediction of the concentration of a single reference standard. In the case of calibration, the confidence may vary according to the underlying concentration. For example, if a calibration were performed between 5 and 15 mM, it is likely that there will be higher confidence in the predicted concentration at 10 mM rather than at 5 mM. It is also interesting to determine the confidence in prediction outside the calibration limits. If the procedure we employ results in a predicted concentration of 18 mM, we may wish to determine what is the uncertainty of this prediction, for example, there may be a 95% confidence limit that the true concentration is between 14 and 22 mM, this provides us with some information about the sample.

There are several different equations, but a common one is to calculate

$$\hat{x} \pm ts \sqrt{\frac{1}{I} + \frac{1}{r} + \frac{(c - \bar{c})^2}{\sum_{i=1}^{I} (c_i - \bar{c})^2}} = \hat{x} \pm l$$

as the limits. Some of the terms in this equation require explanation. The term s is the total root mean square error, which is 1.0229 in our example in Table 2.2. The value of t depends on the number of degrees of freedom for the overall model, which equals the number of experiments minus the number of parameters or $10 - 2$ or 8 in our case, and the confidence limits. It is most common to determine 95% confidence for the predictions; this means that we would expect 19 out of 20 measurements to be within the computed limits if the model is well represented by a normal distribution. To obtain this value, we use a two-tailed t-distribution at 5% confidence because we want both positive and negative bounds. In Excel, we can use the function TINV(0.05,8) for this purpose, the first parameter providing the percentage confidence and the second the number of degrees of freedom, giving a value of 2.306. The value of I is the number of samples in the calibration, and r relates to the number of replicates for a specific measurement. If we are determining the concentration from a single replicate, this is equal to 1; however, sometimes we determine the concentration by averaging several (r) replicates.

The calculation is illustrated in Table 2.3 for both 95% and 50% confidence limits using individual rather than average replicate values at each concentration. As the percentage decreases, the limits get narrower. It can be seen that in our example, all samples are predicted within the 95% limit, but 4 out of the 10 are outside the 50% limit. The observed value of x for the fourth sample is 6.948, but the 50% confidence limits for a concentration of 3 mM are 7.160 and 8.681, so it is below the lowest 50%

TABLE 2.3

95% and 50% Confidence Limits for the Prediction of x from c Using a Two-Parameter Classical Model

				95% Limit			50% Limit	
c	x	$\hat{x}$	l	$\hat{x}-l$	$\hat{x}+l$	l	$\hat{x}-l$	$\hat{x}+l$
1	3.803	3.048	2.695	0.353	5.742	0.825	2.222	3.873
1	3.276	3.048	2.695	0.353	5.742	0.825	2.222	3.873
2	5.181	5.484	2.556	2.928	8.040	0.783	4.701	6.267
3	6.948	7.920	2.483	5.437	10.404	0.761	7.160	8.681
3	8.762	7.920	2.483	5.437	10.404	0.761	7.160	8.681
4	10.672	10.357	2.483	7.874	12.840	0.761	9.596	11.118
4	8.266	10.357	2.483	7.874	12.840	0.761	9.596	11.118
5	13.032	12.793	2.556	10.238	15.349	0.783	12.010	13.576
6	15.021	15.230	2.695	12.535	17.924	0.825	14.404	16.055
6	16.426	15.230	2.695	12.535	17.924	0.825	14.404	16.055

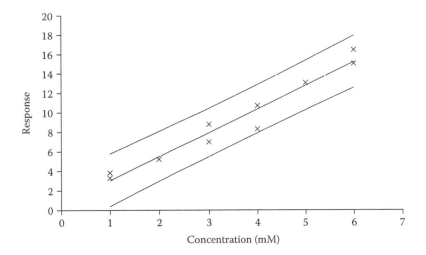

FIGURE 2.2 Data of Table 2.2 presented graphically together with best-fit straight line and 95% confidence bands.

limit. We would expect roughly half the samples to be outside the 50% confidence limit, although on a small dataset, this prediction will not be obeyed exactly; however, the results are promising in this example. The full calibration line together with confidence limits can be presented graphically (Figure 2.2). The graphical representation is useful because it is possible to spot whether there are any obvious outliers or trends. A similar calculation then can be used to determine uncertainties of estimates in concentrations for unknown samples.

ACKNOWLEDGMENTS

David Duewer and Katherine Sharpless of the National Institute of Standards Technology are thanked for helpful comments on the first version of the manuscript.

REFERENCES

1. Natrella, M.G. *Experimental Statistics, NBS Handbook 91, Revised Edition.* Washington, DC: U.S. Dept. of Commerce. 1966.
2. Tobias, P. and Croarkin, C., eds. *NIST/SEMATECH Engineering Statistics Internet Handbook.* Washington, DC: National Institute of Standards and Technology, U.S. Dept. of Commerce. Available at: http://www.itl.nist.gov/div898/handbook/, accessed on July 27, 2011.
3. Massart, D.L., Vandeginste, B.G.M., Buydens, L.M.C., De Jong, S., Lewi, P.J., and Smeyers-Verbeke, J. *Handbook of Chemometrics and Qualimetrics. Part A.* Amsterdam: Elsevier. 1997.
4. Brereton, R.G. *Applied Chemometrics for Scientists.* Chichester: Wiley. 2007.
5. Ellison, S.L.R., Wegscheider, W., and Williams, A. Measurement uncertainty. *Analytical Chemistry.* 69: 607A–613A, 1997.
6. *Guide to the Expression of Uncertainty in Measurement.* Geneva: International Organisation for Standardisation. 1993.
7. Miller, J.C. and Miller, J.N. Basic statistical methods for analytical chemistry. Part 1. Statistics of repeat measurements. *Analyst.* 113: 1351–1356, 1988.
8. Ellison, S.L.R., Rosslein, M., and Williams, A. Quantifying Uncertainty in Analytical Measurement, Eurachem/CITAC Guide. 2000. Available at: http://www.eurachem.org/guides/QUAM2000-1.pdf.
9. Eurachem. Quantifying Uncertainty in Analytical Measurement. Available at: http://www.measurementuncertainty.org/guide/, accessed on July 27, 2011.
10. The NIST Reference on Constants, Units and Uncertainty. Available at: http://physics.nist.gov/cuu/Uncertainty/.
11. Cochran, W.G. *Sampling Techniques.* 3rd ed. London: Wiley. 1977.
12. Lyn, J.A., Ramsey, M.H., Damant, A.P., and Wood, R. Empirical versus modelling approaches to the estimation of measurement uncertainty caused by primary sampling. *Analyst,* 132: 1231–1237, 2007.
13. Gy, P. *Chemometrics Intelligent Laboratory Systems.* 74: 7–60, 2004.
14. Gy, P.M. *Sampling Particulate Material Systems.* Amsterdam: Elsevier. 1979.
15. Miller, J.N. and Miller, J.C. *Statistics and Chemometrics for Analytical Chemistry.* New York: Pearson Education. 2010.
16. Brereton, R.G. *Chemometrics: Data Analysis for the Laboratory and Chemical Plant.* Chichester: Wiley. 2003.
17. Caulcutt, R. and Boddy, R. *Statistics for Analytical Chemists.* London: Chapman and Hall. 1995.
18. Coleman, D. and Vanatta L. Statistics in analytical chemistry—Calibration: Introduction and ordinary least squares. *American Laboratory.* 35: 18, 2003.
19. Miller, J.N. Basic statistical methods for analytical chemistry. Part 2. Calibration and regression methods. *Analyst.* 116: 3–14, 1991.
20. Martens, H. and Naes, T. *Multivariate Calibration.* Chichester: Wiley. 1989.
21. Coleman, D. and Vanatta, L. Statistics in analytical chemistry—Part 4 Calibration: Uncertainty intervals. *American Laboratory.* 35: 60, 2003.

3 Analytical Quality Assurance

Beata Plutowska and Marta Dabrio Ramos

CONTENTS

3.1 INTRODUCTION

Thousands of analytical measurements are performed every day in food laboratories all over the world. The increasing trend in the amount of measurements needed is directly linked to the continuous release of national and international regulations and the growing international trade of foodstuffs. Indeed, compliance with food legislation involves a decision-making process, usually, based on reports of results produced by analytical laboratories. In other words, analytical reports providing results of *poor quality* can trigger misleading conclusions and, subsequently, potentially incorrect managerial decisions. However, what do we think of when referring to quality in the context of analytical chemistry? As a reply to this question, this chapter intends to provide an overview covering basic aspects relevant to quality with a special focus on analytical quality in the field of food analysis.

3.2 NEED AND IMPACT OF THE QUALITY OF ANALYTICAL RESULTS

Although quality has been defined as "the degree to which a set of inherent characteristics fulfills requirements" [1], in the field of analytical chemistry, quality can be ultimately related to the achievement of results with a defined degree of confidence.

In practice, quality is expressed by two complementary characteristics, namely reliability and utility [2]. While reliability involves the delivery of realistic results, which are sufficiently accurate and certain, utility refers to the capability of making a sound managerial decision on the quality and/or safety of a product based on the results. Thus, the reliability of the analytical data is a condition for a correct managerial decision.

At a European level, the regulation for official food control [3] aims to raise standards of food safety and consumer protection. To achieve that, the laboratories analyzing official samples are required to work in accordance with internationally approved procedures or criteria-based performance standards and to use methods of analysis that have, as far as possible, been validated. In other words, in Europe, the official laboratories for food and feed control must operate in accordance with EN ISO/IEC 17025 on *General requirements for the competence of testing and calibration laboratories* [4] be accredited by accreditation bodies which operate according to EN ISO/IEC 17011 on *General requirements for accreditation bodies accrediting conformity assessment bodies* [5, 6]. Both standards reflect the utmost importance given to the quality of analytical results.

Official controls serve the competent authorities of the member states by monitoring the requirements set out in the relevant food and feed laws. As an example, the presence of an increasing number of harmful compounds in foodstuffs is regulated by establishing maximum residue levels or by banning certain substances. The role of the control laboratory is then to assure that the potential exposure hazards are minimized. A good example where food control is paramount is the effect of inappropriate storage conditions of certain foodstuffs possibly resulting in the occurrence of potentially carcinogenic mycotoxins. A batch of imported peanuts checked by appropriate port health authorities and declared to contain mass fractions of aflatoxins above the permitted levels automatically implies the rejection of that particular batch for the food market. This decision, besides the intrinsic benefit for the public health, may have other important economical or even political implications.

An erroneous assessment on the quality of food products can produce consequences possibly causing health risks, economical losses, or illegal practices that could end up in disputes and court trials. Another repeated occurrence in food control is food adulteration, for example, using the blend of oils from various vegetable species of lower cost not in compliance with the label declaration, thus constituting a fraud. Whenever detected and made open to the public, these cases can even affect the trust of the consumers toward the quality of the food they eat.

One main aspect of the reliability of analytical results involves comparability in terms of both the identity and the amount of the target compounds quantified between laboratories and over time [7]. In order to be comparable, results have

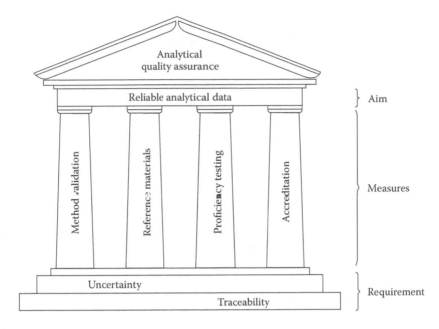

FIGURE 3.1   Concept of analytical quality assurance.

to be traceable to appropriate references and must be reported with a statement of measurement uncertainty. There are different tools at disposal to fulfill both requirements (Figure 3.1). The common usage of reference materials, validation of analytical methods, and participation in interlaboratory comparisons are the most important measures or approaches to achieve reliable chemical measurements. All mentioned and other conditions that are necessary to provide appropriate confidence that the given requirements for quality are satisfied are called the analytical quality assurance (AQA) [8].

3.3   EUROPEAN UNION AND INTERNATIONAL GUIDANCE

Ideally, the quality of analytical results should be assured uniformly at an international level. For this purpose, international organizations such as International Organization for Standardization (ISO), International Union of Pure and Applied Chemistry (IUPAC), or Eurachem have published a number of guidelines and standards that formalize and harmonize the principles of AQA. In Europe, a lot of work in this field is performed by Eurachem, a network of organizations focused on analytical chemistry and quality-related issues. The most widely recognized references are three CITAC/Eurachem guides: on quality in analytical chemistry, on traceability in chemical measurements, and on uncertainty in chemical measurements. These and other relevant guides are listed in Table 3.1. Eurachem has also developed a series of other specific guidelines addressing topics such as method validation and reference materials. Furthermore, additional standards are available from other regulatory bodies, such as standardization agencies and working groups. A selected number

TABLE 3.1
Selected International Guidelines on AQA Issues

Institution	Guide Title	References
Eurachem/CITAC: A Focus for Analytical Chemistry in Europe/Co-Operation on	Guide to Quality in Analytical Chemistry. An Aid to Accreditation	9
	Traceability in Chemical Measurement. A guide to achieving comparable results in chemical measurement	7
International Traceability in Analytical Chemistry	Quantifying Uncertainty in Analytical Measurement	10
	Quality Assurance for Research and Development and Non-routine Analysis	11
Eurachem: A Focus for Analytical Chemistry in Europe	The Fitness for Purpose of Analytical Methods. A Laboratory Guide to Method Validation and Related Topics	12
	The Selection and use of Reference Materials	13
	Selection, use and interpretation of proficiency testing (PT) schemes by laboratories	14
IUPAC: International Union of Pure and Applied Chemistry	Harmonized guidelines for internal quality control in analytical chemistry laboratories	15
	Metrological traceability of measurement results in chemistry: concepts and implementation	16
	Harmonized guidelines for single-laboratory validation of methods of analysis	17
	The international harmonized protocol for the proficiency testing of (chemical) analytical laboratories	18
	Nomenclature for the presentation of results of chemical analysis	19
ISO: International Organization for Standardization	General requirements for the competence of testing and calibration laboratories	4
	Guide to the expression of uncertainty in measurement	20
	Accuracy (trueness and precision) of measurement methods and results	21
	Proficiency testing by interlaboratory comparisons	22
EAL: European Cooperation for Accreditation of Laboratories	The Expression of Uncertainty in Quantitative Testing	23
ILAC: International Laboratory Accreditation Cooperation	Introducing the Concept of Uncertainty of Measurement in Testing in Association with the Application of the Standard ISO/IEC 17025	24
EA/Eurolab/Eurachem: European Co-operation for Accreditation/European Federation of National Associations of Measurement, Testing and Analytical Laboratories/A Focus for Analytical Chemistry in Europe	Use of Proficiency Testing as a Tool for Accreditation in Testing	25

Source: See table for multiple data references.

is adopted by organizations such as FAO/WHO Codex Alimentarius Commission as an effort towards the overall harmonization of quality assurance in the food area [26–28]. The effort is seconded by an increasing number of analytical laboratories having a direct interest in being recognized as qualified and competent for given tasks.

3.4 QUALITY CONTROL AND QUALITY ASSESSMENT

AQA describes the overall planned and systematic measures a laboratory must undertake to ensure that it can consistently perform reliable measurements [9]. The CITAC/ Eurachem *Guide to Quality in Analytical Chemistry* [9] lists a number of components comprising the quality assurance system, such as a suitable laboratory environment; educated, trained, and skilled staff; training procedures and records; properly maintained and calibrated equipment; documented and validated methods; traceability and measurement uncertainty; application of quality control procedures; preventive and corrective actions; proficiency testing (PT); internal audits; complaint procedures; and requirements for reagents, calibrants, measurement standards, and reference materials. All those activities form a flexible scheme of a quality assurance system, most often underpinned by accreditation to European and international standards, such as the ISO/IEC 17025 [4]. Accreditation, as a formal recognition of laboratory competence in carrying out specified measurements, is the last step in the implementation of a quality assurance system.

Quality should be continuously monitored. Two terms are in use to describe the quality assurance system in the analytical laboratory: quality control and quality assessment [11]. The quality control is usually more developed in routine and industrial laboratories. It relates to the quality of results obtained for specific samples or batches of samples, and it is defined as all the practical activities undertaken to control potential errors that may occur during the implementation of an analytical method on a run-by-run basis [29]. In general, the development of a method and the establishment of its performance characteristics do not guarantee obtaining constantly satisfactory results. The internal quality control system incorporates a mechanism for verification of results within acceptable limits, including the use of control charts, analysis of reference materials or reference standards, or analysis of blind samples, blanks, or spiked samples as described in quality control documentation. When results fall outside the acceptable limits, the system is activated for corrective actions. Quality assessment, in turn, is a broader term. It covers all the measures that can be used to evaluate the laboratory performance and to verify that the quality system is both functioning properly and operating within acceptable limits. There are different levels of quality assessment. The evaluation of analytical methods takes place on a small scale every day in the laboratory during data acquisition or optimization of analytical conditions. Also, the method validation is the judgment on method suitability for its intended use. Subsequently, the evaluation of the developed method and laboratory performance can be done, for example, by participation of the laboratory in PT schemes or analyzing certified reference materials (CRMs). Finally, the quality of the whole system should be checked as well, on a regular basis, either by internal or by external audits or reviews. This type of verification is being carried out through the assessment of quality control documents, such as manuals, protocols, standard operational procedures, and validation or calibration reports [30].

The range of available evaluation tools depends mainly on the type of the laboratory. When the laboratory has a formal quality system, the assessment is conducted according to formal procedures, typically in the form of internal and external audits, compulsory validation of analytical methods, and regular participation in PT schemes. However, in the case of research laboratories, where non-routine methods are mainly used, laboratory performance can also be assessed on the basis of the number of patents, publications, and postgraduates and doctors trained, as well as using external parameters, such as the ISI Science Citation Index and even the evaluation of the number and quality of publications, project plans and reports by expert panels [31].

3.5   QUALITY ASSURANCE IN THE TOTAL ANALYTICAL PROCESS

A chemical analysis involves a multistage process with the objective of determining properties such as the identity and quantities of target components in a given sample. The analytical procedure, illustrated as a general flow scheme in Figure 3.2, always begins with the definition of an analytical problem—such as the detection of levels of methylmercury in tuna fish—and ends up with the evaluation of the obtained data and the decision-making process, as to whether or not the fish can be distributed and commercialized for human consumption. The sampling procedure, sample treatment, and analytical measurement complete the so-called total analytical process.

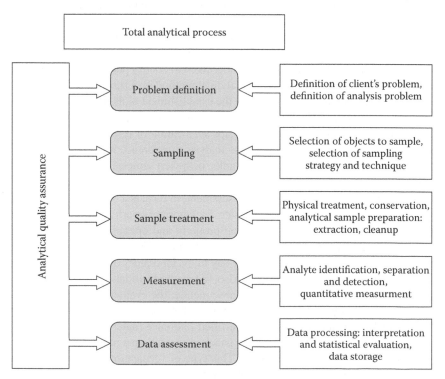

FIGURE 3.2   Flowchart of the total analytical process.

In order to achieve reliable results, every step of the process is covered under the quality assurance system. This involves appropriate methodology characterization, validation, and documentation control. The lack of control over a single stage of the total analytical process may jeopardize the quality of the final results. That can be typically the case for the sampling or sample preparation steps since often they are considered as the weakest links of a method. Likewise, the lack of appropriate experiment planning and final statistical evaluation of the results can squander even a carefully carried out laboratory work. Generally, every analytical process requires an unambiguous aim, scope, and principle of measurement, as well as clear procedural instructions and performance criteria including corrective actions. The requirements and measures described in this chapter are common for them all, independently of the field of application.

3.6 REQUIREMENTS FOR THE QUALITY OF ANALYTICAL RESULTS

According to the IUPAC, the purpose of performing a measurement is to provide information, in the form of a measurement result, on the magnitude of a *measurand* embodied in a specified system. Furthermore, when only the measured quantity value is presented, the associated measurement uncertainty and metrological traceability should be available [16], as illustrated in Figure 3.1.

3.6.1 METROLOGICAL TRACEABILITY

Metrological traceability is defined as "the property of a measurement result whereby the result can be related to a reference through a documented unbroken chain of calibrations, each contributing to the measurement uncertainty" [32]. In practice, establishing traceability can be complex since many different factors—such as the definition of the measurand or the traceability of the calibrators—may influence the traceability chain.

The result can be related to different types of references [32]:

- A measurement unit
- A measurement method
- A measurement standard

The preferred traceability statement refers to the internationally recognized system of units (SI). Establishing traceability to the SI in chemical analysis is possible when all the equipment used in the process are calibrated with standards that are themselves traceable to SI. A simple example illustrating a chain of calibrations is the use of analytical balances [33]. Each balance is regularly calibrated using reference weights that are certified against higher level standards. These are calibrated with national standards, which ultimately are related to the international kilogram standard kept by the International Bureau of Weights and Measures in Paris. In other words, every calibrator in the chain has a quantity value determined by comparing with the preceding calibrator. If the uncertainties of the particular links of this chain are known, then the overall measurement uncertainty of the weight performed can be estimated. When the results are traceable to the same reference, they are comparable, even between different laboratories and at different times. This is, in essence, the meaning of traceability.

Whenever possible, results should be traceable to the measurement unit—in case of chemical measurements—usually to mole or kilogram, as then values are independent of any method or artifact and are universally valid. Unfortunately, in food chemistry this is not always straightforward since the analyte is quite often defined in functional terms, such as extractable fat, unsaturated fat in terms of iodine value, dietary fiber, or protein based on a nitrogen content determination. In such cases, method-defined property values are measured and the reference is closely related to the documented protocol describing the analytical procedure. Method-defined traceability chain is preserved as long as the procedure is strictly followed. The main disadvantage of this approach is that the results are comparable only if obtained using the same method under exactly the same analytical conditions.

The third type of traceability reference, the measurement standards, can be represented by artifacts. For example, when the concentrations of the compounds are calculated directly from the absorbance measurements, the results are traceable to the reference absorbance data. Likewise, in measurements of the amounts of substance based on the biological activity, such as for vitamins or hormones, the so-called International Units defined arbitrarily by the Committee on Biological Standardization of the WHO are often employed [34].

The complexity of the methods used in food analysis—comprising successive analytical steps, such as extraction, clean up, derivatization, separation, and detection—may provoke losses of information due to contamination, interferences, or incomplete recovery, thus generating a bias. The traceability is kept in this case either by relating the results to the method or by evaluating the bias. The latter is possible with the use of measurement standards such as CRMs. Obviously, if the property value of a CRM is traceable to SI, then in principle the measurement can also be traceable to SI. However, when the composition of the sample and the CRM is not sufficiently comparable, the uncertainty of the result may be large [10].

The results of chemical measurements are predominantly calculated indirectly from the measurement of properties other than the one of interest, such as masses, volumes, and concentrations of the samples, reagents, or standards. In order to assure the traceability of the result, all the intermediate measurements involved in the analytical process must also be traceable. Even the quantities of the properties not present in the final equation but used to control the analysis—which can influence the result for the temperature of the extraction—should be taken into consideration [9].

Traceability of the results can be established via [10]

- Traceable standards for calibration of the analytical equipment
- Nonmatrix and matrix CRMs
- Primary method or comparison with the results of primary method
- Standardized or precisely defined procedure

Calibration with appropriate standards is the key operation in establishing traceability. In general, calibration sets up a relation between the quantity values provided by measurement standards and corresponding indications (e.g., of the instrument) and uses this information to obtain a measurement result from the indication [32]. A common way of calibration is to subject a known amount of the measurement standard (e.g., certified standard) to the analytical system and observe the measurement response. The main aspects to be taken into consideration here are the frequency and range of the calibration and the extent of traceability.

The frequency of calibration is influenced by the stability of the instrumentation, whereas the range would depend on the complexity of the system. Usually, each part of the measurement system is calibrated separately and individually contribute to the measurement uncertainty. Separate calibrations are based mainly on the use of gravimetrically prepared pure substances (of a certified purity) or solution reference materials and concern the analytical equipment. However, for many multistep analyses, matrix-matching reference materials are checked through the whole analytical process as it can be the only way to estimate the bias (often presented in terms of recovery) and thus correct the results [9]. When the appropriate matrix reference materials are not available, the evaluation of the difference in response of the measurement system to the standards used and the tested sample can be carried out by spiking the real sample with the standard [7].

Different requirements for calibration depend on whether a relative assessment is made (e.g., repeatability) or the absolute values are to be determined (e.g., for trueness studies). In the former case, traceability to in-house references is typically acceptable, but the reliability of the absolute value determination increases with the grade of the references used up to the international ones. Also, for subsidiary measurements performed to control the analytical process, strict traceability is necessary only when they can significantly influence the result (e.g., thermometers used for the determination of enzymatic activity must be carefully calibrated).

Primary methods or primary reference procedures are methods whose results are directly traceable to the SI without any reference to an external standard of the same quantity [10, 32]. They are considered to have the highest metrological qualities and guarantee the smallest achievable uncertainty of the measurement. Primary methods mainly exist for relatively simple measurements, such as gravimetry, titrimetry, and coulometry of simple homogeneous solutions, and for trace element determinations in such solutions using isotope dilution mass spectrometry [35]. For determination of organic compounds in complicated food matrices, there are no primary reference procedures due to the presence of one or more analytical steps impossible to completely describe and firmly demonstrate their 100% recovery (e.g., involving extraction or chemical reactions) [36].

If the result cannot be obtained by a primary method, compared to the primary method result or in any other way linked to SI, then it can at least be related to the standardized procedure.

Establishing the traceability does not make the results identical, but it allows meaningful comparison of the results by ensuring sameness of measurement units [7]. If one wants to decide that one result is higher or lower than another, with a given confidence level, then a second piece of information is needed—the uncertainty.

3.6.2 Uncertainty of Measurement Result

According to ISO/IEC 17025 [4], an analytical result can be only interpreted if it is expressed in the form of two values:

$$(\text{RESULT}) : (x \pm U)\,(\text{units})$$

$$(3.1)$$

where x is a measured quantity value and U is an uncertainty at a stated level of confidence. A measurement uncertainty is currently defined as "a non-negative parameter characterizing the dispersion of the quantity values being attributed to a measurand, based on the information used" [32]. Strictly, the uncertainty is a quantitative estimation of the range of the values within which the measurement result (e.g., analyte concentration) is expected to lie [9]. In this sense, it does not cast doubt on the result but just the opposite—the uncertainty statement increases the confidence in the validity of the analytical result. For example, if a food regulation states that peanuts must not contain more than 2 μg/kg of aflatoxin B_1 [37], then the producer is obliged to analyze every batch and assure that the legal level is not exceeded. Assume that the result obtained is 1.8 μg/kg. When the uncertainty of this result expressed as a half-range is 0.1 μg/kg, the concentration of the toxin in the product is expected to fall within the interval 1.7–1.9 μg/kg and with a high probability the maximum limit is not exceeded. On the other hand, if the uncertainty is 0.5 μg/kg, then there is no longer sufficient assurance that the peanuts are safe for consumption.

There is a great variety of factors that may contribute to the measurement uncertainty. Eurachem has identified the following potential uncertainty sources [10]:

- Incomplete definition of the measurand
- Sampling and storage conditions of samples
- Instrument effects (e.g., uncertainty of weights and volumetric equipment, instrument resolution, or selectivity)
- Values assigned to reference standards, reagent purity
- Approximations and assumptions incorporated in the measurement procedure (e.g., assumed stoichiometry, values of constants, and reference data)
- Measurement conditions, imperfect evaluation of influence of environmental conditions on the procedure
- Sample effects (e.g., incomplete extraction, matrix effects and interferences, blank correction)
- Operator effects (e.g., contamination during sampling, imperfect sample preparation, personal bias in reading analogue instruments)
- Random variation

Most of these factors are unavoidable during the analytical process. All of them, as error sources, influence the measured quantity value and cause it to deviate from the true value. It is important, however, to differentiate between the terms "uncertainty," "accuracy," and "error." The relationship between these concepts is presented in Figure 3.3. In order to fully estimate the uncertainty of a measurement,

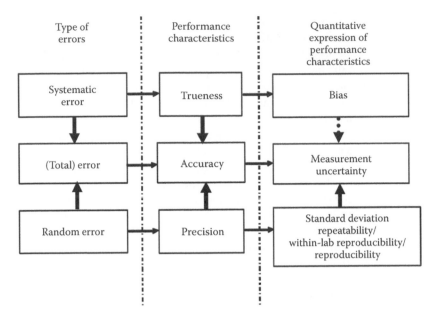

FIGURE 3.3 Relationship between error, accuracy, and measurement uncertainty. (With kind permission from Springer Science+Business Media: *Understanding the Meaning of Accuracy, Trueness and Precision. Accreditation and Quality Assurance*, 12, 2007, 45–47, Menditto, A., Patriarca, M., and Magnusson, B.)

two components have to be taken into consideration: standard deviation and bias. The standard deviation indicates the level of repeatability and/or reproducibility of the measurement, whereas the bias is associated with systematic effects inherent in a method. Both terms are the quantitative expression of two components of error: random error and systematic error. Error is generally related to the accuracy of an analytical method and is defined as a "difference between a result and the true value of the measurand" [39]. In contrast to the uncertainty, the error is a single value. A random error originates from different unpredictable effects influencing the measured value, and it is a measure of the precision of the measurement. Since it is unpredictable, it cannot be compensated, but it can be reduced by increasing the number of replicate analysis [10]. Systematic error, in turn, is a constant part of the total error (or at least varying in a predictable way), thus can be used to correct the result but cannot be reduced by increasing the number of observations. This component of the total error influences the trueness of the measurement. The trueness together with the precision contributes to the accuracy of analytical result, which is one of the method performance characteristics, defined as a "closeness of agreement between a measured quantity value and a true quantity value of a measurand" [32].

As a statistical parameter, uncertainty can never be precisely calculated although it can be estimated to a higher or lower extent. There are many theories described in the literature for estimation of uncertainty [39], but all of them can be grouped into two major strategies: the so-called bottom-up and the top-down approaches.

The bottom-up approach can only be applied if the analytical method is well understood. It is based on identification, quantification, and combination of the potential sources of uncertainty. In order to identify all components contributing to the uncertainty of the measurement, the so-called fishbone diagrams are frequently used [10]. The whole analytical process is investigated as a series of separate actions (e.g., weighing, diluting, extracting), and individual contributions are expressed as standard deviations (standard uncertainties). Then they are combined according to the law of error propagation. The resulting combined uncertainty is finally multiplied by the adequate coverage factor to obtain an expanded uncertainty.

To apply the bottom-up approach, usually, little or even no experimental data are needed because the uncertainty components can be evaluated from the previous information, for example, from the laboratory equipment producer or calibration certificates, or even using the personal judgment of an experienced analyst. An extensive description is available in the *Guide to the Expression of Uncertainty in Measurement* [20]. The bottom-up strategy evaluates the relevance of individual components, thus providing an excellent starting point for potential method improvement. The drawback is that it is a very time-consuming process for analyses that are seldom performed and—due to the frequent complexity of the mathematical models—prone to mathematical errors and to overlooking important effects.

The top-down approach applies the opposite principle. In this case, the estimation is based on the method performance data such as validation or interlaboratory comparison results for uncertainty estimation. Therefore, the main difference compared with the former strategy is that this approach assumes that the analytical process is a "black box" and measures the combined influence of many intermediate effects by means of overall method performance parameters such as repeatability, reproducibility, and trueness [40]. Just as in the bottom-up approach, all the uncertainty contributions have to be expressed as standard deviations, combined together, and multiplied by an appropriate coverage factor. More detailed information on uncertainty estimation together with numerous examples can be found in the Eurachem guide [10].

3.7   MEASURES FOR ACHIEVING THE QUALITY OF ANALYTICAL RESULTS

As previously mentioned, different means are available for demonstrating the excellence of analytical data, including the use of validated analytical methods, participation in PT schemes, and introduction of internal and external quality control systems (Figure 3.1). These measures have been extensively described in the literature and therefore essential notions will be included in this section, where special attention is given to one particular measure serving as a link to all, that is, the use of reference materials.

3.7.1   REFERENCE MATERIALS IN THE TOTAL ANALYTICAL PROCESS

Reference materials are materials or substances that are strictly characterized for one or more specified property in a qualitative and/or quantitative way [41]. In chemistry, these properties are usually the identity (i.e., chemical structure) and chemical composition in terms of quantities of particular compounds [35]. Chemical reference materials are most often encountered in the form of pure chemical substances or corresponding solutions of these substances and the so-called matrix reference materials prepared from unspiked or spiked real-life samples or as synthetic mixtures [42]. Reference materials available in the form of pure substances or solutions are well characterized for chemical purity and/or concentration and thus are mainly used for calibration purposes. Matrix reference materials either are based on real materials containing the analytes of interest in their natural form and content or can be prepared from synthetic mixtures and mimic the real sample as closely as possible [13]. Depending on the intended use, they can be characterized for the composition of specified major, minor, or trace constituents. The first group includes reference materials for nutritional properties of food products such as milk powder, flour, or meat, which are well characterized in terms of total fat, protein, and carbohydrate content, and the second group can be represented by reference materials characterized for the content of fat and water-soluble vitamins in foodstuffs. The most numerous are reference materials certified for trace compounds. Various impurities such as polychlorinated biphenyls, pesticides, veterinary drug residues, polycyclic aromatic hydrocarbons in meat or fish tissues, or mycotoxins in cereals, nuts, or milk decrease the quality and health safety of food. For this reason, the importance of reliable determination of those compounds in food products is very high and so is the demand for relevant reference materials.

Reference materials often allow the control and assessment of the whole analytical process. One of their uses is the calibration of measurement equipment. For this purpose, pure substances and their solutions are commonly used. However, matrix reference materials are usually not recommended for calibration due to relatively high uncertainties of certified values. They might be also required in the case of matrix-dependent methods employing solid sample techniques such as atomic absorption spectrometry or x-ray fluorescence [43]. Another important application is the establishment of traceability of the results in the laboratory. When the CRM is analyzed simultaneously with the sample using the same procedure, the result of the measurement of the unknown sample can be linked to an internationally accepted reference point [44].

The main reasons for using CRMs are usually a need for proving trueness of a method and for estimating measurement uncertainty. Other significant applications of reference materials are the validation of analytical procedures and subsequent regular control of validated methods used routinely in a laboratory. The use of reference materials is also necessary in laboratories that aspire to obtain or have already obtained accreditation according to ISO 17025 and is common in PT schemes organized among laboratories.

The "reference material" is a general term, and different materials of unequal quality exist. Materials of the highest metrological order are accompanied by the written statement on the established property values and their uncertainty and traceability. They are called "certified reference materials" [45]. Whenever possible and economically justified, it is advisable to use CRMs on a routine basis. Using CRMs offers significant advantages over laboratory-prepared standards. They have been tested for homogeneity and stability so can be easily switched from one bottle to the other and even opened are usually more stable than ordinary laboratory samples. There are many analytical tasks, however, in which CRMs are replaceable by non-CRMs, which are also called as "quality control materials" and represented by different kinds of working standards, such as previously analyzed homogenous samples or pure compounds and their solutions. For example, when only the level of variation of trueness or precision in time is being monitored by a comparison of a series of measurement results and a well-defined certified value and uncertainty of a CRM is not essential, usually a noncertified material is sufficient [46]. This case concerns especially the measurements being repeated routinely over extended period (e.g., for internal quality control purposes) or laboratories with a wide range of activities, taking into account limited amount and relatively higher cost of CRMs in comparison with non-CRMs or even in-house reference materials.

Depending on the particular use, the quality and the source, the reference materials are often named in different ways. Materials for interlaboratory studies are called Proficiency Testing Materials, while materials for laboratory method development and validation are called Laboratory Reference Materials. A third group includes materials for statistical quality control purposes, called Laboratory Control Materials. The reference materials can be commercially produced or prepared in-house, and they can be certified or noncertified. It should be borne in mind that all of them fulfilling the characteristics of adequate homogeneity and stability belong to the same family [47, 48]. The only limitation is that the conception of the reference material cannot be implemented to materials where either the matrix or the analyte is unstable [11].

3.7.2 REFERENCE MATERIALS IN METHOD VALIDATION

Validation is the confirmation by examination and the provision of objective evidence that the specified requirements are adequate for an intended use [32]. Let us imagine that a food manufacturer is interested in monitoring the levels of acrylamide generated during the processing of commercial toasted bread. Then accordingly, the laboratory develops or implements an analytical method for the quantification of the potential carcinogenic compound that needs to be properly characterized. The extent to which the validation is required is set by the intended use and the field of the application. A standardized approach given by the ISO 17025 [4] particularly points out the necessity to evaluate the performance of the method by one or more of the following techniques:

- Calibration using reference standards or RMs
- Comparison of results achieved with other methods
- Interlaboratory comparisons
- Systematic assessment of the factors influencing the result
- Assessment of the uncertainty of the results based on scientific understanding of the theoretical principles of the method and practical experience

Furthermore, the international standard requires establishing and assessing a range and accuracy of the values obtainable from the validated method according to the intended use. To achieve that, different parameters are investigated including the uncertainty of the results, detection limit, selectivity of the method, linearity, limit of repeatability and reproducibility, robustness against external influences, and/or cross-sensitivity against interference from the matrix of the sample object.

The role of the CRM in validation can be adjusted to the different needs or possibilities. In the acrylamide example, an appropriate CRM, as a stable and homogeneous material, is an ideal choice for the assessment of most of the parameters mentioned above, although not strictly necessary and always affordable. Whenever accuracy plays a role, the CRM becomes indispensable. That is the case for the trueness estimation during validation [46]. A potential bias of the method can be detected by the replicate analysis of a CRM using the methodology under validation. The average measurement result obtained ($\bar{x}$) is compared with the "true" value (μ), in this case the certified value, taking into account the uncertainties of the certified value and that associated to the measurement. In practice, it is simply expressed as follows:

$$\left| \bar{x} - \mu \right| < k \sqrt{\left(u_m^2 + u_{CRM}^2 \right)} \tag{3.2}$$

where u_{CRM} is the standard uncertainty of the certified value, u_m is the standard measurement uncertainty (ideally, replicate analyses performed on different days to achieve intermediate precision conditions), and k is the coverage factor, typically $k = 2$ to cover a confidence interval of around 95%. The fulfillment of the condition indicates no statistical significant bias of the measurements [49].

For the given example, the validation of an acrylamide method, the certified value for acrylamide in toasted bread in the reference material is 425 ± 29 ng/g. Since a coverage factor of 2 was employed, u_{CRM} is estimated to be $29/2 = 14.5$ ng/g. The analysis of six replicates of the CRM spread over 3 days resulted in 452 ± 38 ng/g, thus $u_m = 38/\sqrt{6} = 15.5$ ng/g. Therefore, the difference between the certified and measured value $| 452 - 425 | = 27$ ng/g) is lower than the combined uncertainties with a coverage factor of $2(2\sqrt{(15.5^2 + 14.5^2)} = 42.4$ ng/g), meaning that no significant bias is detected.

Whenever appropriate CRMs are not available, the assessment of the trueness is approached by conducting spiking experiments and recovery studies.

3.7.3 REFERENCE MATERIALS IN INTERNAL QUALITY CONTROL

The aim of internal quality control, as a continuation of method validation, is the continuous monitoring of the total analytical process in order to ensure that the results obtained are acceptable and fit for their intended purpose [15]. This is done in two ways: the analysis of control materials to monitor trueness and replication to monitor precision [29]. Typically, control materials such as CRMs, spiked samples, blank, and blind samples are analyzed in replicates and subjected to exactly the same sample treatment as test samples and are measured in the same analytical sequence. By comparing the results obtained using the analytical method with the reference value of the material, any bias can be detected and appropriate corrective actions can be taken [50]. The approach for statistical interpretation of obtained results depends on the type and frequency of analysis. In measurements carried out on a routine basis, control charts are often established [9, 15].

Different types of control materials may be used to detect different types of variation within the process. Reference materials analyzed at intervals may indicate drift in the system, and analysis of blank reference materials may indicate contamination during the sample preparation stage and/or instrument contribution besides that of the analyte [9, 51]. Regardless of the type of control materials, they should be always traceable to either an appropriate reference method, a CRM, or another well-characterized material. A very important issue here is that the control materials and the materials used for calibration should not come from the same source [15]. Using quality control material prepared from the same standard solution of analyte would not allow the detection of any inaccuracy originating from the improper preparation of the stock solution.

3.7.4 REFERENCE MATERIALS IN EXTERNAL QUALITY ASSESSMENT

The "external quality assessment" is a general term used for all periodic demonstrations of a laboratory's analytical performance and technical competence in terms of ability to achieve continually reliable results. There are two complementary measures that laboratories can use: a so-called third-party assessment and regular participation in PT schemes [14, 52].

A third-party assessment is a physical inspection of the laboratory's activities and documents to ensure that it is competent to carry out specified activities and/or its quality system conform to certain standards. The objective of a third-party assessment is to obtain an attestation—certification or accreditation [52]. Accreditation, as mentioned before, enhances the confidence of the customer in the competence of the laboratory delivering the analytical data on the basis that the given laboratory was able to convince a panel of experts that it has qualified staff and necessary facilities to correctly carry out specified calibrations and tests or types of calibrations and tests [53]. However, it does not directly concern the quality of the results in contrast to the PT.

The main objective of PT is to assess the accuracy of the results obtained by participating laboratories. This is achieved by the distribution of one or more blind samples to several laboratories for the determination of given compounds by an independent scheme's coordinator. Each laboratory uses either its own or imposed method of analysis and returns the results to the organizer. At the end, test results are compared with each other and the participants are provided with a report illustrating the statistical interpretation of the data, from which the laboratory's performance can be evaluated.

A PT is only one of the different types of interlaboratory comparison presented in Figure 3.4. Their common feature, however, is the comparison of test results obtained by analysis of test samples. To enable reliable comparability of the analytical data between different laboratories, the test materials used in such interlaboratory exercises have to be sufficiently homogenous and at least display short-term stability [44]. The fact that the analyte level is unknown at the time of analysis makes

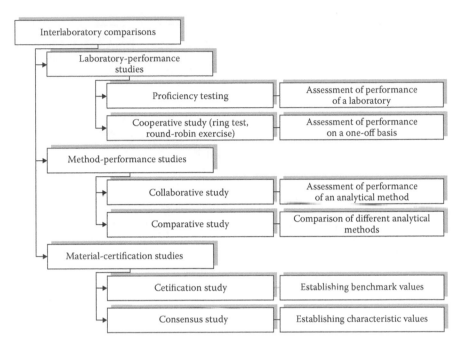

FIGURE 3.4 Types of interlaboratory comparisons. (Adapted from *Eurachem Guide: Selection, Use and Interpretation of Proficiency Testing (PT) Schemes by Laboratories, 2000.* Available at: http://www.eurachem.org/guides/ptguide2000.pdf; van Reeuwijk, L.P., and Houba, V.J.G. *Guidelines for Quality Management in Soil and Plant Laboratories.* FAO Soils Bulletin—74, 1998. Available at: http://www.fao.org/docrep/W7295E/w7295e00. HTM.)

the PT an independent tool for demonstration of various aspects of analytical proficiency [51]. For example, testing the reference material using different methods enables the detection of errors related to this procedure and the way a method is applied by a laboratory, while using the same predefined method, the performance criteria, such as repeatability and reproducibility, between different laboratories may be compared [36].

The use of CRMs in PT schemes has many advantages over other types of material. They help to assess not only precision performance between laboratories but also systematic errors and trueness. Figure 3.5 illustrates the exemplary results of a PT scheme on the determination of the arsenic content in tuna fish. Since in this scheme a CRM was used and the true value was known, it was possible to identify the laboratories that reported correct values (among the scattered pool of results provided by the participants). Such unsatisfactory results are still being obtained relatively often in the area of food testing [56], which proves a need for continuous monitoring of the laboratories and analytical methods' performance by means of international interlaboratory comparison programmes like Food Analysis Performance Assessment Scheme (http://www.fapas.com/) or International Measurement Evaluation Programme (http://irmm.jrc.ec.europa.eu/html/interlaboratory_comparisons/).

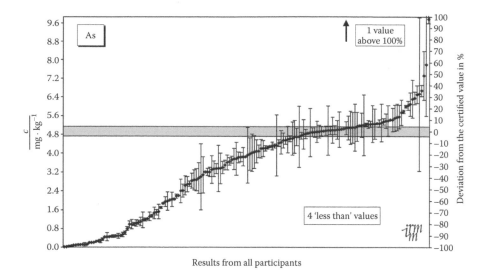

FIGURE 3.5 An example of an IMEP-20 proficiency PT scheme results of determination of arsenic content in tuna fish on a CRM; certified value: 4.93 ± 0.21 mg/kg (Reprinted from Aregbe, Y., Harper, C., Verbist, I., Van Nevel, L., Smeyers, P., Quétel, C., and Taylor, P.D.P. *Trace Elements in Tuna Fish*. IMEP-20 Report to Participants, 2004. With permission. Available at: http://irmm.jrc.ec.europa.eu/html/interlaboratory_comparisons/imep/imep-20/EUR21018EN.pdf.)

3.7.5 AVAILABILITY OF REFERENCE MATERIALS FOR FOOD ANALYSIS

The main limitation in common use of reference materials in food analysis laboratories is their restricted availability considering the wide range of possible analytes and/or matrices [44]. Obviously, in order to obtain reliable results, the composition of the reference material should be as similar as possible to that of a sample of interest in terms of level (e.g., concentration) of the analytes, composition of the matrix, physical form, stability and available quantity of the material, and acceptability of the uncertainty of the certified value [46]. In practice, it is rarely possible to obtain exactly matching reference materials and thus the relevance of the differences between the sample and CRM has to be evaluated and the best available material for the particular measurement has to be chosen. For example, when the concentration of the given analyte in the CRM is more than 10 times higher than its concentration in the sample, then the usefulness of such a reference material is questionable [51]. It is even more difficult to evaluate discrepancies related to sample matrix. For example, when the CRM is available in freeze-dried or lyophilized form and fresh foodstuff samples are to be analyzed [50].

Information about reference materials is available from a number of sources including the Internet. Currently, the biggest global-scale Internet database for a selection of reference materials provided by the Federal Institute for Materials Research and Testing is COMAR (http://www.comar.bam.de). It contains thousands of CRMs produced worldwide by about 220 producers in 25 countries. Information on many aspects of reference materials including a search engine can also be found at the webpage of the Virtual Institute on Reference Materials (http://www.virm.net).

Reference materials can also be searched in online catalogues of producers, e.g. members of the European Reference Materials (http://www.erm-crm.org): the Institute for Reference Materials and Measurements (irmm.jrc.ec.europa.eu), LGC Ltd. (http://www.lgc.co.uk) and the Federal Institute for Materials Research and Testing (http://www.bam.de). Other important non-European producers are: the National Institute of Standards and Technology (http://www.nist.gov), the Korea Research Institute of Standards and Science (http://www.kriss.re.kr), or the National Metrology Institute of Japan (http://www.nmij.jp).

REFERENCES

1. ISO 9000:2005. Quality management systems—Fundamentals and vocabulary.
2. van Zoonen, P., Hoogerbrugge, R., Gort, S.M., van de Wiel, H.J., van't Klooster, H.A. Some practical examples of method validation in the analytical laboratory. *Trends in Analytical Chemistry*. 1999, 18, 584–593.
3. Regulation (EC) No 882/2004 of the European Parliament and of the Council of 29 April 2004 on Official Controls Performed to Ensure the Verification of Compliance with Feed and Food Law, Animal Health and Animal Welfare Rules. OJ L 191, 28.5.2004. Available at: http://eurlex.europa.eu/LexUriServ/LexUriServ.do?uri=CONSLEG:2004 R0882:20060525:EN:PDF (01.08.2011).
4. ISO/IEC 17025:2005. General requirements for the competence of testing and calibration laboratories.
5. ISO/IEC 17011:2004. Conformity assessment—General requirements for accreditation bodies accrediting conformity assessment bodies.
6. Commission Regulation (EC) No 1029/2008 of 20 October 2008 Amending Regulation (EC) No 882/2004 of the European Parliament and of the Council to Update a Reference to Certain European Standards. OJ L 278, 21.10.2008. Available at: http://eur-lex.europa. eu/LexUriServ/LexUriServ.do?uri=OJ:L:2008:278:0006:0006:EN:PDF (01.08.2011).
7. Eurachem/CITAC Guide. Traceability in Chemical Measurement. A Guide to Achieving Comparable Results in Chemical Measurement, 2003. Available at: http://www.eura-chem.org/index.php/publications/guides (01.08.2011).
8. ISO 8402:1994. Quality assurance and quality management—Vocabulary.
9. CITAC/Eurachem Guide. Guide to Quality in Analytical Chemistry. An Aid to Accreditation, 2002. Available at: http://www.eurachem.org/index.php/publications/ guides (01.08.2011).
10. Eurachem/CITAC Guide CG 4. Quantifying Uncertainty in Analytical Measurement, 2000. Available at: http://www.eurachem.org/index.php/publications/guides (01.08.2011).
11. Eurachem/CITAC Guide. Quality Assurance for Research and Development and Non-Routine Analysis, 1998. Available at: http://www.eurachem.org/index.php/publications/ guides (01.08.2011).
12. Eurachem Guide. The Fitness for Purpose of Analytical Methods. A Laboratory Guide to Method Validation and Related Topics, 1998. Available at: http://www.eurachem.org/ index.php/publications/guides (01.08.2011).
13. Eurachem Guide EEE/RM/062rev3. The Selection and Use of Reference Materials, 2002. Available at: http://www.eurachem.org/index.php/publications/guides (01.08.2011).
14. Eurachem Guide. Selection, Use and Interpretation of Proficiency Testing (PT) Schemes by Laboratories, 2000. Available at: http://www.eurachem.org/index.php/publications/ guides (01.08.2011).

15. Thompson, M., WR. IUPAC Technical Report: Harmonized guidelines for internal quality control in analytical chemistry laboratories. *Pure and Applied Chemistry*. 1995, 67, 649. Available at: http://media.iupac.org/publications/pac/1995/pdf/6704x0649.pdf (01.08.2011).

16. De Bièvre, P., Dybkaer, R., Fajgelj, A., Hibbert, D.B. IUPAC Provisional Recommendations: Metrological Traceability of Measurement Results in Chemistry: Concepts and Implementation, 2009. Available at: http://old.iupac.org/reports/provisional/abstract09/debievre_prs.pdf (01.08.2011).

17. Thompson, M., Ellison, S.L.R., Wood, R. IUPAC Technical Report: Harmonized guidelines for single-laboratory validation of methods of analysis. *Pure and Applied Chemistry*, 2002, 74, 835. Available at: http://media.iupac.org/publications/pac/2002/pdf/7405x0835.pdf (01.08.2011).

18. Thompson, M., Wood, R. IUPAC Technical Report: The international harmonized protocol for the proficiency testing of (chemical) analytical laboratories. *Pure and Applied Chemistry*, 1993, 65, 2123. Available at: http://media.iupac.org/publications/pac/1993/pdf/6509x2123.pdf (01.08.2011).

19. Currie, L.A., Svehla, G. IUPAC Recommendations: Nomenclature for the Presentation of Results of Chemical Analysis, 1994. Available at: http://media.iupac.org/publications/pac/1994/pdf/6603x0595.pdf (01.08.2011).

20. ISO/IEC Guide 98–3:2008. Uncertainty of measurement—Part 3: Guide to the expression of uncertainty in measurement (GUM:1995).

21. ISO 5725. Accuracy (trueness and precision) of measurement methods and results, 1994 (Parts 1–4, 6), 1998 (Part 5).

22. ISO/IEC Guide 43. Proficiency testing by interlaboratory comparisons, 1997 (Part 1,2).

23. EAL-G23. The Expression of Uncertainty in Quantitative Testing, 1996. Available at: http://www.lysconsultores.com/Descargar/EAL-G23.pdf (01.08.2011).

24. ILAC-G17:2002. Introducing the Concept of Uncertainty of Measurement in Testing in Association with the Application of the Standard ISO/IEC 17025. Available at: http://www.ilac.org/guidanceseries.html (01.08.2011).

25. Eurachem/Eurolab/EA Guide EA-03/04. Use of Proficiency Testing as a Tool for Accreditation in Testing, 2001. Available at: http://www.eurolab.org/docs/use-of-pt.pdf (01.08.2011).

26. CAC/GL 54–2004. Guidelines on measurement uncertainty.

27. CAC/GL 59–2006. Guidelines on estimation of uncertainty of results.

28. CAC/GL 49–2003. Harmonized IUPAC guidelines for single-laboratory validation of methods of analysis.

29. Taverniers, I., De Loose, M., Van Bockstaele, E. Trends in quality in the analytical laboratory. II. Analytical method validation and quality assurance. *Trends in Analytical Chemistry*. 2004, 23, 535–552.

30. Prosek, M., Golc-Wondra, A., Krasnja, A. Quality assurance systems in research and routine analytical laboratories. *Accreditation and Quality Assurance*. 2000, 5, 451–453.

31. Robins, M.M., Scarll, S.J., Key, P.E. Quality assurance in research laboratories. *Accreditation and Quality Assurance*. 2006, 11, 214–223.

32. ISO/IEC Guide 99:2007. International vocabulary of metrology—Basic and general concepts and associated terms (VIM) and JCGM 200:2008: International vocabulary of metrology—Basic and general concepts and associated terms (VIM). Available at: http://www.bipm.org/utils/common/documents/jcgm/JCGM_200_2008.pdf (01.08.2011).

33. Bulska, E., Świtaj-Zawadka, A., Spójność pomiarowa. In: *Ocena i kontrola jakości wyników pomiarów analitycznych*. Wydawnictwa Naukowo-Techniczne, 2007, Warszawa, 73–89.

34. 55th WHO Technical Report. WHO Expert committee on biological standardization, *WHO Technical Report Series*, 932, 2006. Available at: http://whqlibdoc.who.int/trs/WHO_TRS_932_eng.pdf (01.08.2011).

35. Zschunke, A. The role of reference materials in analytical chemistry. *Accreditation and Quality Assurance*. 2003, 8, 247–251.

36. Quevauviller, P., Donard, O.F.X. Stated references for ensuring traceability of chemical measurements for long-term environmental monitoring. *Trends in Analytical Chemistry*. 2001, 20, 600–613.

37. Commission Regulation (EC) No 1881/2006 of 19 December 2006 Setting Maximum Levels for Certain Contaminants in Foodstuffs. OJ L 364/5, 20.12.2006. Available at: http://eur-lex.europa.eu/LexUriServ/LexUriServ.do?uri=OJ:L:2006:364:0005:0024:EN:PDF (01.08.2011).

38. Menditto, A., Patriarca, M., Magnusson, B. Understanding the meaning of accuracy, trueness and precision. *Accreditation and Quality Assurance*. 2007, 12, 45–47.

39. Taverniers, I., Van Bockstaele, E., De Loose, M. Trends in quality in the analytical laboratory. I. Traceability and measurement uncertainty of analytical results. *Trends in Analytical Chemistry*. 2004, 23, 480–490.

40. Kuselman, I., Shenhar, A. Uncertainty in chemical analysis and validation of the analytical method: Acid value determination in oils. *Accreditation and Quality Assurance*. 1997, 2, 180–185.

41. ISO Guide 35. Reference materials—General and statistical principles for certification, 2006.

42. ILAC-G9. Guidelines for the Selection and Use of Reference Materials, 2005. Available at: http://www.ilac.org/guidanceseries.html (01.08.2011).

43. Emons, H. Measurement standards in chemistry. Prerequisites for achieving reliable analytical results. *GIT Laboratory Journal*. 2007, 3–4, 42–44.

44. Emons, H., Held, A, Ulberth, F. Reference materials as crucial tools for quality assurance and control in food analysis. *Pure and Applied Chemistry*. 2006, 78, 135–143.

45. ISO Guide 30:1992/Amd 1:2008. Revision of definitions for reference material and certified reference material.

46. ISO Guide 33. Uses of certified reference materials, 2000.

47. Emons, H., Linsinger, T.P.J., Gawlik, B.M. Reference materials: Terminology and use. Can't one see the forest for the trees? *Trends in Analytical Chemistry*. 2004, 23, 442–449.

48. Emons, H. The 'RM family'—Identification of all of its members. *Accreditation and Quality Assurance*. 2006, 10, 690–691.

49. Linsinger, T. Comparison of a measurement result with the certified value. ERM Application Note 1, 2005. Available at: http://www.erm-crm.org/ERM_products/application_notes/Pages/index.aspx (01.08.2011).

50. Walker, R., Lumley, I. Pitfalls in terminology and use of reference materials. *Trends in Analytical Chemistry*. 1999, 18, 594–616.

51. Jorhem, L., Engman, J., Sundström, B., Nilsson, A. Evaluation of measurement data for Cd, Cr and Pb in certain uncontaminated foodstuffs published in surveys: Analytical quality vs. uncertainty of measurements. *Accreditation and Quality Assurance*. 2006, 10, 647–658.

52. Linsinger, T.P.J., Bernreuther, A., Corbisier, P., Dabrio, M., Emteborg, H., Held, A., Lamberty, A., Lapitajs, G., Ricci, M., Roebben, G., Trapmann, S., Ulberth, F., Emons, H. Accreditation of reference material producers: The example of IRMM's Reference Materials Unit. *Accreditation and Quality Assurance*. 2007, 12, 167–174.

53. Laboratory Accreditation in Developing Economies. Tested Once—Accepted Everywhere. UNIDO Working Paper No. 2, 2003. Available at: http://www.unido.org/fileadmin/import/55885_02ILAC_UNIDO_PUBLICATION.PDF (01.08.2011).

54. van Reeuwijk, L.P., Houba, V.J.G. Guidelines for Quality Management in Soil and Plant Laboratories. *FAO Soils Bulletin*—74, 1998. Available at: http://www.fao.org/docrep/W7295E/W7295E00.htm (01.08.2011).

55. Aregbe, Y., Harper, C., Verbist, I., Van Nevel, L., Smeyers, P., Quétel, C., Taylor, P.D.P. Trace Elements in Tuna Fish. IMEP-20 Report to Participants, 2004. Available at: http://www.irmm.jrc.be/interlaboratory_comparisons/imep/imep20/eur21018en.pdf (01.08.2011).

56. Kaarls, R. Metrology, essential to trade, industry and society. *Accreditation and Quality Assurance*. 2007, 12, 435–437.

4 Analysis of Drinking Water

Marek Biziuk and Malgorzata Michalska

CONTENTS

Water is a substance necessary for human well-being. Daily we drink 2–3 L of water. It is the source and basis of life on the earth, but at the same time, it can be a serious hazard to our health and life. Intensive industrialization has resulted in an increased input of toxic inorganic and organic compounds, drastically reducing the quality of surface waters, which are a source of drinking water for a large part of the world's population. Groundwater supply is diminishing and, in addition, anthropogenic groundwater pollution has become a fact of life. The majority of anthropogenic water pollutants are toxic and dangerous not only to humans but also to animals and plants. These pollutants should be determined and monitored in tap and surface water. Water pollutants can be classified as physical, biological, radioactive, inorganic, or organic. This chapter discusses only the biological and chemical organic and inorganic water pollution.

4.1 ORGANIC POLLUTANTS

Most organic compounds in water, such as humic and fulvic acids or peptides, are of natural origin. These compounds are predominantly nontoxic, but they can be precursors of toxic compounds in the process of water treatment. Anthropogenic organic compounds present the main hazard to life and health. The number of known organic compounds is now estimated to be about 16 million, 2 million of which are produced by synthesis alone. At present, about 70,000 organic compounds are commercially available, with an annual global production of 100–200 million tons. More than 30% of all organic compounds produced end up in the environment, including tap water.[1-10] In some water supplies, over 700 chemical compounds have been detected, including more than 600 organic compounds, many of which are biologically active.

Anthropogenic organic compounds occur in drinking water at relatively low levels, but even at these levels, they are dangerous. This low concentration causes, in the majority of cases, the need for isolation and preconcentration of the analytes from the complex water matrix. Sample pretreatment, the initial stage of an analytical procedure, becomes the essential stage in environmental analysis. According to participants in the survey taken by LC-GC International,[11] among analytical laboratories, the sample pretreatment takes up 61% of total time of analysis and is the source of 30% of the errors generated during sample analysis. The above data clearly indicate that sample pretreatment, including isolation and preconcentration of the analytes, is an essential step of trace analysis. The methods of isolation and preconcentration of organic compounds from water are closely associated with the type of analytes, their volatility, polarity, stability, water solubility, solubility in organic solvents, and so on. A variety of techniques of isolation and preconcentration of the analytes from water have been developed.[3,5,9,10,12-19] The most popular methods are solvent extraction or liquid–liquid extraction (LLE), solid-phase extraction (SPE), solid-phase microextraction (SPME), and techniques utilizing the distribution of solute among the liquid and the gaseous phase (headspace, purging). Less commonly used methods include supercritical fluid extraction, freezing out, lyophilization, vacuum distillation, steam distillation, and membrane processes (reverse osmosis, ultrafiltration, and dialysis).[3,5,9,10,12-19] For the final determination, gas chromatography with specific detectors such as electron capture detector (ECD), nitrogen–phosphorus detector (NPD), flame photometric detector (FPD), atomic emission detector, or electrolytic conductivity detector (ELCD), or universal methods such as flame ionization detection (FID) or mass spectrometry (MS) are used. Liquid chromatography methods with ultraviolet (UV), diode-array (DAD), electrochemical, or fluorescent detectors are also often used. A schematic diagram of the utilization of various isolation techniques for the determination of organic compounds in water is shown in Figure 4.1.[9] In Figure 4.1, two main ways for the determination of organic compounds in water are shown: determination of individual compounds and determination of total parameters. Total parameters including total organic carbon or halogen (TOC, TOX) and dissolved and suspended organic carbon or halogen are used to characterize the content of organic compounds in water. Other parameters are defined in terms of the method of isolation of an organic fraction from water: volatile, purgeable, extractable, and adsorbable organic carbon or halogen. However, total parameters measuring the carbon content in an organic fraction are not particularly suitable as an estimate of anthropogenic water pollutants and their hazard to human health because a decisive majority of organic compounds in water are biogenic.

The two main methods of isolation and preconcentration of organic contaminants from water are LLE and SPE, and although both are simple and do not require

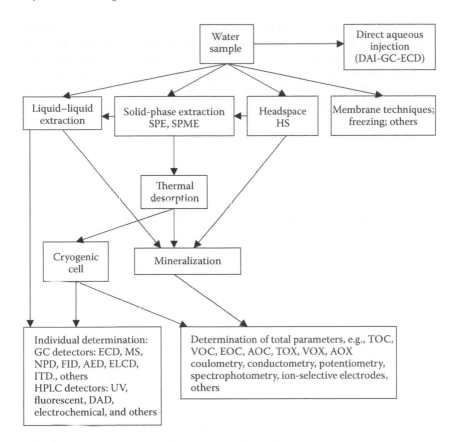

FIGURE 4.1 Classification of techniques for the isolation and determination of organic compounds in water.

sophisticated apparatus, they have a number of drawbacks. LLE is one of the oldest methods for isolation of organic compounds from aqueous phase. It is based on a favorable partition coefficient of the analytes between an aqueous phase and an organic solvent with which the analytes are extracted. This principle also determines the applicability of LLE. Organic solvents used for the extraction have to be immiscible with water (i.e., relatively nonpolar), and the analytes have to dissolve better in the solvent than in water. The most favorable case for the analyst is a close-to-100% transfer of the analytes into the organic phase. This can be achieved by proper selection of the solvent, by adjusting the pH of the aqueous phase, or by adjusting its ionic strength. If the extraction into the organic phase is not close to 100%, a correction factor has to be introduced or an internal standard with properties similar to those of the analytes has to be added to the sample. This does not assure reliability of the results because of the error in determining the correction factor and the effect of the matrix and temperature on the partition coefficient. The extraction proper is carried out by adding an organic solvent immiscible with water to the aqueous sample (typically in a 1:10 to 1:100 ratio) and vigorously stirring or shaking the two phases. More-volatile compounds [e.g., trihalomethanes (THMs), other volatile halocarbons, and volatile hydrocarbons] evaporate during these operations, with a consequent loss of recovery and lower results. Some compounds are poorly extracted from the water phase (e.g., ketones and alcohols), or a partition coefficient is strongly dependent on matrix (diversified). The LLE requires the use of large volumes of expensive, and possibly toxic, solvents. After SPE (except with the methods using a thermal desorption for volatile compounds), solvents are also used, but not in such large quantities. Solvents then are evaporated and after analysis, they create the problem of storage of hazardous waste. Working with these solvents may require the use of personal protection devices, such as respirators. The formation of emulsion is yet another problem with LLE. The solvents and sorbents used for extraction should be of very high purity, free from traces of the analytes and other potential interference. Otherwise, a large background is obtained precluding the determination.

For volatile compounds, headspace techniques[4,9,10,13,14,19] can be used. The analytes in a liquid sample are, thus, determined by measuring their concentration in the gaseous phase that is in thermodynamic equilibrium with the sample. The methods are rapid and straightforward. They eliminate the effects of matrix components, which could interfere with the determination and contaminate the GC injector, detector, and column. The problems associated with the blank contributed by the solvents in LLE methods, or by the sorbents in SPE methods, are also eliminated. Hence, headspace techniques are suitable for the analysis of samples with a high content of inorganic compounds (e.g., seawater, juices, wastewater), high–molecular weight organic compounds (e.g., polymers, humic substances), and nonhomogeneous mixtures (e.g., blood, other physiological fluids, colloids, wastewater), which would otherwise require complex sample purification and treatment procedures. These techniques are limited only to the volatile compounds, and a partition coefficient is also strongly dependent on matrix (unknown) and affects the accuracy. Additionally, some compounds belonging to priority pollutants, listed by World Health Organization (WHO) as the most dangerous pollutants, are not purgeable from the water phase (e.g., acrylonitrile, propionitrile).

The only method for the determination of organic compounds in water that avoids the isolation and preconcentration step uses direct injection of an aqueous sample onto a gas chromatographic column (DAI-GC-ECD).

4.2   VOLATILE ORGANOHALOGEN COMPOUNDS

Volatile organohalogen compounds are used mainly as solvents, cleaning and degreasing agents, blowing agents, polymerization modifiers, and heat-exchange fluids. As wastes, they find their way into lakes and rivers and then into seas and oceans. Their concentrations in water and air are variable and depend on atmospheric conditions due to the washing by rain and evaporation from water during long periods of warm weather. It is estimated that the annual global production of organohalogen solvents alone amounts to several million tons.[1] One of the most important sources of organohalogen compounds, particularly volatile ones, is water disinfection by chlorination.[1,2,4,9,19–23] The actual disinfecting agent is hypochlorous acid formed in the course of disproportionation reaction that takes place when chlorine dissolves in water. During chlorination, humic and fulvic compounds (the so-called precursors), which are harmless and occur naturally in water, are converted into organohalogen compounds, which are dangerous to human life and health. The largest group of compounds formed during chlorination is the THMs. These include chloroform (the most abundant compound), bromodichloromethane, dibromochloromethane, and tribromomethane. Organobromine compounds are formed when the water being chlorinated contains a large amount of bromides or when the chlorine used for disinfection is contaminated with bromine. Hypobromous acid formed in the reaction of bromide ions with hypochlorous acid reacts with an organic matrix about 200 times faster than hypochlorous acid.[22] The amount and kind of organohalogen compounds formed depend on the water pH, the amount of chlorine used, and the content of organic matrix (TOC) in chlorinated water. THMs are not the only organohalogen compounds formed in the course of chlorination. Other volatile organochlorine compounds, such as tetrachloromethane, chloroethylene, 1,1-dichloroethylene, 1,1,2-trichloroethylene, tetrachloroethylene, 1,1,1-trichloroethane, and 1,2-dichloroethane, are also commonly found in chlorinated water. In addition, chlorination of humic substances and organic water pollutants yields a variety of other derivatives, of which over 100 have already been identified, including chlorinated acetone, chlorinated acetonitrile, chloropicrin, chloral, chloroacetic acids, chlorinated ethers, chlorophenols, and chlorinated ketones. Koch[24] estimated that among organohalogen compounds formed during chlorination of water, 77% are THMs, 15% are haloacetic acids, 3% are halonitriles, 4% are trichloroacetaldehyde hydrates, and 1% are the remaining compounds. Chlorination causes very toxic, mutagenic, and carcinogenic volatile organohalogen compounds to be present in tap water at relatively high levels (100 µg/L and more). WHO, European Union (EU), and most countries, including Poland, introduced maximum admissible concentration (MAC) for some organohalogen compounds present in tap water. These compounds include THMs (trichloromethane, bromodichloromethane, dibromochloromethane, tribromomethane), tetrachloromethane, 1,2,2-trichloroethene, 1,1-dichloroethene, 1,1,2,2-tetrachloroethene, 1,2-dichloroethane, and 1,1,1-trichloroethane. The MAC varied from 0.3 µg/L for 1,1-dichloroethene (WHO, Norway) to 100 µg/L (EU, WHO, Environmental Protection Agency [EPA], United Kingdom)and to 350 µg/L (Canada) for the sum of THMs. Typical concentration of trichloromethane (chloroform), the compound most frequently present in the water treated by chlorination, varied from 1 to 30 µg/L; 30 µg/L is the MAC in the directives of WHO, Poland, and Great Britain for chloroform in the tap water. DAI-GC-ECD is a suitable method for determination of organohalogen compounds in this range of concentration. This technique also avoids all problems related to the hot vaporizing injection technique by introduction of the sample directly into the oven-thermostated column inlet. However, aqueous samples, particularly environmental ones, have a very complex matrix and a large amount of inorganic salts and organic compounds (humic acids and other contaminants). These compounds can rest on the top of the column and cause additional retention or deactivation of the column. Some low-volatile compounds can be eluted from the column over a long period of time. After direct introduction of aqueous samples, during evaporation, a large volume of water vapor is formed. At 20°C, from 1 µL of water, 60 mL of carrier gas saturated with water vapor is produced. A large volume of water can deactivate the stationary phases and the layer and change the chromatographic condition of the column. Another problem is retention of water on the chromatographic column. It is, particularly in the packed column, a tedious and time-consuming process. Water also has high surface tension (poor wetting properties) and poor properties concerning solvent effects. Most of the detectors used in chromatographic determination are not compatible with water, especially the flame-based detectors or infrared (IR) or other spectroscopic detectors. Flame in an FID or a thermoionic NPD can be put out by water. On the other hand, in the case of IR detector, a measurement cell, made very often from KBr, can be diluted in water. Relatively resistant to the presence of water are ECDs or ion trap detectors.

The DAI method requires a special injector allowing cold on-column injection and special capillary columns. The nonvaporizing, septumless, cold on-column injector with secondary cooling (Grob type) fulfills nearly all requirements for a good introduction of sample onto a chromatographic column. The construction of such an injector produced by Carlo Erba is shown in Figure 4.2.[9] The analyzed aqueous sample (1–2 μL) is introduced directly to the oven-thermostated capillary gas chromatographic column inlet using a syringe with outer diameter of the needle smaller than the inner diameter (ID) of the column. The secondary air cooling of the inlet of the column prevents sample evaporation in the syringe needle (including its tip). The evaporation starts after the deposition of the sample on the column wall. The sample is mechanically transferred instead of being evaporated. So, the high-boiling

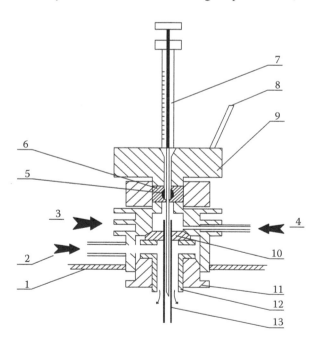

FIGURE 4.2   Schematic diagram of nonvaporizing, septum-less, cold on-column injector: (1) oven wall; (2) secondary air cooling; (3) principal air-cooling system; (4) carrier gas inlet; (5) stainless steel rotating valve; (6) valve seal; (7) syringe; (8) valve lever; (9) upper part of injector; (10) column seal; (11) bottom part of injector; (12) cooling jacket; and (13) capillary column.

components do not remain in the syringe needle. The secondary cooling system is switched on only before and during the sample injection. The primary cooling of the injector body works permanently. The secondary cooling enables rapid cooling of the bottom part of the injector attached to the oven, preventing discrimination of compounds at the syringe tip. It also allows the introduction of a large volume of sample, which permits a decrease in the detection limits. The sample is injected at an injector temperature lower than the solvent boiling point, but at an oven temperature higher than the solvent boiling point. Nonvolatile inorganic and organic compounds can gradually accumulate at the column inlet, which will deteriorate the resolving power of the column. To prevent this, readily exchanged, deactivated, noncoated pre-columns (2 m × 0.32 mm) connected to the analytical column through a zero dead-volume fitting are used. Following a specified number of analyses, the precolumn can be completely exchanged or a part of it can be broken off.

The precolumns play another important role. They serve as the so-called retention trap, narrowing chromatographic bands of the analytes. The sample volume during fast injection spreads in the column inlet, flooding it and causing "band broadening in the space," so it is necessary to reconcentrate the broad inlet band. The band broadening in the space depends on the temperature, the speed of injection, and the injected volume. In the precolumn, all compounds present in the gaseous phase migrate with the same speed due to lack of any interaction. In the main column coated with a stationary film of nonpolar phase, the analytes interacting with the stationary phase are focused, and water, which does not react with the stationary phase, is drained out freely. To prevent retention of water, the precolumn and the analytical column should be deactivated. The film of the stationary phase should also be immobilized by cross-linking. During the chromatographic process, water is eluted before the analytes because of the use of an inert column coated with a thick layer (approximately 5 μm) of nonpolar stationary phase, which results in rapid and complete elution of water and sufficient retention of the analyzed organohalogen compounds. Methyl silicones (DB-1) are used as nonpolar stationary phases. To improve separation, the methyl silicone stationary phases modified with vinyl groups (PS-255—1% vinyl groups) are applied. Capillary columns 30 m long with an ID of about 0.5 mm are used. Longer columns provide an improved resolution but at the expense of sensitivity. Isothermal runs at about 102°C–105°C are employed. As a result of the high sensitivity of the ECD employed, the stationary phase in the analytical column is immobilized through cross-linking and chemical binding to the column wall.

The DAI-GC-ECD method proves to be a good and very useful method for determination of volatile organohalogen compounds in tap and swimming pool water because of its simplicity (no isolation and preconcentration steps are necessary), repeatability, reduction of the possibility of sample contamination by solvent or sorbent used, reduction of possible losses of volatile compounds during additional steps, and low detection limits (0.002–0.1 μg/L; only for dichloromethane, 1 μg/L) depending on the percentage of halogen in the compound. The detection limit of the method is related to the amount of the analyte in the sample injected onto the column and the volatility of the analytes. On the basis of presented investigations, Biziuk and Czerwiński[25] elaborated the Polish Standard Method (PN-C-04549–1) for determination of dichloromethane, trichloromethane, tetrachloromethane, bromodichloromethane, dibromochloromethane, tribromomethane, trichloroethylene, tetrachloroethylene, 1,1,1-trichloroethane, and 1,1,2,2-tetrachloroethane in water by gas chromatography using direct injection of the sample.

In the method routinely used in the laboratory of SAUR-Neptun Gdańsk (water supply and sewage system for Gdańsk) for the determination of volatile and semivolatile organohalogen compounds in tap water, a column of Rtx-Volatiles (60 m × 0.32 mm ID) coated with a 3-μm film and an oven temperature program (104°C for 1 min and then to 160°C at 2.5°C/min) was used. The exemplary chromatogram for 15 compounds obtained in this laboratory is shown in Figure 4.3.

The use of the DAI-GC-ECD method for samples with very complex matrices containing many inorganic salts or nonvolatile organic compounds, as well as for low concentration of analytes, creates many problems. Only small amounts of the samples (2 μL) are introduced into the column, which causes a high detection limit. The detection limit of the DAI-GC-ECD method for volatile organohalogen compounds is sufficient for tap and swimming pool water treated by chlorination and for polluted surface waters but insufficient for unpolluted underground or surface

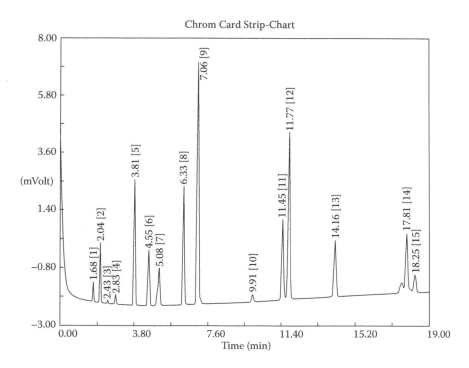

FIGURE 4.3 A sample chromatogram of volatile organohalogen compounds (DAI-GC-ECD—temperature program): (1) 1,1-dichloroethene; (2) dichloromethane; (3) 1,2-dichloroethene; (4) 1,1-dichloroethane; (5) trichloromethane; (6) 1,1,1-trichloroethane; (7) tetrachloromethane; (8) trichloroethene; (9) bromodichloromethane; (10) 1,1,2-trichloroethane; (11) tetrachloroethene; (12) dibromochloromethane; (13) 1,1,1,2-tetrachloroethane; (14) tribromomethane; and (15) 1,1,2,2-tetrachloroethane.

waters. DAI-GC-ECD cannot be performed for complex matrices such as water rich in humic acids, body fluids, mineral waters, soft drinks, beers, and juices. Inorganic salts and nonvolatile organic compounds deposited in the precolumn retain water, causing broadening of the water peak and an increase in the retention time for analytes. These contaminants can be slowly transferred to the main column and to the detector, affecting the peak shape, deactivating the column, and changing the chromatographic condition of the column and the detector. These drawbacks can be eliminated by applying a combination of thin-layer headspace (TLHS) technique with the DAI method. This combination allows the preconcentration of analytes in the aqueous phase and a change of the matrix from a complex mixture of inorganic and organic compounds for a solution of analytes in pure water. The countercurrent TLHS method introduced by Kozlowski and coworkers[4,9,14,26 38] is based on the passing of volatile compounds from a thin layer of analyzed aqueous sample to a stream of countercurrently flowing stream of purified air. The main part of the apparatus for TLHS is the vertically positioned, spirally wound thermostated glass tube (2 m × 7 mm ID) shown in Figure 4.4. The analyzed sample is introduced on the top of the tube and flows down in the form of thin film, from which the analytes together with water vapors are transferred to the gaseous phase and transported for the final determination. This approach can be used as a different method for determination of volatile organic compounds: for determination of total parameters and of individual compounds and for preparation of the sample for DAI analysis.[33–38] In

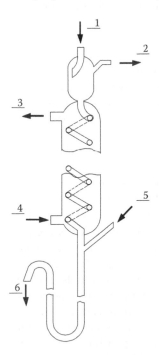

FIGURE 4.4   Column for countercurrent TLHS in a spiral tube: (1) sample inlet; (2) outlet of the gas stream containing the liberated analytes (to the combustion tube); (3 and 4) outlet and inlet of water from a thermostat, respectively; (5) inlet of purified air; and (6) sample outlet.

such a case, two TLHS columns have to be used. In the first column, thermostated in high temperature (up to 90°C), the analyzed compounds are transferred together with water vapors to the stream of air; in the second column, they are cooled to 10°C–11°C, and column water vapors are condensed together with the analytes. The condensation leaving the second column is analyzed using DAI-GC-ECD. There are very important advantages of using such a procedure: the compounds of interest are isolated from the complex aqueous matrix and preconcentrated in pure water and it allows analysis of volatile organohalogen compounds in beverages[36] (mineral waters, juices, soft drinks, etc.) and in biological fluids (urine).

Numerous studies have dealt with the LLE with pentane of volatile organohalogen compounds, the products of chlorination of humic acids in the course of water treatment.[4,9] Besides pentane, other solvents such as diethyl ether, hexane, methylcyclohexane, isooctane, dichloromethane, or other solvents and their mixtures have been used.[4,9] In the majority of cases, the isolated volatile organohalogen compounds were separated and determined by GC with ECD. The cleaned extracts were also used to determine extractable organic halogen, following their combustion and coulometric determination of the resulting halide ions.[4,9]

4.3 VOLATILE HYDROCARBONS

For the volatile organic compounds (volatile organohalogen compounds and volatile hydrocarbons), the best methods of determination in water or, generally, liquid phase (juices, soft drinks, etc.) are based on the headspace techniques (static headspace, stripping, purge and trap, closed-loop stripping analysis).[4,9,13,14,19] Figure 4.5 presents the classification of headspace methods used for the determination of total parameters and individual compounds.

The simplest version of headspace technique is static headspace.[4,9,13,14,19] It involves the analysis of the gaseous phase that is in thermodynamic equilibrium, at a constant temperature T, with the analyzed liquid sample. The equipment necessary to carry out static headspace analysis is very simple and enables automation of the sample injection into the GC column. The design of headspace accessories has been described in a number of studies,[4,9,13,14,19] and the accessories are now available commercially from numerous sources, such as Perkin Elmer (HS-6B; HS-100; HS-101), Hewlett-Packard (HP-19395A), Carlo Erba (currently Fisons, HS-250), and DANI (HSS-3950). Detection limits of static headspace GC depend primarily on the sensitivity of the GC detector used, the boiling point of the analyte, and the analyte's distribution coefficient between the liquid and the gaseous phase. Using the FID and injecting a 1 mL gaseous sample, a detection limit of the order 1–10 µg/L is obtained. For organohalogen compounds, the detection limit can be lowered to 0.1 µg/L by using the ECD. Besides increasing the temperature, the distribution coefficient value can be increased for certain compounds by as much as 2 orders of magnitude by salting out the liquid sample (mostly with NaCl solution) or by varying its pH, thus improving the detection limit. The static headspace technique is suitable for volatile compounds with boiling points up to 200°C. The major advantages of static headspace GC include simplicity and rapidity, high sensitivity and precision, ability to remove most of the solvent and inorganic and nonvolatile organic

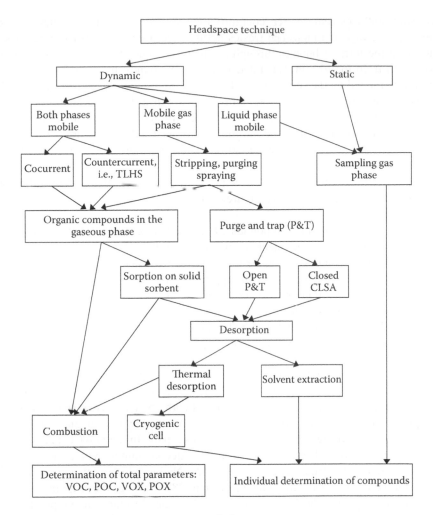

FIGURE 4.5　Classification of headspace techniques.

compounds from the sample injected into a gas chromatograph, and the possibility of automation of the final determination.

Headspace techniques employing immobile liquid phase and mobile gaseous phase fall into two categories. A stream of gas can pass over the surface of the analyzed liquid sample or it can bubble through the sample (stripping and purging). The term "immobile liquid phase" applies to the liquid sample that is not being replenished because in both categories, the liquid phase is vigorously stirred either by a magnetic stirrer or by a gas stream. A substantially increased interface is achieved by passing a stream of gas in the form of tiny bubbles through the liquid phase. For the determination of volatile hydrocarbons (such as heptane, benzene, isooctane, nonane, decane, undecane, xylene, and toluene) and chlorobenzene, purge and trap technique, sorption on solid sorbent (Tenax TA), thermal desorption, and GC-FID final determination can be used.[4,10,13–17,39–42] The presence of volatile hydrocarbons indicates anthropogenic water contamination by crude oil–like compounds of municipal sewage and industrial wastes nature. The stream of purified purge gas is passed in the form of bubbles through an analyzed aqueous layer at a flow rate of 30–40 mL/min for 10 min. Liberated volatile compounds are dried by the Nafion drier and trapped on the solid sorbent Tenax TA. Subsequently, after turning on the six-port switching valve, analyzed compounds are desorbed in a stream of helium (20 mL/min) at a temperature of 250°C for 2 min and thereafter are analyzed by GC-FID. The volume of the used purging device is 20 mL, the volume of the water sample is 10 mL, and the mass of the sorbent layer is 120 mg. The detection limit of this method is about 0.05 µg/L.

The purge technique can also be used in a different way. To this end, the following approaches can be employed (see Figure 4.2):

*   The liberated organic compounds are trapped in a cryogenic cell, converted to liquid nitrogen, and injected into a GC column.
*   The purged compounds can be trapped directly inside a cooled GC column [purge with whole-column cryotrapping (P/WCC)].
*   The liberated compounds can be adsorbed on a solid sorbent using an open purge and trap system.
*   The liberated compounds can be trapped on a solid sorbent using a closed-loop stripping system (closed-loop stripping analysis).

In the first approach, the freezing of the purged compounds takes place in a capillary tube cooled with liquid nitrogen. Water is removed using a drier (a piece of Nafion tubing passing through a container filled with a 5-Å molecular sieve). After the trapping process is completed, the cryogenic cell is rapidly heated, and the released compounds are introduced into a GC column as a plug. In the P/WCC method, volatile organic compounds are stripped from the aqueous sample with a stream of gas and, after removing water vapor (e.g., by using Nafion tubing), introduced directly into a GC column cooled with liquid nitrogen to 280°C. When the stripping/trapping process is completed, the GC column is rapidly heated to the starting point of the temperature program, and a normal GC analysis is performed.

Headspace techniques with mobile liquid and gaseous phases involve the transfer of volatile organic compounds from the mobile aqueous phase to the gaseous phase moving in countercurrent or in cocurrent.[4,9,14,26–38] The techniques have been used extensively in the removal of volatile organic compounds from tap water and especially toxic organohalogen compounds formed in the water-treatment process. The removal is usually accomplished in stripping towers or columns in which the water being purified and a stream of air run countercurrently, and the volatile organic compounds are stripped from the aqueous phase. The devices used for the determination of volatile organic compounds in water are based on similar principles. The TLHS method involves the transfer of volatile organic compounds from a thin layer of the aqueous phase flowing through a spirally wound, thermostated glass tube to a stream of purified air flowing countercurrently.[4,9,26–38] A diagram of such a device, patented by Kozlowski and coworkers, is shown in Figure 4.4. The air used as a stripping gas is prepurified by passing it through the Körbl catalyst heated to

500°C, which removes organic and organosulfur or organohalogen compounds. The analytes released from water are introduced through the air stream into the final determination system.

4.4 NONVOLATILE ORGANIC COMPOUNDS (PESTICIDES, POLYCYCLIC AROMATIC HYDROCARBONS, AND POLYCHLORINATED BIPHENYLS)

Pesticides are particularly important pollutants among organic compounds because of their common use, persistence in the environment, and toxicity. They are predominantly anthropogenic. Pesticides increase crop yields by reducing the amount of crop that is lost to pests and helping control diseases transmitted by insects. Pesticides are necessarily toxic as they are used to kill pests in agriculture, industry, and households. The present global annual production of pesticides is estimated at several hundred thousand tons. Pesticides in soil and surface water can enter drinking water system. Also, other toxic pollutants such as polychlorinated biphenyls (PCBs), polycyclic aromatic hydrocarbons (PAHs), and phenols can enter the environment, including tap water. Recently, some pesticides that are persistent and nonbiodegradable have been abandoned, and the most dangerous ones have been banned in most countries (dichlorodiphenyltrichloroethane and other organochlorine pesticides). For humans, the major route of exposure to these pollutants is the gastrointestinal system, mainly via food (because of bioaccummulation and biomagnification in the food chain) and drinking water. Pesticides, PCBs, and some PAHs are carcinogenic, mutagenic, and teratogenic and cause cardiovascular and neurological diseases.[5]

For determination of pesticides, PCBs, and PAHs in water, sample pretreatment step includes the analytes' isolation, enrichment, and purification of extracts (removal of matrix components). LLE has been used extensively in the isolation of semivolatile and nonvolatile compounds from the aqueous phase. Examples of such applications include the determination of triazine and phenylcarbamate herbicides following their extraction with dichloromethane, trichloromethane, or a mixture of benzene and ethyl acetate or the determination of halogenated pesticides and PCBs following their extraction with hexane, dichloromethane, freon TF, diethyl ether, naphtha ether, ethyl acetate, pentane, tetrachloromethane, trichloromethane, or a mixture of solvents.[4,10,13] The extract obtained during an LLE usually has to undergo a further cleanup. Typically, water is removed by the addition of anhydrous sodium sulfate. Solvent evaporation is performed in a rotary evaporator, in a Kuderna-Danish evaporator, or in a needle evaporator in a stream of inert gas.

To clean up the extract and to isolate the analyte fraction, columns with the following packings are used:

Silica gel is most widely used for the extract cleanup and the removal of strongly-basic compounds.

Most frequently used neutral alumina (pH 6.9–7.1) is employed to isolate hydrocarbons, esters, aldehydes, ketones, lactones, quinones, alcohols, and weak organic acids and bases. Basic alumina (pH 10–10.5) is used to isolate compounds that are unstable in acidic medium, and acidic alumina (pH 3.5–4.5) is used to isolate organic acids and inorganic compounds.

Florisil (magnesium silicate) is widely used to clean up natural samples to determine trace amounts of pesticides.

Following sample cleanup and chromatographic separation, nitrogen- and phosphorus-containing organic compounds are usually determined using the NPD, and organohalogen compounds are determined using the ECD. GC-MS is also used extensively for the identification and determination of analytes. The LLE methods, although simple and do not require sophisticated apparatus, have a number of drawbacks, which are mentioned above.

Solvent microextraction is an interesting new variant of LLE, one that is much less labor intensive than LLE or SPE. It is realized by immersing an organic solvent drop in an aqueous sample. It is intensified by using a magnetic stirrer. SME has many advantages compared with the classical LLE: it largely minimizes the problems associated with large volumes of expensive and toxic solvents of high purity (generating hazardous waste), the formation of emulsion, and the effect of solvent background on the results of analyses. The SME process is rapid, not requiring the use of a separatory funnel, and the construction of the extraction vessel is simple.

Sorption on solid sorbents is currently the most common technique for the isolation of organic compounds from water.[5,9–18] The process is based on partition of organic compounds dissolved in water between a solid sorbent and the aqueous phase. The retention of substances on solid adsorbents is predominantly due to van der Waals forces. The technique has many advantages:

- It allows isolation and preconcentration on the sorbent bed of both volatile and nonvolatile organic compounds present in water, which enables the determination of both individual analytes and total parameters (TOC, TOX, TOS, and so on); water and inorganic compounds are minimally retained and readily removed from the bed during washing and drying, following the sorption step.
- For properly selected adsorbents, the partition coefficients of the analytes between the sorbent and water approach infinity.
- Wettability of a sorbent by water results in a satisfactory transport of the analytes toward the sorbent surface.
- Sorbents do not chemically react with the preconcentrated analytes and can be readily regenerated, which extends their lifetime to several years.
- Compared with LLE, the problems associated with using large volumes of expensive and toxic solvents of high purity—generating hazardous waste, the formation of emulsions, and the effect of solvent background on the results of analyses—are largely minimized.
- The analytes trapped on a sorbent can be transported and stored, allowing one to perform the final determination at the most convenient time.
- The enrichment process is rapid and straightforward and can be readily automated.

Preconcentration can be carried out by passing the analyzed aqueous sample through a column packed with a sorbent or in a batch mode by agitating the aqueous sample with a sorbent added to it and then separating the sorbent by filtration. Following preconcentration of the analytes, further analysis can be carried out as follows:

- The whole sorbent bed can be combusted and the total parameters of water pollution determined (in the case of carbon sorbents only).
- The trapped analytes are solvent extracted, and following the extract cleanup, individual analytes are determined by chromatography, or the extract is combusted and the total parameters are determined.

In analytical practice, a variety of adsorbents have been used for the isolation of organic compounds from water, and new sorbents are being developed that are better suited for specific matrices, methods of final determination, or selective isolation of a particular group of compounds. The most common kinds of sorbents, along with their functional groups, matrices, and possible mechanisms of interactions with the analytes, have been tabulated in the literature. Powdered activated charcoal and granulated activated charcoal have been used for the longest time for the isolation of organic compounds from water. This sorbent has been used frequently because of its high sorption capacity, thermal stability, and the possibility of combustion of the entire sorbent bed. The drawbacks of activated charcoal include incompleteness of sorption and desorption, the possibility of reactions of the sorbent with the analytes, and a relatively high background.

Amberlite XAD macroporous resins have found use in the isolation of organic compounds from water. Their basic physical properties have been described elsewhere. Like activated charcoal, XAD resins also have to be pretreated prior to use in trace analysis. XAD-2 and XAD-4 (styrene-divinylbenzene copolymers) have been used most frequently because of their hydrophobicity and lack of adsorption of inorganic ions. Solvent extraction is the most common method of desorption of the trapped analytes from the sorbent bed. The solvents used for desorption include diethyl ether, ethyl acetate, methanol, acetone, acetonitrile, cyclohexane, n-hexane, dichloromethane, ethanol, or a mixture of two solvents. Solvent extraction of the sorbent bed is usually carried out by passing the solvent through the column packed with the sorbent, but it is also performed in a Soxhlet extractor. The extracts are often pretreated before the final chromatographic analysis by removing water using anhydrous sodium sulfate or by freezing out, by evaporating the solvent, by replacing the solvent with another solvent, or by fractionating the analytes in the extract using columns packed with Sephadex, Zorbax-CN, and Spherisorb-CN. Final determination of the analytes are typically performed by gas chromatography with flame ionization or mass spectrometric detection, but selective detectors such as ECD and ELCD for organohalogen compounds, NPD for organonitrogen and organophosphorus compounds, and FPD for organosulfur compounds are also used. High-performance liquid chromatography (HPLC) with the UV detection has recently found some applications in this area.

Chemically modified silica gels are relatively new sorbents used for the isolation of organic compounds from the aqueous phase. Such sorbents are prepared by chemically bonding organic groups to the surface of silica gel. Depending on the organic group bonded to the surface, the modified sorbent is then used for the isolation of selected classes of organic analytes, which is an obvious advantage of this approach. Most often, commercially available prepacked sorption tubes are used. The procedures using sorption tubes packed with modified silica gels usually involve the following steps:

- Selection of the kind and amount of a sorbent, depending on the concentration level of the analytes and interferents, the matrix composition, and the sample volume available. Most commonly, sorption tubes packed with 1 mL of the sorbent are used, but the tubes containing 3 or 6 mL of the sorbent can also be used.
- Conditioning of sorption tubes in order to activate the sorbent bed. The method of conditioning depends on the kind of sorbent used. After this step, the sorbent bed should not be allowed to dry.
- Preconcentration of the analyte(s) on the sorbent bed. The sample volume may vary from milliliters to liters. In order to improve the selectivity of enrichment of the analytes, pH of the sample can be adjusted, inorganic salts or organic solvents can be added to the sample, and the analytes can be converted into derivatives that are better adsorbed on the sorbent or easier to determine. An internal standard is usually added to the sample prior to the preconcentration step.
- Washing the sorbent bed to remove impurities weakly retained on the bed.
- Extraction of the trapped analytes with a suitable solvent (0.2–2 mL).
- Regeneration of the sorbent bed aimed at a removal of the remaining impurities.

The obtained extract can be treated further by fractionation, solvent evaporation, or solvent exchange, depending on the method of final determination. Final determination is carried out by gas chromatography, HPLC, or supercritical fluid chromatography.

The most commonly used sorbents from this group are silica gels with bonded octadecylsilyl (ODS) or octylsilyl groups. Applications of these sorbents include the following:

- Determination of organophosphorus and organonitrogen pesticides using GC-NPD
- Determination of organochlorine pesticides and PCBs using GC-ECD
- Determination of a variety of pesticides and other compounds using GC-MS
- Determination of a wide spectrum of compounds using HPLC with UV, DAD, and fluorescence detection

Another approach to using chemically modified silica gel is the extraction disk technique, in which sorption tubes are replaced by disks with diameters 25, 47, or 90 mm and thickness of 0.5 mm made of polytetrafluoroethylene and sorbent particles. The disks with the ODS phase are most often used, but the disks with the octylsilyl phase are also common. A sorbent makes up about 90% of the mass of a disk, and in the case of disks 47 mm in diameter, the sorbent mass is about 500 mg. The diameter of sorbent particles is approximately 8 mm, and the pore size is about 0.6 mm. The analytical procedure involving extraction disks is analogous to that utilizing sorption tubes and includes the same steps. A number of procedures using extraction disks have been developed, including those for the determination of dioxins, PAHs, PCBs, phthalates, organochlorine pesticides, phenols, and explosives in water and in wastewater. The advantages of using extraction disks include an increased efficiency of sorption or desorption compared with sorption tubes due to (1) a decrease in particle diameter of sorbents; (2) better packing of the sorbent particles, which reduces the mean free path of the analyte molecule to the sorbent particle; and (3) a decrease in linear velocity of the analyzed sample with the simultaneous reduction in its flow rate as a result of a considerable increase in the flow cross section. The extraction disk technique is relatively recent; it was first used in 1990. However, its advantages—the simplicity and low cost—have caused a growing interest among analysts and increased application, despite high blank values and batch-to-batch variation of the sorbents.

The advantages of extraction disks and extraction columns are connected in a new SPE technique—Speedisk and Speedisk Column. It is a compromise between column (high recovery, large volume of sample, and low detection limit) and disks (possibility of high flow of the sample and shorter time of analysis).

Although SPE has a number of attractive features, it also has a number of limitations such as low recovery and plugging of the cartridge or blocking of the pores in the sorbent by solid components, which results in small breakthrough volumes, high blank values, and batch-to-batch variations of the sorbents. Because SPE involves a multistep procedure, including concentration of the extract, it is limited to the analytes with a relatively low volatility, with boiling points above those of the solvent. One solution to these disadvantages is to improve the geometry of the sorbent by coating it on a fiber or a wire made of fused silica. The cylindrical geometry of such an SPME system allows rapid mass transfer during sorption and desorption, prevents plugging, eliminates the use of solvents, and facilitates the introduction of the analytes into analytical instruments.[43,44] Commercial SPME devices are manufactured by Supelco (Bellefonte, Pennsylvania, USA). In such a device, the fused-silica fiber coated with a suitable GC stationary phase is connected to a stainless steel tube to increase the mechanical strength of the fiber assembly. The stainless steel tubing is then contained in a specially designed syringe. During SPME, the fiber is first withdrawn into the syringe needle and then lowered into the sample vial. The fiber coating is exposed for a predetermined time to extract analytes from the sample. Once sampling is completed, the fiber is directly transferred into a GC injector. Analytes are thermally desorbed from the fiber coating and quantitatively determined by GC. SPME consists of two processes: partitioning of analytes between the coating and the sample and desorption of preconcentrated analytes into an analytical instrument, typically a gas chromatograph. Thus far, SPME applications have focused on extracting organic compounds from various matrices, such as air, water, and soil. SPME retains all the advantages of SPE, such as simplicity, low cost, ease of automation, and on-site sampling. At the same time, it eliminates the drawbacks of SPE, such as plugging and the use of solvents. No special thermal desorption module is required, and no modification of gas chromatographs is needed. SPME completely eliminates organic solvents from extraction and injection, and it combines both processes into a single step. The geometry of SPME allows placement of the sorbent into a sample (gas or liquid) or the headspace above the sample to extract the analytes. By sampling from headspace, very complex matrices can be analyzed.

Methods of isolation of organic compounds from the aqueous phase using membranes can also be used to isolate organic and inorganic analytes from water. Separation techniques employing membranes for the preparation of samples for chromatographic analysis have been introduced quite recently. In the past, the membrane techniques have found wider applications in industrial processes, such as water deionization. Due to the progress in polymer research, the use of membranes in analytical chemistry has been increasing. Briefly, a membrane is a selective barrier between two phases. The separation process is based on the ability of a membrane to allow a much more rapid transfer of some components from the donor to the acceptor phase than others although the membrane is never an ideal semipermeable barrier. Classification of membranes can be based on a number of criteria, such as their origin (biological or synthetic), structure, applicability, or the mechanism of separation.

4.5 DETERMINATION OF OTHER ORGANIC COMPOUNDS

Phenols can be determined in drinking water after sorption on solid sorbent (C-18 cartridge).[45] Before sorption, each column has to be conditioned with 5 mL of methanol and then with 10 mL of 0.01 M HCl containing 0.5% isopropanol. One liter of water sample is adjusted to pH 2 with HCl and 0.5% of isopropanol, and 40 g of NaCl is added. Phenols are eluted from the sorbent bed with 1 mL of methanol. Twenty microliters of the eluate is analyzed directly by HPLC. The column is packed with LiChrospher RP-18e. A Merck-Hitachi liquid chromatograph equipped with UV-visible detector Type L 4250 can be used. All phenols are detected at 280 nm except pentachlorophenol, which is detected at 300 nm. The detection limit of this method is about 0.5 μg/L.

Haloacetic acids, formed mainly during water chlorination, can be determined after extraction with methylbutyl ether (MTBE) from the water sample adjusted to pH 0.5 with sulfuric acid, derivatization with diazomethane, and final determination using GC-ECD.[45]

MX-3-chloro-4(dichloromethyl)-5-hydroxy-2(5H)-furanone, formed mainly during water chlorination, can be determined in drinking water after extraction using solid sorbents Amberlit XAD-4 and XAD-8, elution with ethyl acetate, solvent evaporation, derivatization with methanol, and final determination using gas chromatography-high-resolution mass spectrometry.[45]

Aldehydes and ketones, formed mainly during water ozonation, can be determined after derivatization with pentafluorobenzyl hydroxylamine (PFBOA), 2,4,6-trichlorophenylhydrazine (TCPH), 2,3,4,5,6-pentafluorophenylhydrazine, or 2,4-dinitrophenylhydrazine (DNPH); extraction with hexane; extract cleanup with sulfuric acid; drying with anhydrous sodium sulfate; and final determination using GC-ECD (for TCPH, PEPH, and PFBOA), LC-UV, or LC-MS (for DNPH).[45]

Haloacetonitriles, formed mainly during water chlorination, can be determined in drinking water after extraction with MTBE from the water sample (pH = 6–7) and final determination using GC-ECD.[45]

4.6 DETERMINATION OF INORGANIC COMPOUNDS

Some of the inorganic compounds in drinking water are necessary for life—for example, calcium, sodium, potassium, and magnesium. Some of these elements are toxic when concentrations are high (e.g., iron, manganese, and fluorides), and some are not needed by organisms (e.g., cadmium and lead). In underground water, used as a source for drinking water, we can often find too high natural concentrations of manganese, iron, fluorides, ammonia, and sulfurretted hydrogen. Lead and asbestos can enter drinking water from lead and asbestos pipes used in water supply systems in the past. During chlorination or ozonation, bromates can be formed. In treated surface water, all elements and inorganic substances present in nature and introduced by industry are found. The most dangerous are aluminum, antimony, arsenic, asbestos, barium, boron, cadmium, chromium, cobalt, copper, cyanides, lead, manganese, mercury, nickel, silver, tin, vanadium, and zinc. All these elements and substances and their maximum acceptable concentrations are listed in the WHO, EPA, and European Community (EC) standards and have to be determined in drinking water.

Heavy metals comprise one of the most dangerous groups of anthropogenic environmental pollutants. This is because of their toxicity, bioaccumulation, persistence in the environment, and biomagnification in the food chain. Metals can be transported with water and air; it can endanger humans through the air they breathe and the water they drink. Cadmium toxicity can be acute (nausea, vomiting, salivation, diarrhea, liver injury, and muscular cramps) or chronic (proteinuria, renal dysfunction, or kidney damage).[8] Mercury binds to the sulfhydryl groups in many proteins and gets deposited mainly in the kidney and brain. In the case of acute poisoning, mercury causes nausea, vomiting, pharyngitis, bloody diarrhea, nephritis, anuria, and hepatitis, followed by death from gastrointestinal and/or kidney lesions. In the case of chronic toxicity, mercury is carcinogenic for animals and humans and attacks the nervous system (Minamata disease) and brain.[8] Chronic poisoning by lead produces the following symptoms: loss of appetite, constipation, metallic taste, anemia, weakness, insomnia, muscle and joint pains, renal dysfunction, and damage to the peripheral nervous system. To prevent human intake of pollutants, heavy metals should be monitored in drinking water. Flame atomic emission spectrometry (FAES), flame atomic absorption spectrometry, electrothermal atomic absorption spectrometry (ETAAS), inductively coupled plasma, and electrochemical methods [anodic stripping voltammetry (ASV)] have been used for final analysis.[2,3,7,8,10,12,14,17,46–51]

In FAES, the sample is aspirated as a mist of fine droplets into a flame in which the solvent is evaporated, and the residual solutes are then vaporized and partially converted to their constituent atoms. Because of the high temperature of the flame, these atoms are excited and emit light at a wavelength characteristic of the element. Radiation of the wavelength of interest is isolated using a monochromator, and its intensity is measured using a detector photoelectrically. The intensity of the emitted radiation is a measure of the concentration of the determined analyte. ETAAS combines the advantages of high sensitivity and low sample volume requirements. The detection limit is 2–3 orders of magnitude better than those of atomic absorption spectrometry. Spectrometric measurements can be performed with apparatus equipped with a graphite tube. Samples of 5–50 µL are introduced into the graphite tube and heated with a temperature program, and free atoms are formed. Free atoms can absorb the characteristic wavelength sent by an external light source (e.g., hollow-cathode lamp). For elements in Groups IV, V, and VI of the periodic table (antimony, bismuth, germanium, selenium, tellurium, tin, and lead), hydride generation atomic absorption spectrometry can be used. Volatile hydrides are formed in the water phase and purged with an inert gas to the atomizer. For mercury, the best method is based on cold vapor generation. Mercury salts are reduced in solution to the free element and purged with air, argon, or nitrogen to the cuvette; at room temperature, absorption can be measured at the wavelength characteristic for mercury.

ASV[50,51] is a method characterized by compact equipment and relatively low cost. In the case of aqueous samples, analyte preconcentration is not required. Voltammetric measurements are performed with a microprocessor voltamperometric analyzer equipped with three electrodes, a computer, and a monitor displaying graphically and numerically a run of analysis. Very low concentrations can be determined using ASV method because of the concentration of metal in the mercury film. The limit of detection for all analyzed metals is very low (about 0.01 µg/L); in fact, it is much lower than that obtained for atomic absorption methods. However, in the case of ASV, only the dissolved metals (in the form of ions) are detected. Quantitative analysis is performed by addition of standards (5 µL, concentration 100 mg/L) directly to the electrochemical cell containing an analyzed sample. The results are calculated by comparing the signals of the real samples and that of the samples with addition of standards. This method eliminates the influence of matrix on the results.

4.7 MICROBIOLOGICAL RESEARCH METHODS FOR ANALYSIS OF DRINKING WATER

Drinking water, in order to be healthful, should be distinguished by the appropriate chemical and biological qualities. From the hygienic-sanitary point of view, water must not contain harmful living or dead organisms.

4.8 CHARACTERISTICS OF SOME BACTERIA OCCURRING IN AQUEOUS ENVIRONMENT

Surface and underground waters are a natural environment for different living microorganisms. Generally, bacteria in the aqueous environment could be divided into three groups based on their origin: water bacteria, soil bacteria, and bacteria of effluent origin. The water bacteria typically live and reproduce in water. They occur both in water and on the surface of the tapping zone. The majority are psychrophilic bacteria, whose optimal growth temperature is about 20°C. According to their food requirements, water bacteria are chemotrophic, photoautotrophic, or heterotrophic organisms. Morphologically, they resemble soil bacteria, whose cells acquire spherical, cylindrical, or screw-like forms. Among this group are filiforms and stylicforms. The majority of these bacteria are active, motile with bilateral symmetry or have peritrichous flagella.[52] The water group comprises purple sulfur bacteria (Thiorhodaceae), purple nonsulfur bacteria (Athiorhodaceae), iron-oxidizing bacteria (*Gallionella*, *Leptothrix*), nitrifying bacteria (*Nitrosomonas*, Nitrobacter), sulfite-reducing bacteria (*Beggiatoa*, *Thiobacillus*), and heterotrophics (*Pseudomonas*, *Achromobacter*, *Spirochaeta*). In bottom sediments, cellulolytic, hydrogen, and methane bacteria could be found. Water bacteria are considered harmless to humans.[53]

Soil bacteria are flushed into water during rainfalls. Soil is their natural environment; in water, their population declines because of the self-cleaning of water. Most of them are aerobic, spore-forming bacilli belonging to the family Bacillaceae. Cellulolytic, decayed, nitrifying, and denitrifying bacteria belong to this group. Because their optimal living temperature is about 37°C, most are mesophilic bacteria. In waters heavily polluted with organic material, soil bacteria may exist for a longer time. In general, soil bacteria are harmless to humans, but they could cause an unpleasant smell and aftertaste in water.

Effluent microorganisms are mainly heterotrophic, parasitic species living on vegetables. Their main representatives are *Escherichia coli*, *Aerobacter aerogenes*, *Pseudomonas fluorescens*, and *Proteus vulgaris*. All are mesophilic (i.e., living at 37°C). Besides the bacteria mentioned above, water may also contain pathogenic bacteria, which get there with the wastewater. Municipal effluents and industrial water carry intestinal tract bacteria (coliform bacteria, fecal streptococci), pathogenic bacteria (rods associated with typhoid fever and dysentery: *Vibrio cholerae*), and viruses (polio virus, echo virus) (see Table 4.1 for a list of pathogenic bacteria found in water). Microorganisms pathogenic for humans do not multiply in water, and their survival depends on several factors. First, it depends on the individual resistance of pathogenic bacteria and viruses to changes in thermal conditions and

TABLE 4.1

Some Important Pathogenic Bacteria Found in Water and the Diseases They Cause[54]

No.	Pathogenic Bacteria	Diseases
1	Rod-shaped *Salmonella*	Abdominal typhoid fever, paratyphoid fever, salmonellosis
2	Rod-shaped *Shigella*	Bacterial dysentery
3	Curved rods *Vibrio cholerae*	Cholera
4	*Francisella tularensis*	Tularemia
5	Spirochaeta from *Leptospira* genus	Spirochetal jaundice, leptospirosis
6	Rod-shaped *Escherichia coli*	Acute dysentery, intestinal infections
7	*Legionella*	Acute pneumonia, Legionnaires' disease
8	Aerobic, spore-forming bacilli, *Bacillus*	Corneal abrasion, ankylosing spondylitis, feed poisoning, anthrax
9	Rod-shaped *Mycobacterium*	Tuberculosis

level of water pollution. It has been determined that pathogenic organisms live longer in clear water than in polluted ones.

Drinking water should fulfill the following requirements:

- It should be clear, odorless, colorless, and refreshing in taste.
- It should not contain pathogenic bacteria; animal parasites and their larvae or eggs; toxic compounds; and excessive quantities of calcium, magnesium, iron, or manganese compounds.
- It should be easily accessible, be always of good quality, and occur in appropriate amounts.
- It should be permanently protected from contamination.
- It should contain substances that are necessary for human life in the proper amounts.

The danger of humans contracting infections from water makes bacteriologists and hygienists carry out permanent sanitary control of drinking water. They use simple methods that are able to indicate the presence of pathogenic organisms because the indirect methods are laborious and very time-consuming. Currently, in routine studies, intermediate methods based on indicative bacteria are used. The organism indicative of contamination has to comply with the following conditions:

- It has to be a permanent inhabitant of the intestine and occur in large numbers.
- It should not be a pathogenic bacterium.
- It should not form spores.
- It should not multiply in water.
- It should be easy to detect.
- It should live longer than pathogens in the natural environment (water, soil).
- It should get removed during the water conditioning in a similar way as the pathogen organisms.[54]

The main indicating organisms of water fecal contaminations are *E. coli*, lacto fermenting bacteria, coliform bacteria, thermotolerant *E. coli* types, fecal enterococcus, and *Clostridium* bacteria. Several of the above-mentioned indicator conditions are fulfilled by *E. coli* and to a lesser extent by coliform bacteria and thermotolerant *E. coli* types. *Escherichia coli* belong to the family Enterobacteriaceae. They are small, nonspore-forming, gram-negative, and rod-shaped bacteria and live in the large intestine of humans and animals, where their density reaches a maximum of 10^9 cells per gram of fecal matter. These bacteria are grown at temperatures of 44°C–45°C on complex culture media. Usually, they ferment lactose and mannitol with the formation of acid and aldehyde and produce indole from tryptophan. *Escherichia coli* possess two enzymes: b-galactosidase and b-glucoronidase; but they do not form oxidase and do not hydrolyze urea. Lacto-fermenting bacteria form colonies in aerobic conditions at 36°C–38°C on culture medium with lactose, with the formation of acid. Coliform bacteria are nonspore-forming, oxidase-negative, gram-negative, rod-shaped bacteria that ferment lactose at 35°C–37°C to form acids and aldehydes within 24–48 hours. Thermotolerant *E. coli* types are coliform bacteria, which ferment lactose at 44°C–45°C. *Enterobacter*, *Citrobacter*, and *Klebsiella* also belong to this group. Coliform bacteria do not have to be directly connected with fecal contamination or with the occurrence of pathogens in drinking water. They could be present in droppings, water rich in nutrient substances, soil, and decayed residues of vegetables. These bacteria cannot be present in conditioned drinking water. The determination of their presence in water suggests improper conditioning of water, secondary contamination, or excessive content of nutrient substances in conditioned water. Coliform bacteria could be exploited as an indicator of effectiveness of water conditioning. Frequency of indicative bacteria is determined to set the coliform titer—the lowest volume of water (mL) in which these bacteria are detected. According to ISO standards, the final result is given as an index of coliform bacteria or *E. coli* in 100 mL of water tested. Fecal enterococci are spherical or oval, catalase-negative, gram-positive bacteria, occurring as short chains. They are able to reduce 2,3,5-triphenyltetrazolic chloride to formazine and hydrolyze Askulin at 44°C. They express D Lancefield group D antigen. The term fecal enterococcus refers to the bacteria that occur in droppings of humans—for example, *Streptococcus faecalis, S. faecium, S. avium,* and *S. gallinarum*. Two subspecies *S. faecalis liquefaciens* and *S. faecalis zymogenese* belong to fecal enterococcus, but their validity as subspecies is questionable. Fecal enterococci that are not found in human fecal matter but found in droppings of animals do not belong to the fecal enterococcus group. In the external environment, they die very quickly, faster than many pathogenic bacteria. The presence of fecal enterococcus in water indicates fresh contamination of water, so it also proves the potential menace of pathogenic bacteria. Because these bacteria are resistant to drying, they could be helpful in routine control of water quality carried out after building new water supply systems or after their repair. After *E. coli*, fecal enterococcus is the second best indicative organism of contamination. *Clostridium* bacteria are sulfite-reducing, gram-positive, spore-forming bacilli bacteria, which can survive for a long time—even for many years—in soil and water. Usually, they are found in droppings, but they can derive from other sources as well. These bacteria are grown at 37°C. Replication of this species indicates a very "old" water contamination, after death of all pathogenic bacteria. The spores of *Clostridium* bacteria are resistant to disinfection; therefore, their presence in disinfected water may point out a shortcoming in water conditioning. These bacteria cannot be used for routine monitoring of water quality or detection of fresh contamination because they might be present for a long time after the contamination and far away from the place of this contamination.

In sanitary analysis of drinking water, besides determining the presence of indicative bacteria, the general number of psychrophilic and mesophilic bacteria in 1 mL of water is determined. Culturing of psychrophilic bacteria is carried out on a solid agar culture medium for 72 hours at 22°C—optimum conditions for these bacteria. The large number of these bacterial colonies indicates the inflow of organic substances to water, which creates favorable conditions for the growth of saprophytic bacteria. Culturing of mesophilic bacteria is performed on a solid agar culture medium for 24 hours at 37°C. Thermal optimum for these bacteria is the temperature of the human body—37°C. The mesophilic bacterial culture consists of effluent bacteria and some soil bacteria. The detection of these bacteria in water is indicative of contamination of water with domestic sewage and industrial waters.

REFERENCES

1. de Kruijf, H.A.M. and Kool, J.H., eds. *Organic Micropollutants in Drinking Water and Health.* Elsevier, Amsterdam, 1985.
2. Dojlido, J. *Chemia wód powierzchniowych.* Wydawnictwo Ekonomia i Środowisko, Bialystok, 1995.
3. Hellmann, H. *Analysis of Surface Waters.* Ellis Horwood, New York, 1987.
4. Biziuk, M. and Przyjazny, A. Methods of isolation and determination of volatile organo-halogen compounds in natural and treated waters. *J. Chromatogr. A.* 733, 417, 1996.
5. Biziuk, M., et al. Occurrence and determination of pesticides in natural and treated waters. *J. Chromatogr. A.* 754, 103, 1996.
6. Moore, J.W. and Ramamoorthy, S. *Organic Chemicals in Natural Waters. Applied Monitoring and Impact Assessment.* Springer Verlag, New York, 1984.
7. Nemerow, N.L. *Stream, Lake, Estuary and Ocean Pollution.* Van Nostrand Reinhold Co., New York, 1985.
8. Moore, J.W. *Inorganic Contaminants of Surface Water. Research and Monitoring Priorities.* Springer-Verlag, New York, 1991.
9. Biziuk, M., Methods to isolate and determine organochlorine compounds and selected pesticides in natural and treated waters. *Zeszyty Naukowe Politechniki Gdańskiej, Nr 513, Chemia XXXI,* Politechnika Gdańska, Gdańsk, 1994.
10. Nollet, L.M.L. *Handbook of Water Analysis.* Marcel Dekker, New York, 2000.
11. Majors, R.E. An overview of sample preparation. *LC-GC Intl.,* 4(2), 10, 1991.
12. Fresenius, W., Quentin, K.E., and Schneider, W., eds. *Water Analysis.* Springer Verlag, Berlin, 1988.
13. Namieśnik, J., et al. Isolation and preconcentration of volatile organic compounds from water. *Anal. Chem. Acta.* 237, 1, 1990.
14. Namieśnik, J. and Jamrógiewicz, J., eds. *Fizykochemiczne metody kontroli zanieczyszczeń środowiska.* WN-T, Warszawa, 1998.
15. Dressler, M. Extraction of trace amounts of organic compounds from water with porous organic polymers. *J. Chromatogr.* 165, 167, 1979.
16. Poole, S.K., et al. Sample preparation for chromatographic separation: An overview. *Anal. Chim. Acta.* 236, 3, 1990.
17. Leenheer, J.A. *Water Analysis.* Academic Press, Oxford, 1984.
18. Starostin, L. and Witkiewicz, Z. Environmental water samples preparation for chemical analysis. *Chem. Anal. (Warsaw).* 39, 263, 1994.
19. Biziuk, M. Gas chromatography by direct aqueous injection in environmental analysis. In *Encyclopedia of Analytical Chemistry,* Mayers, R.A., ed. Wiley, Chichester, 2000, vol. 3, pp. 2549–2587.
20. de Leer, W.B. *Aqueous Chlorination Products: The Origin of Organochlorine Compounds in Drinking and Surface Waters.* Delft University Press, Delft, 1987.
21. Jolley, R.L., et al., eds. *Water Chlorination. Environmental Impact and Health Effects.* Ann Arbor Science Publishers Inc., Ann Arbor, 1980.
22. Rebhun, M., Manka J., and Zilberman, A. Trihalomethane formation in high-bromide Lake Galilee water. *Research and Technology, JAWWA.* 1988, 84–89.
23. Montgomery, J.M., ed. *Water Treatment: Principles and Design.* John Wiley and Sons, New York, 1985.
24. Koch, B. and Krasner, S.W. *Occurrence of Disinfection By-products in a Distribution System.* Proc. Water Qual. Technol. Conf. AWWA, Los Angeles, June 18–22, 1989, AWWA, Denver, 1989, 1203–1230.
25. Biziuk, M. and Czerwiński, J. Polish Standard Method-PN-C-04549–1. Determination of dichloromethane, trichloromethane, tetrachloromethane, bromodichloromethane, dibromochloromethane, tribromomethane, trichloroethylene, tetrachloroethylene,

1,1,1,-trichloroethane, 1,1,2,2-tetrachloroethane in water by gas chromatography using direct injection of the sample. Polish Committee for Standardization, Warsaw, 1998.

26. Kozlowski, E., Sieńkowska-Zyskowska, E., and Biziuk M. Countercurrent thin-layer head-space as a new approach to continuous analysis of volatile organic compounds in water. *Chem. Anal. (Warsaw)*, 28, 817, 1983.

27. Kozlowski, E., Sieńkowska-Zyskowska, E., and Górecki T. Continuous flow thin-layer head-space (TLHS) analysis. I. Conductometric determination of volatile organic halogens (VOX) in tap water. *Fresenius J. Anal. Chem.* 339, 19, 1991.

28. Kozlowski, E., Górecki, T., and Sieńkowska-Zyskowska, E. Continuous flow thin-layer head-space (TLHS) analysis. II. Spectrophotometric determination of volatile organic halogens (VOX) in water. *Fresenius J. Anal. Chem.* 339, 882, 1991.

29. Kozlowski, E., Górecki, T., and Sieńkowska-Zyskowska, E. Continuous flow thin-layer head-space (TLHS) analysis. III. Potentiometric determination of volatile organic halogens (VOX) in water. *Fresenius J. Anal. Chem.* 340, 454, 1991.

30. Kozlowski, E., Górecki, T., and Sieńkowska-Zyskowska, E. Continuous flow thin-layer head-space (TLHS) analysis. IV. Indirect potentiometric determination of volatile organic halogens (VOX) in water. *Fresenius J. Anal. Chem.* 340, 773, 1991.

31. Kozlowski, E., Sieńkowska-Zyskowska, E., and Górecki, T. Continuous flow thin-layer head-space (TLHS) analysis. V. Comparison of the conductometric and indirect potentiometric methods of volatile organic halogens (VOX) determination in tap water. *Fresenius J. Anal. Chem.* 342, 20, 1992.

32. Kozlowski, E., Górecki, T., and Sieńkowska-Zyskowska, E. Continuous flow thin-layer head-space (TLHS) analysis. VI. Comparison of the results of determination of volatile organohalogen compounds in tap water in the form of a group parameter (VOX) with partial speciation by DAI-ECD capillary gas chromatography (GC)^2. *Fresenius J. Anal. Chem.* 342, 401, 1992.

33. Kozlowski, E., Kolodziejczyk, A., and Kozlowski, J. Continuous flow thin-layer head-space (TLHS) analysis. VII. General equation of simultaneous separation and precon-centration of volatile analytes in liquids. *Chem. Anal. (Warsaw).* 38, 315, 1993.

34. Kozlowski, E. and Polkowska, Z. Continuous flow thin-layer head-space (TLHS) analy-sis. IX. Isolation of trihalomethanes from complex mixtures with autogenous generation of a liquid sorbent stream, *Chem. Anal. (Warsaw).* 41, 173, 1996.

35. Polkowska, Z. and Kozlowski, E. Continuous flow thin-layer head-space (TLHS) analy-sis. X. Preconcentration of volatile analytes in autogenously generated stream of liquid sorbent. *Chem. Anal (Warsaw).* 41, 183, 1996.

36. Polkowska, Z., et al. Volatile organohalogen trace analysis in beverages by thin-layer headspace enrichment and electron-capture gas-chromatography. *Int. J. Food Sci. Techn.* 31, 387, 1996.

37. Polkowska, Z., et al. Comparison of the new version of thin-layer headspace analysis with autogenous generation of the liquid sorbent (TLHS-DAI-GC-ECD) with liquid–liquid extraction for the determination of volatile organohalogen compounds in liquid samples. *Process Control and Quality.* 11, 1, 1998.

38. Polkowska, Z., et al. Theoretical principles of thin layer headspace analysis (TLHS). *Toxicol. Environ. Chem.* 68, 1, 1999.

39. Janicki, W., et al. Thermal desorber with an intermediate sorbent trap. *Chem. Anal.* 37, 599, 1992.

40. Janicki, W., et al. A simple device for permeation removal of water vapour from purge gases in the determination of volatile organic compounds in aqueous samples. *J. Chromatogr.* 654, 279, 1993.

41. Zygmunt, B. Determination of trihalomethanes in aqueous samples by means of purge-and-trap system with on-sorbent focusing coupled to gas chromatography with electron-capture detection. *J. Chromatog. A.* 725, 157, 1996.

42. Biziuk, M., et al. Determination of volatile organic compounds in water intakes and tapwater by purge and trap and direct aqueous injection-electron capture detection techniques. *Chem. Anal. (Warsaw)*. 40, 299, 1995.
43. Pawliszyn, J. *Solid Phase Microextraction—Theory and Practice.* Wiley, New York, 1997.
44. Pawliszyn, J. *Application of Solid Phase Microextraction.* The Royal Society of Chemistry, Cambridge, 1999.
45. Dojlido, J., ed. *By-products of Water Disinfection.* Polskie Zrzeszenie Inzynierów i Techników Sanitarnych, Warszawa, 2002.
46. Hewitt, C.N., ed. *Instrumental Analysis of Pollutants.* Elsevier, London, 1991.
47. Hunt, D.T.E., and Wilson, A.L. *The Chemical Analysis of Water.* Royal Society of Chemistry, London, 1988.
48. Willard, H.H., et al. *Instrumental Methods of Analysis.* Wadsworth, Belmont, 1981.
49. Taylor, L.R., Pap, R.B., and Pollard, B.D. *Instrumental Methods for Determining of Elements.* VCH, New York, 1994.
50. Bersier, P.M., Howell, J., and Bruntlett, C. Advanced electroanalytical techniques versus atomic absorption spectrometry, inductively coupled plasma atomic emission spectrometry and inductively coupled plasma mass spectrometry in environmental analysis. *Analyst.* 119, 219, 1994.
51. Wang, J. *Stripping Analysis. Principles, Instrumentation and Applications.* VCH, Deerfield Beach, FL, 1985.
52. Rheinheimer, G. *Mikrobiologia wód.* Panstwowe Wydawnictwo Rolnicze i Lesne, Warszawa, 1977.
53. Pawlaczyk-Szpilow, M. *Mikrobiologia Wody i Ścieków.* PWN, Warszawa, 1980.
54. Nawrocki, J. and Bitozor, S. *Uzdatnianie wody: Procesy chemiczne i biologiczne.* PWN, Warsawa, Poznań, 2000.

5 Analysis of Proteins, Peptides, and Amino Acids in Foods

Edward Kolakowski

CONTENTS

5.1 BASIC RECOMMENDATIONS FOR THE ANALYSIS OF PROTEINS, PEPTIDES, AND AMINO ACIDS

To prevent undesirable changes in proteins and peptides, samples should be collected as quickly as possible and immediately placed in an isolating medium at a low temperature (0°C–4°C). When isolating proteins for further analysis (e.g., electrophoresis), the isolating medium must contain substances that prevent bacterial contamination (e.g., 0.1 mM sodium azide), protect [e.g., by adding 2 mM ethylenediaminetetraacetic acid (EDTA)] the proteins from chelating metals, and prevent protein degradation caused by endogenous proteases by, for example, adding phenylmethylsulfonyl fluoride (PMSF) to 0.1 mM final concentration, PMSF acting as a serine protease inhibitor.

5.2 PROTEIN ISOLATION AND PURIFICATION

Isolation and purification is aimed at separating proteins from a certain material in a possibly homogeneous form or in a group of proteins suitable for further analyses. The first stage usually involves obtaining a solution of the protein to be isolated. To this end, tissues and cells have to be disintegrated by homogenization or osmolysis so that the solvent is capable of accessing the proteins. Homogenization is carried out in high-speed (10,000–15,000 rpm) blenders equipped with sharp blades.

Osmolysis starts after a biological material has been placed in a hypotonic solution, such as water or sucrose-free buffer, which penetrates the cytosol, causing cells to swell and rupture. Unnecessary cell organelles can then be removed from the material treated this way. Membrane proteins require a detergent (e.g., Triton X-100) for separation; the detergent damages the lipid bilayer and releases the integral membrane proteins. Techniques used for isolation and purification of proteins are based on protein solubility, molecular weight, charge, and/or protein-specific binding affinity.

5.2.1 EXTRACTION OF PROTEINS

Native skeletal muscle proteins can be divided, based on their solubility in aqueous solvents, into three basic groups:

1. *Sarcoplasmic proteins* are soluble in solutions of low salt concentration (I, 0.1) (e.g., globulin X, albumins, myoglobin, and the so-called myogen fraction that includes most glycolytic enzymes). They are most often extracted with phosphoric buffer of ionic strength I 5 0.05 (e.g., 15.5 mM Na_2HPO_4 13.38 mM KH_2PO_4, pH 7.3). As these proteins have been sometimes extracted with pure water, they are commonly known as water-soluble proteins.

2. *Myofibrillar proteins* are soluble in solutions of neutral salts of ionic strength 0.5 (in practice, between 0.7 and 1.5) and often known as salt-soluble proteins. Generally, the 5% NaCl solution buffered to pH 7.0–7.5 (optimally 7.3) by using 0.02–0.003 $NaHCO_3$ is regarded as the best solvent. To model conditions of myofibrillar protein extraction for industrial operations (e.g., mincing during preparation of sausage emulsion), lower NaCl concentrations (2.5%–3.0%) are recommended (Kolakowski and Szybowicz 1976).

   Actomyosin can be isolated from myofibrillar protein extract by diluting it with distilled water (volumetric ratio of 1:10–1:12) and keeping the mixture overnight in a refrigerator. The precipitated actomyosin can be separated by centrifugation; the supernatant contains sarcoplasmic proteins and nonprotein nitrogen compounds.

3. *Connective tissue proteins* are composed mostly of collagen and elastin. These proteins are soluble neither in neutral salt solutions of ionic strength 0.5 nor in 0.05 N solutions of NaOH and HCl; they are known as stromal proteins.

Myofibrillar proteins that, due to effects of denaturing factors (e.g., frozen storage), have lost their original property to dissolve in neutral salt solution of ionic strength between 0.05 and 1.5 form a group differing in its solubility-related properties. They are commonly known as insoluble proteins. However, these proteins are well soluble in 0.05 N NaOH or 0.05 N HCl. So, if a precipitate obtained from a salt-soluble protein extraction is additionally extracted by using 0.05 NaOH or 0.05 N HCl, stromal proteins will remain in the precipitate and the solute will contain the so-called nonsoluble proteins (i.e., proteins not soluble in salt solutions).

Extractability of plant α-globulins as a function of sodium chloride concentration is similar to that of the muscle myofibrillar proteins. The values taper to a fairly constant level beyond 0.8 M NaCl and remain at that level up to 2.0 M NaCl concentration. The pH range of maximum extractability increases with increasing NaCl concentration. Salts of divalent cations such as $MgCl_2$ and $(NH_4)_2SO_4$ are usually more effective for globular protein extractability than salts of monovalent cations such as NaCl, NH_4Cl, and KCl. In standardized procedures of plant protein solubilization, NaCl at pH 9.0 (which yields a fraction defined as globulins), water (albumins), 70% ethanol (prolamins), and 0.05 M CH_3COOH (glutelins) are used.

Extraction of milk proteins, which are water soluble, most often boils down to separation of emulsified fat and separation of protein from liquid milk. To achieve this, different physical (such as ultracentrifugation, membrane microfiltration, and electrodialysis) and chemical (such as isoelectric point precipitation and addition of enzymatic coagulants) techniques are used.

The protein (nitrogen) content in protein extract is most often determined with the Kjeldahl or biuret technique.

5.2.1.1   Special Solutions Used for Protein Extraction

Special extraction solutions disrupt certain types of chemical bonds and make it possible to selectively extract protein or to solubilize those fractions that have lost their solubility in neutral salt solutions. To disrupt hydrogen bonds and hydrophobic interactions, 6–8 M urea or 1% sodium dodecyl sulfate (SDS) is used. Acetic acid (0.5 M) is applied to extract soluble collagen from the residue obtained after alkali extraction of meat. Another way to extract collagen from the residue after myofibrillar protein extraction is to reextract the residue with 5% trichloroacetic acid (TCA) at 90°C for 30 minutes. The collagen content in the extract is calculated from the amount of hydroxyproline (HOP).

Proteins that are to be used in immunochemical assays require special extraction methods. On the one hand, the antigenic properties of those proteins have to be preserved as much as possible; on the other, a possibly complete extraction has to be achieved to perform quantitative immunoassays. In the case of salt solution extraction, denaturation of protein is minimal, and native epitopes can be easily recovered. Proteins derived from processed products, and also from wheat, barley, rice, and soybean, in which hydrophobic interactions usually occur, require the use of 2–9 M urea or 1%–5% SDS or SDS-dithiothreitol (DTT) followed by dialysis to remove the reagents and reconstitute the native epitopes. Extraction of gluten proteins from flour or whole meal requires the absence of reducing agents; the best results are obtained with 4% SDS in Tris–HCl buffer. The best solution of gliadin is arrived at with 2 M urea in carbonate buffer of pH 9.6.

5.2.2 DEPROTEINATION PROCEDURES

Deproteination is one of the basic procedures in the analyses of food products. Deproteination of samples and separation of the "nonprotein" from the "protein" fraction usually requires one of the following five methods to be employed: chemical precipitation, precipitation by thermal treatment, ultrafiltration, dialysis, or purification by column chromatography.

5.2.2.1 Chemical Deproteination

Chemical deproteination is most widely used in the analysis of food products because an appropriate precipitating agent can be selected without any difficulty and the procedure is simple to perform

Organic solvents (such as ethanol, acetone, and acetonitrile) affect the spatial structure of a protein molecule, weaken its hydrophobic interactions, and directly react with charged groups on the protein molecule's surface. This damages the water coat of the molecule and leads to protein denaturation. The complete denaturation, however, occurs only at high concentrations of the precipitating agents (65%–80%), at an elevated temperature (20°C–30°C), and after an extended time of precipitation (24 hours). Ethanol precipitation (final concentration of 80%) is particularly recommended for free amino acid assays as a sample can be easily concentrated during chloroform rinsing of alcohol extracts. It is recommended that the protein residue be precipitated in an ethanol extract and that saturated alcohol zinc chloride (500:0.5 v/v) solution be added to the extract. Acetone with 1% HCl brings about complete deproteination as early as after 15–30 minutes; precipitation with pure acetone requires a longer time, usually from 3 hours to overnight. Acetonitrile used in 66.6% concentration as a deproteinating agent for fresh pork muscle and dry-cured ham produced recoveries of free amino acids in excess of 97% (Aristoy and Toldrá 1991).

Acid precipitation is most efficient if used with undigested or unhydrolyzed protein. If digested protein is used, incomplete precipitation may occur, whereas a too high acid concentration may cause a partial protein hydrolysis. For this reason, different materials usually require different reagent concentrations.

Sulfosalicylic acid (SSA) was reported as producing especially poor recovery of aspartic and glutaminic acids, histidine, arginine, valine, isoleucine, leucine, and lysine. Curiously, SSA is one of the most common precipitants used with biological materials and cheese. Another relevant fact is the abnormally high proline (around 150%) and glycine (around 110%) levels in the filtrate (Aristoy and Toldrá, 1991).

Tungstic acid precipitation is the result of loss of arginine, histidine, and lysine. These dibasic amino acids form protein–tungstate complexes. Phosphotungstic acid, too, has been observed to produce very poor recoveries of arginine, ornithine, and lysine, as well as poor recoveries of aspartic and glutaminic acids, asparagine, glutamine, and tryptophan.

Deproteination with 1% picric acid produces rather good recoveries of α-amino acids except for Trp, which is decomposed, but the entire procedure is little complicated. On the other hand, the advantage of picric acid is its better (compared with TCA) ability to precipitate polypeptides. Some losses of Trp occur also during precipitation with perchloric acid. However, the latter has an advantage of being susceptible to removal from the extract by precipitating it with potassium ions.

TCA is a deproteinating agent most commonly used in food analysis. It acts relatively rapidly, the filtrate produced being clear and easily separable from the precipitate; TCA is also less expensive than other precipitants. An almost complete deproteination of fish flesh homogenates was demonstrated to occur at TCA concentrations as low as 3%. However, higher concentrations (5%) are safer to use, whereas partial protein hydrolysis was observed at concentrations exceeding 12% (Kolakowski, 1973). The nonionic detergent Triton X-100 is recommended as a coprecipitating carrier to precipitate protein in lipid-rich samples. In practice, TCA does not precipitate amino acids (recoveries proved higher than 90%), peptides, polypeptides, and mucoproteins. To analyze peptides in deproteinated fish meat extracts, Kolakowski (1973) proposed to apply a two-stage precipitation with TCA (used initially at 1% and then at 5% concentration) to prevent coprecipitation of peptides with the precipitated proteins.

5.2.2.2 Deproteination by Thermal Treatment

A protein denatured thermally at an isoelectric point is nonsoluble. Acidification of water, used as a heating medium, to isoelectric point of the protein present in the sample greatly facilitates and accelerates deproteination. Acetic acid (0.12% final concentration) is used most frequently. Proteins differ in the temperature of their thermal denaturation; the process may start within a temperature range of 80°C–100°C and is completed after 3–5 minutes of heating. Only few proteins, such as gelatin and ribonuclease, are able to withstand a short period of heating at those temperatures. Some proteases and peptidases are also thermoresistant, which are capable of enduring heating for some time and which may hydrolyze proteins at certain optimal temperatures and distort the actual protein to nonprotein fraction ratio. Therefore, it is recommended that thermal deproteination be performed by homogenization of a sample with very hot (95°C–100°C) water so that the resultant temperature is not lower than 80°C and is maintained at that level for 5 minutes.

5.2.3 Filtration

Filtration paper is practicable only for homogenates obtained from high volumes of acid precipitate solutions (e.g., 5% TCA), relative to the sample weight (10 > 1). The medium hard filtration paper ensures a sufficiently high rate of the operation and does not require any suction pressure. On the other hand, gravitational filtration of protein extracts through filter paper is very difficult. As a rule, the filtration takes too long, and the filter pore size gradually diminishes due to clogging, which may also affect the filtrate composition. Filtration through plastic (polyvinyl chloride, polyamide) filters of more uniform and stable pores, which perform better when used with the Büchner and Hartley funnels, is not practicable in routine assays either and therefore is frequently replaced by cold centrifugation.

Vacuum filtration of protein hydrolysates containing high concentrations of acid or alkali is enhanced by sintered-glass filters, most frequently G-3 and G-4 (pore sizes 15–40 μm and 5–15 μm, respectively). The required suction pressure on the order of 2.7–8.0 hPa (20–60 mm Hg) is produced with a water pump or a rotary vacuum pump with vapor trap. Filters G-0 and G-1 (pore sizes 150–200 μm and 90–150 μm, respectively), attached to suitably large cases, may be used for vacuum filtration of tissue homogenates obtained at high sample-to-precipitant ratios (>1.0) when a high extract concentration of the substance under study is required. As pores of those filters are relatively large and clog easily on filtration, it is recommended that a wet filter paper circle be placed on the top. The G-5 filters (pore size 1–2 μm) are so fine that they may be used to filter out bacteria.

5.2.4 ULTRAFILTRATION

Ultrafiltration through polysulfone or polyamide membranes is most often used for extraction, deproteination, separation, cleaning-up, and concentration. Membranes used for ultrafiltration are characterized by their molecular weight cutoff (MWCO). The MWCO is not precise because it depends on solute rejection. Solute rejection depends on solute size, geometry, and flexibility, as well as on solute–membrane interactions, operating conditions, and the state of the apparatus used. In deproteination, the 10,000-M cutoff membrane is most often used. Molecular weights of the substances to be separated by ultrafiltration should differ by at least an order of magnitude. Filters are prewashed by sonication in distilled water to remove any contaminant. Filtration is most often induced by using centrifugation or pressure. This technique is regarded as one of the best procedures because it is fast, the protein is not denatured during operation, and no amino acid losses are incurred.

5.2.5 CENTRIFUGATION

Centrifugation is widely used to separate tissue homogenates into the supernatant and semisolid residues (pellets). Efficiency of centrifugation in removing dispersed particles and suspensions from tissue protein extracts depends primarily on the g value, centrifugation time, and density of particles. At $200 \times g$, the nuclei, intact cells, connective tissue, sarcolemmal sheaths, and myofibrils are sedimented after 10–15 minutes of centrifugation. Sarcosomes sediment after centrifugation for 10–20 minutes at 2,000–10,000 × g. Lysosomes and "heavy" microsomes need 20–30 minutes of centrifugation at 10,000–25,000 × g; microsomes, polysomes, ribosomes, and sarcolemma fragments sediment after 120–180 minutes of centrifugation at 25,000–100,000 × g.

5.2.6 DENSITY GRADIENT CENTRIFUGATION

Stepwise density gradient centrifugation and constant density gradient centrifugation are used. Density gradient–producing substances employed most frequently include glycerine, sucrose, cesium chloride (CsCl), and Ficoll®. They are easily soluble in water and chemically neutral, compared with proteins. Ficoll is a synthetic, high–molecular weight (M_r 5 400,000) copolymer of sucrose and epichlorohydrin. Solvent densities commonly range between those only slightly higher than 1.0 and 1.80.

The stepwise density gradient (usually from 5%–20% to 10%–60%) centrifugation is a layer separation method, in which density gradient is maintained by gravitational fields. Sucrose or glycerine gradients can be prepared in a simple device consisting of two interconnected vessels, one filled with a denser and the other with a low-density solution of a gradient-producing substance. The solutions are mixed in the outlet with a magnetic stirrer and are collected directly into the centrifuge tube placed in the rotor. The gradient solution concentration is selected, appropriately to the required (de Duve et al. 1959). The protein sample is carefully (to avoid vortices) placed onto the gradient solution surface. Sucrose gradient centrifugation proceeds for 3–4 hours at 0°C and 25,000 rpm. The centrifuge should be stopped without braking. The gradient fractions are collected by siphoning them off or by allowing them to drip through a needle inserted into the bottom of the tube.

The constant density gradient centrifugation, called cesium chloride gradient centrifugation, requires no prior gradient preparation. The gradient is formed and maintained in the centrifugal field during sufficiently long centrifugation. Prior to centrifugation, centrifuge tubes are filled with a homogeneous fluid containing the protein solution and CsCl. Concentration of the latter is selected so that it is within the density range of the macromolecules to be separated. The formula for calculating the amount of CsCl necessary to obtain the required initial density was given by Vinograd and Hearst (1962). During centrifugation (60 hours at 30,000 rpm), the macromolecules begin to migrate toward sites in the gradient that correspond to their apparent densities.

Density gradient centrifugation is particularly recommended for separation of ribonucleic acids from proteins in tissue extracts.

5.2.7 SALTING-OUT

Proteins are usually precipitated from solutions by using salts containing polyvalent anions, such as ammonium sulfate, sodium sulfate, magnesium sulfate, sodium citrate, and ammonium phosphate; polyvalent salts are more efficient than univalent ones. Ammonium sulfate is used most often because it is very well soluble in water, whereby high salt concentrations can be obtained (saturated solutions are those of 3.9 M at 0°C and 4.1 M at 25°C). Concentrated ammonium sulfate solutions are acidic (pH is approximately 5.5) because of the hydrolysis of the salt, and the pH should be adjusted by adding NH_4OH or H_2SO_4 before use. The most widely used method of ammonium sulfate protein salting involves addition of the salt *in substantia* or in a saturated solution. A useful compilation of the properties of ammonium sulfate solution and the *in substantia* form for preparation of solutions of known fraction saturation are given in Table 5.1. Mixing equal amounts of protein solution and saturated solution of ammonium sulfate (50% saturation) usually suffices to salt out most globular proteins; albumin proteins require a much higher (exceeding 80%) saturation. It should be remembered that individual proteins, depending on their structure and properties, require specific concentrations of ammonium sulfate; the concentrations have to be determined experimentally. The differences between proteins make it possible to use a "reverse" method for fractionating proteins that were salted out with ammonium sulfate. The protein precipitated at 80% $(NH_4)_2SO_4$ saturation, packed into a column (e.g., Sephadex G-200), can be fractionated after elution with linear $(NH_4)_2SO_4$ gradient from 80% to 0% or 80% to 40% saturation, buffered at pH 7.

TABLE 5.1
Preparation of Ammonium Sulfate Solution of Known Percent Saturation

Initial Concentration of Ammonium Sulfate (%)	Final Concentration of Ammonium Sulfate (%) Amount (g) of Solid-State Ammonium Sulfate Added to 1 L Solution																
	10	**20**	**25**	**30**	**33**	**35**	**40**	**45**	**50**	**55**	**60**	**65**	**70**	**75**	**80**	**90**	**100**
0	56	114	144	176	196	209	243	277	313	351	390	430	472	516	561	662	767
10		57	86	118	137	150	183	216	251	288	326	365	406	449	494	592	694
20			29	59	78	91	123	155	189	225	262	300	340	382	424	520	619
25				30	49	61	93	125	158	193	230	267	307	348	390	485	583
30					19	30	62	94	127	162	198	235	273	314	356	449	546
33						12	43	74	107	142	177	214	252	292	333	426	528
35							31	63	94	129	164	200	238	278	319	411	506
40								31	63	97	132	168	205	245	285	375	465
45									32	65	99	134	171	210	250	339	431
50										33	66	101	137	176	214	302	392
55											33	67	103	141	179	264	353
60												34	69	105	143	227	314
65													34	70	107	190	275
70														35	72	153	237
75															36	115	198
80																77	157
90																	79

5.2.8   DIALYSIS AND ELECTRODIALYSIS

Dialysis is usually applied to separate macromolecules from ions. It can also be used to concentrate protein solutions by removing water under vacuum and to remove low–molecular weight compounds. The potential for binding of free amino acids and peptides to native proteins is rather small, so it is possible to separate those compounds by using molecular sieving membrane technique based on the molecular size differences. The prerequisite is the selection of a membrane with an appropriate pore size. For instance, the nominal MWCO is 200, 5,000, and 30,000 in Bio-Fiber 20 (cellulose acetate), Bio-Fiber 50 (cellulose), and Bio-Fiber 80 (cellulose acetate), respectively (Zabin 1975). A simple method of dialysis, involving polyethylene glycol, to concentrate protein-containing fluids was developed by Kohn (1959).

Electrodialysis substantially reduces the duration of a time-consuming operation. Problems encountered with negative electric charge of synthetic membranes can be overcome by reversing the direction of dialysis from time to time.

All the dialysis techniques involve losses of the compounds studied, particularly the low–molecular weight ones. For this reason, the best results are obtained in desalting and separation of high–molecular weight proteins.

5.2.9   PURIFICATION AND GROUP FRACTIONATION BY
TRAP COLUMNS AND CHROMATOGRAPHY

An increasing amount of various ready-to-go materials and accessories is commercially available. Those products include, among others, different types of short columns for cleanup, preconcentration, desalting and buffer exchange, and group separation.

Solid-phase extraction is a more efficient method of sample preparation than the traditional method of liquid–liquid extraction. Its purpose is twofold: (1) sample cleanup (to eliminate impurities and/or to isolate the component of interest from the matrix) and (2) sample concentration (to selectively concentrate components of interest prior to the analysis). Commercially available column processors (Bakerbond Spe, Speedisk, and others) consist of several small columns (most often 1- to 6-cm^3 volume) filled with different phases (adsorption or reversed phase, cation and anion exchanger, hydrophilic or hydrophobic, molecular sieve) involving wide-pore sorbents (10–50 μm), with high particle surface area and adsorption capacity, which allows one to characterize the protein by a rough estimate of its isoelectric point (pI), hydrophobicity, and molecular weight. At the same time, after each step, the cleaning-up/desalting and concentration of protein sample is achieved. Positive pressure processor ensures identical flow rate in all the columns and enables reproducible extraction of proteinaceous substances at a small-scale regime.

Fast desalting columns can be used in any application, including extract desalting, which requires group separation of substances of high and low molecular weights. Most convenient are fast desalting columns filled with Sephadex G-25 Superfine that separate larger proteins or polypeptides from all molecules under its 5000 MW exclusion limit. The technique is relatively rapid as each run takes 1–4 minutes, depending on the sample volume and flow rates.

5.3 PROTEIN HYDROLYSIS FOR DETERMINING AMINO ACID COMPOSITION

For hydrolysis, the following procedure is usually followed:

1. Hydrolysis with 6 M hydrochloric acid at 100°C for 24 and 48 hours: Once hydrolysis is complete, hydrochloric acid is distilled off by heating to a temperature not higher than 40°C, and the hydrolysate is dissolved in buffer or water before amino acids are assayed. Results obtained from the 24- and 48-hour hydrochloric acid hydrolysis are next extrapolated to time zero to obtain values pertaining to the state before the decomposition caused during the acid hydrolysis, except for Cys, Val, Met, Ile, and Trp. During the 24-hour hydrolysis, Val and Ile are not released quantitatively; the correct values for these are obtained from the 48-hour hydrolysis.
2. Hydrolysis with 5 M sodium hydroxide for 16 hours at 110°C, for an accurate assay of Trp and Leu: After cooling, the hydrolysate is neutralized by hydrochloric acid and brought to the desired volume with distilled water.
3. Performic acid oxidation and subsequent hydrolysis with 6 M hydrochloric acid, for accurate assay of Cys and Met: Performic acid is prepared by adding 0.5 mL of 30% H_2O_2 to 4.5 mL of 80% formic acid and heating for 3 minutes at 50°C in water bath. A mixture of performic acid (0.4 mL) and sample (12 mg) is heated for 15 minutes at 50°C in water bath and then evaporated to dryness in a rotary vacuum still. The residue is then hydrolyzed with 5 mL of 6 M hydrochloric acid for 22 hours, following procedure 1.

There are also other hydrolysis techniques that produce lower amino acid losses. For example, an assay carried out with 3-N-mercaptoethanesulfonic acid at 110°C for 22 hours allows recovery of all amino acids, including tryptophan, with yields higher than those obtained with toluenesulfic acid or HCl + thioglycolic acid (Sheppard 1976).

5.4 ANALYTICAL METHODS USED IN PROTEIN, PEPTIDE, AND AMINO ACID ANALYSIS

5.4.1 KJELDAHL METHOD

The Kjeldahl method is a major technique in food protein analysis and is also used as a reference in comparisons against all other methods. When properly conducted, the Kjeldahl assay ensures high precision and good reproducibility of results. During a routine assay, the mean error does not exceed 1%.

The major difference between the very numerous Kjeldahl nitrogen assays lies in the catalyst used for mineralization and in the technique of ammonia distillation and titration. Regardless of different options, the basic Kjeldahl procedure consists of five stages:

1. Preparation and weighing of a sample
2. Addition of reagents
3. Mineralization

4. Steam distillation of ammonia
5. Titration

The sample should be as homogeneous as possible and may vary in size from 0.2 to 2.0 g, depending on the protein content in a product under assay. At a protein content of 5% –25%, an optimal sample weight is 0.5–1.0 g.

The Kjeldahl procedure requires that three types of reagents be added to the sample, each serving a certain function during mineralization:

1. Sulfuric acid: It causes oxidation of organic compounds to carbon dioxide and water and induces breakdown and transformation of nitrogen compounds to free ammonia, which is subsequently bound into ammonium sulfide.
2. Salts (K_2SO_4, Na_2SO_4): It is necessary to raise the mineralization temperature.
3. Catalysts to accelerate oxidation: These include metals, Hg, Cu, Se; oxides, CuO, HgO, P_2O_5, TiO_2; peroxides, H_2O_2; salts, $CuSO_4$, Hg_2SO_4.

Mercury and mercuric oxide are the most efficient catalysts available; their application, however, has drastically dwindled due to their adverse environmental effects. In addition, those catalysts share a drawback because mercury complexes with ammonia, which restrains distillation of the latter. To decompose the ammonia–mercury complex and also to include nitrate and nitrite, sodium thiosulfate or salicylic acid is added to the sample.

Mineralization takes longer time in the presence of copper as a catalyst than in the presence of mercury. Copper compounds, however, are less environmentally harmful and thus are used more often to replace mercury catalysts, also in the official descriptions of the approved procedures [International Organization for Standardization and Association of Official Analytical Chemists (AOAC)]. The most efficient and environmentally neutral copper catalyst is $CuSO_4$.

Selenium is a catalyst that has to be used in small and repeated doses. At larger doses and during longer mineralization, it may cause losses of nitrogen and its incomplete recovery during distillation, for which reason caution is recommended.

Reagents belonging to groups 2 and 3 are prepared as special tablets or mixtures, all containing K_2SO_4 or Na_2SO_4 and various catalysts. For example, tablets manufactured by Tecator contain K_2SO_4 and one of the five different catalysts, and the widely known selenium mixture contains Na_2SO_4 (90%), Hg_2SO_4 (7%), $CuSO_4$ (1.5%), and Se (1.5%).

Mineralization is used to transform all the nitrogen in the sample into ammonium ions. The presence of sulfuric acid only restricts mineralization temperature to that of sulfuric acid boiling point (338°C). Addition of a salt (K_2SO_4 or Na_2SO_4) raises the boiling point and reduces the decomposition time for organic compounds and facilitates decomposition of hard-to-mineralize substances. Boiling point of sulfuric acid increases almost linearly with the K_2SO_4/H_2SO_4 (g/mL) ratio to reach 383°C at the ratio of 1:1; at the ratio of 2:1, a temperature as high as 450°C is achieved. However, addition of K_2SO_4 not only results in elevated boiling point but also consumes sulfuric acid in potassium hydrogen sulfate formed.

A food product may contain mineralization-resistant compounds; nitrogen in these compounds is not assayable with the Kjeldahl method. Such compounds include nitrates, nitrites, alkaloids, pyridine, chinoline derivatives, triazoles, pyrazolones, aminopyrine, and antipyrine. To subject such compounds to the Kjeldahl assay, samples containing them have to be additionally treated with appropriate reducing agents, such as chromium, zinc, iron (VI) sulfate, salicylic acid, and sucrose. Reduction can be performed before or during mineralization. Relevant procedures are described in detail by appropriate directives and regulations (AOAC 955.04, 970.02, 978.02, and 970.03).

The samples that contain lipids in addition to protein are more difficult to mineralize and are more prone to foaming than the protein- or carbohydrate-rich ones. To prevent foaming, boiling rods or chips can be used, a few drops of hydrogen peroxide or octanol can be added, or a special antifoaming emulsion can be applied. Under normal conditions, antifoaming substances can be omitted when two-phase heating is applied. At the initial phase, until the sample is carbonized, heating should be carried out carefully, the sample being gradually heated up to about 200°C. The second phase, mineralization proper, begins by gradually raising the temperature to about 420°C and involves heating over the time necessary for the sample to become completely clear. Once mineralization is completed, sample-containing flasks should be cooled down to room temperature.

The contents of a receptacle into which ammonia has been distilled are titrated with hydrochloric acid if boric acid is used or with sodium hydroxide if hydrochloric acid is used. When titrating, it is convenient to use the Tashiro indicator (0.2 g of methyl red and 0.19 g of methylene blue dissolved in 100 mL of ethanol) or a conductometer.

5.4.2 Ultraviolet Absorption

Proteins and peptides containing aromatic amino acids, particularly tyrosine and tryptophan, can be successfully assayed with the ultraviolet absorption technique. However, nucleic acids exhibit a strong ultraviolet absorption within the same wavelength range, but their absorption is stronger at 260 nm than at 280 nm, whereas the absorption pattern of proteins is reverse. Therefore, by determining optical density of an appropriately diluted protein solution at 280 and 260 nm, interference of nucleic acids can be, to some degree, eliminated.

Kalckar (1947) proposed a simplified formula to calculate the protein concentration from a spectrophotometric assay; his equation has the following form:

$$P_c \, (\text{mg/mL}) = 1.45 \, D_{280} - 0.74 \, D_{260}$$

Whitaker and Granum (1980) proposed a direct method for protein assay, based on the difference between absorbances at 235 and 280 nm. Protein concentration (P_c) is calculated using the formula: P_c (mg/mL) = 1.45 D_{280} − 0.74 D_{260} where 2.51 is the difference between the average extinction coefficients ($E^{0.1\%}$) measured at 235 and 280 nm. The main advantage of the method is the lack of interference by nucleic acids as they have essentially identical absorbance at 280 and 235 nm.

Protein estimation with the ultraviolet absorption method is very simple and rapid but produces a considerable error when mixtures containing more than 20% nucleic acids or very turbid solutions are used. This method is particularly recommended for comparative assays of large series of samples with relatively low protein concentrations or when high contents of ammonium salts rule out other analytical techniques (Layne 1957). The method is frequently used for an assay of the proteolysis rate and to determine peptide and protein contents in chromatographic effluents.

5.4.3   2,4,6-TRINITROBENZENE 1-SULFONIC ACID-SPECTROPHOTOMETRIC METHOD

2,4,6-Trinitrobenzene 1-sulfonic acid (TNBS) reacts only with primary amino groups of amino acids, peptides, and proteins under slightly alkaline conditions (pH 8–9); the reaction is terminated by reducing the pH.

Those findings aided in developing the TNBS-spectrophotometric method for determination of amines, amino acids, and peptides (Satake et al. 1960). The authors quoted used the following procedure: 1.0 mL of sample solution to be analyzed (at a concentration of 0.01–0.87 µmol/mL), 1.0 mL of 4% sodium bicarbonate, and 1.0 mL of 0.1% TNBS were mixed; kept in dark at 40°C for 2 hours; and the optimal density was measured at 340 nm after acidification with a defined volume of 1 N HCl. Under these conditions, the kinetics of color development for various amines, amino acids, peptides, and proteins are practically identical (Satake et al. 1960; Mokrasch 1967), except for two peptides, carnosine and glutathione, which produce optical density values higher by about 2.5% compared with the final concentration of N.

The α- and ϵ-amino groups in the sample amino acids were found to be involved in the second-order reaction, an unprotonated amino group being the reactive species (Freedman and Radda 1968). The SH group of cysteine reacts with TNBS; ammonia does not form a complex with TNBS, but urea slightly delays the TNBS reaction with free amino groups. On the other hand, a high (8 M) concentration of urea eliminates differences in kinetic reactivity of amino groups of different kinds of proteins. In the analytical procedure, it is important that constant temperature, pH, and reaction time be maintained.

The method of Satake et al. underwent numerous modifications and improvements in order to make it applicable in, among other things, coestimation of amines, amino acids, and protein mixtures (Mokrasch 1967); determination of available lysine in proteins (Kakade and Liener 1969); determination of protein amino groups unreactive toward TNBS using SDS (Habeeb 1966); and determination of the protein directly in 6N HCl hydrolysates (Gehrke and Wall 1971). Other valuable modifications were aimed at determining the degree of hydrolysis of food protein hydrolysates (Adler-Nissen 1979) and assessing protein degradation in meat (Madovi 1980) and cheese (Clegg et al. 1982) with the TNBS method.

In numerous food products, results obtained with the TNBS method were found to correlate well with those produced by the AOAC Kjeldahl technique. Sensitivity of the TNBS method is a few times higher than that of the ultraviolet absorption. The major advantage of the TNBS method, however, lies in its high specificity and low sensitivity to interfering substances.

5.4.4 Biuret Method and Its Modifications

5.4.4.1 Classic Procedure

The biuret reaction is based on the formation of a purple complex between copper ions and amide linkages or peptide bonds of the protein in strongly alkaline solutions. Each copper ion chelates four to six peptide nitrogens, depending on the protein (Strickland et al. 1961). That reaction affects all compounds in which a molecule contains two –CO-NH– groups interconnected in one of the following ways:

- Directly (e.g., in oxalic acid diamide, $H_2N-CO-CO-NH_2$)
- Via nitrogen (in biuret)
- Via a carbon atom (e.g., in malonic or succinic acid diamides and in protein hydrolysis products)

Free amino acids and dipeptides produce no positive biuret reaction, as opposed to tri- and oligopeptides and proteins. Proline peptides do not form ultraviolet-absorbing complexes with copper, whereas prolyl peptides form such complexes. This explains the finding that gelatin (which has a high proportion of proline and HOP residues) is less reactive with copper than are other proteins. Casein and myofibrillar proteins, too, produce a color that is less intensive than that formed, at identical protein concentrations in a sample, by albumin and sarcoplasmic proteins.

It has been proposed to use glycol, tartrate, and citrate to increase the stability of such biuret reagents. Gornall et al. (1949) optimized the composition of the tartrate-stabilized reagent and developed a simple procedure for the determination of serum total protein, albumin, and globulin.

The biuret reaction is regarded as simple and rapid. The main disadvantage is its relatively low sensitivity and being subjected to interference when the protein sample contains various thiols, high concentrations of deoxyribonucleic acid (DNA), saccharides, or lipids. For this reason, the biuret reaction has undergone a number of modifications, discussed below.

5.4.4.2 Colorimetric Determination of Protein in the Presence of Thiols

The thiol (β-mercaptoethanol [ME], β-mercaptoethylamine, glutathione) interference can be prevented in most cases by using a biuret reagent that is chelated with EDTA, but in samples containing DTT, it is also necessary to add iodoacetamide prior to addition of the biuret reagent (Chromy et al. 1974; Pelley et al. 1978).

5.4.4.3 Colorimetric Determination of Protein in the Presence of TCA-Soluble Interference Substances

The method involves protein precipitation, separation of the precipitate from the filtrate, dissolution of the precipitate in sodium or potassium hydroxide solution, and determination of protein with the biuret reaction. Depending on their composition, the proteins are totally precipitated by adding 5%– 25% TCA.

5.4.4.4 Colorimetric Determination of Protein in Lipid-Rich Samples

Colorimetric protein assays in samples of a high lipid- or phospholipid-to-protein ratio produce highly turbid extracts. Hence, before the reaction is performed, lipids have to be removed from such samples—for example, by using an acetone:petroleum ether (1:1) mixture extraction (Beyer 1983). The lipid-free samples are subsequently solubilized with 10% sodium deoxycholate or sodium hydroxide and treated with the biuret reagent.

5.4.4.5 Colorimetric Determination of Protein in the Presence of Reducing Saccharides

Reducing sugars interfere with the biuret reaction. Therefore, hydrogen peroxide is added to the acidic supernatant (TCA extract) to oxidize the sugars. In addition, a 10-fold increase in the concentration of the biuret reagent (15 g $CuSO_4$ 5 H_2O, 60 g K, Na-tartrate, $4H_2O$ in 1 L 7.5 N NaOH) improves sensitivity of the assay and allows it to be applied to peptide bond determination in food suspensions (Hung et al. 1984).

5.4.4.6 Spectrophotometric Determination of Protein in the Presence of Deoxyribonucleic Acid

Measuring ultraviolet absorption of the complex formed by protein and copper in strongly alkaline copper sulfate solutions is usually 10–15 times more sensitive than the standard biuret reaction and makes it possible to determine protein even in very dilute solutions, on the order of 0.01–0.1 mg/mL.

Ellman (1962) proposed to measure the absorption of the copper–protein complex at 263 nm, using a copper sulfate concentration of 0.04% and sodium hydroxide concentration of 2N. The choice of wavelength in the case of DNA-containing protein solutions is limited by the very high absorption of DNA. For this reason, Itzhaki and Gill (1964) developed a procedure based on the measurement of absorption at 310 nm, which can safely be used in the presence of DNA concentrations as high as 0.7 mg/mL in the final reaction mixture.

5.4.5 THE POPE AND STEVENS METHOD

5.4.5.1 The Classic Pope and Stevens Procedure

Free amino acids and peptides form well-soluble complexes with poorly soluble copper phosphate. Once the excess nonsoluble copper phosphate is separated by centrifugation or filtration, the amino acid–bound copper in a clear filtrate or supernatant is determined iodometrically. Such complexes are strongly blue colored, their maximum absorbance falling at 620 nm. The complex color intensity depends on the concentration and type of an amino acid involved. In most cases, two molecules of an amino acid react stoichiometrically with one cupric ion to form a blue-colored copper salt. However, histidine forms a salt at a ratio of two cupric ions to three histidine molecules rather than at the normal ratio. The copper salt of cystine is much more insoluble than salts formed by the remaining amino acids. For this reason, not all amino acids produce comparable color, although for most of them the results ranged from 98% to 107% of the expected values (Pope and Stevens 1939).

Schroeder et al. (1950) applied copper phosphate that had been washed free of excess phosphate and suspended in borax and obtained more satisfactory and reproducible results. Three amino acids (HOP, leucine, and threonine) produced a higher but consistent result (102%), whereas histidine and cystine could not be determined satisfactorily.

The Pope and Stevens method is fairly specific with respect to free amino acids, but the reaction is produced also by peptides and proteins present in extracts of biological materials and in protein hydrolysates. Tokarczyk and Kolakowski (2003) demonstrated oligopeptides to be involved in quantitative determination of amino acids by iodometric titration of their copper salts, the involvement being dependent on both molecular weight and amino acid composition. Dipeptides containing Trp or Tyr produce results similar to those obtained with free tyrosine and tryptophan. Tripeptides containing no aromatic amino acids (e.g., triglycine and α-alanyl-glycyl-glycine) produce results close to those of the major group of free amino acids; glutathione, both reduced and oxidized, produces results similar to those obtained with tetrapeptides (about 30% of those for Glu). Polypeptides (e.g., bacitracin) interfere very weakly (about 15%, compared with Glu), whereas proteins of molecular weight exceeding 14 kDa virtually do not interfere in determination of α-amine nitrogen.

5.4.5.2 Spectrophotometric Modification of the Pope and Stevens Method

Spies and Chambers (1951) developed a simple and rapid spectrophotometric method for determining amino acids and peptides. The method is based on the color obtained when the combined copper is converted to a copper salt of alanine. They tested the procedure when studying protein hydrolysis rate. Color intensities of 16 amino acids were found to range from 79.8% for glycine to 125.9% for HOP when analyzed without alanine added. When analyzed with alanine, the range was 98.2%–103.6% and the overall deviation from 100% was 11.2%, a degree of accuracy satisfactory for most analytical purposes. The amount of copper bound to histidine and some other amino acids was strongly dependent on the pH of the solution. Therefore, careful control of pH was essential. Lactic acid, glucose, galactose, sucrose, inositol, mannitol, and ammonium chloride produced no color with copper and would not therefore interfere.

5.4.6 Formol Titration

The amino acid amine group produces a methylene derivative of the amino acid with formaldehyde and loses its cationic properties, whereas the carboxyl group in the methylene amino acid is unblocked and can be titrated with an alkali solution in the presence of phenolphthalein or thymolphthalein as indicators. The number of carboxyl groups released is equivalent to the number of amino acids bound to formaldehyde. Thus, the titration outcome makes it possible to determine the amount of amino acids in the solution assayed. The method is not specific as—besides free amino acids—peptides and proteins, too, react with neutral formaldehyde. Moreover, proline produces an unstable binding with formaldehyde, which suppresses the titration efficiency; the efficiency is increased by tyrosine, which contains a phenol group. The formol titration technique cannot be used in the presence of ammonium salts as those, while reacting with formalin, produce urotropin, and an acid equivalent is released. In laboratory practice, this rapid and simple procedure is frequently used as an accessory technique during enzymatic digestion of meat protein, although some authors recommend it for determining the protein content, particularly of milk and meat protein.

5.4.7 Alcohol Titration

Carboxyl groups of amphoteric amino acids, peptides, and proteins can be determined by titrating in aqueous solutions without formalin. This is the principle of a convenient and simple procedure of amino acid determination in deproteinated extracts or filtrates—the alcohol titration micromethod. The sample solution pH is adjusted to 6.5 by adding a base or an acid with as little dilution as possible. After an indicator solution has been added, the mixture is titrated with alcoholic HCl until the color turns pink. Then, 9–10 volumetric units of absolute alcohol per 1 volumetric unit of the amino acid solution are added. The indicator turns yellow. The solution should be retitrated, with standardized alcoholic (90%) KOH solution, until it turns pink again. The difference between the blank and the experimental value is equivalent to the amount of carboxyl groups in the dissociated substance (Levy 1957).

The preliminary neutralization of a sample to pH 6.5 is related to the stoichiometric point of amino acid, except for the imidazole group of histidine, which is a drawback of the method: it becomes virtually useless at large amounts of imidazole. The guanidine group of arginine will not be included in the titration because it is still positively charged.

Similar is the principle of the acetone titration method (Levy 1957). Both methods are particularly useful to study kinetics of enzymatic processes controlled by protease effects on proteins and peptides.

5.4.8 Lowry Method

The method developed by Lowry et al. (1951) is based on a color reaction produced by the Folin-Ciocalteu reagent with peptide bonds of proteins and peptides and aromatic amino acids in appropriate alkaline medium.

The reaction proceeds via two basic stages:

1. The biuret reaction, in which a copper ion (Cu^{2+}) complexes with protein and peptides containing at least two peptide bonds
2. Reduction of phosphomolybdic acid (the Folin reagent) through the Cu^{2+}– protein complex to molybdenum blue

The Folin-Ciocalteu reagent reacts with a number of other chemicals, such as some free amino acids, thiol compounds, sucrose and other saccharides, lipids and fatty acids, amines and trimethylamine oxide, chelating substances such as EDTA, some buffers, and reagents. An exhaustive list of the many compounds known to interfere with the Lowry assay was given by Peterson (1979).

To eliminate or restrict interference of accompanying compounds during the Lowry assay of protein, the technique underwent a number of modifications, including heating of a sample before and after treating it with the Folin-Ciocalteu reagent, addition of perhydrate, chloramine-T, SDS, extraction with organic solvents to remove lipids, and so on, which was reviewed in more detail in a previous study (Kolakowski 2001). However, the most useful seems to be the modifications involving a parallel measurement of sample absorbance in the absence (A_1) and in the presence (A_2) of a copper ion and determination of "protein" from the difference $A_2 - A_1$ (Rieder 1966; Potty 1969), and initial protein precipitation with TCA (Bennett 1967; Polacheck and Cabib 1981) or deoxycholate-TCA (Peterson 1977) and assaying it after solubilization in leach solution. To assay peptides soluble in TCA extracts of muscles, Kolakowski et al. (2000) used a procedure similar to that mentioned above, assuming $A_2 - A_1 = A_{PEP}$ to correspond with "peptides" and A_1 to "tyrosine," although tryptophan—when treated with the Folin-Ciocalteu reagent—produces a pronounced color, a weak color being produced by phenylalanine, histidine, lysine, and taurine and by other TCA-soluble substances. Of the 57 interfering compounds examined, it was only the reduced glutathione that was shown to reduce A_{PEP}, while A_1 includes also tryptophan-containing oligopeptides (Kolakowski et al. 2000).

In practice, however, the original Lowry method is most frequently used. Its popularity is due chiefly to its high sensitivity, 100 times that of the biuret method and 10 times that of the spectrophotometric assay.

5.4.9 BICINCHONINIC METHOD

Bicinchoninic acid (2,2'-diquinolyl-4,4'-dicarboxylic acid) is a highly specific chromogenic reagent for Cu(I) that forms a purple complex at the absorbance maximum at 562 nm. The reaction was used in assaying protein (Smith et al. 1985) because it reduces one alkaline Cu(II) to Cu(I). At room temperature, the color of the reaction complex intensifies slowly over 21 hours, and hence incubation of the mixture at 37°C or 60°C reduces the assay time to 30 minutes or to 10–15 minutes, respectively. The absorbance is directly proportional to the protein concentration although some substances such as DTT, glycine, and detergents may interfere during assays in an appropriate high concentration (Brown et al. 1989). As the technique is fairly sensitive, it can be used to measure very dilute protein concentrations and be applied to very small sample volume. In tube assays, the technique accurately measures protein concentrations of 10–30 mg/mL. For this reason, the technique may be regarded as an alternative to the Folin-Ciocalteu reagent in the Lowry assay. Variations in the color yield reflecting the protein in a sample follow a pattern similar to that of the Lowry method. The order of increasing color formation is as follows: gelatin, avidin, BSA, immunoglobin G, chymotrypsin, insulin, and ribonuclease (Owusu-Apenten 2002).

5.4.10 DYE-BINDING METHODS

When reacting, under specific conditions, with dyes containing acid sulfonic groups ($-SO_3H$), the functional groups of proteins, particularly the basic groups in the side chains of arginine, lysine, and histidine, induce a color reduction in the dye solution in proportion to the protein content of food. Organic dyes bind with proteins by ionic or electrostatic interactions although van der Waals forces participate as well (Owusu-Apenten 2002). The reaction is optimal in a strongly acidic medium and produces soluble or insoluble complexes (Kolakowski 1974). Thus, the concentration of protein in a sample may be easily determined by measuring absorbance of the dye solution before and after protein addition or after separation of the insoluble dye–protein complex by centrifugation or filtration.

Several dyes have been used as reagents to determine the protein content; among them, Amido Black 10B, Coomassie Brilliant Blue G-250, Orange G, and Acid Orange 12 seem to be most popular in food analysis. The techniques applied most frequently are those of Bradford and Udy.Bradford's technique involves using Coomassie Brilliant Blue G-250 (C.I. 42655) to produce a sparingly soluble complex with proteins (Bradford 1976), which can be precipitated at low temperature (8°C), or by using TCA at room temperature; calcium phosphate for protein coprecipitation may also be used in the assay. The other technique, originally developed by Udy (1971), calls for using Acid Orange 12 to obtain a protein–dye precipitate. So far, the two techniques have undergone numerous modifications and improvements. The techniques are recommended mainly for milk protein analysis (AOAC Method 967.12) and routine analyses of food products in which the protein contents are very low (beer, wine, fruit and vegetable juices, etc.).

Dye-binding methods are very simple and rapid, particularly when applied as a one-step procedure (Anon. 1968), convenient for routine use, highly sensitive, and inexpensive. Color usually develops in less than 5 minutes and remains stable for at least 0.5–1 hour. An important advantage of the dye-binding method is the lack of interference from many compounds known to interfere with the Lowry method; there is also no need for skillful manipulation and use of corrosive reagents of the Kjeldahl procedure. The method's drawback, however, lies in a different affinity of a dye for several pure proteins that react or bind differently, according to their structural characteristics, which virtually requires that the method be adjusted to a given protein each time it is applied, thereby ruling out its use as a universal procedure for total protein assay. For example, some proline-rich proteins (e.g., collagens, histones) stain red with Coomassie Brilliant Blue R-250 and show absorption maxima different from those of the conventional proteins. Small peptides that react with the Lowry reagents remain undetected when dye binding is applied. A list of interfering substances in the dye-binding procedures, compiled from relevant references, was published by Owusu-Apenten (2002).

5.4.11 Ninhydrin Method

Oxidative deamination of amino acids in acid solution with ninhydrin leads to the formation of ammonia and the reduction of ninhydrin to hydrindantin. Ammonia condenses then with hydrindantin to form diketohydrindylidenediketohydrindamine (DYDA) with a characteristic purple color. Ammonia produces DYDA with ninhydrin only in the presence of a reducing agent capable of producing hydrindantin. In the method developed by Moore and Stein (1954), the agent is stannous chloride; potassium cyanide, more stable than stannous chloride, was used in the procedure of Yemm and Cocking (1955). Standara et al. (1999) proposed to reduce ninhydrin by sodium borohydride, which has a number of advantages compared with the classic reducing agents. Organic solvents such as ethanol, dioxan, methyl cellosolve, pyridine, and phenol were found to accelerate color development to varying degrees, depending on the solvent concentration. Acting on that premise, Lee and Takahashi (1966) simplified the procedure of Yemm and Cocking (1955) by eliminating the reducing agent and by using glycerol (60%) as the organic solvent.

In spite of being rapid and sensitive, the ninydrin method has its disadvantages, which should be borne in mind when the method is to be applied. The stoichiometric reaction concerns most of the amino acids with a quantitative yield of DYDA, although cystine indicates that only one half of the amino groups react, and cysteine does not form DYDA, producing a yellow product similar to that formed by proline. Tryptophan, too, produces a clearly weaker color with the ninhydrin reagent than most amino acids. On the other hand, the DYDA yield from tyrosine and phenylalanine is largely dependent on the relative proportion of water to methyl cellosolve, whereas the higher yield in the case of lysine is probably brought about by the partial reaction of the terminal amino group (Yemm and Cocking 1955). Apart from amino acids, the positive color ninhydrin reaction is produced also by ammonium salts, amino sugars, and ammonia. Various cations, particularly Cu^{2+} and Fe^{3+}, drastically inhibit color development (D'Aniello et al. 1985), while microgram quantities of Mn^{2+}, Fe^{2+}, and Mo^{2+} increase the intensity of the color developed by amino acids reacting with ninhydrin (Singh et al. 1978). The intensity of reaction with peptides and proteins depends on the number of amino groups available. It is also important that optimal reaction conditions in the procedure used be maintained, particularly a constant pH (5.5), temperature (100°C), and reaction time (12–15 minutes).

The ninhydrin reaction is one of the most commonly used methods for determination of free amino acids in deproteinated tissue extracts.

5.4.12 HYDROXYPROLINE DETERMINATION

Because HOP occurs only in collagen, determining its content allows us to indirectly determine the total amount of collagen (after the sample has been hydrolyzed in 6 N chloride containing 0.75% $SnCl_2$ under reflux) or products of its hydrolysis, soluble in deproteinated extracts (TCA, hot water, etc.).

Most of the published HOP determination procedures involve oxidation of the amino acid to pyrrole-2-carboxylic acid or pyrrole and then formation of chromophore with p-dimethylaminobenzaldehyde (Prockop and Undenfriend 1960). Oxidation of HOP with hydrogen peroxide is somewhat less reproducible than oxidation with chloramine-T (N-chloro-p-toluenesulfonamide sodium salt). Optical density of the solution in which color has developed is measured at 557–565 nm against a blank.

5.4.13 N-AMIDE DETERMINATION

A complete deamidation of a food protein can be achieved by the conventional method of hydrolyzing the protein with 2N HCl at 100°C for 3–6 hours, depending on the type of product sample. The amount of ammonia generated during the reaction is used to calculate the amide content of the protein. For example, the amide nitrogen content of soybean protein is about 11% of the total nitrogen. Those proteins richer in their N-amide concentration are, as a rule, more susceptible to deamidation or transglutaminase-induced cross-linking.

5.4.14 CHROMATOGRAPHY

5.4.14.1 Filter Paper Chromatography

This old technique is still widely used, both in its one- and two-dimensional varieties, for qualitative amino acid analysis. Apart from the classic solvents—phenol/water (8:2) and collidine/lutidine (1:1, water saturation)—a number of other mixtures are used, for example, n-butanol/acetic acid/water (4:1:1), n-butanol/butyric acid/acetic acid/water (5:0.5:7.5), and others, particularly for separation of protein hydrolysate components. Repeated separation on the same chromatogram allows one to separate a mixture of amino acids to several fractions; the most frequently encountered difficulties involve separation of alkaline amino acids from one another, as well as separation of leucine and isoleucine.

The structure and chromatographic mobility of amino acids are rather closely interrelated. The amino acids possessing hydrophobic side chains show a higher affinity to the mobile organic phase, and hence R_f increases with increasing hydrophobicity of an amino acid (Gly, Ala, Val, Leu). Amino acids with hydrophilic side chains show a higher affinity to the aqueous stationary phase and produce lower R_f values than amino acids with identical carbon skeletons, but with nonsubstituted side chains.

After the development of ion-exchange filter paper, the potential of filter paper chromatography has been greatly expanded.

Following separation, individual amino acids are identified and quantitatively determined with specific color reactions, ninhydrin being the most frequently used reagent. Red complexes obtained by an additional spraying of the chromatogram with dilute Cu(II) solution (e.g., copper nitrate) are relatively persistent as well. They can be eluted from the filter paper and used in semiquantitative analysis of amino acids; however, an error on the order of 20%–50% is involved.

5.4.14.2 Thin-Layer Chromatography

Separation is carried out on a plate of glass, an aluminum foil, or a plastic foil coated with a thin layer of a vector. Vectors such as silica gel, diethyloaminoethyl (DEAE)-cellulose, and cellulose MN 300, as well as various starch and dextran gels, are used most often.

In terms of its utility, thin-layer chromatography (TLC) is greatly superior to filter paper chromatography, as it allows one to (1) reduce the separation time considerably; (2) use a wider array of adsorbents, adjusted to the composition of a sample; (3) obtain smaller and more contained patches (without the so-called tails); and (4) increase sensitivity. For these reasons, TLC is used not only in qualitative analysis but also in quantitative analysis of amino acids and peptides. The colored derivatives of amino acids and peptides can be quantified spectrophotometrically following elution from the plates.

The modern high-performance TLC, involving commercially precoated, high-efficiency plates, reduces development time still further (3–20 minutes), pushes down detection limits (0.1–0.5 ng for absorption and 5–10 pg for fluorescence), and allows a higher number of sample lanes per plate (Sherma 1994).

5.4.14.3 Gel-Filtration Column Chromatography

Commercially available gel-filtration media are mainly marketed as Sephadex G (cross-linked dextrans), Bio-Gel P (polyacrylamide), Bio-Gel A (agarose chains), Sepharose 2B (agarose gels), and Superdex (dextran covalently binding to cross-linked agarose beads). By selecting a suitable type and exclusion limit of a gel, practically all proteins and products of their hydrolysis can be fractionated by their molecular weight; moreover, the eluent may be selected almost at will. To fractionate proteins of high molecular weight (e.g., myofibrillar proteins), usually Sephadex G-150, Sephadex G-200, and columns that are not too long (optimum length of 45 × 2.5 cm) due to a slow flow are used. Separation of three major groups is possible; the peaks emerging on polyacrylamide gels are sharper than those obtained on dextran gels. Sephadex gel grades G-75 and G-50 are sufficient to separate sarcoplasmic proteins; the number of peaks produced under optimal conditions (50 × 2.5 cm column) range from 5 to 8. Peptides and amino acids are separated on Sephadex G-25, Sephadex G-15, and Sephadex G-10; the necessary column length then reaches 80–100 cm although it can be, in extreme cases, as long as 150 cm. The number of groups separated increases from about 5 to more than 10 as the gel grade decreases, and the column is elongated because aromatic amino acids (tyrosine, tryptophan, and phenylalanine) have a tendency to produce separate peaks.

By using media with a highly porous matrix, stabilized by covalent cross-links (Sephacryl, Superdex), a more accurate and legible protein separation can be obtained in a much shorter time than when using the conventional dextrans (Sephadex G). The structure of those special gels is unaffected by dissociating agents (urea, guanidine-HCl) and detergents (such as SDS) used in protein analysis.

The gel-filtration column chromatography is a relatively simple method, and when the column is coupled with a flow-through spectrophotometer and a computer-monitoring system, the method is very convenient and useful. On the other hand, the disadvantage of the gel-filtration method is the prolonged retention of aromatic amino acids and peptides in the column, whereby they are not eluted in the order of decreasing molecular weight. The use of the phenol/acetic acid/water (1:1:1, w/v/v) mixture as eluent eliminates absorption of aromatic amino acids and peptides on Sephadex G-25 and allows a more accurate determination of molecular weight of peptides (Carnegie 1965). However, the eluent is not universally applicable.

Gel-filtration chromatography was successfully used to directly separate proteins from other substances present in food extracts; it was also applied as a convenient technique for preliminary fractionation of peptides prior to their more in-depth analysis by TLC peptide mapping, ion-exchange column chromatography, capillary electrophoresis (CE), and other methods.

5.4.14.4 Ligand-Exchange Chromatography on Cu^{2+}-Sephadex

The simple gel filtration is unreliable because single amino acids show a separation behavior different from that of the amino acids in mixtures with peptides. Thus, isolated peptide fractions often contain different amounts of free amino acids as impurities. The way to eliminate those disadvantages is to use a Sephadex column loaded with Cu(II). The method is based on the charge differential, at alkaline pH, between Cu(II) complexes of amino acids and peptides, while Sephadex behaves as a solid ligand. α-Amino acids that form copper complexes having a stability comparable to that of copper–Sephadex complexes are retained on the column. Peptides form strong complexes, stripping copper from the column due to the chelate effect; therefore, they are retained in small amounts only. The column is eluted with borate buffer (pH 11), and the fractions collected may be analyzed colorimetrically, but the method is mostly applicable when copper is removed from eluents with the chelating ion-exchanger Dowex A-1. The peptide fractions are completely free of α-amino acids. Only β-alanine and α-amino butyric acid are eluted together with the peptides (Rothenbühler et al. 1979).

Ligand-exchange chromatography was successfully used to, among other things, determine low–molecular weight peptides in food hydrolysates.

5.4.14.5 Ion-Exchange Chromatography

This technique separates molecules based on their net charge. Negatively or positively charged functional groups are covalently bound to a solid-support matrix producing a cation or an anion exchanger. When a charged molecule is applied to an exchanger of opposite charge, it is adsorbed, whereas uncharged molecules or molecules of the same charge do not bind. The ion exchangers most suitable for amino acids are the following Dowex resins: Dowex 50 × 8 and Dowex 50 × 4, containing 8% and 4% transverse bonds with divinylbenzene, respectively, as well as fine-grained sulfone polystyrene resins (Amberlit IR-120). The automatic amino acid analyzer works by simultaneously using two columns; the larger column is used to separate acidic and neutral amino acids eluted with 0.2 M sodium citrate, initially at pH 3.25 (acidic amino acids) followed by at pH 4.24 (neutral amino acids), and the smaller column is applied to separate basic amino acids, eluted with 0.35 M sodium citrate at pH 5.28. The amino acids separated are derivatized during the postcolumn reaction with the suitably prepared ninhydrin reagent and analyzed photometrically at 550 nm; 440 nm is used for proline. The conventional amino acid analysis, based on ion-exchange chromatography followed by derivatization with ninhydrin, is widely used to analyze the composition of amino acids in protein hydrolysates and to determine free amino acids in extracts of biological materials. Although sensitivity is not usually a problem in the amino acid analysis, the replacement of ninhydrin in the postcolumn detection system by fluorescent reagents such as fluoroscamine or o-phthalaldehyde considerably improves the sensitivity and provides another major advancement.

The automated amino acid analysis has been recently substantially improved. The development of single-column amino acid analyzers coupled with application of microbead resins improves buffer compositions, and the high-pressure chromatographic technique reduces the duration of the analysis from several hours to 90 minutes or less.

The ion exchangers most useful in protein analysis are those having a hydrodynamically appropriate matrix of large beads, adapted to fast flow; they have a high capacity and are practically unaffected by changes in the ionic strength or pH of the eluent (i.e., methacrylate). Ion-exchange chromatography of proteins is based on differences between their overall charge. The protein of interest must have a charge opposite to that of the functional group attached to the resin in order to bind. The proteins of a negative net charge are separated using resins with DEAE or trimeth-ylammonium groups (QAE and Q), charged positively, such as DEAE cellulose, DEAE Sepharose, DEAE Sephadex, or QAE Sephadex. This type of ion-exchange chromatography is called "anion-exchange chromatography." The proteins of a positive net charge are separated using carboxymethyl (CM) or sulfopropyl (SP) groups, charged negatively, such as CM cellulose, CM Sepharose, CM Sephadex, or SP Sephadex. This type of ion-exchange chromatography is called "cation-exchange chromatography." Because interaction is ionic, binding must take place under low ionic conditions. Elution of protein is achieved by increasing the ionic strength (NaCl gradient) of eluent, which disrupts the ionic interaction, and by increasing (cation exchange column) or decreasing (anion-exchange chromatography) the pH of buffer. Depending on the nature of the resins, they are equilibrated with an anion (i.e., Cl^-) or with a cation (i.e., Na^+), which forms anionic bonds with the functional groups.

Ion-exchange chromatography is used in many standard procedures for separation and purification of proteins. Ion-exchange types with designations A-25 and C-25 should be used for separation of molecules with molecular weights below 30,000, and types A-50 and C-50 for the molecular weights in the range of 30,000–200,000.

5.4.14.6   High-Performance Liquid Chromatography

High-performance liquid chromatography (HPLC) is based on the principle of separation of components of a mixture in relatively short columns (15–30 cm) packed with very uniform microparticulate silica (3–10 μm), chemically modified to obtain a surface coating with different functional groups. The use of high pressure and automatic equipment enables an efficient and fast separation of components. Generally, two chromatographic techniques are used in HPLC: (1) normal phase (polar stationary phase and nonpolar mobile phase) and (2) reversed phase (nonpolar stationary phase and polar mobile phase).

In the case of foods, most analyses are carried out using the reversed-phase chromatography. The stationary phases for this type of HPLC are prepared by covalently binding hydrophobic ligands such as C4-, C8-, C18-alkyl chains or aromatic functions to the surface of a rigid siliceous or polymeric support. Due to the strong hydrophobic character of the stationary phase, proteins and peptides are bound very strongly from an undiluted aqueous mobile phase so that their dilution requires the use of hydroorganic eluents. The separation of peptides and protein in reversed-phase chromatography is typically carried out by gradient elution with increasing concentration of an organic modifier such as acetonitrile, methanol, tetrahydrofuran, and isopropanol. The mobile phase usually contains low levels of trifluoroacetic or phosphoric acids, the role of which is to protonate the residual silanol groups at the surface of the siliceous support and the carboxyl groups of the elutes and to form ion pairs with the charged amino groups of the substances to be separated.

Some amino acids (Phe, Tyr, Trp) can only be sufficiently accurately determined by HPLC directly, without derivatization involving UV absorption, fluorescence, or electrochemical detection. For general determination, however, derivatization of amino acids before analysis is universally used for detection and quantitation. Derivatization improves separation, extinction coefficient, and detectability. The reagents most often used for amino acid derivatization include dansyl chloride,

2,4-dinitrophenol, fluorescamine, and *o*-phthalaldehyde. Due to the high rate of reaction and its high sensitivity, the latter is particularly recommended for precolumn derivatization with fluorescence detection of amino acids and peptides.

Reversed-phase chromatography usually results in a very thorough separation; its disadvantage, however, is the difficulty in predicting retention times for unknown compounds. Generally, retention increases with size and hydrophobicity of the molecules. However, in the case of larger peptides and proteins, the number and location of hydrophobic patches at the molecule surface becomes predominant.

5.4.14.7 Affinity Chromatography

Separation and purification of specific proteinaceous substances is usually time-consuming and poorly effective. Affinity chromatography successfully eliminates these drawbacks. It is a highly specific process of bioselective adsorption and subsequent recovery of a compound from an immobilized ligand. A selective ligand is generally immobilized on a beaded and porous matrix that may be in the form of a column packing or batchwire adsorption medium. In practice, the column is filled with the affinity chromatography resin, subsequently coated with the more or less contaminated protein solution. The component showing affinity to the resin is retained, whereas the accompanying substances pass the column unhampered. The component bound to a complex is then eluted by appropriately adjusting the pH or salt concentration in the eluent. The column can be reused.

Historically, cyanogen bromide–activated matrices have been very popular. At present, there are many more activated resins with coupling chemistries suitable for stable attachment to most types of potential ligands (Harakas 1994).

5.4.14.8 Gas Chromatography

Amino acids, due to their low volatility, cannot be directly determined with gas chromatography. They have to be transformed beforehand into suitable volatile derivatives. The derivatives used most frequently include methyl, isopropyl, pentyl, and 2-octyl esters of *N*-trifluoroacetyl amino acids; *N*-heptafluorobutyryl isobutyl ester derivatives; methyl esters of *N*-isobutyl amino acids or N-2,4-dinitrophenyl amino acids; *n*-propyl or *n*-pentyl esters of *N*-acetyl amino acids; and degradation products of reaction with ninhydrin, lithium-aluminium hydroxide, iodobenzene, or hypobromide. Techniques of preparing volatile amino acid derivatives were described in numerous reviews and original papers. Separation of ester derivatives as a liquid (stationary) phase usually involves polyester (PEGA), ethylene polyoxide, or polysilanes (SE-30). The mobile phase is a nitrogen, helium, or hydrogen jet. Gas chromatography allows one to detect 10–14 pmol methyl ester of a dinitrophenyl amino acid. This high sensitivity is ensured by flame-ionization detection and the high electronic affinity of the dinitrophenyl group.

The GC/MS method proved very useful in the analysis and identification of dipeptides present in biological mixtures.

5.4.15 Electrophoresis

Electrophoresis is one of the methods most useful in separation of biological macromolecules. The most commonly used techniques in the analysis of food proteins and products of their hydrolysis are the SDS-polyacrylamide gel electrophoresis (SDS-PAGE) (including isoelectric focusing [IEF]) and two-dimensional SDS-PAGE; electrophoresis on starch and agarose gels, filter paper, and cellulose acetate membranes is common as well, albeit at a smaller scale. On the other hand, CE is gaining importance. Only the most popular and useful electrophoretic techniques used in food protein analysis are described below.

5.4.15.1 SDS-Polyacrylamide Gel Electrophoresis

The anionic detergent SDS, which binds to almost all proteins in essentially identical amounts (1.4 g of SDS per 1 g of protein), minimizes the influence of charge on electrophoretic mobility of proteins to such a degree that they will migrate in the polyacrylamide gel exclusively as a function of their molecular weights. At a sample SDS concentration higher than that given above, all the proteins take up a rigid-rod conformation, which masks differences in total overall charges between them. Larger molecules have a lower net mobility. Proteins having an extreme charge (e.g., histones and prolamines) or those that do not assume a rigid-rod conformation after complexing with SDS (e.g., collagen molecules), as well as low–molecular weight proteins (where the shape is closer to a sphere than to a rigid rod), show no direct relationship between their net mobility and molecular weight. This anomaly can be eliminated by using a calibration curve plotted from different standards. Practically, the sample SDS concentration is 0.5%–2%, depending on the amount of protein involved. The electrophoretic mobility of the SDS–protein complex becomes independent of SDS concentration only at about 0.1%. Therefore, the presence of SDS is important not only in the sample but also in polyacrylamide gels (buffers) during electrophoresis.

SDS has an additional advantage of facilitating sample solubilization because the detergent disrupts electrostatic linkages and hydrophobic forces while leaving disulfide bonds intact. If the individual subunits of the protein have to be separated, a reducing agent such as ME or DTT should be added to the sample to reduce and prevent disulfide linkages. A sample of protein(s) in a medium containing SDS and either ME or DTT is heated for 3–5 minutes at 100°C for SDS–protein binding and good solubilization, which simultaneously prevents any possible enzymatic activity and bacterial contamination. After the solution has been filtered (preferably when hot) and cooled down to room temperature, 10%–40% sucrose or glycerol is added to increase the density, along with about 0.001% of bromophenol blue or pyronin Y, which act as markers during sample migration.

To avoid overlapping of protein bands and to obtain a legible electrophoretic separation, a preliminary fractionation of protein is recommended (e.g., with the extractability method). For muscle protein analysis, good results are obtained when sarcoplasmic proteins are first extracted with a low–ionic strength (I, 0.05) solution, the pellet is rinsed with that buffer, recentrifuged, and the residue is extracted in SDS buffer or SDS–urea buffer solution, with addition of ME or DTT (disulfide reducing conditions) or without such addition (nonreducing conditions) to extract myofibrillar proteins (Kolakowski 1988). To guard against the effects of enzymatic activity, sarcoplasmic protein extraction should be carried out at low temperature, preferably after an appropriate inhibitor (EDTA, PMSF, STI, iodoacetamide, etc.) has been added to the buffer solution. SDS should be carefully added to the sarcoplasmic protein extract, preferably in the form of concentrated solution rather than as a powder, with continuous mixing until an appropriate concentration of the detergent in the sample is obtained; further on, the analysis proceeds as described above.

Polyacrylamide gels are applicable to the separation of proteins within a wide range of molecular weights (MW < 1000 to MW > 10^6). In practice, 3%–7%, 7%, or 7%–30% of acrylamide concentrations are used to separate 1–2×10^6, 10^4–10^5, and 2×10^3 – 10^4 molecular weights, respectively. Gels containing 2% – 4% of acrylamide are used as stacking gels or to separate cytoskeletal proteins (titin, nebulin), lipoproteins, and protein polymers emerging during covalent cross-linking of protein.

The basic procedures of protein separation by SDS-PAGE are based on the classic studies of Shapiro et al. (1967) and Laemmli (1970). The procedures were adapted to separation of low–molecular weight proteins (polypeptides of molecular weights from 1,400 to 25,000) by, among others, Swank and Munkers (1971) and Hashimoto et al. (1983). Coomassie Brilliant Blue R-250 or Coomassie Brilliant Blue G-250 (both marketed under various names) are the most commonly used stains for protein and can be detected down to about 0.1 µg protein in a typical zone; staining with colloidal silver or gold, however, can be 10–100 times more sensitive. However, it should be remembered that the classic SDS-PAGE is not applicable to the analysis of peptides of molecular weight below 1000 or some alkaline proteins (e.g., protamine sulfate) as they are easily washed out of the gel during acid fixing, staining, and destaining.

SDS-PAGE is an excellent method to study protein fraction composition and to compare it in different foods; to study proteolysis of myofibrillar proteins during postmortem storage, marinading, salting, fermentation, and thermal treatment; to evaluate cytoskeletal protein degradation; to detect polymers emerging during covalent cross-linking of proteins (e.g., by transglutaminase); and used in numerous other applications.

5.4.15.2 Isoelectric Focusing in Polyacrylamide Gel

IEF involves a mixture of ampholytes (Ampholine, Bio-Lyte, Pharmalyte, Servalit, Resolyte) of different isolelectric points (pI), which, in the electric field, produce a pH gradient ranging 3–0. Different protein fractions migrate as dictated by their molecular charges until they reach a position corresponding to their individual isolelectric point. To facilitate movements of protein molecules, particularly in the pH region close to their isoelectric point, it is important to use gels with a minimal sieving effect. This is achieved by using a low concentration of a monomer or a high bis-to-acrylamide ratio. Typically, polyacrylamide gels of the total acrylamide concentration of 5%, containing 4% of bis cross-linker, are made up in water rather than in the buffer solution, and ampholytes are added to produce a concentration of 1%–2%. IEF in polyacrylamide gels can be carried out in gel rods or in gel slabs. The remaining part of the procedure is identical to that of SDS-PAGE.

IEF is most commonly applied to identify animal species and food products—raw and cooked meat of fish and shrimp, surimi from different fish, beef mixed with pork, etc.—from specific protein bands.

5.4.15.3 Two-Dimensional Electrophoresis

Two-dimensional electrophoresis allows one to separate very complex protein mixtures into individual fractions, visualized as spots or smudges, for which reason it is classified among the highest resolution methods. It is very useful as an accompanying technique in unidirectional electrophoresis for corroborating an individual nature or complexity of bands.

Two-dimensional electrophoresis can be limited by the quality of the sample. Protein precipitation is often employed prior to electrophoresis to selectively separate proteins in the sample from contaminating substances and to obtain a sufficient concentration of protein in sample. The precipitation procedure should not introduce ions that interfere with electrophoresis. The purified protein sample should be fully solubilized in urea, a nonionic detergent, buffer, and reducing agent.

Two-dimensional electrophoresis is most often applied in either of two versions:

- The first- and second-dimension separations are carried out with the same system (e.g., SDS-PAGE).
- The second-dimension separation is performed with a system different from the first (e.g., IEF and SDS-PAGE).

Two-dimensional electrophoresis technique has found a number of applications in food analysis, including characterization of raw materials such as wheat flours, bovine caseins, and arctic fish muscle and the study of protein changes during washing of minced fish. It has also proved very useful in monitoring of proteolysis during cheese ripening or fish fermentation.

5.4.15.4 Capillary Electrophoresis

CE is a technique utilizing a silica capillary tube normally filled with a buffer solution, although for certain purposes this may be internally coated or filled with a gel matrix. The ends of the tube are placed in electrolyte reservoirs, and a high voltage is applied across these two solutions. This causes migration of charged species, and the bands produced are detected and measured online, providing immediate results as with a chromatogram. The process allows separations that with conventional electrophoretic techniques would take hours but can be achieved within minutes.

CE is not a single technique but consists of a set of related techniques with distinct mechanisms of separation, which differ greatly in terms of their utility in the analysis of amino acids, peptides, or proteins.

5.4.15.4.1 Capillary Zone Electrophoresis

Capillary zone electrophoresis (CZE) employs a single-buffer system in a free solution. Separations are based on differences in mass-to-charges ratios of sample components. CZE is particularly successful in separation of peptides, including compounds with slight structural differences, that cannot be separated by other methods. The separation is based on differences in the electrophoretic rate of molecules in a uniform separation buffer. Very distinct separation is obtained both with acidic (pH 2.5) and basic (pH 10) buffers. At low pH, peptides migrate as cations toward the cathode. In high-pH buffers, peptides migrate as anions, but the strong electroendoosmotic flow (EOF) induces migration in the same direction. CZE enables both a direct analysis of peptides according to their molecular mass (Grossman et al. 1989) and an analysis after modifying their mobility and making it possible to manipulate the selectivity of the separation (Cifuentes and Poppe 1997).

Protein separation with CZE is rendered difficult due to the strong absorption of proteins to ionic or hydrophobic sites on the capillary surface, which causes local variations in the rate of EOF as separation progresses. For this reason, the most successful protein separations are achieved using coated capillaries and buffer additives to suppress adsorption and using a sieving medium in which the migration rate depends on molecular size.

5.4.15.4.2 Dynamic Sieving Capillary Electrophoresis

Polyacrylamide gel–filled capillaries are not useful for protein separations because the strong UV absorbance of the gel restricts their use to longer wavelengths where proteins have poor detection sensitivity. Dynamic sieving capillary electrophoresis (DSCE) employs a polymer solution as the sieving agent. This solution is introduced into the capillary by pressure and is replenished between each analysis. The UV transparency of the polymer solution is a major advantage for separations of proteins by sieving because detection in the low UV region greatly increases sensitivity. DSCE can be used for separation of native proteins but cannot be used for molecular weight determination because of mixed electrophoretic and sieving contributions to migration velocity and polymer–protein interactions. In contrast, DSCE can be useful for molecular weight determination of SDS–protein complexes because all proteins have the same mass-to-charge ratio after complexation of the surfactant.

5.4.15.4.3 Micellar Electrokinetic Chromatography

The separation medium is a buffer containing a surfactant at concentration higher than the critical micelle concentration. Surfactant micelles constitute a pseudophase into which analyte molecules are partitioned by hydrophobic interactions with the lipophilic micelle interior. EOF is used as a pump to transport micelles and the bulk electrolyte toward the detection point. In the most popular embodiment of micellar electrokinetic chromatography (MEKC), SDS is used as the surfactant. Since the SDS micelles are anionic, they migrate electrophoretically in a direction opposite to that of EOF, extending the separation range. Because the magnitude of EOF is greater than the electrophoretic velocity of the micelles, all the species eventually move through the detector. MEKC separations have been successfully used in, among other things, amino acid assays. As most of them lack chromophores, amino acids are most often used as dansyl, dabsyl, phenylthiohydantoinic (PTH), or fluorenylmethoxycarbonyl derivatives to increase reaction sensitivity.

5.4.15.4.4 Capillary Isoelectrophoretic Focusing

Capillary isoelectrophoretic focusing is performed by filling the capillary with a mixture of ampholytes and the sample and then forming a pH gradient. This makes it possible not only to quantitatively separate unknown compounds but also to characterize them in terms of their isoelectric point.

5.4.16 IMMUNOCHEMICAL METHODS

The methods are based on antigen–antibody interactions and are gaining importance. They are very useful, primarily, in detecting allergenic properties of proteins, identification of electrophoretic fractions of proteins, detecting heat treatment and bacterial contamination, and protein structure control. Enzyme-linked immunosorbent assay and Western blotting are the immunochemical tests most widely applied in food analysis. The first is applied to detect and quantitate specific proteins present in very small amounts; the other serves to find a required protein in a complex mixture. These techniques are very sensitive (they are capable of detecting an antigen present in quantities lower than 10 ng/mL) and relatively rapid and simple.

5.4.16.1 Enzyme-Linked Immunosorbent Assay

The basic part of the assay kit is a polystyrene multiwell plate, with each well coated by an antigen. The samples are placed in the wells. If they contain the antigen recognized by an appropriate antibody, the antigen is bound. Subsequently, the wells are rinsed to remove the excess of unbound protein, and the secondary antibody, recognizing another epitope of the same protein antibody, is applied. The secondary antibody is connected with an enzyme that transforms a colorless or nonfluorescent substrate into a colored or fluorescent product. The color or fluorescence intensity is then measured for each sample and serves as a basis from which the amount of the antigen present in the sample is determined.

5.4.16.2 Western Blotting

A sample is subjected to SDS-PAGE, whereby proteins are separated by their molecular weights. Then the gel is placed on a nitrocellulose or polyvinylidene difluoride sheet, or between such sheets, and the protein is transferred from the gel to a membrane by the Southern method or electroblotting. The blotted membrane is then probed with an antigen-specific antibody (primary antibody), followed by a secondary antibody, which binds to the primary one. The secondary antibody usually has a conjugated enzyme attached. The membrane is then incubated with a substrate solution, and the conjugated enzyme produces—with the chromogenic substrate—a dark or fluorescent band at the site of the antigen.

5.4.17 OTHER ANALYTICAL METHODS

In addition to the methods described above, numerous other procedures and techniques are used in the food protein analysis as auxiliary or complementary techniques and also as rapid analytical methods. Some of them merit at least a brief mention here.

Turbidimetric techniques are rapid and convenient, but they yield different values with different proteins and do not differentiate between proteins and acid-insoluble compounds such as nucleic acids.

Mass spectrometry is used mainly for identifying protein subunits and to determine molecular weight of proteins and peptides from different sources. The method is highly sensitive. The molecular weight of a protein can be determined with a precision better than 0.01%.

Infrared spectroscopy (IRS) is based on using the wavelength region that coincides with oscillatory–rotational movements of atoms in a given molecule. The emerging spectra are specific for a given atom arrangement. For example, the range of 1430 to 910 cm^{-1} is particularly rich in absorption bands associated with oscillation of bonds such as C–C, C–N, C–O, and C–H. Although complex arrangements do not produce pure spectra, the range mentioned, frequently called the dactyloscopic (fingerprint) interval, will always produce clear differences and may be used both in qualitative and in quantitative analyses. Identification of substances, determination of their purity, structural analysis, and following changes in chemical processes are the major IRS applications in qualitative analysis. In quantitative assays, IRS is more and more frequently applied in rapid, multicomponent (protein, fat, water, saccharides) analyses of food products with automated and computer-controlled equipment. The analysis is not overly accurate because the presence of accompanying substances often distorts the picture of near-infrared spectra of protein and all the other substances assayed.

Nuclear magnetic resonance (NMR) uses magnetic nuclei or unpaired electrons occurring inside a molecule to unravel its structure, to follow intramolecular dynamic processes, and to observe chemical reactions. NMR methods are at present regarded as most important for studying protein macromolecule structure (particularly in liquids). Additional advantages include ease of sample preparation, a small amount (10^{-6} to 10^{-2} g) necessary, short time needed to produce a spectrum (usually a few minutes), and low cost of a routine spectrum. The fact that a spectrum provides abundant structural information and its interpretation is relatively simple is the most important advantage of the method.

Differential scanning calorimetry is one of the most effective methods to follow thermodynamics of protein stability during denaturation; it is also a powerful tool for assessing structural changes associated with chemical modification and effects of different physical and chemical laboratory and food processing operations. The use of thermal analysis techniques in the food protein analysis and in food research has been reviewed by Biliaderis (1983) and by Ma and Halwalkar (1991).

N-terminal amino acid analysis is applied using standard Edman chemistry. Dinitrofluorobenzene or dansyl chloride forms stable covalent bonds with N-terminal amino acid of a protein or a peptide. The derivatives of the N-terminal amino acid obtained, released during acidic hydrolysis, which simultaneously splits all the remaining amino acids, can be identified chromatographically (e.g., using HPLC) by reference to appropriate standards.

Protein sequencing is used to determine the sequence of amino acids in a protein or a peptide. Edman degradation, involving sequential removal of marked amino acid rests from the N-terminal of a protein or a peptide, is used; peptide bonds between the remaining amino acids are not disrupted. There are two basic ways of identifying N-terminal amino acids. The first involves binding an appropriate substitute (e.g., 1-fluoro-2,4-dinitrobenzene,5-Dimethylaminonaphthalene-1-sulfonyl chloride) to the N-terminal amino acid. After acidic or enzymatic hydrolysis, the amino acid marked by the "substitute" can be detected spectrophotometrically, fluorometrically, or isotopically and identified using chromatography. The other way to determine the N-terminal amino acid involves binding the amino acid rest occurring at the end of a protein, phenylisothiocyanate, to an α-amine group; the result is the formation of a phenylisothiocarbamoyl derivative released from the protein in a weakly acidic medium in the form of a PTH amino acid derivative. The peptide, shortened by one amino acid rest, can be subjected to another cycle of marking and splitting.

Sequence information may also be obtained from the carboxyl terminal of proteins or peptides. The carboxyl-terminal degradation (C-alkylation) method, which activates the C-terminal only once at the start of the chemistry cycle, prevents the detection of background amino acids and increases the accuracy of sequence calling.' Degradation is achieved by both chemical and enzymatic techniques. Enzymatic hydrolysis involves specific exopeptidases that release a single amino acid from the polypeptide chain terminal: aminopeptidases release the amino acid from the N-terminal, while carboxypeptidases release it from the C-terminal.

The Edman degradation method has been automated and improved so that it is possible to elucidate a sequence of about 50 amino acids from the N-terminal of a protein with a picomole amount of the initial material. Amino acid sequence analysis in high molecular–weight proteins usually requires that they be preliminarily degraded into smaller fragments of 20–100 rests; the fragments are then separated and sequenced.

Proteomics is a complete system that includes sample preparation, two-dimensional electrophoresis, imaging, spot picking, digestion, spotting, mass spectrometry, and bioinformatics. Modern proteomics solves many problems by accumulating databases. Automated monoisotopic peptide mass fingerprinting is a powerful analytical strategy for identification of a known protein. If the protein class or species of interest are not well represented in the databases or the proteins are extensively posttranslationally modified, it is possible to obtain complementary MS-MS data from an electrospray (electrospray ionization) mass spectrometer. The analysis of nongel-separated protein mixtures is made possible through the use of trypsin digestion, followed by two-dimensional HPLC separation on the Micromass CapLC™ system. This has been optimized to provide automated high-resolution separations using strong cation exchange in the first dimension, followed by reversed-phase separation in the second dimension. This makes it possible to collect both qualitative and relative quantitative information on protein complexes.

REFERENCES

Adler-Nissen, J. 1979. Determination of the degree of hydrolysis of food protein hydrolysates by trinitrobenzenesulfonic acid. *J. Agric. Food Chem.* 27: 1256–1262.

Anon. 1968. *Bio-Rad Protein Assay, Bulletin 1069.* Bio-Rad Laboratories, Richmond, CA.

Aristoy, M. C. and Tolda, F. 1991. Deproteination technique for HPLC amino acid analysis in fresh pork muscle and dry-cured ham. *J. Agric. Food Chem.* 39: 1792–1795.

Bennett, T. P. 1967. Membrane filtration for determining protein in presence of interfering substances. *Nature (London).* 213: 1131–1132.

Beyer, R. E. 1983. A rapid biuret assay for protein of whole fatty tissues. *Anal. Biochem.* 129: 483–485.

Biliaderis, G. G. 1983. Differential scanning calorimetry in food research—A review. *Food Chem.* 10: 239–265.

Bradford, M. 1976. A rapid and sensitive method for quantitation of microgram quantities of protein utilizing the principle of protein dye-binding. *Anal. Biochem.* 72: 248–254.

Brown, R. E., Jarvis, K. L., and Hyland, K. J. 1989. Protein measurement using bicinchoninic acid: Elimination of interfering substances. *Anal. Biochem.* 180: 136–139.

Carnegie, P. R. 1965. Estimation of molecular size of peptides by gel filtration. *Biochem. J.* 95: 9P.

Chromy, V., Fischer, J., and Kulhanek, V. 1974. Re-evaluation of EDTA-chelated biuret reagent. *Clin. Chem.* 20: 1362–1363.

Cifuentes, A. and Poppe, H. 1997. Behavior of peptides in capillary electrophoresis: Effect of peptide charge, mass and structure. *Electrophoresis* 18: 2362–2376.

Clegg, K. M., Lee, Y. K., and McGilligan, J. F. 1982. Technical note: Trinitrobenzenesulfonic acid and ninhydrin reagents for the assessment of protein degradation in cheese. *J. Fd. Technol.* 17: 517–520.

D'Aniello, A., D'Onofrio, G., Pischetola, M., and Strazzullo, L. 1985. Effect of various substances on the colorimetric amino acid-ninhydrin reaction. *Anal. Biochem.* 144: 610–611.

de Duve, C., Berthet, J., and Beaufay, H. 1959. Gradient centrifugation of cell particulates. Theory and applications. *Progr Biophys Biophys Chem.* 9: 325–335.

Ellman, G. L. 1962. The biuret reaction: Changes in the ultraviolet absorption spectra and its application to the determination of peptide bonds. *Anal. Biochem.* 3: 40–48.

Freedman, R. B. and Radda, G. K. 1968. The reaction of 2,4,6,trinitrobenzenesulfonic acid with amino acids, peptides and proteins. *Biochem. J.* 108: 383–391.

Gehrke, C. W. and Wall, L. L. 1971. Automated trinitrobenzene sulfonic acid method for protein analysis in forages and grain. *J. AOAC.* 54: 187–191.

Gornall, A. G., Bardawill, C. J., and David, M. M. 1949. Determination of serum proteins by means of the biuret reaction. *J. Biol. Chem.* 177: 751–766.

Grossman, P. D., Colburn, J. C., and Lauer, H. H. 1989. A semiempirical model for the electrophoretic mobilities of peptides in free solution capillary electrophoresis. *Anal. Biochem.* 179: 28–33.

Habeeb, A. F. S. A. 1966. Determination of free amino groups in proteins by trinitrobenzenesulfonic acid. *Anal. Biochem.* 14: 328–336.

Harakas, N. K. 1994. Protein purification process engineering. Biospecific affinity chromatography. *Bioprocess Technol.* 18: 259–316.

Hashimoto, F., Horigome, T., Kanbayashi, M., Yoshida, K., and Sugano, H. 1983. An improved method for separation of low-molecular-weight polypeptides by electrophoresis in sodium dodecyl sulfate polyacrylamide gel. *Anal. Biochem.* 129: 192–199.

Hung, N. D., Vas, M., Cseke, E., and Szabolcsi, G. 1984. Relative tryptic digestion rates of food proteins. *J. Food Sci.* 49: 1535–1542.

Itzhaki, R. F. and Gill, D. M. 1964. A micro-biuret method for estimating proteins. *Anal. Biochem.* 9: 401–410.

Kakade, M. L. and Liener, I. E. 1969. Determination of available lysine in proteins. *Anal. Biochem.* 27: 273–280.

Kalckar, H. M. 1947. *J. Biol. Chem.* 167: 461 (citation according to to Layne, 1957).

Kohn, J. 1959. A simple method for the concentration of fluids containing protein. *Nature* 183(4667): 1055.

Kolakowski, E. 1973. *Determination of fish meat peptides generated in technological processes* (in Polish with English summary), Wydawnictwo Uczelniane Akademii Rolniczej w Szczecinie, Rozprawy No. 32, pp. 1–101.

Kolakowski, E. 1974. Determination of peptides in fish and fish products. Part 1. Application of Amido Black 10B for determination of peptides in trichloroacetic acid extracts of fish meat. *Nahrung.* 18: 371–383.

Kolakowski, E. 1988. Comparison of krill and Antarctic fish protein solubility. *Z. Lebensm. Untersuch und Forschung.* 185: 420–424.

Kolakowski, E., 2001. Protein determination and analysis in food systems. In *Chemical and Functional Properties of Food Proteins.* Z. E. Sikorski, ed., Chap. 4, pp. 57–112, Technomic Publishing, Lancaster/Basel.

Kolakowski, E., Bednarczyk, B., and Nowak, B. 2000. Determination of protein hydrolysis products by a modified Lowry method. In *Food in the Course of Scientific Expansion: Potential, Expectation, Prospects* (in Polish). 31st Scientific Session of Polish Academy of Sciences, Poznań, September 14–15, 2000, p. 1: 119–120.

Kolakowski, E. and Szybowicz, Z. 1976. Application of the protein extractability index for estimation of fish meat technological value (in Polish). *Przem. Spoz.* 30: 97–99.

Laemmli, U. K. 1970. Cleavage of structural proteins during the assembly of the head of bacteriophage T4. *Nature.* 227: 680–685.

Layne, E. 1957. Spectrophotometric and turbidimetric methods for measuring proteins. In *Methods in Enzymology.* S. P. Colowick and N. O. Kaplan, eds., Vol. 3, pp. 447–454, Academic Press, New York, London.

Lee, Y. P. and Takahashi, T. 1966. An improved colorimetric determination of amino acids with the use of ninhydrin. *Anal. Biochem.* 14: 71–77.

Levy, M. 1957. Titrimetric procedures for amino acids (formol, acetone and alcohol titration). In *Methods in Enzymology.* S. P. Colowick and N. O. Kaplan, eds., Vol. 3, pp. 454–458, Academic Press, New York, London.

Lowry, O. H., Rosebrough, N. J., Farr, A. L., and Randall, R. J. 1951. Protein measurement with the Folin phenol reagent. *J. Biol. Chem.* 193: 265–275.

Ma, C.-Y. and Halwalkar, V. R. 1991. Thermal analysis of food proteins. *Adv. Food Nutr. Res.* 35: 317–366.

Madovi, P. B. 1980. Changes in the free $-NH_2$, free $-CO_2H$ and titratable acidity of meat proteins. *J. Fd. Technol.* 15: 311–318.

Mokrasch, L. C. 1967. Use of 2,4,6-trinitrobenzenesulfonic acid for the coestimation of amines, amino acids, and proteins in mixtures. *Anal. Biochem.* 18: 64–71.

Moore, S. and Stein, W. H. 1954. A modified ninhydrin reagent for the photometric determination of amino acids and related compounds. *J. Biol. Chem.* 211: 907–913.

Owusu-Apenten, R. K. 2002. *Food Protein Analysis. Quantitative Effects on Processing.* Marcel Dekker, New York, Basel, pp. 169–194.

Pelley, J. W., Garner, Ch.W., and Little, G. H. 1978. A simple rapid biuret method for the estimation of protein in samples containing thiols. *Anal. Biochem.* 86: 341–343.

Peterson, G. L. 1977. A simplification of the protein assay method of Lowry et al. which is more generally applicable. *Anal. Biochem.* 83: 346–356.

Peterson, G. L. 1979. Review of the Folin phenol protein quantitation method of Lowry, Rosebrough, Farr and Randall. *Anal. Biochem.* 100: 201–220.

Polacheck, I. and Cabib, E. 1981. A simple procedure for protein determination by the Lowry method in dilute solutions and in the presence of interfering substances. *Anal. Biochem.* 117: 311–314.

Pope, C. G. and Stevens, M. F. 1939. Determination of amino nitrogen using a copper method. *Biochem. J.* 33: 1070–1077.

Potty, V. H. 1969. Determination of proteins in the presence of phenols and pectins. *Anal. Biochem.* 29: 535–539.

Prockop, D. J. and Undenfriend, S. 1960. A specific method for the analysis of hydroxyproline in tissues and urine. *Anal. Biochem.* 1: 228–239.

Rieder, H. P. 1966. Eine neue Modifikation der Cu-Folin-Methode zur Bestimmungs des Totalproteins im Liquor cerebrospinalis. *Klin. Wschr.* 44: 1036–1040.

Rothenbühler, E., Waibel, R., and Solms, J. 1979. An improved method for the separation of peptides and a-amino acids in copper-sephadex. *Anal. Biochem.* 97: 367–375.

Satake, K., Okuyama, T., Ohashi, M., and Shinoda, T. 1960. The spectrophotometric determination of amine, amino acid and peptide with 2,4,6,-trinitrobenzene 1-sulfonic acid. *J. Biochem.* 47: 654–660.

Schroeder, W. A., Kay, L. M., and Mills, R. S. 1950. Quantitative determination of amino acids by iodometric titration of their copper salts. *Anal. Chem.* 22: 760–763.

Shapiro, A. L., Vinulea, E., and Maizel, J. V. 1967. Molecular weight estimation of polypetide chains by electrophoresis in SDS-polyacrylamide gels. *Biochem. Biophys. Res. Commun.* 28: 815–820.

Sheppard, R. C. 1976. *Amino-acids, Peptides, and Proteins*, Vol. 7. The Chemistry Society, Burlington House, London.

Sherma, J. 1994. Modern high performance thin-layer chromatography. *J. AOAC Intern.* 77: 297–306.

Singh, J. V., Khanna, S. K., and Singh, G. B. 1978. Effect of manganese on ninhydrin color development by amino acids. *Anal. Biochem.* 85: 581–585.

Smith, P. K., Krohn, R. I., Hermanson, G. T., Mallia, A. K., Gartner, F. H., Prevenzano, M.D., Fujimoto, E. K., Goeke, N. M., Olsen, B. J., and Klenk, D. C. 1985. Measurement of protein using bicinchoninic acid. *Analyt. Biochem.* 150: 76–85.

Spies J. R. and Chambers D. C. 1951. Spectrophotometric analysis of amino acids and peptides with their copper salts. *Biochem. J.* 191: 787–797.

Standara, S., Drdák, M., and Veselá, M. 1999. Amino acid analysis: Reduction of ninhydrin by sodium borohydride. *Nahrung*. 43: 410–413.

Strickland, R. D., Freeman, M. L., and Gurule, F. T. 1961. Copper binding by protein in alkaline solution. *Anal. Chem.* 33: 545–552.

Swank, R. T. and Munkres, K. D. 1971. Molecular weight analysis of oligopeptides by electrophoresis in polyacrylamide gel with sodium dodecyl sulfate. *Anal. Biochem.* 39: 462–477.

Tokarczyk, G. and Kolakowski, E. 2003. Specificity of the Pope and Stevens method for amino nitrogen determination. *Proceedings of the 34th Scientific Session of Polish Academy of Sciences*, Sept. 10–11, 2003, p. 2.

Udy, D. C. 1971. Improved methods for estimation of protein. *J. Am. Oil Chem. Soc.* 48: 29A–33A.

Vinograd, J. and Hearst, J. E. 1962. Fortschrifte Chemie organ. Naturstoffe, 20, 273 (citation acc. to: Chandra, P., Appel, W.: *Methoden der Molekularbiologie für Biochemiker, Mediziner und Biologen*, Gustav Fischer Verlag, Stuttgart, 1973)

Whitaker, J. R. and Granum, P. E. 1980. An absolute method for protein determination based on differences in absorbance at 235 and 280 nm. *Anal. Biochem.* 109: 156–159.

Yemm, E. W. and Cocking, E. C. 1955. The determination of amino acids with ninhydrin. *Analyst*. 80: 209–213.

Zabin, B. A. 1975. Hollow-fiber separation devices and processes. In *Methods of Protein Separation*, Vol. 1, N. Catsimpoolas, ed., pp. 239–254, Plenum Press, New York and London.

6 Extraction and Analysis of Food Lipids*

Robert A. Moreau and Jill K. Winkler-Moser

CONTENTS

6.1 INTRODUCTION: A DESCRIPTION OF THE COMMON TYPES OF LIPIDS FOUND IN FOODS

Lipids are often defined as a group of biomolecules that are insoluble in water and soluble in organic solvents such as hexane, diethyl ether, or chloroform. W.W. Christie, an international authority on lipids, defines them as follows: "Lipids are fatty acids and their derivatives and substances related biosynthetically or functionally to these compounds."[1] Sterols, tocopherols, and carotenoids are also common components of lipid extracts. Methods for the analysis of the sterols are described in this chapter (tocopherol analysis is covered in Chapter 9 and carotenoid analysis in Chapter 10). Triacylglycerols (TAGs) (Figure 6.1a) are the main storage lipid (storing energy and carbon skeletons) in plants and animals. TAGs include fats (which are solid at 20°C) and oils (which are liquid at 20°C). In general, most fats are found in animal tissues, and most oils are found in plant tissues. Since fats and oils comprise about 40% of

---

* Mention of a brand or firm name does not constitute an endorsement by the US Department of Agriculture above others of a similar nature not mentioned.

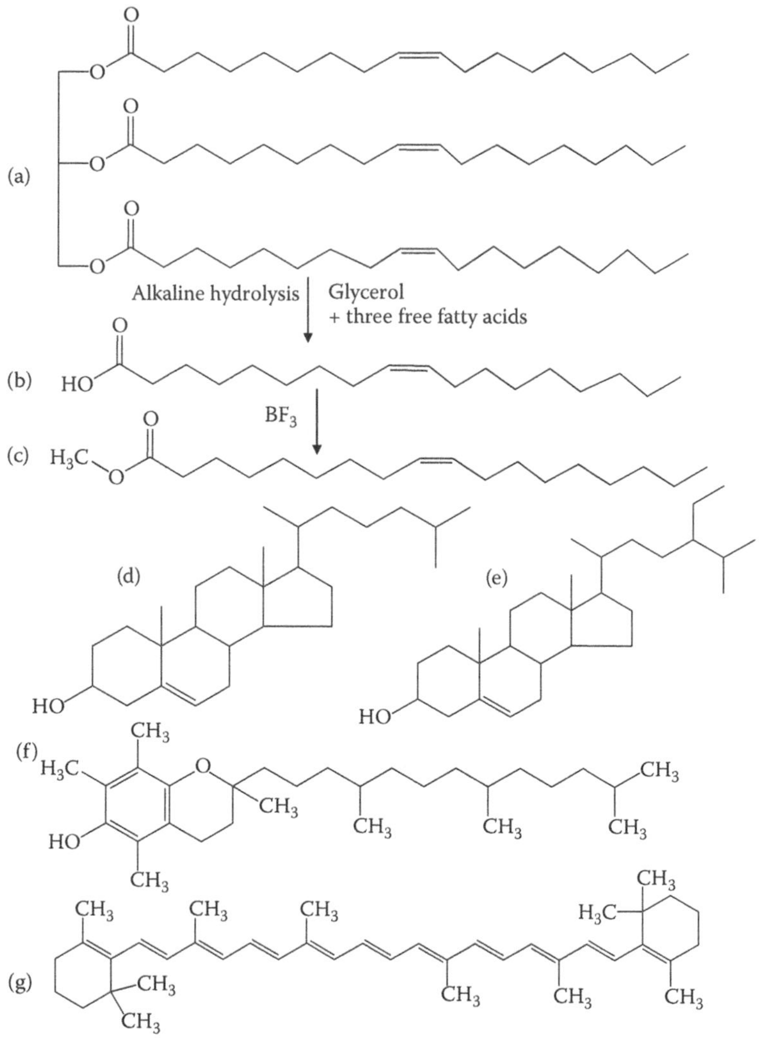

FIGURE 6.1 Nonpolar lipids commonly found in foods: (a) Triolein (a common triacylglycerol [TAG] molecular species), (b) oleic acid, (c) oleic acid methyl ester, (d) Cholesterol (the common sterol in animals), (e) Sitosterol (a plant sterol or phytosterol), (f) Vitamin E (α-tocopherol), and (g) beta carotene (a carotenoid). The arrows show how a common lipid such as a TAG can be hydrolyzed to yield free fatty acids and these can then be converted to methyl esters for GC analysis.

the dietary calories consumed in Western Europe and North America, an understanding of their composition is essential to the field of nutrition.[1] "Crude fat" is a compositional term that includes all of the nonpolar lipids that are extractable with diethyl ether (primarily TAGs, but also other nonpolar lipids such as waxes, sterols, free fatty acids, and tocopherols).

TAGs (Figure 6.1a) are a lipid class that can contain an assortment of different molecules, each with a different molecular weight determined by the specific combination of three fatty acids (Figure 6.1b) in the molecule. For example, the most abundant TAG molecular species in olive oil is triolein, consisting of a glycerol backbone with each of the three hydroxyls of glycerol esterified to oleic acid (a fatty acid with 18 carbons and a double bond between Carbons 9 and 10).

For food compositional analysis, the levels of "saturated," "monounsaturated," and "polyunsaturated" fats are often measured. Saturated fats are those that contain no carbon–carbon double bonds (mainly palmitic and stearic acids). Monounsaturated fatty acids have one double bond (primarily oleic acid), and polyunsaturated fatty acids (PUFA) have two or more double bonds (primarily linoleic and linolenic acids). For this type of analysis, the ester bonds of the TAGs (and the other extractable lipids) are hydrolyzed, and the free fatty acids are converted to fatty acid methyl esters (FAMES) (Figure 6.1c). The FAMES are then analyzed via gas chromatography (GC) (see the section "Methods for the Quantitative Analysis of Fatty Acids, Trans Fatty Acids, and Sterols via Gas Chromatography"). In general, animal fats are high in saturates (TAGs containing saturated fatty acids), and vegetable oils can either be high in monounsaturates (olive and canola oils) or high in polyunsaturates (corn and sunflower oils).

The other common nonpolar lipids include waxes (long chain alkanes, fatty alcohols, and esters of fatty acids and fatty alcohols), diacylglycerols, free fatty acids, and sterols. In animal tissues, cholesterol (Figure 6.1d) is the most abundant sterol, and in plant tissues, several types of plant sterols (phytosterols, Figure 6.1e) are often found. The exact proportions of these phytosterols are often species-specific. Sterols can exist either with a free hydroxyl group (in which case they are usually incorporated in biological membranes along with phospholipids) or as esters, bound to common fatty acids (in which case they are often associated with TAGs). In plants, a large proportion (the specific proportion is usually species-specific and tissue-specific) of the sterols can also be found as glycosides, usually bound to glucose. Vitamin E (α-tocopherol, Figure 6.1f, and other tocopherol vitamers) and carotenoids (Figure 6.1g) are also minor components of most lipid extracts. Other than pointing out that tocopherols and carotenoids are components of lipid extracts, methods for their analyses are not covered in this chapter, but are discussed in Chapters 9 and 10 of this volume, which focus on the analysis of vitamins and carotenoids, respectively.

Phospholipids (Figure 6.2a) are the main structural lipid component of animal and plant tissues. Five major types of phospholipid classes are commonly found, each containing a distinct phosphoryl "head group," including phosphorylcholine, phosphorylethanolamine, phosphorylinositol, phosphorylglycerol, and phosphorylserine. Phosphatidylcholine is a phospholipid with a phosphorylcholine head group. Like TAGs, phospholipids have a glycerol backbone, and each phospholipid class (e.g., phosphatidylcholine) can include various molecular species (various combinations of two fatty acids). Although phospholipids are the main structural lipids in biomembranes, cholesterol and plant sterols are required to stabilize membranes. Other major membrane components include sphingomyelin (in animals) (Figure 6.2b), glycosphingolipids (in both plants and animals) (Figure 6.2c), and galactolipids (only in plants) (Figure 6.2d).

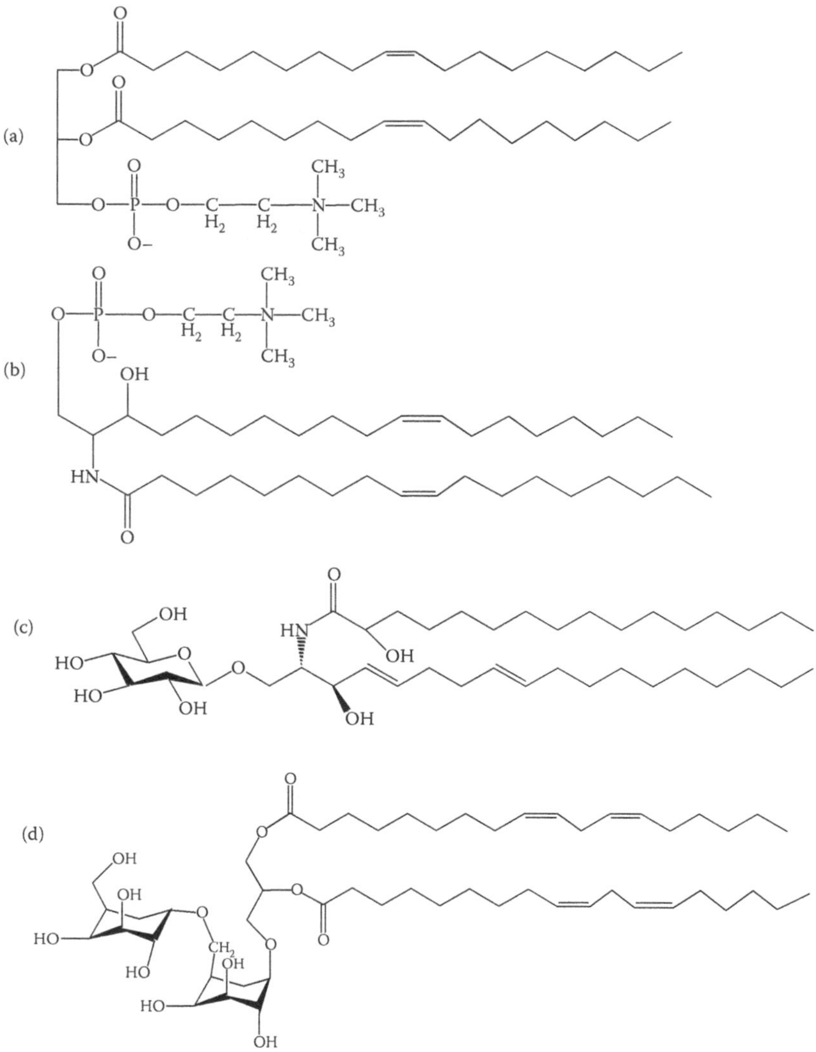

FIGURE 6.2 Polar lipids commonly associated with biological membranes and found in foods: (a) Phosphatidylcholine (a glycerolphospholipid), (b) sphingomyelin (a sphingophospholipid), and (c) glucocerebrocide (a glycosphingolipid) and (d) digalactosyldiacylglycerol (a plant galactolipid).

Recently, a new comprehensive nomenclature system has been proposed.[2] It includes eight major types of lipids (Table 6.1). It was created to provide a systematic way to catalog lipid analytical data for the new field of "lipidomics" that is described in detail at the end of this chapter. Using this system, each lipid can be categorized by a code—the first two letters are the code of the Fixed Database designation (LM = Lipid MAPS), the next two letters are the two-letter category code from Table 6.1, followed by a two-digit class code, a two-digit subclass code, and finally a unique four-character identifier. Using

TABLE 6.1

A Classification System for Lipids

Category	Abbreviation	Examples
Fatty acyls	FA	Fatty acids, fatty alcohols, wax esters
Glycerolipids	GL	Monoacylglycerols, diacylglycerols, triacylglycerols, monogalactosyldiacylglycerol
Glycerophospholipids	GP	Phosphatidylcholine, phosphatidylethanolamine, lyso-phosphatidylcholine
Sphingolipids	SP	Cardiolipin, ceramides, cerebrocides
Sterol lipids	ST	Cholesterol, sitosterol, estrogens, bile acids, hopanoids
Prenol lipids	PR	Monoterpenes, ubiquinone, vitamins E and K, dolichol
Saccharolipids	SL	Aminosugars, Lipid X, Lipid IV_A
Polyketides	PK	Aflatoxin, B1

Source: Fahy, E., Subramanianm, S., Brown, H.A., et al., *J. Lipid Res.,* 46, 839–861, 2005.

this nomenclature system, arachidonic acid is classified as LMFA01030001 and has the systematic name of 5Z,8Z,11Z,14Z-eicosatetraenoic acid. This classification provides a database framework for up to 1.68 million lipid structures. The LIPID MAPS Structure Database (LMSD) is a relational database encompassing structures and annotations of biologically relevant lipids. As of November 2009, LMSD contains 21,715 unique lipid structures, making it the largest public lipid-only database in the world.[3]

6.2 METHODS FOR THE DETERMINATION OF THE TOTAL FAT CONTENT IN FOODS

In the United States and many other countries, all packaged retail foods are required to carry a nutrition information label (Table 6.2).[4] More detailed information about the composition of lipids and various other nutrients in foods can be found at the USDA Nutrient Database[5] and other Web sites. One of the main nutrients on a nutrition label is the "total fat" (sometimes called "crude fat"), which is usually expressed as grams of fat per serving (i.e., per tablespoon for oils or fats, butter, margarine, and spreads) or grams of fat per 100 g (or 100 ml) in many European countries and Australia. In the United States, for foods containing more than 0.5 g fat per serving, the label must also contain a breakdown of the grams of fat from saturated and trans-unsaturated fats, while the breakdown of the grams of monounsaturates and polyunsaturates can be voluntarily listed.[4] These numbers are obtained by hydrolyzing the sample and analyzing the individual fatty acids (see the section "Methods for the Quantitative Analysis of Fatty Acids, Trans Fatty Acids, and Sterols via Gas Chromatography"). Although cholesterol does not contribute to the caloric content, it still must be listed on the nutrition label due to concerns about the relationship between dietary cholesterol and the risk for developing cardiovascular heart disease.

TABLE 6.2

Lipid Information on Food Labels

	Olive Oil[a]	Corn Oil[a]	Corn Oil Margarine[a]	Promise Activ® Light Margarine[b]	Lard[a]
Fat	100	100	80.5	35 (5 g)	100
Polyunsaturates	8.4	58.7	18.0	11 (1.5 g)	11.2
Monounsaturates	73.7	24.2	45.8	18 (2.5g)	45.1
Saturates	13.5	12.7	13.2	7 (1.0 g)	39.2
Trans fat	0	0	0	0	0
Cholesterol	0	0	0	<0.035 (<5 mg)	0.095
Phytosterols (plant sterols)	0.221	0.968	0.571	14.0 (1.0 g)	0
Moisture	0	0	15.7	~60	0

Note: Lipid information is expressed in grams of each type of lipid per 100 g of food; numbers in parentheses are grams per 1 tablespoon (14 g) serving.

Source: a Adapted from USDA Nutrient Database for Standard Reference, Release 15. Nutrient Data Laboratory Home Page, http://www.nal.usda.gov/fnic/foodcomp/Data/SR15/sr15.html.

b From package label.

To measure total fat, various methods have been approved by the regulatory agencies of most major countries (Table 6.3). Most older methods involve solvent extraction using either continuous or semicontinuous methods and gravimetric mass measurement of the lipid residue. In 1879, Franz von Soxhlet designed the first automatic extraction apparatus,[6] which was named as Soxhlet extractor after its inventor, and in the intervening years, several modifications of this apparatus (Soxhtherm, Soxtec™, Butt, Goldfisch, Bailey-Walker, Rohrig, etc.) have been developed. The original Soxhlet and Butt-type extractors are not conducive to the simultaneous analysis of multiple samples because the glassware is somewhat cumbersome, and the extraction times can vary from several hours to overnight. However, automated systems such as the Soxtec™ instrument (FOSS, Eden Prairie, MN) and Soxtherm (Gerhardt Gmbh, Bonn, Germany) allow up to 42 samples to be extracted in a day.[7] Several official methods, such as the Association of Official Analytical Chemists (AOAC) 2003.05, 2003.06, and 991.36, and International Organization for Standardization (ISO) 1444:1996, have been developed for the use of these instruments.[8–12] The majority of dietary lipids that contribute to fat calories in foods are TAGs; so for total fat determination,[6–12] most methods use nonpolar solvents because they selectively extract mainly nonpolar lipids. Ethyl ether (diethyl ether) is often the solvent of choice as it is relatively nonpolar and extracts the nonpolar lipids (e.g., TAGs, sterols, and tocopherols) well while poorly extracting the polar lipids (e.g., glycolipids and phospholipids). Petroleum ether and hexane are other commonly used solvents.[6–12]

With most solvent extraction methods, it is important to include a drying step because moisture will interfere with the ability of solvents to penetrate foods and to efficiently extract all lipids.[13,14] In addition, the extraction efficiency will be facilitated by reducing the food particle size either before extraction or during the solvent

TABLE 6.3

Methods for Total Fat Determination in Foods

Method	Description	Solvent	Apparatus	References
AOCS Method Ac 3–44	Total fat in soybeans	Petroleum (Pet) ether	Butt extractor	8
AOCS Method Aj 4–89	Total fat in corn germ	Hexane	Spex ball mill	8
AACC Method 30–10	Crude fat in flour, bread, and baked cereal products	Pet ether	Rohrig or Mojonnier extractor	9
DGF Standard Method B 1 5	Total fat	Petroleum benzene	Soxhlet	10
ISO Standard Number 659	Total fat	Pet ether	Soxhlet	11
AOAC Procedure 920.39	Crude fat in food	Anhydrous ethyl ether	Soxhlet	12
AOAC Procedure 954.02 Fat hydrolysis	Fat in baked products and pet foods	Acid hydrolysis and ethyl ether extraction	Majonnier	12
AOAC Official Method 991.36	Fat in meat and meat products	Pet ether	Extraction system (such as Soxtec™)	12
AOCS Method Am 5–04	Oil and fat in foods	Pet ether (or hexane, or diethyl ether)	Ankom X-10 or X-15	8,15
AOCS Method Am 3–96	Oil in oilseeds by supercritical fluid extraction	Supercritical CO_2 (± 15% ethyl alcohol)	Supercritical fluid extractor	8
AOAC Official Method 989.04	Fat in raw milk	None, acid hydrolysis	Babcock milk test bottle	12
AACC Method 58–19	Total fat, saturates, mono and polyunsaturates in cereals by GC	Acid hydrolysis, ethyl ether/petroleum ether	GC	9
AOCS Method Ak 5–01	Oil in oilseeds by NMR	None	NMR spectrometer	8
ISO Method 10565	Oil and moisture measurement in oilseeds by NMR	None	NMR spectrometer	11
AACC Method 39–20	Oil and protein in soybeans	None	NIR spectrometer	9

extraction through milling, grinding, or blending. In certain foods, such as dairy products, seafoods, or meat, lipids are either bound to proteins or emulsified; so some methods, such as AOAC Methods 948.15 and 954.02,[12] use either acid or alkali to dissolve proteins and disrupt emulsions before solvent extraction.

Another automated alternative to the Soxhlet type of extraction methods is based on the filter bag technology developed by Ankom Technology Inc. (Macedon, NY). This alternative method was approved by American Oil Chemists' Society (AOCS) as an Official Procedure, Am 5–04.[15] Unlike conventional methods, this filter bag method allows for batch extraction of 10–15 samples at a time and precise control, and thus offers features of faster determination and a higher level of automation. Methods have also been developed to use supercritical fluid extraction, rather than solvent extraction, for total fat determination.[8] These methods have been compared with solvent extraction, and in most cases, the results are in very close agreement.

Several nongravimetric methods are also commonly used for fat-content analysis. In the Babcock (AOAC methods 989.04 and 964.12) or the Gerber (AOAC method 2000.18, ISO 2446:2008) methods[11,12] for determining total fat content, foods are treated with a strong acid to hydrolyze proteins and to melt and release the lipids, and then the fat content is determined volumetrically. The Babcock method and the modifications thereof are commonly used for analyzing fat content in milk and seafood.[12] In addition, there are several methods (e.g., AACC 58–19, AOAC 996.06) that allow determination of total fat as well as saturated, monounsaturated, and PUFA content by GC: after adding an internal standard, food samples are subjected to acid or alkaline hydrolysis, followed by extraction of lipids and quantitative GC analysis of FAMES.[9,12]

Nuclear magnetic resonance (NMR) spectroscopy methods[8,11] have also been developed for total fat and moisture determination of a number of foods. Careful calibration is required, and the most accurate measurements are obtained when the instrument is calibrated with the food matrix being measured.

Near infrared (NIR) spectroscopy[9] has been used as a nondestructive method to measure total fat. NIR determination of total fat also requires careful calibration and, consequently, is most accurate if the NIR instrument is calibrated with the particular food matrix being measured.

6.3   METHODS FOR THE EXTRACTION OF TOTAL LIPIDS IN FOODS (FOR SUBSEQUENT ANALYSIS)

The extraction of total lipids in foods is often considered to be a simple procedure. However, because of the diversity of biochemical matrices occurring in natural and prepared foods, and the diversity of lipid classes, achieving complete or near-complete extraction can be a challenging task.

The extraction methods presented in the Section 6.2 (both using organic solvents and supercritical CO_2) are designed to efficiently extract TAGs and other nonpolar lipids from foods that have low moisture content. To extract total lipids (nonpolar and polar), more polar solvents must be used as extractants (Table 6.4). In addition, since many foods from plant and animal sources contain high levels of water, extraction methods need to take into consideration the presence of endogenous water.

The most popular method for total lipid extraction is the Folch Method[16] (Table 6.4). Lipids are extracted from a sample (1 g) with 20 mL of chloroform–methanol (2:1 by volume). The endogenous water in the sample forms a third solvent in the monophasic extraction mixture. After extraction, one-fifth volume of water or saline solution (0.29%) is added, and the mixture is shaken and then equilibrated. After equilibration, two phases will form, with the bottom phase comprising chloroform–methanol–water, 86:14:1, plus virtually all of the lipids. The upper phase consists of the same solvents

TABLE 6.4

Methods for the Routine Extraction of Lipids in Foods for Using the Extract for Subsequent Analysis of Individual Lipid Components

Method	Solvents	Types of Lipid Extracted	References
Soxhlet-related methods	Hexane or petroleum ether	Nonpolar lipids	Table 6.3
Supercritical fluid extraction	CO_2, sometimes with ethyl alcohol	Nonpolar lipids	8
Folch	Chloroform, methanol, and water	Nonpolar and polar lipids	16, 17
Bligh and Dyer	Chloroform, methanol, and water	Nonpolar and polar lipids	18
Hara and Radin	Hexane, Isopropanol, and water	Nonpolar and polar lipids	22
Accelerated solvent extraction	Hexane	Nonpolar lipids	19
	Ethanol or isopropanol	Polar and nonpolar lipids	

in the proportion 3:48:47, respectively, and contains most of the nonlipid components. This method works best if the final proportion of the three solvents, chloroform–methanol–water, is 8:4:3. A modified Folch procedure, with more experimental details, was subsequently published.[17]

The Bligh and Dyer method[18] was developed to extract total lipids from fish, but it has been employed for the extraction of lipids from numerous animal and plant tissues. The method specifies that the tissue be extracted at a ratio of 1:2:0.8, chloroform–methanol–endogenous water. After extraction of the monophasic mixture, one part of chloroform and one part of water (0.88% potassium chloride) are added and mixed, and after equilibration (which can be accelerated by centrifugation), the lipids are recovered in the lower phase.

For the extraction of lipids from large numbers of samples, an instrument called Accelerated Solvent Extractor (ASE®) has been marketed by Dionex (Sunnyvale, CA, USA) since about 2001. It is useful for the extraction of lipids from solid and semisolid samples (1–100 g) using water or organic solvents at accelerated temperatures (up to 200°C) and pressures (up to 1500 psi). A comparative study examined the types of lipid extracted from ground corn and ground oats using an ASE®, with hexane, methylene chloride, isopropanol, and ethanol.[19]

Since some plant tissues contain very active lipolytic enzymes, the lipid degradation that they sometimes cause can be minimized by macerating the plant tissue in hot isopropanol before chloroform–methanol extraction.[20] A variation of this method was recently developed for the routine extraction of a large number of leaf samples for lipid class analysis via high-performance liquid chromatography (HPLC).[21] Because of the health concerns of working with chloroform, methods employing less toxic solvents, such as hexane–isopropanol,[22] have been developed. The hexane-isopropanol extraction method was employed in the 1980s for many studies, but it is our opinion that when using new types of tissues, this method should be carefully evaluated (preferably by comparing the composition of extracts with those of the Folch[16] or Bligh and Dyer[18] extracts) before it is routinely employed.

6.4   METHODS FOR THE SEPARATION AND QUANTITATIVE ANALYSIS OF INTACT LIPID CLASSES

Lipid class separation is often required to prepare samples for subsequent fatty acid analysis. An accurate quantitative analysis of lipid classes can also provide valuable structural and physiological information. Some relatively simple methods have been developed to isolate lipid classes from total lipid extracts using "open-column" (low-pressure) chromatography. Christie[1] described a method to separate nonpolar lipid classes (hydrocarbons, cholesterol esters, TAGs, cholesterol, diacylglycerols, and monoacylglycerols) from a lipid extract using an open column of silicic acid and eluting with step fractions of 0%–100% diethyl ether in hexane (Table 6.5). Similar methods for nonpolar and polar lipid fractionation have been published using adsorbents such as Florisil™ and DEAE Cellulose.[1]

Prepacked solid-phase extraction (SPE) columns, packed with a variety of sorbents, can be purchased. Kaluzny et al.[24] published a method to sequentially separate

TABLE 6.5
Methods for the Separation and Quantitative Analysis of Intact Lipid Classes

Method	Stationary Phase	Mobile Phase	Lipid Classes Separated	References
Open-column LC	Silica	Hexane-diethyl ether	HC, CE, TAG, Ch, DAG, MAG	1
Solid-phase extraction	Silica	Eight solvent mixture steps	CE, TAG, Ch, DAG, FFA, PL	24
TLC—nonpolar lipids	Silica	Hexane, ether, formic acid—80:20:2	CE, TAG, FFA, Ch, DAG, MAG	1
TLC— polar lipids	HP-Silica	Diisobutyl ketone, acetic acid, water— 40:25:3:7	MGDG, SG, PE, DGDG, PC, PI	1
HPLC— nonpolar lipids	Diol	Hexane, acetic acid, isopropanol	See Figure 6.3	Figure 6.3
HPLC— polar lipids	Diol	Hexane, acetic acid, isopropanol	See Figure 6.3	Figure 6.4
HPLC—FFA and FAMES	Ultrasphere ODS	Methanol, water, acetic acid, gradient	Common FFAs and FAMES	31
HPLC—hydroxy and epoxy fatty acids	Silica	Hexane, isopropanol, acetic acid, gradient	Common hydroxyl and epoxy-free fatty acids and FAMES	32

Note: The methods for the separation and quantitative analysis of intact lipid classes including free and hydroxyl fatty acids are low-pressure open-column chromatography (LC), SPE, thin-layer chromatography (TLC), and high-performance liquid chromatography (HPLC).

Abbreviations: CE, cholesterol esters; Ch, cholesterol; CSE, cinnamic acid steryl esters; DAG, diacylglycerols; DGDG, digalactosyldiacylglycerol; FAMES, fatty acid methyl esters; FFA, free fatty acids; FS, free sterol; HC, hydrocarbons; MAG, monoacylglycerols; MGDG, monogalactosyldiacylglycerol; PE, phosphatidylethanolamine; PC, phosphatidylcholine; PI, phosphatidylinositol; PG, phosphatidylglycerol; PL, phospholipids; SE, sterol esters; SG, sterol glycosides; TAG, triacylglycerols.

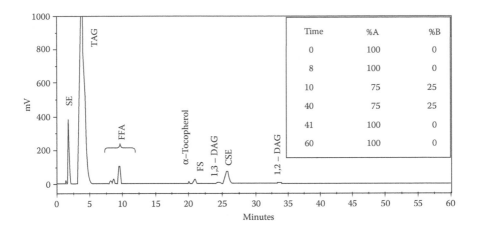

FIGURE 6.3 HPLC chromatogram of the separation of nonpolar lipids from corn fiber oil (100 µg). The column was a LiChrosorb 5 DIOL, 3 × 100 mm, flow rate 0.5 ml/min. The inset table shows the proportions of the two gradient components at each step point. A = 1000:1 hexane:acetic acid, B = 100:1 hexane:isopropanol. For abbreviations, see Table 6.5.

ten lipid classes using a silica SPE cartridge. Various techniques for the SPE fractionation and/or purification of lipids were compared.[25]

Thin-layer chromatography (TLC) has long been a popular technique for the prefractionation or separation of lipid classes. Today, most analysts use precoated TLC plates for convenience and to ensure a high degree of reproducibility. Although it has been suggested by some that HPLC will replace TLC, some important advantages of TLC include the following: (1) it can be employed without having to invest in costly instruments, and (2) selective spray reagents can provide very valuable structural information. Common TLC methods for the separation of nonpolar and polar lipids are listed in Table 6.5. Although most analysts use one-dimensional TLC with one development step, it may sometimes be necessary to use two-dimensional TLC (with a different solvent mixture for each dimension) to enhance the separation of certain mixtures of lipid classes. For some applications, it may be possible to separate both nonpolar and polar lipids in the same dimension by using two development steps in the same direction, with two different solvent mixtures.

HPLC methods for lipid analysis began to be published in the 1980s. UV detection (at 205–210 nm) is sufficiently sensitive to detect unsaturated lipids, but quantitative detection with a UV detector is very difficult because the detector response is proportional to the total number of carbon–carbon double bonds in a lipid. Also, saturated lipids are not detected via UV detection. In the mid 1980s, two different types of "mass detectors" were invented: the flame ionization detector (FID) and the evaporative light scattering detector (ELSD). The FID was only marketed for several years. Several manufacturers have continued to improve and perfect ELSDs, and these detectors have become a very valuable tool for lipid analysts. Several HPLC-ELSD methods have been developed for the quantitative analysis of nonpolar (Figure 6.3) and polar lipids (Figure 6.4). Common HPLC methods for the analysis of lipid classes were reviewed.[26]

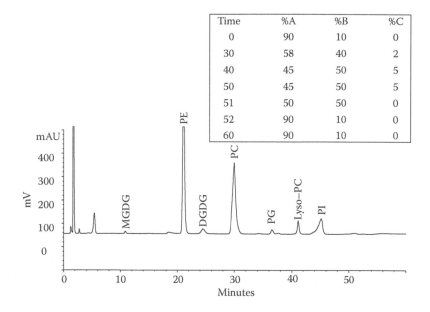

Time	%A	%B	%C
0	90	10	0
30	58	40	2
40	45	50	5
50	45	50	5
51	50	50	0
52	90	10	0
60	90	10	0

FIGURE 6.4 HPLC chromatogram of the separation of polar lipids from commercial soy lecithin (50 μg). The column was a LiChrosorb 5 DIOL, 3 × 100 mm, flow rate 0.5 ml/min. The inset table shows the proportions of the three gradient components at each step point. A = 1000:1 hexane:acetic acid, B = isopropanol, C = water. For abbreviations, see Table 6.5.

Recently, a new type of mass detector for HPLC, a Charged Aerosol Detector (CAD), was developed by ESA (Chelmsford, MA, USA) and was evaluated for the analysis of lipids using common normal-phase and reverse-phase HPLC methods.[27,28] Using a normal-phase HPLC system for the analysis of nonpolar lipids, the minimum limits of detection using HPLC-CAD were found to be as low as 1 ng.[27]

Although GC is usually not employed for lipid class analysis, several methods have been developed to separate nonpolar lipid classes (Table 6.6). Most of these methods employ very high temperatures, and some require derivatization (trimethylsilylation and others) of the sample to lower the boiling point of the lipids.[29] Since most of these methods require very high temperatures, care must be taken to ensure that lipids are not degraded during chromatography.

6.5 METHODS FOR THE QUANTITATIVE ANALYSIS OF FATTY ACIDS, TRANS FATTY ACIDS, AND STEROLS VIA GAS CHROMATOGRAPHY

Determination of the fatty acid composition of TAGs is one of the most common methods of analysis for the characterization of food lipids. Listing of the contents of saturated and trans-unsaturated fatty acids is mandatory on food nutrition labels.[4] In addition, there is increasing interest in the bioactive properties of individual fatty acids, especially omega-3 and omega-6 fatty acids as well as conjugated linoleic

TABLE 6.6

Methods for the Quantitative Analysis of Fatty Acid Methyl Esters (FAMES) and Other Lipids by GC

GC Column	Gradient Temperatures (°C)	Types of FAMES and Other Lipids Separated	References
CP-Wax 52 CB	170–220	Erythrocyte FAMES	1
SP-2340	130–250	Corn fiber oil FAMES	41
CP-Sil 88, high polarity	177–220	Separates some cis and trans isomers of hydrogenated oils.	1
CP-Sil 84	30–180	Milkfat FAMES, especially good for short chain acids, separates 4:0 to 18:3	1
SP-2560 or CP-Sil 88 (100 m length)	180 (isothermal)	cis, trans, saturated, monounsaturated, polyunsaturated FAMES in vegetable and nonruminant animal oils and fats	8
SAC-5	250–265	Free plant sterols	41
RTX-5, HP-5	300 (isothermal)	Sterol TMS derivatives	38
RTX-5w/INTEGRA	70–275		42
SPB-1701	250–280		43
SE-54 (high temperature limit)	150–340	CE, Ch, TAG, DAG, ceramides	29

Note: For abbreviations, see Table 6.5.

acid; so methods for their quantitative analysis are becoming more important.[30] By far, the most common procedures for the analysis of fatty acid composition involve the hydrolysis of fatty acids from the glycerol backbone, conversion of the fatty acids to FAMES, and separation and quantitation of FAMES by GC.[30] HPLC methods have additionally been developed to separate (using UV detection) and quantitatively analyze (using ELSD detection) free fatty acids and FAMES,[31] but most analysts agree that GC methods are superior. In contrast, the analysis of FAMES of hydroxy and epoxy fatty acids can be achieved by GC methods,[2] but HPLC methods are usually considered to be superior.[32]

For accurate analysis of fatty acid composition, lipids must first be completely extracted from a food product. Intact TAGs can be extracted from foods using the methods described in the sections "Methods for the Determination of the Total Fat Content in Foods" and "Methods for the Extraction of Total Lipids in Foods (for Subsequent Analysis)", or food samples (and lipids contained therein) can be first hydrolyzed either with acid or base, followed by extraction of the free fatty acids using lipid extraction methods.[30] The AOAC Official Method 996.06 uses the latter approach.[12] Either way, care should be taken to avoid oxidative degradation of PUFA; therefore, an antioxidant such as pyrogallol or butylated hydroxytoluene may be added before the extraction or hydrolysis.[33] Several hydrolysis-methylation and transesterification methods are commonly used to prepare FAMES,[34,35] but each of these methods has limitations, and care must be taken to ensure complete conversion of fatty acids to FAMES. Intact triacylglycerols are typically converted to FAMES using either acid- or alkali-catalyzed transesterification, although some methods have separate hydrolysis and methanolysis steps; free fatty acids are typically methylated using acid catalysts. The most common acid catalysts are BF_3 in methanol and HCl in methanol, while sodium methoxide and potassium hydroxide in methanol are the most common base catalysts.[34]

Numerous GC methods for FAMES analysis have been published (Table 6.6). Most modern methods employ capillary columns, whereas earlier methods (before approximately mid-1980s) employed packed glass columns. Both sensitivity and resolution were greatly improved after capillary column technology was perfected. Simple mixtures of FAMES, or those that contain epoxy or hydroxy FAMES, can be adequately separated on a nonpolar column.[1] FAME mixtures that contain a diverse mixture of fatty acids (including very short, very long, or very unsaturated mixtures of FAMES) require the use of polar columns (e.g., CP-Sil 88, BPX 70, SP-2340) ranging in length from 15 to 60 m to achieve optimal separation. On these types of columns with a 100% cyanopropylsilicone stationary phase, fatty acids elute in order of chain length, degree of unsaturation and, to some extent, by the location of double bonds along the hydrocarbon chain. High-speed GC methods by which FAMES composition of a sample can be determined in just a few minutes[34] have also been described.

Trans fatty acid analysis is challenging because of the number of positional isomers of both cis and trans fatty acids that are present in hydrogenated oils as well as in milk products.[35] Some of the polar columns described above can separate a few cis and trans isomers of some fatty acids,[1] but when used for the analysis of partially hydrogenated oils or natural milk fatty acids, considerable overlap of the various cis and trans isomers is seen. Current methods, such as AOCS Official Method Ce 1h-05 and AOAC 996.06,[8,12] utilize longer (100 m) polar (SP-2340, SP-2560) columns and isothermal oven temperatures to obtain much better separation. Generally, trans-positional isomers will elute before cis isomers; however, there is still some overlap of cis and trans isomers of oleic acid.[36]

Cholesterol and plant sterols can also be quantitatively analyzed by GC. Plant sterols and stanols are well known for their ability to lower blood cholesterol by blocking readsorption of cholesterol from the gut.[37] The Food and Drug Administration approved a health claim for foods containing at least 0.65 g or 1.7 g per serving of plant sterol or plant stanol esters, respectively,[4] yet as part of the requirements for that health claim, food processors must indicate on the food package the quantity (in grams) of phytosterols in a serving of the food.[4] Therefore, quick and accurate methods for phytosterol quantitation in foods are needed.

Several procedures exist for the analysis of endogenous cholesterol and sterols in food and plant tissues, such as AOCS Official Method Ch 6–91 and AOAC 994.10.[8,12] Laakso[38] adapted traditional methods for analysis of phytosterols in functional foods, such as the Promise Activ® spread in Table 6.2, where phytosterol content is much higher (up to 8%) than that found endogenously in food and plant tissues (≤1%). Analysis of sterols usually begins with the addition of an internal standard and a saponification step using alkaline (0.5–2.0 N KOH) conditions to hydrolyze sterol esters as well as TAGs. As with FAMES analysis, the starting material can be either a lipid extract or a food.[39] However, some foods, such as cereals, require acid hydrolysis (4–6 M HCl in methanol) before saponification to release sterols from the food matrix and to hydrolyze sterol glycosides.[40] Free sterols are then extracted using a solvent such as ether or hexane, and interfering lipids (mainly fatty acid soaps) are removed with water washes. Free sterols can then be quantitatively analyzed by GC without derivatization[41]; however, most analysts prefer to derivatize free sterols to obtain better peak response to the FID detector so that lower column temperatures can be used, thus minimizing the possibility of the degradation of sterols during chromatography.[40,42,43] Sterols can be either acetylated or silylated with trimethylsilane (TMS); however, TMS derivatives are more commonly used for analysis of food and oil sterols.[44] Most researchers use capillary dimethylpolysiloxane columns of slight polarity such as SAC-5, PTE-5, or DB-5 or mid-polarity capillary columns such as DB-1701 or SPB-1701 for the separation of plant sterols.[44] The major plant sterols (sitosterol, campesterol, stigmasterol, brassicasterol, stigmastanol, campestanol) are typically identified by comparing their retention time with that of commercially available standards. For functional foods with phytosterols added in high percentages, these plant sterols will be all that will need to be identified.[38] However, most of the less common phytosterols are not available commercially, and when analyzing endogenous phytosterol content of some foods, GC-MS analysis may be necessary to confirm the identity of these phytosterols.[45] Since many sterols have same molecular weight, sometimes a comparison of the relative (to either sitosterol or the internal standard) retention times and mass spectral patterns of the sterols to be identified to that of known sterols in the literature is required for peak identification.

6.6 METHODS FOR THE QUANTITATIVE ANALYSIS OF THE MOLECULAR SPECIES OF VARIOUS COMMON CLASSES OF LIPIDS VIA HPLC

Several reverse-phase HPLC methods (Table 6.7) have been developed for the separation of molecular species of TAGs.[46] Before injection into the HPLC, pure TAGs must first be prepared by TLC or HPLC lipid class methods. If the lipid mixtures are extremely complex (such as fish oil), the TAGs are sometimes prefractionated into several fractions using silver ion chromatography (which subfractionates lipids based on the total number of carbon–carbon double bonds), and then several fractions are each injected into the HPLC to generate several simpler molecular species chromatograms.[1]

Several methods have also been developed to separate molecular species of the various classes of phospholipids.[47] After purification of an individual phospholipid class (such as phosphatidylcholine), the molecular species of the given phospholipid class are similarly separated via reverse-phase HPLC.

Methods have also been developed for the separation of molecular species of other lipid classes such as steryl esters,[48] plant galactolipids,[49] plant acylated steryl glucosides,[49] and plant steryl glucosides.[50]

TABLE 6.7
Reverse-Phase HPLC Methods for the Quantitative Analysis of the Molecular Species of Various Common Classes of Lipids

Lipid Class	Column	Mobile Phase	Lipid Source	References
Triacylglycerols	Inertsil ODS	Acetonitrile–dichloromethane gradient	Corn oil	46
Phospholipids	Bondpack C18	Methanol, water, acetonitrile, isocratic—90.5:7:2.5	Plant phosphatidylcholine	47
Phytosterol esters	Zorbax ODS	Acetonitrile, isopropanol, isocratic—60:40	Corn oil	48
Plant galactolipids	Luna C18	Methanol, water, isocratic—98:2	Bell peppers	49
Plant acylated steryl glycosides	Luna C18	Methanol, ethanol isocratic—3:2	Bell peppers	49
Plant steryl glucosides	Prevail C18 3μ	Methanol, water isocratic—96:4	Biodiesel	50

6.7   LIPIDOMICS—AN EMERGING FIELD OF LIPID ANALYSIS

Lipidomics is a new area of research, which has developed over the last 10 years, following the rapid emergence of the two major "omics" fields: genomics, which began in the 1980s, and proteomics, which began in the 1990s. What is lipidomics? Various experts have different definitions. According to Spener and Lagarde,[51] it is "The full characterization of lipid molecular species and of their biological roles with respect to expression of proteins involved in lipid metabolism and function, including gene regulation." Molecular species are individual lipid molecules such as triolein in olive oil and PC 36:4 (phosphatidylcholine containing fatty acids with a total of 36 carbons and 4 double bonds) in cell membranes. In contrast, a lipid class is a group of lipid molecular species with similar characteristics (e.g., TAGs, phosphatidylcholine, and free sterols). The new lipid nomenclature system that was developed for the field of lipidomics was described in the section "A Description of the Common Types of Lipids Found in Foods."[2] Some experts feel that lipidomics is a high-tech name for lipid profiling or high-throughput lipid profiling.[52] Although traditional analytical methods such as GC and HPLC can be used to generate data for lipidomics, most of the current lipidomic data is being generated using powerful mass spectrometry instruments (usually MS/MS), and the data is analyzed and archived using specialized software, similar to that used for genomics and proteomics.

6.8   CONCLUSION

Numerous methods are available for lipid analysis. As with any tool, the analyst must choose the most appropriate tool for each application. Many lipid analysis techniques have benefited from the advances in electronics and computer science (especially the areas of data acquisition, data analysis, and interpretation), but some "low-tech" analytical methods, such as SPE and TLC, can provide valuable information at modest costs.

REFERENCES

1. Christie, W.W. *Lipid Analysis*, 3rd edn. Oily Press, Bridgewater, UK, 416 pp, 2003.
2. Fahy, E., Subramanianm, S., Brown, H.A., Glass, C.K., Merrill, A.H. Jr., Murphy, R.C., Raetz, C.R.H., Russell, D.W., Seyama, Y., Shaw, W., Shimizu, T., Spener, F., van Meer, G., VanNieuwenhze, M.S., White, S.H., Witztum, J.L., and Dennis, E.A. A comprehensive classification system for lipids. *J. Lipid Res.* 46, 839–861, 2005.
3. Lipid, MAPS. 2009. http://www.lipidmaps.org/data/structure/index.html.
4. FDA. *Code of Federal Regulations, Title 21, Part 101, Food Labeling*. Silver Spring, MD, 2008.
5. Anonymous. *USDA National Nutrient Database for Standard Reference, Release 15*. 2003. Nutrient Data Laboratory Home Page: http://www.nal.usda.gov/fnic/foodcomp/Data/SR15/sr15.html.
6. Soxhlet, F. Die gewichtsanalytische bestimmung des milchfettes. *Polytechnisches J.* 232, 461–465, 1879.
7. Matthaus, B. and Bruhl, L. Comparison of the different methods for the determination of oil content in oilseeds. *JAOCS.* 78, 95–102, 2001.
8. Firestone, D., Ed. *Official Methods and Recognized Practices of the American Oil Chemists' Society*, 6th edn. AOCS Press, Champaign, IL, 2009.
9. AACC. *Approved Methods of the AACC*, 9th edn. American Association of Cereal Chemists, St. Paul, MN, 1995.
10. DGF. *Deutsche Einheitsmethoden zur Untersuchung von Fetten, Fettproduckten, Tensiden und verwandten Stoffen*. Wissenschaftliche Verlagsgesellschaft, Stuttgart, 1998.
11. International Organization for Standardization. *Oilseeds*. International Organization for Standardization, Geneva, 1988. http://www.iso.org.
12. AOAC International. *Official Methods of Analysis*, 17th edn. AOAC International, Gaithersburg, MD, 2000.
13. Min, D.B. Crude fat analysis. In *Introduction to the Chemical Analysis of Foods*, Nielson, S.S., Ed. Jones and Bartlett, Boston, MA, 181–192, 1994.
14. Shahidi, F. and Wanasundara, P.K.J.P.D. Extraction and analysis of lipids. In *Food Lipids*, 3rd edn., Akoh, C.C. and Min, D.B., Eds. CRC Press, Boca Raton, FL, 125–156, 2008.
15. AOCS Official Procedure Am 5–04, Rapid determination of oil/fat utilizing high temperature solvent extraction. http://www.ankom.com/09_procedures/CrudeFat_0504_013009.pdf
16. Folch, J., Lees, M., and Stanley, G.H.S. A simple method for the isolation and purification of total lipids from animal tissues. *J. Biol. Chem.* 226, 497–509, 1957.
17. Ways, P. and Hanahan, D.J. Characterization and quantification of red cell lipids in normal man. *J. Lipid Res.* 5, 318–328, 1964.
18. Bligh, E.G. and Dyer, W.J. Rapid method of total lipid extraction and purification. *Can. J. Biochem. Physiol.* 37, 911–917, 1959.
19. Moreau, R.A., Powell, M.J., and Singh, V. 2003. Pressurized liquid extraction of polar and nonpolar lipids in corn and oats with hexane, methylene chloride, isopropanol, and ethanol. *J. Am. Oil Chem. Soc.* 80, 1063–1067, 2003.
20. Nichols, B.W. Separation of the lipids in photosynthetic tissues: Improvement in analysis by thin-layer chromatography. *Biochim. Biophys. Acta.* 70, 417–425, 1963.
21. Christie, W.W., Gill, S., Nordback, J., Itabaski, Y., Sanda, S., and Slabas, A.R. New procedures for rapid screening of leaf lipid components from *Arabidopsis*. *Phytochem Anal.* 9, 53–57, 1998.

22. Hara, A. and Radin, N.S. Lipid extraction of tissues with a low-toxicity solvent. *Anal. Biochem.* 90, 420–426, 1978.
23. Wu, X.Y., Moreau, R.A., and Stumpf, P.K. Studies of biosynthesis of waxes by developing jojoba seed. III. Biosynthesis of wax esters from acyl-CoA and long chain alcohols. *Lipids.* 16, 897–902, 1981.
24. Kaluzny, M.A., Duncan, L.A., Merritt, M.V., and Epps, D.E. Rapid separation of lipid classes in high yield and purity using bonded phase columns. *J. Lipid Res.* 26, 135–140, 1985.
25. Ruiz-Gutierrez, V. and Perez-Camino, M.C. Update on solid-phase extraction for the analysis of lipid classes and related compounds. *J. Chromatogr. A.* 885, 321–341, 2000.
26. Moreau, R.A. Quantitative analysis of lipids by HPLC with a flame ionization detector or an evaporative light scattering detector. In *Lipid Chromatographic Analysis*, Shibamoto, T., Ed. Marcel Dekker, New York, 251–272, 1994.
27. Moreau, R.A. The quantitative analysis of lipids via HPLC with a charged aerosol detector. *Lipids.* 41, 727–734, 2006.
28. Moreau, R.A. Lipid analysis via HPLC with a charged aerosol detector. *Lipid Technol.* 21, 191–103, 2009.
29. Kuksis, A., Myher, J.J., and Geher, K. Quantitation of plasma lipids by gas–liquid chromatography on high temperature polarizable capillary columns. *J. Lipid Res.* 34, 1029–1038, 1993.
30. Ruiz-Rodriguez, A., Reglero, G., and Ibañez, E. Recent trends in the advanced analysis of bioactive fatty acids. *J. Pharm. Biomed. Anal.* 51, 305–326, 2010.
31. Lin, J.T., McKeon, T.A., and Stafford, A.E. Gradient reversed-phase high performance liquid chromatography of saturated, unsaturated, and oxygenated fatty acids and their methyl esters. *J. Chromatogr. A.* 699, 85–91, 1985.
32. Gerard, H.C., Moreau, R.A., Fett, W.F., and Osman. Separation and quantification of hydroxy and epoxy fatty acids by high-performance liquid chromatography with an evaporative light scattering detector. *J. Am. Oil Chem. Soc.* 69, 301–304, 1992.
33. House, S.D., Larson, P.A., Johnson, R.R., DeVries, J.W., and Martin, D.L. Gas chromatographic determination of total fat extracted from food samples using hydrolysis in the presence of antioxidant. *J. AOAC Int.* 77, 960–964, 1994.
34. Ichihara, K., Shibihara, A., Yamamoto, K., and Nakayama, T. An improved method for rapid analysis of the fatty acids of glycerolipids. *Lipids.* 31, 525–539, 1996.
35. Mossoba, M.M. and McDonald, R.E. Methods for trans fatty acid analysis. In *Food Lipids*, 3rd edn., Akoh, C.C. and D.B. Min, D.B., Eds. CRC Press, Boca Raton, FL, 157–202, 2008.
36. Kramer, J.K.G., Blackadar, C.B., and Zhou, J. Evaluation of two GC columns (60-m Supelcowax 10 and 100-m CP Sil 88) for analysis of milkfat with emphasis on CLA, 18:1, 18:2, and 18:3 isomers, and short- and long-chain FA. *Lipids.* 37, 823, 2002.
37. Gylling, H. and Miettinen, T.A. The effect of plant stanol- and sterol-enriched foods on lipid metabolism, serum lipids and coronary heart disease. *Ann. Clin. Biochem.* 42, 254–263, 2005.
38. Laakso, P. Analysis of sterols from various food matrices. *Eur. J. Lipid Sci. Technol.* 107, 402–410, 2005.
39. Toivo, J., Lampi, A.-M., Aalto, S., and Piironen, V. Factors affecting sample preparation in the gas chromatographic determination of plant sterols in whole wheat flour. *Food Chem.* 68, 239–245, 2000.
40. Moreau, R.A., Whitaker, B.D., and Hicks, K.B. Phytosterols, phytostanols, and their conjugates in foods: Structural diversity, quantitative analysis, and health-promoting uses. *Prog. Lipid Res.* 41, 457–500, 2002.
41. Moreau, R.A., Singh, V., Nunez, A., and Hicks, K.B. Phytosterols in the aleurone layer of corn kernels. *Biochem. Soc. Trans.* 28, 803–806, 2000.

42. Moreau, R.A., Lampi, A-M., and Hicks, K.B. Fatty acid, phytosterol, and polyamine conjugate profiles of edible oils extracted from corn germ, corn fiber, and corn kernels. *J. Am. Oil Chem. Soc.* 86, 1209–1214, 2009.
43. Winkler, J.K., Rennick, K.A., Eller, F.J., and Vaughn, S.F. Phytosterol and tocopherol components in extracts of corn distiller's dried grains. *J. Agric. Food Chem.* 55, 6482–6486, 2007.
44. Abidi, S.L. Chromatographic analysis of plant sterols in foods and vegetable oils. *J. Chromatogr. A.* 935, 173–201, 2001.
45. Beveridge, T.H.J., Li, T.S.C., and Drover, J.C.G. Phytosterol content in American ginseng seed oil. *J. Agric. Food Chem.* 50, 744–750, 2002.
46. W.C. Byrdwell, W.E. Neff, and List, G.R. Triacylglycerol analysis of potential margarine base stocks by high-performance liquid chromatography with atmospheric pressure chemical ionization mass spectrometry and flame ionization detection. *J. Agric. Food Chem.* 49, 446–457, 2001.
47. Demandre, C., Tremolieres, A., Justin, A., and Mazliak, P. Analysis of molecular species of plant polar lipids by high performance and gas liquid chromatography. *Phytochemistry.* 24, 481–485, 1985.
48. Billheimer, J.T., Avart, S., and Milani, B. Separation of steryl esters by reversed-phase liquid chromatography. *J. Lipid Res.* 24, 1646–1651, 1983.
49. Yamauchi, R., Aizawa, K., Inakuma, T., and Kato, K. Analysis of molecular species of glycolipids in fruit pastes of red bell pepper (*Capsicum annuum* L.) by high-performance liquid chromatography-mass spectrometry. *J. Agric. Food Chem.,* 49, 622–627, 2001.
50. Moreau, R.A., Scott, K.A., and Haas, M.J. The identification and quantification of steryl glucosides in precipitates from commercial biodiesel. *J. Am. Oil Chem. Soc.* 85, 761–770, 2008.
51. Spener, F. and Lagarde, M. What is lipidomics? *Eur. J. Lipid Sci. Technol.* 105, 481–482, 2003.
52. Christie, W.W. Lipidomics – A personal view. Lipid Technol. 21, 58–60, 2009.
53. USDA Nutrient Database for Standard Reference, Release 15. Nutrient Data Laboratory Home Page, http://www.nal.usda.gov/fnic/foodcomp/Data/SR15/sr15.html.

7 Advanced Analysis of Carbohydrates in Foods

Miguel Herrero, Alejandro Cifuentes,
Elena Ibáñez, and Maria Dolores del Castillo

CONTENTS

7.1   INTRODUCTION

Carbohydrates are one of the most abundant and diverse class of organic compounds occurring in nature. Chemically they are composed of carbon, hydrogen, and oxygen in the ratio $C_n:H_{2n}:O_n$. Food carbohydrates include a wide range of macromolecules, which can be classified according to their chemical structure into three major groups: (1) low-molecular-weight mono- and disaccharides, (2) intermediate molecular weight oligosaccharides, and (3) high-molecular-weight polysaccharides. They can also be classified as simple or complex carbohydrates. Simple carbohydrates are monosaccharides and disaccharides, whereas complex carbohydrates are made up of many monosaccharides such as starches and fiber (polysaccharides).

Carbohydrates are a major source of energy in the human diet with intakes ranging from 40% to 80% of total energy requirements (Muir et al. 2009). Carbohydrates constitute the main source of energy for all body functions, particularly brain functions, and are necessary for the metabolism of other nutrients. Other important effects of carbohydrates on human physiology are satiety and gastric emptying; control of blood glucose, insulin metabolism, and serum cholesterol; and influencing colonic microflora and gastrointestinal processes, such as laxation and fermentation (Muir et al. 2009).

Cereals, vegetables, fruits, rice, potatoes, legumes, and flour products are the major sources of carbohydrates. Thus, naturally occurring sugars are consumed as part of a healthy diet. Monosaccharides, disaccharides, and polysaccharides are present in all vegetables (Hounsome et al. 2008). Carbohydrates are synthesized in all green plants and in the body. They are either absorbed immediately or stored in the form of glycogen. They can also be manufactured in the body from some amino acids and the glycerol component of fats. Moreover, sugars can be added to foods during processing or preparation, mainly to enhance food sensorial quality (Murphy and Johnson 2003).

Nutritionists divide food carbohydrates into two classes: (1) available carbohydrates or carbohydrates readily utilized and metabolized, including mono-, di-, or polysaccharides such as glucose, fructose, lactose, dextrin, and starch and (2) unavailable carbohydrates or carbohydrates not directly utilized, but instead broken down by symbiotic bacteria yielding fatty acids and thus not supplying the host with carbohydrates. The second class includes structural polysaccharides of plant cell walls and many complex polysaccharides such as cellulose, pectin, and β-glucans (Cui 2005). In other words, available carbohydrates are those that are hydrolyzed by enzymes of the human gastrointestinal system, whereas the unavailable ones (sugar alcohols, many oligosaccharides, and nonstarch polysaccharides) are those that are not hydrolyzed by endogenous human enzymes, but they can be fermented by microorganisms in the large intestine to varying extents and then absorbed (Hounsome et al. 2008).

Sugars (glucose, sucrose, fructose, lactose, and maltose), sugar polyols (sorbitol and mannitol), oligosaccharides (galactooligosaccharides [GOS] and fructooligosaccharides [FOS]), and polysaccharides (starch and nonstarch polysaccharides) have been described as the major classes of carbohydrates relevant to human nutrition. The fermentable short-chain carbohydrates such as oligo-, di-, and monosaccharides and polyols that can be poorly absorbed by the small intestine are called FODMAPs, and they have different effects on gastrointestinal health. FODMAPs can be found in a wide variety of foods (Muir et al. 2009).

Detailed information related to the chemistry of food carbohydrates can be found elsewhere (Cui 2005). In Sections 7.1.1–7.1.4, only key characteristics of the food carbohydrates relevant to human nutrition are provided together with the importance of developing powerful and informative techniques for their adequate analysis.

7.1.1 Simple Carbohydrates: Monosaccharides and Disaccharides

Monosaccharides require no digestion and can be absorbed directly into the blood stream. All the monosaccharides can be synthesized by the body (Hounsome et al. 2008). The basic carbohydrates are the monosaccharide sugars, of which glucose, fructose, and galactose are the most important from a nutritional point of view. Structure of fructose is shown in Figure 7.1. Fructose, a major component of fruits, fruit juices, honey, and corn syrup (Park et al. 1993), is one of the principal FODMAPs of the Western diet (Gibson et al. 2007).

Disaccharides are composed of two monosaccharides. Nutritionally, the important disaccharides are sucrose, a dimer of glucose and fructose; lactose, a dimer of glucose and galactose; and maltose, a dimer of two glucose units. Lactose has also been classified as FODMAP (Muir et al. 2009).

7.1.2 Sugar Polyols

Sugar alcohols, also called sugar polyols, are obtained by reduction of the corresponding aldoses and ketoses. Sugar polyols have generated much interest as food additives since they can be used as low-calorie sweeteners. Figure 7.2 shows structure of mannitol and sorbitol, two sweeteners very commonly used in the food industry. Although few data on levels of polyols in foods have been published so far, it is known that mannitol is present in several vegetables such as celery, carrot, parsley, pumpkin, onion, endive, and asparagus (Hounsome et al. 2008); sorbitol is present in many fruits and it is also used as sweetener. Some authors have suggested the

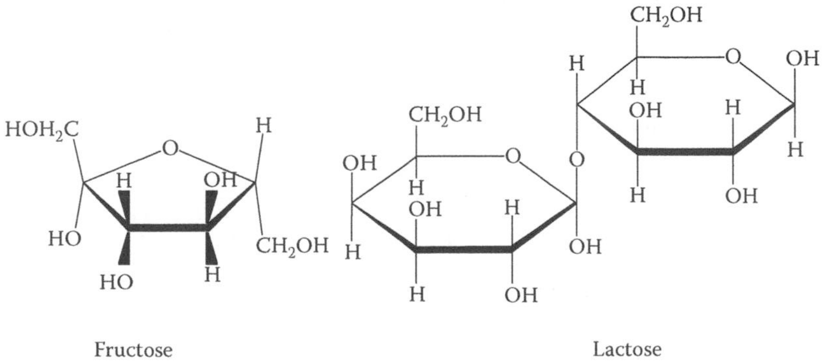

Fructose   Lactose

FIGURE 7.1 Example of monosaccharide and disaccharide structures.

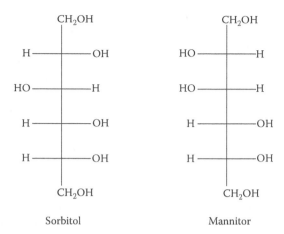

FIGURE 7.2   Example of the chemical structure of two sugar polyols.

use of mannitol and sorbitol as indicators for authenticity or adulteration in foods (Martínez-Montero et al. 2004).

7.1.3   OLIGOSACCHARIDES

Numerous oligosaccharides, comprising three to five monosaccharide units, occur in foods, and they are in general nondigestible. Oligosaccharides can also be defined as those carbohydrates formed by 2–10 basic sugar molecules. Fructans and FOS (e.g., nystose, kestose; see Figure 7.3) and GOS (e.g., raffinose, stachyose) are considered FODMAPs and prebiotic carbohydrates (Muir et al. 2009). FOS and nonstarch polysaccharides play an important role as dietary fiber (Hounsome et al. 2008).

Garlic, Jerusalem artichokes, and onions contain high levels of fructans (Muir et al. 2007). The Jerusalem artichoke has been found to have the highest concentration of FOS in cultured plants. FOS have been used in dietary supplements in Japan since the 90s, and they are now becoming increasingly popular in Western cultures for their prebiotic effects. FOS serve as a substrate for microflora in the large intestine, increasing the overall gastrointestinal tract health. Moreover, several studies have found that FOS and the polysaccharide inulin promote calcium absorption in both animal and human gut. The intestinal microflora in the lower gut can ferment FOS, resulting in the production of gases and the reduction of pH that increases calcium solubility and bioavailability.

7.1.4   POLYSACCHARIDES

Glycans is a general term given to polysaccharides in which large numbers of monosaccharides are naturally joined by O-glycosidic linkages. Polysaccharides are condensation polymers in which glycosidic linkage is formed from the glycosyl moiety of hemiacetal, or hemiketal, and a hydroxyl group of another sugar unit, acting as an acceptor molecule or aglycone (see Figure 7.4). Polysaccharides may be linear or

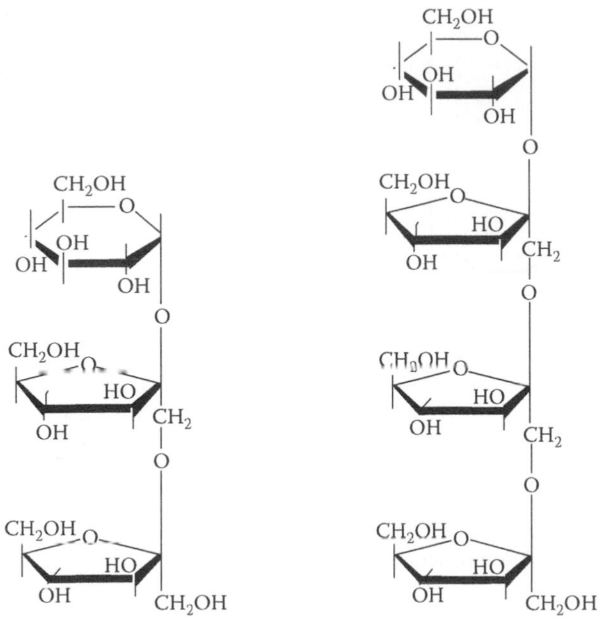

FIGURE 7.3 Example of the chemical structure of two fructooligosaccharides.

branched (see Figure 7.4), and there are two types of polysaccharides: homopolysaccharides and heteropolysaccharides. A homopolysaccharide has only one type of monosaccharide repeating in the chain, whereas a heteropolysaccharide has two or more types of monosaccharides. Unbranched polysaccharides contain only alpha 1,4 linkages, while some branched polysaccharides are linked to a molecule via alpha 1,4 and to another one via alpha 1,6 glycosidic bonds. These macromolecules have large physiological interest, and they affect both food quality and nutrition.

7.1.5 CONJUGATED CARBOHYDRATES

Carbohydrates can also be attached to other different compounds such as proteins, lipids, and phenols. Thus, glycans covalently bound to these compounds (glycoproteins, glycolipids, glycophenols) play a pivotal role on their bioactivity, and they are, for instance, involved in cell signaling or bioavailability. Glycation and deglycosylation of proteins have been proposed as strategies for modulating the immunoreactivity of key food allergens (van de Lagemaat et al. 2007; Amigo-Benavent et al. 2009) or for providing new technological attractive properties to proteins (Corzo-Martinez et al. 2010).

From the information given in this section, it can be concluded that the development of new analytical approaches for separation, identification, and quantification of carbohydrates is of huge importance in many research fields. Thus, knowledge on the quality and quantity of carbohydrates in fruits, vegetables, cereals, and many

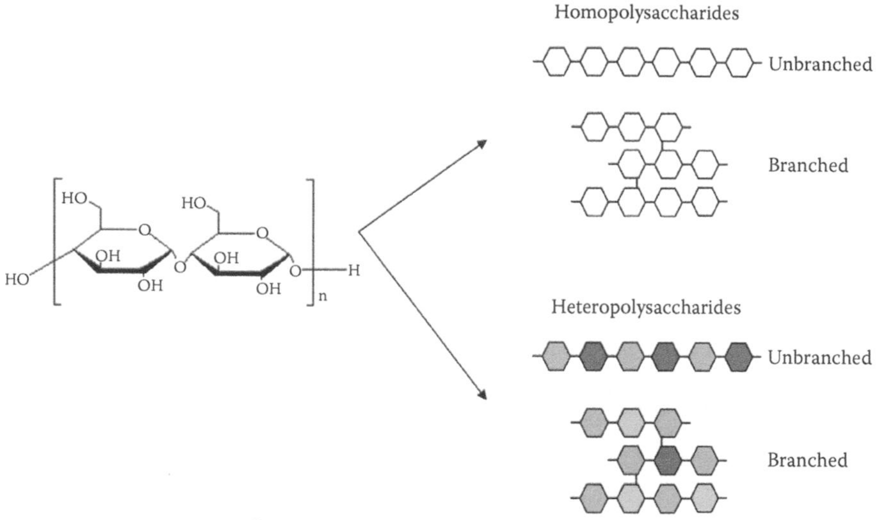

FIGURE 7.4 Typical basic structure of a polysaccharide.

other foods is essential to characterize them, determining important food properties such as flavor, maturity, quality, authenticity, and storage conditions. Moreover, a better knowledge on the composition of carbohydrates is mandatory to assess and understand their role in several functions in the human body, an issue of crucial interest in the production of healthier foods.

7.2   SAMPLE PREPARATION FOR CARBOHYDRATES ANALYSIS

A critical step when analyzing carbohydrates in food matrices is separation of carbohydrates from the rest of the main components, such as lipids and proteins, which can somehow interfere with their determination and quantification. Once isolated, carbohydrates can be either analyzed directly or subjected to some other additional treatments, such as hydrolysis and/or derivatization, to favor their subsequent analysis.

7.2.1   EXTRACTION AND FRACTIONATION

The separation of carbohydrates from other food components is generally performed using some kind of extraction or cleanup step prior to their analysis. Nevertheless, in some cases the sample obtained is still too complex to be easily analyzed, and a further fractionation step is necessary to obtain a particular carbohydrate fraction. In this section, the methods used for extraction and fractionation of carbohydrates are described.

7.2.1.1   Filtration

Usually, when studying the carbohydrate fraction of a beverage or a liquid sample (e.g., wine, milk, whey), a simple filtration and dilution step could be enough for sample preparation. Even ultrafiltration and filtration based on membrane technologies have been used to concentrate carbohydrates (Kamada et al. 2002). Nevertheless, when the original matrix is too complex or the amount of different class of carbohydrates is very different, more sophisticated sample preparation procedures are needed. It is interesting to mention that a filtration step can be used as a cleanup step before using any of the procedures explained below.

7.2.1.2 Traditional Extraction Methods: Liquid–Liquid Extraction, Solid-Phase Extraction

Nowadays, the so-called traditional extraction techniques, such as liquid–liquid extraction and solid-phase extraction (SPE), have been widely applied to the extraction of carbohydrates from food matrices prior to analysis. SPE has been used to extract carbohydrates from wine, beer, or honey, among others (Weston and Brocklebank 1999). The nature of the solid column packing and the solvent used for elution is selected according to the nature of the sample to be analyzed. C_{18} cartridges are probably the most used (Megherbi et al. 2008) although others such as C_8 (Kitahara and Copeland 2004), porous graphitic carbon (Arias et al. 2003), or even cartridges based on strong cation exchange (Castellari et al. 2000) are also used. SPE has shown its capability in fractionating complex samples; for instance, active carbon and celite columns were used to fractionate monosaccharides and oligosaccharides from honey using different ethanol/water ratios to obtain a selective elution (Weston and Brocklebank 1999). Although the results using these techniques might be satisfactory, some important shortcomings, such as large volume of organic solvents needed, have brought about an increase in the use of other extraction techniques.

7.2.1.3 Supercritical Fluid Extraction

Supercritical fluid extraction (SFE) is based on the use of fluids above their critical pressure and temperature. Under these conditions, supercritical fluids acquire properties between liquids (similar solubility) and gases (similar diffusivity). CO_2 is the most commonly used fluid, due to its moderate critical temperature and pressure, along with other important advantages such as the easiness to modify the selectivity of the extraction by varying the operating pressure, the possibility to obtain fractions through different depressurization steps, and the attainment of an extract free of solvent, since CO_2 is released as a gas when pressure is decreased under ambient conditions. However, its use in carbohydrate analysis has been partly hampered by its low polarity, which makes difficult the solubilization of these compounds. In order to overcome this problem, different amounts of organic solvents are used as modifiers. SFE has been used several times to fractionate different carbohydrates that could be used as prebiotics in the food industry (Montañés et al. 2007; Montañés et al. 2006). In these works, the authors demonstrated that the selectivity was mainly influenced by the cosolvent composition, whereas other parameters such as temperature, pressure, and amount of polar cosolvent had an effect on carbohydrates recovery. It is expected that other applications of this technique to the extraction and fractionation of carbohydrates can be developed following similar methods.

7.2.1.4 Pressurized Liquid Extraction

Pressurized liquid extraction (PLE) is based on the use of solvents at high temperatures and pressures enough to maintain their liquid state during the whole extraction process. As a result of these experimental conditions, the extraction processes are faster, produce higher yields, and consume lesser amounts of solvents compared to the more traditional extraction techniques. Due to these advantages, PLE is being widely used to extract sample components from food and natural products (Mendiola et al. 2007). Different solvents, including water, organic solvents, and their mixtures can be used. As for carbohydrates fractionation, few works have been published dealing with the selective extraction of lactulose from its mixtures with lactose using a 70:30 ethanol/water mixture at 40°C for 30 minutes (Ruiz-Matute et al. 2007a). Also, enriched fractions of di- and trisaccharides were obtained from honey using PLE combined with the use of activated charcoal (Ruiz-Matute et al. 2008). In this case, honey was adsorbed onto activated charcoal and packed into a PLE extraction cell that was submitted to two different extraction cycles. The first one using 1:99 ethanol/water for 5 minutes, and the second using 50:50 ethanol/water for 10 minutes. This PLE procedure can easily be adapted for fractionating carbohydrates from other sources.

7.2.1.5 Field Flow Fractionation

This technique is based on the separation of (high molecular weight) sample components being pumped by a laminar flow inside an open channel in which a perpendicular field (electric, thermal, magnetic, or gravitational) is applied. The separation or fractionation takes place based on the differences in the mobility of the components as a result of the applied field. For carbohydrates, this technique is often used to fractionate large polysaccharides that could not be otherwise fractionated, such as cellulose, starch, and pullulan, among others.

7.2.1.6 Chromatography-Based Methods

Different techniques based on chromatographic procedures have been employed to fractionate and separate carbohydrates. Size-exclusion chromatography (SEC) is one of the most popular techniques for the determination of molecular mass distributions of (relatively large) carbohydrates (Eremeeva 2003). SEC has been widely used to fractionate and separate oligo- and polysaccharides, including the analysis of prebiotic galactosaccharides and analysis of carbohydrates in rice, sorghum, wheat, or coffee, among others (Hernandez et al. 2009; Arya and Mohan Rao. 2007).

Another chromatography-based technique broadly applied to fractionate carbohydrates is ion-exchange chromatography (IEC). IEC is mostly suited for oligo- and monosaccharides fractionation. Both anion- and cation-exchange resins can be used depending on the particular application. The normal IEC approach is to use this technique as analytical step rather than for fractionation, and it will be discussed in more detail in Section 7.3.

The last technique included in this group is simulated moving bed chromatography. This technique has been widely employed in the sugar industry for large-scale palatinose-trehalose fractionation. Two streams of products (raffinate and extract) are obtained from the continuous countercurrent separation of a sample, eluted from a series of different adsorbents. These adsorbents can be based on ligand-exchange chromatography or on SEC. This technique has also been employed to fractionate lactose from a complex mixture of human milk oligosaccharides (HMO) (Geisser et al. 2005).

7.2.1.7 Membranes and Other Methods

The use of membranes is not new in carbohydrates sample preparation. In fact, the fractionation of sugars in sugarcane has been performed by dialysis, while the use of membranes combined with ultrafiltration and nanofiltration systems has also allowed the separation and fractionation of carbohydrates according to their molecular mass (Godshall et al. 2001).

Another technique worth to mention is carbon fractionation. This technique has been extensively used in official methods for many years, and it involves adsorption of carbohydrates in an activated charcoal column and a later elution with different proportions of ethanol (1% for monosaccharides, 5% for disaccharides, and 50% for oligosaccharides) (AOAC Official method of analysis n954.11) to fractionate them selectively according to their degree of polymerization. The fractionation with active charcoal has been shown to be particularly useful to obtain monosaccharides from very complex food samples (Morales et al. 2006). This technique may also be used in combination with other sample preparation processes or fractionation steps (e.g., PLE) in order to allow a more convenient characterization of carbohydrates in a subsequent analytical phase (Ruiz-Matute et al. 2008).

Other methods employed in carbohydrates sample preparation include the use of molecularly imprinted polymers (MIPs). MIPs have been proved to be useful for the recognition of different saccharides or for the epimeric differentiation of disaccharides. Although the potential of this technology is remarkable, its use in food science is still not very extended due to the difficulty in obtaining selective and robust MIPs in a reproducible way.

7.2.2 CHEMICAL TREATMENTS

Two chemical procedures are commonly used in the sample preparation step of food carbohydrates in order to facilitate their subsequent analysis: hydrolysis and derivatization.

7.2.2.1 Hydrolysis

Frequently, prior to a chromatographic analysis, a hydrolysis step is carried out of large carbohydrates to obtain fractions of low-molecular-weight mono- or oligosaccharides that can be analyzed more easily. Under heating and in the presence of a strong acid, the glycosidic bonds can be cleaved, being the most commonly employed acids trifluoroacetic acid, hydrochloric acid, or even sulfuric acid. The treatment conditions, mainly concentration of acid, temperature, and time, have to be optimized to achieve a total hydrolysis of the polysaccharides contained in the sample. At the same time, these conditions should not be too strong because they can produce the sample degradation, since it is well known that during this procedure, the released monosaccharides are prone to degradation. As an example, typical conditions applied to process wheat flour include the use of 2 M HCl at 100°C for 90 minutes. Polysaccharides can be also submitted to partial hydrolysis before analysis. Due to the differential susceptibility to hydrolysis of the different glycosidic linkages, partial degradations can be obtained at controlled conditions in order to produce mixtures of different mono- and oligosaccharides whose analysis could be useful to determine the polysaccharide structure. Recently, microwave irradiation has been employed as a tool to reduce the process time for carbohydrates hydrolysis (Corsaro et al. 2004).

7.2.2.2   Derivatization

Given their intrinsic properties, carbohydrates can be difficult to analyze by different separation techniques due to their low volatility, lack of chromophore, or very similar polarity. Thus, if analyzed by gas chromatography (GC), carbohydrates have to be derivatized so that their volatility is increased. Monosaccharides can be analyzed by GC as alditol acetates obtained by reduction and acetylation, while oligosaccharides can be analyzed by GC as trimethylsilyl or trifluoroacetyl ethers.

If capillary electrophoresis (CE) or high-performance liquid chromatography (HPLC) are the techniques of choice, care must be taken about the sensitivity required, since the vast majority of intact carbohydrates can only absorb ultraviolet (UV) light below 195 nm or can be detected using amperometric or refractive index detectors. Therefore, to increase their detection sensitivity, UV-absorbing or fluorescent labels are used in different derivatization processes. The reaction may take place before or during the analysis, being the most common procedure precolumn derivatization (Lamari et al. 2003). The most used approach implies the use of reductive amination. These procedures are based on the condensation of a carbonyl group with primary amines to produce a Schiff base that is reduced to give a *N*-substituted glycosil amine. 2-Aminopyridine of trisulfonates is among the reagents more commonly used. In addition, considering that most carbohydrates are neutral and hydrophilic compounds, derivatization not only enhances sensitivity but also causes polarity and electrical charge variations that can make easier their separation by CE or HPLC. Extensive reviews on this topic can be found in the literature (Lamari et al. 2003; Suzuki et al. 2003).

7.3   ADVANCED ANALYSIS OF CARBOHYDRATES IN FOOD

Spectroscopic techniques, separation techniques, and their multiple combinations are the main tools used to analyze carbohydrates in foods although some other approaches based on the use of sensors or the new glycomics platforms are already of importance in this field. The papers cited in this section have been selected on the basis of their novelty and relevance, trying to provide an updated overview on food carbohydrate analysis.

7.3.1   SPECTROSCOPIC TECHNIQUES

For years, spectroscopic techniques have been used for qualitative and quantitative analysis of foods and food ingredients using mainly UV-visible, fluorescence, infrared , Raman, atomic absorption, atomic emission, electron spin resonance, nuclear magnetic resonance (NMR), and mass spectrometry (MS) techniques (Ibañez and Cifuentes 2003).

Nowadays, there is a clear trend to use high-resolution NMR and MS instruments for the development of different foodomics approaches (Cifuentes 2009) including metabolomics. In these metabolomics approaches, it is essential to detect and identify compounds (metabolites) such as carbohydrates, amino acids, vitamins, hormones, flavonoids, phenolics, and glucosinolates and their role in the nutritional and sensorial quality of foods, including other important topics such as plant and animal growth, plant development, stress adaptation, or defense. NMR and MS techniques provide impressive possibilities in food carbohydrates analysis, and many structural studies involve the application of both NMR and MS for identification and characterization of food components.

7.3.1.1 Nuclear Magnetic Resonance

Because the structure of the carbohydrate will determine its function, the use of NMR is becoming crucial given that NMR may provide information related to the structure, purity, and safety of carbohydrate samples. NMR has been extensively used for identification of food carbohydrates and related compounds, including newly designed and natural carbohydrates (Hounsome et al. 2008). Currently, oligosaccharides naturally comprising the food matrices and those newly designed are of great interest because of their numerous applications (sweetener, health promoting, etc.) in the food industry. Nowadays, there is an increasing interest for the search of natural, low-cost, bioactive compounds to be used as health-promoting food ingredients and functional foods. FOS are the most popular oligosaccharides worldwide; however, there is a lack of information related to the properties of other non-inulin-type fructans. Recently, the application of H^1 NMR along with enzymatic hydrolysis allowed the identification of fructans in milling fractions of two wheat cultivars (Haska et al. 2008). Arabinoxylan, an important non-digestible component in cereal grains with beneficial health effects, was found to be a main component of the dietary fiber of all fractions. Complementary characterization of the fructans, such as determination of molecular weight, involved the use of other techniques for structural analysis such as high-performance anion-exchange chromatography with pulse amperometric detection and MALDI-TOF-MS (Haska et al. 2008).

Synthetic carbohydrates can also be used in the formulation of specialty foods for people with particular health concerns. The presence of glucose and lactose, for instance, may be risky for diabetics and lactose intolerants. As an example, new FOS prepared from sucrose using fungal fructosyltransferase have been identified using NMR spectroscopy and LC-MS (Mabel et al. 2008). In that work, the detection and quantification of monosaccharides and disaccharides in the FOS preparations to be used as nonnutrient sweeteners was also undertaken. In addition, the spectral data proved the absence of any toxic microbial metabolites in the carbohydrate preparation certificating its safety as sweetener with potential benefits to diabetics (Mabel et al. 2008).

NMR has been also successfully used for structural characterization of trisaccharides obtained by enzymatic transglycosylation of lactulose. The authors use either one-dimensional (^{1}H and ^{13}C TOCSY) or two-dimensional (gCOSY, TOCSY,

ROESY, gHSQC, and gHMBC) NMR techniques. The structure of two trisaccharides obtained by enzymatic treatment of lactulose was completely elucidated, and full proton and carbon assignment was carried out. Based on the structural data derived from NMR analysis, the authors proposed the enzymatic transgalactosylation as a new route for prebiotic synthesis (Martínez-Villaluenga et al. 2008).

Very complex molecules containing carbohydrates like coffee melanoidins have also been characterized by NMR. Melanoidins contribute to color and flavor of coffee and are believed to have physiological activity. The application of two-dimensional NMR technique permitted to obtain novel information supporting the idea that coffee melanoidins are based on condensed polyphenols. Data are in opposition to those previously published indicating that they can be formed by low-molecular-weight chromophores bound to polysaccharides or proteins (Gniechwitz et al. 2008). It is believed that the spectrometric approaches, and particularly NMR, are very important for the analysis of carbohydrates linked to proteins and other physiological relevant food components.

Other complex carbohydrates like polysaccharides have been identified by ^{1}H and ^{13}C NMR. For instance, major polysaccharides of hazelnut cell walls have been determined by NMR (Dourado et al. 2003); also, composition of polysaccharides in unripe tomato and a medicinal mushroom has been also recently characterized using different NMR experiments (Chandra et al. 2009).

Various papers related to carbohydrate metabolomic studies in vegetable and animal foods have been also reported (Hounsome et al. 2008; Karakach et al. 2009).

7.3.1.2 Mass Spectrometry

MS is a powerful analytical technique used to identify chemical compounds based on their different mass-charge ratio (m/z). In this section, only the application of direct MS analysis to food carbohydrates is described. The use of mass detectors coupled to different separation techniques will be discussed in following sections.

The chemical compound has to be ionized before entering the MS analyzer; electron impact ionization, chemical ionization, electrospray ionization (ESI), plasma desorption ionization, and matrix laser desorption ionization (MALDI) are the most commonly used ionization techniques for the analysis of organic compounds such as food carbohydrates. High-quality reviews providing huge and detailed information related to the state of the art of MS analysis of simple, complex, and conjugated carbohydrates have been published (Zaia 2004; Harvey 2008).

Historically, the structural analysis of carbohydrates has been a difficult task to be undertaken due to their multiple combinations, complex nature, and low volatility; consequently, the functional implications of the carbohydrates are poorly known. At present, MS is being considered crucial for a fast characterization of structures and functions of these macromolecules under the so-called field of glycomics.

Selection of the right matrix and ionization technique are critical for obtaining good structural results by MS (Lastovickova et al. 2009). For instance, MALDI-TOF-MS is considered to be more sensitive for detecting larger oligosaccharides, while ESI-MS seems to be more useful for structural studies. MS techniques have been used for the analysis of isolated carbohydrates important in food chemistry. Applications of MS to the analysis of oligosaccharides in various foods were revised by Careri et al. (2002). MALDI-TOF-MS has been applied for characterization of oligosaccharides in vegetable samples (Jerusalem artichoke, red onion, and glucose syrup from potatoes); once the spectrometric conditions were optimized, the degree of polymerization of the oligosaccharides in each food was established. The authors found MALDI-TOF-MS to be more sensitive for detection of oligosaccharides than the chromatographic methods used for the same purpose, underlying the importance of MS for characterization of food carbohydrates (Stikarovska and Chmelik 2004).

According to the tremendous impact of both NMR and MS on glycomics and food chemistry, a large number of NMR and MS application for the analysis of food carbohydrates is expected.

7.3.2 Gas Chromatography

GC has been extensively applied to determine carbohydrates in foods. Considering the high diversity of carbohydrates in terms of structure, size, functionality, and low volatility, GC has been mainly used to study mono-, di-, and trisaccharides. However, oligosaccharides with degree of polymerization up to 7 have been successfully determined by GC with programmed temperatures reaching 360°C as maximum temperature (Montilla et al. 2006a). Trimethylsilyl oximes and ethers are possibly the most used derivatives for the GC analysis of carbohydrates. Nevertheless, acylated derivatives are preferred to detect closely related species, such as aldohexoses (mannose, galactose, glucose) and sugar alcohols (mannitol, galactitol, sorbitol), which only differ in one or more hydroxyl groups.

Generally, for the separation of sugars by GC, dimethyl polysiloxane stationary phases with a given percentage of polar phenyl groups are used, with final programmed temperatures reaching typically 300°C–320°C. Among the detection methods, flame ionization detector (FID) is the most commonly used, although MS is used, providing several advantages over the traditional FIDs, mainly, a higher degree of certainty in the identification of carbohydrates due to the determination of the molecular mass and fragmentation pattern that can also be used in the characterization of unknown carbohydrates. A common approach involves both a GC-MS method to identify carbohydrates and a GC-FID to quantify. This approach was used to identify and quantify minor carbohydrates in carrots and coffee. In fact, using the MS detector, it was possible to identify for the first time trimethylsilyl oximes of scyllo-inositol and sedoheptulose in the chemical composition of carrot, together with other two known minor carbohydrates, mannitol and myo-inositol (Soria et al. 2009).

Honey is a carbohydrate-rich food whose composition depends on its floral origin. The study of carbohydrates contained in honey as well as their relationship has been used to assess the particular composition of unifloral honey and their origin (de la Fuente et al. 2007). Statistical analyses, such as principal component analysis, have demonstrated their usefulness for correct sample assignment (Terrab et al. 2001; Cotte et al. 2004).

The use of GC-MS has also been shown to be valuable to identify unknown carbohydrates. The disaccharides composition in several honey samples was determined in different analytical columns. The use of two different columns allowed studying the elution behavior of an unknown disaccharide. This was then compared with other commercial standards. Combining all these information and its fragmentation in the MS, the presence of inulobiose could be confirmed (Ruiz-Matute et al. 2007b).

Another interesting example is the determination of lactulose; this disaccharide is formed by lactose isomerization under heat treatment in commercial milk. Several studies have demonstrated that its different concentrations depending on the type and intensity of heat treatment; therefore, this compound can be considered a marker of heat treatment processes in sterilized and ultra-high-temperature (UHT) treated commercial milk. To determine lactulose, different methods have been described; among them, the GC analysis of their corresponding trimethylsilyl ethers separated on a 50% diphenyl/50% dimethylsiloxane capillary column. Using this protocol, it was possible to quantify lactulose on different treated milks (UHT, sterilized, pasteurized, condensed, and powder milk) (Montilla et al. 2005). The authors demonstrated that the amount of lactulose quantified on UHT milk was significantly higher than those determined in pasteurized samples, thus, demonstrating that this compound could serve as heat treatment marker in commercial samples (Montilla et al. 2005). In this regard, carbohydrate determination can also be used for detecting food adulterations (Cotte et al. 2003).

Maltulose, another disaccharide resulting from maltose isomerization during heat treatment has also been suggested as a possible marker of heat treatment. In fact, the maltose/maltulose ratio measured after GC analysis was suggested as a possible indicator to assess heat treatment during manufacture and storage of infant formulae (Morales et al. 2004). Difructose anhydrides have also been used as quality markers in sugar-rich foods such as honey and coffee (Montilla et al. 2006b). These compounds are produced by condensation of two fructose molecules during caramelization reactions that take place during heating. These compounds were analyzed by GC because it is possible to statistically differentiate between samples treated with different heating treatments (Montilla et al. 2006b).

In conclusion, although GC provides excellent results for the analysis of food carbohydrates, the usefulness of this technique is limited by the required sample preparation steps and its application to relatively small carbohydrates. Other techniques, such as CE and HPLC, have been developed to overcome these limitations.

7.3.3 High-Performance Liquid Chromatography

HPLC is now the main analytical technique used for the analysis of carbohydrates. Thus, although less sensitive than GC, HPLC is faster, does not necessarily require sample derivatization, allows the analysis of larger carbohydrates, and is much more versatile due to the different separation mechanisms and detectors used.

7.3.3.1 HPLC Separation Mechanisms

The separation mechanism can be selected depending on the matrix to be analyzed, the particular carbohydrates of interest, and the instrumental availability. The most used HPLC methods for the analysis of carbohydrates are IEC, SEC, and partition (normal or reversed-phase) chromatography.

Ion-exchange chromatography can be performed using anionic- or cationic-exchange resins. Particularly, high-performance anion-exchange chromatography (HPAEC) has found great success in the analysis of carbohydrates (Cataldi et al. 2000). To perform the separation, quaternary ammonium resins are commonly used under highly alkaline conditions to allow the ionization of carbohydrates although different types of commercial columns are available nowadays. Mobile phase has to be carefully selected since it will strongly influence not only the selectivity of the separation but also the total analysis time. In addition to mono- and disaccharides, oligosaccharides can also be analyzed using HPAEC. Different methods have been developed to separate oligosaccharides according to their degree of polymerization. For instance, oligosaccharides from honey samples and corn syrups have been studied using a gradient from 1 M sodium acetate to deionized water, maintaining 10% 1 M sodium hydroxide constant during the whole analysis (Morales et al. 2008). Using this method, different malto-oligosaccharides (degree of polymerization from 1 to 16) were adequately separated, as it can be appreciated in Figure 7.5. Oligo- and polysaccharides can also be analyzed using HPAEC after the sample hydrolysis in order to simplify the complex carbohydrates profile to a more simple mono- and disaccharides composition.

Cationic metal ions, such as Ca^{2+} or Ag^+, have also been used to separate mono- and oligosaccharides from different food matrices. The retention mechanism in this case relies on the formation of weak complexes between the metal ions and the hydroxyl groups of carbohydrates. This particular separation method, also called ligand-exchange chromatography, allows the complete separation of monosaccharide enantiomers, such as glucose, mannose, or galactose. Also, sugar alcohol can be separated. However, this technique has several shortcomings: columns are not very robust as the metals can be eluted from the resins when analyzing complex samples, and the resolution of oligosaccharides is often poor due to the very weak interaction with the metal ions.

Another separation mechanism widely used for food carbohydrates separation is SEC. As already mentioned in Section 7.2, SEC might be also used to fractionate carbohydrates. Moreover, technological developments have allowed manufacturing analytical columns that enable SEC to be used as a HPLC-based procedure. SEC is based on the permeation of substances through resins with a particular pore size and structure. Polysaccharides, being large molecules, cannot penetrate into the three-dimensional network inside the resins and consequently are eluted first. The rest of the components will be eluted according to the decrease in molecular size. Thus, monosaccharides with small size will interact more with the resins and will take longer time to elute. This technique has been applied, for instance, to the chemical characterization of arabinoxylans from cereals (Hartmann et al. 2005). Arabinoxylans are related to different organoleptic parameters of bread and several technological aspects in bread making. SEC was used together with a UV detector to determine the mass distribution of wheat and rye arabinoxylans during bread-making process.

The last separation mechanism that has to be mentioned is partition chromatography, either normal or reversed-phase separations. The notable improvements and developments made in the field of column technology are increasing the number of applications based on the use of this HPLC mechanism. In this regard, the new

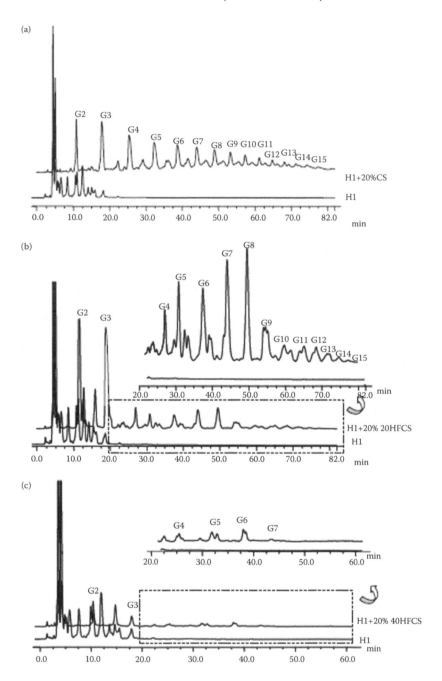

FIGURE 7.5 HPAEC-PAD chromatographic profiles of oligosaccharide fraction of a honey sample H1 and H1 adulterated with (a) 20% of corn syrup, (b) 20% of high-fructose corn syrup, and (c) 40% of high-fructose corn syrup. (Reprinted from *Food Chem*, 107, Morales, V., Corzo, N., Sanz, M. L. HPAEC-PAD oligosaccharide analysis to detect adulterations of honey with sugar syrups. 922–928, Copyright (2008), with permission from Elsevier.)

stationary phases composed of spherical silica particles bonded with trifunctional amino propylsilane are finding many applications in carbohydrate analysis. This special case of normal-phase separation is called "hydrophilic interaction liquid chromatography" (HILIC). The functionalized amino columns have shown to be more resistant than the first commercial amino-bonded silica gel columns that were used to separate mono- and disaccharides. The column was eluted with acetonitrile/water (75:25, v/v) for the analysis of fermentable oligo-, di-, and monosaccharides and polyols, including glucose, fructans, FOS, GOS, and different sugar polyols in several vegetables and fruits (Muir et al. 2009). The same mobile phase was employed with other column type, based on secondary and tertiary amines, for the analysis of oligosaccharides from chickpeas (Xiaoli et al. 2008). More standard C_{18} RP-HPLC columns have also been used although the results obtained are not, in general, as satisfactory as with the amino columns. Also in the latter case, derivatization of the carbohydrates may be necessary.

7.3.3.2 HPLC Detectors

When analyzing carbohydrates by HPLC, one of the parameters to be considered is the detection mode since its choice will greatly influence the sample preparation to be performed. In fact, if a photometric detection is going to be used (UV-visible or fluorescence), it will be necessary to submit the carbohydrates to a derivatization step in order to enable their detection and/or to increase sensitivity. Derivatization reactions of carbohydrates can be divided into two main groups: the first involves the introduction of chromophores reacting with hydroxyl groups and the second involves the reaction with amines at the carbonyl group. One of the most used derivatizing reagents to confer UV absorbance is 1-phenyl-3-methyl-5-pyrazolone. This compound can react with carbohydrates under mild conditions without the common requirements of an acidic catalyst, which avoids degradation and unwanted isomerization providing a sensitive UV detection at 245 nm (Zhang et al. 2003). Other reagents used are *p*-nitrobenzoyl chloride (absorbance at 260 nm) and phenylisocyanate (detection at 240 nm).

Although precolumn derivatization reactions are the most commonly used, postcolumn derivatization reactions have also been described. Benzamidine was used as derivatizing agent to produce a fluorescent complex in the presence of reducing carbohydrates in alkaline medium and high temperatures. This reaction was implemented for the simultaneous determination of mono- and oligosaccharides in different food materials by HPLC (Kakita et al. 2002). The fluorescence detection was performed using 288 nm as excitation wavelength and 470 nm for emission, resulting in LOD of 1.78 and 2.59 pmol for D-glucose and maltohexaose, respectively. A similar strategy was followed to detect reducing sugars and phosphorylated sugars in chicken meat. In this case, tetrazolium blue was the chosen reagent, which conferred absorbance at 550 nm (Aliani and Farmer 2002).

Considering the problematic photometric detection and in order to be able to avoid laborious derivatization steps, other detection systems have been traditionally employed in food carbohydrate analysis. In fact, refractive index detection has been widely used; this detector is not selective, is incompatible with gradient elution, gives low sensitivity, and is quite susceptible to flow rate and temperature changes. Still, it has been broadly applied to different matrices, such as powder milk, cider, or meat products, among others (Chavez-Servin et al. 2006).

Evaporative light scattering detectors (ELSD) have also been utilized to improve sensitivity. This detector is for nonvolatile components and provides approximately the same response for every component. Moreover, ELSD allows gradient elution since the solvent is evaporated just before detection. This kind of detector has been employed together with HPLC and a silica gel column for the direct analysis of carbohydrates in food (Wei and Ding 2002). The presence of soluble sugars in onion was also confirmed by ELSD coupled to size exclusion and amino columns (Davis et al. 2007). Moreover, ELSD has also been proved to be useful as detector in combination with HPAEC separations of carbohydrates (Muir et al. 2009).

Although ELSD and refractive index detectors have found many applications in food carbohydrate analysis, the most widely used detection method has been electrochemical detection. Pulsed amperometric detection (PAD) is considered the ideal detector for HPLC separations of carbohydrates. PAD is suitable for those compounds containing functional groups that are oxidizable at the detection voltage applied. Moreover, electrochemical detection in PAD works better in alkaline medium, which makes it perfectly suitable to be coupled to HPAEC carbohydrate separation. When using PAD, the potential waveform has to be adequately optimized for the detection of carbohydrates. The most frequently used waveform implies three steps, each lasting a given time (60–300 ms): detection, involving the direct oxidation of the electrode surface; cleaning, using higher potentials in order to fully oxidize the electrode surface; and regeneration, eliminating the previous gold oxide film by using a very low potential and making the electrode ready for the next detection step. Both potential and time interval have to be set to attain the best possible results in each application. In the analysis of underivatized carbohydrates, compared to the above-mentioned refractive index and ELSD detectors, PAD provides with better sensitivities reaching LODs below 10 μg/L (Cataldi et al. 2000).

In spite of the good results obtained using PAD detectors in the analysis of food carbohydrates, the use of MS detectors together with HPLC has also been increasing. The use of HPLC-MS in carbohydrate analysis has several advantages because it can corroborate saccharides identification and help characterize unknown carbohydrates, reducing the necessary time for isolation and purification. HPAEC has been successfully coupled to MS detection for the analysis of food carbohydrates. This coupling is not easy, given the technological incompatibility that poses the typical alkaline conditions of HPAEC with volatilization and ionization prior to MS typically using an electrospray interface (ESI). The salts inhibit the correct formation of the electrospray due to their low volatility. Therefore, when operating at these conditions, a desalting device has to be employed between the HPAEC column and the ESI-MS detector. Moreover, to further increase carbohydrates signal in the MS analyzer, a solution of lithium chloride is added to the eluent in order to produce Li^+ (or chloride) adducts. Using this approach, several carbohydrates of different nature were correctly detected in diverse food and beverages allowing the determination of high-molecular-mass carbohydrates (>2000 Da) with a single quadrupole analyzer (Bruggink et al. 2005).

HPLC can also be combined with isotope ratio MS for the analysis of sugars. This technique can measure the abundance of ^{13}C in different compounds. High-fructose corn syrup and honey were found to have significantly different natural abundance of ^{13}C. Therefore, this technique was proved to be useful to detect adulteration of honey with low amounts of corn syrup (Abramson et al. 2001).

Other detection methods have also been applied to a less extent by coupling to HPLC to determine carbohydrates. For instance, HPLC has been coupled to Fourier transform infrared spectrometry to detect different carbohydrates in food samples. Fructose, glucose, maltose, and sucrose were determined and quantified in beverages using this procedure (Kuligowski et al. 2008). Inductively coupled plasma-atomic emission spectrometry (ICP-AES) has also been coupled to HPLC to be used as a carbon detector. In this case, the use of organic compounds in the mobile phases should be avoided to prevent interferences in the detection. Under these conditions, glucose, fructose, sucrose, sorbitol, and lactose, together with other carboxylic acids and alcohols, were detected and quantified. Results were comparable, in terms of LODs, to those obtained with refractive index and UV–visible detectors (Paredes et al. 2006).

In summary, HPLC has been widely employed to analyze carbohydrates in foods combining different mechanisms and detectors (Martínez-Montero et al. 2004; Soga 2002). These applications include control of manufacturing processes, control of products, detection of adulterations, and authenticity determination, among others (Megherbi et al. 2009).

7.3.4 CAPILLARY ELECTROPHORESIS

CE has been already demonstrated to be a very useful analytical technique in food analysis (Herrero et al. 2009; García-Cañas and Cifuentes 2008). Under the conditions typically employed in CE, charged compounds are separated in dissolution according to their effective electrophoretic mobilities under the influence of an electric field. Regarding the analysis of carbohydrates by CE, the main drawbacks to be solved are similar to those mentioned for HPLC; that is, most of these compounds are neutral and lack chromophores. In spite of these limitations, carbohydrates have directly been separated without derivatization using UV detection at 195 nm or indirect UV detection. By using indirect detection, it was possible to improve the low response of these compounds at 195 nm and to detect fructose, glucose, maltose, maltotriose, galactose, lactose, and sucrose in cereal (after ultrasonication) and dairy (after dilution) samples with relatively low LODs (Jager et al. 2007).

Modification of the carbohydrate structure through derivatization procedures can improve both sensitivity and selectivity of the CE separation. Among the derivatizing compounds used in carbohydrate analysis to allow UV detection, phenylethylamine (200 nm), 4-aminobenzoic acid ethyl ester (306 nm), or 6-aminoquinolone (245 nm) can be cited as examples. p-Nitroaniline was used to allow UV detection of different hexoses in powdered milk and rice syrup using a blue light-emitting diode as light source and detection at 406 nm. The use of this reagent was justified considering its high solubility in water and molar absorptivity. The detection limits could be lowered down to 1.1 μM using a boric acid buffer at pH 9.7 (Momenbeik et al. 2006). Analysis of carbohydrates in beverages has also been described using *p*-aminobenzoic acid as derivatizing agent; the reaction took place at 40°C for 1 hour in the presence of the derivatizing agent and 20% acetic acid. The products of this reaction were separated with borate buffer (pH 10.2) in less than 12 minutes. Glucose, maltose, and maltotriose were identified in different beverages (Cortacero-Ramirez et al. 2004). Fluorescence labels that allow using laser-induced fluorescence detectors have also been employed for carbohydrate analysis. 9-Aminopyrene-1,4,6-trisulfonic acid can be used to detect oligosaccharides with different degrees of polymerization (Khandurina and Guttman 2005).

Neutral mono- and disaccharides can form enediolate anions due to a cascade reaction comprising ionization, mutarotation, enolization, and isomerization in aqueous alkaline solution (Rovio et al. 2007). Using this approach, 11 different neutral carbohydrates were detected at 270 nm after reaction in an alkaline electrolyte composed of 130 mM sodium hydroxide and 36 mM disodium hydrogen phosphate hydrate. Detection limits as low as 0.02 mM could be obtained. This method was successfully applied to determine target carbohydrates in fruit juices and cognac, with only a dilution step as sample preparation (Rovio et al. 2007).

The interest of on-column reaction derives from the search of more straightforward methods that do not imply extensive and labor-intensive sample preparation procedures. An on-capillary cleanup has also been proposed for the analysis of carbohydrates in juices. This procedure was based on the partial filling of the capillary with carboxylated single-walled carbon nanotubes that are able to efficiently retain interferences from the matrix allowing a more precise identification and quantification of carbohydrates (Morales-Cid et al. 2007).

Although difficult to implement, CE coupled to MS has been used to analyze carbohydrates in red and white wines (Klampfl and Buchberger 2001). In this case, volatile organic bases, such as diethylamine, were combined with ESI-negative ionization mode. These conditions allowed the detection of deprotonated carbohydrates. Fifteen different carbohydrates could be separated and quantified, obtaining LODs ranging 0.5–3.0 mg/L. Results were in agreement with those previously published using other analytical techniques. Other detection methods that have been used together with CE to detect underivatized carbohydrates are electrochemical detection and Fourier transform infrared (FTIR) detection. In the latter case, an original approach was followed to couple these two techniques (Kolhed and Larlberg 2005). The separation took place in high alkaline conditions in order to have the carbohydrates deprotonized, and a FTIR spectrometer was adapted as online detector. Under these conditions, sucrose, galactose, glucose, and fructose were detected and quantified in orange juice. In addition, their spectra could be recorded and the compounds accordingly identified. The LODs obtained ranged between 0.7 and 1.9 mM for the studied carbohydrates, more sensitive than those obtained from a more common UV detector (Kolhed and Larlberg 2005).

There is a clear trend in CE toward the development of CE analysis in microchips. In this case, the separation is performed in a miniaturized device that allows the separation in a narrow channel typically made of glass, poly(dimethylsiloxane), or other polymer-made plate. The separations on these chips have several advantages, such as low reagents and sample consumption and high speed of analysis, that is, high throughput. A silicone/quartz chip was used combined to an amperometric detector to analyze carbohydrates in acacia honey samples, with LOD as low as 90 amol (Zhai et al. 2007). A typical electropherogram obtained using this device under optimized conditions can be observed in Figure 7.6. An increase in the number of applications of chips for the detection of carbohydrates in food analysis can be expected in the near future.

7.3.5 MULTIDIMENSIONAL TECHNIQUES, GLYCOMICS AND OTHERS

In this section, multidimensional and hyphenated techniques for the analysis of carbohydrates together with new approaches for glycomics are presented.

Sample complexity has led to the development of new hyphenated methods to solve in a fast and automatic way the complete analysis of carbohydrates. A typical example is the sample preparation by microdialysis coupled online to HPAEC, and integrated pulsed electrochemical detection-mass spectrometry (IPED-MS). HPAEC-IPED-MS was used to determine oligosaccharides during the online monitoring of hydrolysis of wheat starch (Torto et al. 1998). One of the key steps was the selection of a desalting device based on a single cation-exchange membrane that is able to exchange sodium ions by hydronium ions, allowing an easy coupling to a MS detector. Several food applications have already been described using the same approach (Okatch et al. 2003).

Although to a less extent, multidimensional techniques have also been developed for the analysis of carbohydrates. A comprehensive two-dimensional GC with time-of-flight MS has been used to analyze metabolites (amino acids, carboxylic acids, and carbohydrates) in plant samples such as basil, peppermint, and sweet

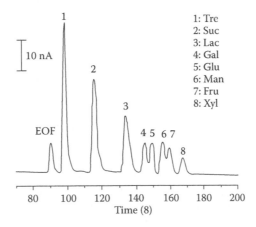

FIGURE 7.6 Separation of carbohydrates (125 µM each) under optimized conditions in a miniaturized device with amperometric detection. Tre, trehalose; Suc, sucrose; Lac, lactose; Gal, galactose; Glu, glucose; Man, mannose; Fru, fructose; Xyl, xylose. (Reprinted with permission from Zhai, C., Li, C., Qiang, W., Xei, J., Yu, X., Ju, H. 2007. Amperometric detection of carbohydrates with a portable silicone/quartz capillary microchip by designed fracture sampling. *Anal. Chem.* 79:9427–9432. Copyright 2007 American Chemical Society.)

herb stevia (Pierce et al. 2006). As for HPLC, the use of an hyphenated RP-HPLC/ HILIC approach has been proposed to analyze different model samples, including sugars. Even if the application is not focused on the separation of carbohydrates from food samples, the setup might be suitable for other complex samples (Louw et al. 2008). In that contribution, authors described a simple approach to couple HILIC to RP-HPLC through the use of a 2-mm I.D. column in the first dimension and a 4.6-mm I.D. in the second dimension combined with the addition of an excess of mobile phase containing high acetonitrile content (i.e., ≥80%) to the mobile phase eluting from the RP-HPLC column via a T-piece. Combination of both mechanisms allowed the elution of compounds widely different in their structure and properties (Louw et al. 2008).

Glycomics, that is, the systematic study of all glycan structures of a given (sample) cell type or organism. Until recently, the complexity of glycan structures was a barrier to understanding the relationship between structure and function of glycans. The development of new analytical tools based on the use of micro- and nanochips coupled to biosensors and/or MS along with HPLC mapping methods has increased the knowledge on the role played by glycans in different biological functions and on their mechanisms of action. Takahashi et al. (1995) developed a three-dimensional sugar mapping technique using three different HPLC columns (diethylaminoethyl, octadecylsilica, and amide) to separate pyridyl-2-aminated derivatives of oligosaccharides. Data in three dimensions, corresponding to the elution in the three HPLC columns, were used to map the position of several N-glycans and thus to identify their structure. Data on more than 500 different glycans, mainly those derived from glycoproteins, are available in the web application GALAXY (Takahashi and Kato 2003) including MS data. Recently, data on anionic oligosaccharides (Yagi and Kato 2009) and on sulfated N-glycans (Yagi et al. 2005) have also been reported, widening the applicability of the technique.

Microarray technologies have emerged as an important analytical tool in glycobiology, complementing other techniques such as MS (Haslam et al. 2006). One typical approach of these microarrays is the use of lectin-based microarrays; lectins are carbohydrate-binding proteins that can help efficiently profile glycan variations in a wide range of biological systems (Gemeiner et al. 2009). More information on microarray-based analysis of glycans can be found in this review paper (Hsu and Mahal 2009). Other approaches have been used, for instance, to analyze the degree of glycosylation of proteins using a chip-based method based on the energy transfer between quantum dots (QDs) and gold nanoparticles (AuNPs). Basic principle relies on modulation in the energy transfer efficiency between the lectin-modified QDs and carbohydrate-conjugated AuNPs by glycoproteins. The chip-based format enables more reliable analyses with no aggregation of the nanoparticles, requiring smaller amount of reagents. Moreover, the energy transfer mechanism allows the development of a simple and highly sensitive method for the detection of glycoproteins, in contrast to the chip-based systems employing a single fluorophore (Kim et al. 2009a). In terms of improving detection of glycated proteins, a system based on the use of a protein chip together with a plasmon resonance imaging biosensor has been developed to determine the presence of advanced glycation end products, implicated in diabetic complications, in serum samples from diabetic rats (Kim et al. 2009b).

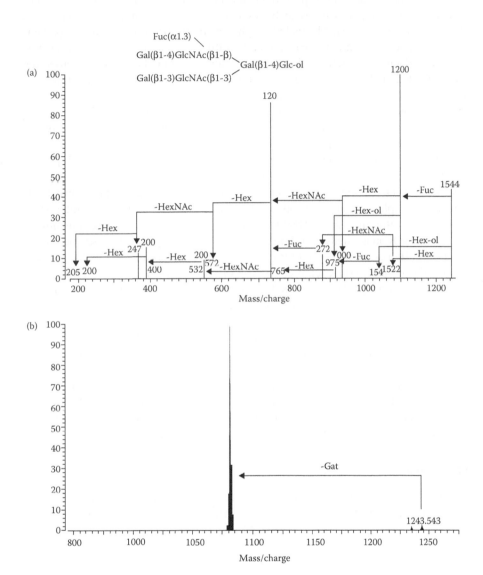

FIGURE 7.7 (a) Infrared multiphoton dissociation spectrum of m/z 1244 from a HPLC fraction in positive mode [M + Na]⁺ and (b) MALDI-FTMS spectrum of m/z 1244 in digestion mixture with recombinant β(1–3)-galactosidase. The loss of a β(1–3)-galactose during digestion is consistent with the structure of 3'-fucosyllacto-N-hexaose. (Reprinted with permission from Ninonuevo, M. R., Park, Y., Yin, H., et al. 2006. A strategy for annotating the human milk glycome. *J. Agric. Food Chem.* 54:7471–7477. Copyright 2006 American Chemical Society.)

Finally, micro- and nanochips have been used, coupled to MS, as a tool for structural analysis of carbohydrates in biological samples. Advantages of chip-based MS over the conventional MS include high sensitivity for minute samples, prevention of sample loss and extensive sample handling, increased ionization efficiency, high reproducibility, high-quality spectra, superior stability of the electrospray over extended analysis time, and possibility of high-throughput MS screening and sequencing (Zamfir et al. 2005). Recently, the integration of various analytical steps onto chips and the direct hyphenation to MS have given rise to a new tool for glycan studies: the lab-on-a-chip MS glycomics. The use of chip devices for MS studies of glycans follows basically two different trends: the direct infusion of the sample mixture for ionization through the chip emitters and the use of integrated chips able to deliver in a single device various sample treatment procedures such as chromatographic or electrophoretic separation, purification, enzymatic reaction, derivatization, and sample ionization (Bindila and Peter-Katalinic 2009). One interesting application is the analysis of HMO mixture by chip-LC in direct conjunction to orthogonal TOF mass analyzer (Ninonuevo et al. 2006). Figure 7.7 shows the typical spectra of oligosaccharides used for identification. The chromatographic separation was conducted on a HPLC-chip consisting of an integrated sample loading structure, an LC column packed with graphitized carbon media, and nanoelectrospray tip (Ninonuevo et al. 2005). A total of 150 distinct neutral and acidic species expressing a high structural variety were found, including novel glycan structures. The potential of the HPLC-chip/MS approach was further explored for differential profiling of HMO in five different randomly selected individuals to study person-to-person variability of the oligosaccharide expression.

These technologies (microarrays, chips, and their couplings with biosensors and MS) are expected to find interesting applications in the food science and technology domain, more probably within the glycomics area, although some other interesting applications within the foodomics field cannot be discarded.

ACKNOWLEDGMENTS

This work was supported by AGL2008–05108-C03–01 (MICINN) and CSD2007–00063 FUN-C-FOOD projects (Programa CONSOLIDER, MEC). M.H. would like to thank Spanish Science and Innovation Ministry (MICINN) for a "Ramón y Cajal" contract.

REFERENCES

Abramson, F. P., Black, G. E. and Lecchi, P. 2001. Application of high-performance liquid chromatography with isotope-ratio mass spectrometry for measuring low levels of enrichment of underivatized materials. *J. Chromatogr. A* 913: 269–273.

Aliani, M. and Farmer, L. J. 2002. Postcolumn derivatization method for determination of reducing and phosphorylated sugars in chicken by high performance liquid chromatography. *J. Agric. Food Chem.* 50: 2760–2766.

Amigo-Benavent, M., Athanasopoulos, V. I., Ferranti, P., Villamiel, M. and del Castillo, M. D. 2009. Carbohydrate moieties on the in vitro immunoreactivity of soy beta-conglycinin. *Food Res. Int.* 42: 819–825.

Arias, V. C., Castells, R. C., Malacalza, N., Lupano, C. E. and Castells, C. B. 2003. Determination of oligosaccharide patterns in honey by solid-phase extraction and high-performance liquid chromatography. *Chromatographia* 58: 797–801.

Arya, M. and Mohan Rao, L. J. 2007. An impression of coffee carbohydrates. *Crit. Rev. Food Sci. Nutr.* 67: 47–51.

Bindila, L. and Peter-Katalinic, J. 2009. Chip-Mass spectrometry for glycomic studies. *Mass Spectrom. Rev.* 28: 223–253.

Bruggink, C., Maurer, R., Herrmann, H., Cavalli, S. and Hoefler, F. 2005. Analysis of carbohydrates by anion exchange chromatography and mass spectrometry. *J. Chromatogr. A* 1085: 104–109.

Careri., M. Bianchi, F. and Corradini C. 2002. Recent advances in the application of mass spectrometry in food-related analysis. *J. Chromatogr. A* 970: 3–64.

Castellari, M., Versari, A., Spinabelli, U., Galassi, S. and Amati, A. 2000. An improved HPLC method for the analysis of organic acids, carbohydrates, and alcohols in grape musts and wines. *J. Liq. Chromatogr. Related Technol.* 23: 2047–2056.

Cataldi, T. R. I., Campa, C. and De Benedetto, G. E. 2000. Carbohydarate analysis by high-performance anion-exchange chromatography with pulsed amperometric detection: The potential is still growing. *Fresenius J. Anal. Chem.* 368: 739–758.

Chandra, K., Ghosh, K., Ojha, A. K. and Islam, S. S. 2009. A protein contaning glucan from an edible mushroom. Termitomyces microcarpus. *Nat. Prod. Commun.* 4: 553–556.

Chavez-Servin, J. L., Romeu-Nadal, M., Castellote, A. I. and Lopez-Sabater, M. C. 2006. Evolution of free mono- and di-saccharide content of milk-based formula powder during storage. *Food Chem.* 97: 103–108.

Cifuentes, A. 2009. Food Analysis and Foodomics. *J. Chromatogr. A* 1216:7109.

Corsaro, A., Chiacchio, U., Pistara, P. and Romeo, G. 2004. Microwave-assisted chemistry of carbohydrates. *Current Org. Chem.* 8: 511–538.

Cortacero-Ramirez, S., Segura-Carretero, A., Cruces-Blanco, C., Hernainz-Bermudez, M. and Fernandez-Gutierrez, A. 2004. Analysis of carbohydrates in beverages by capillary electrophoresis with precolumn derivatization and UV detection. *Food Chem.* 87: 471–476.

Corzo-Martinez, M., Moreno, F. J., Villamiel, M. and Harte, F. M. 2010. Characterization and improvement of rheological properties of sodium caseinate glycated with galactose, lactose and dextran. *Food Hydrocolloids* 24: 88–97.

Cotte, J. F., Casabianca, H., Chardon, S., Lheritier, J. and Grenier-Loustalot, M. F. 2004. Chromatographic analysis of sugars applied to the characterization of monofloral honey. *Anal. Bioanal. Chem.* 380: 698–705.

Cotte, J. F., Casabianca, H., Chardon, S., Lheritier, J. and Grenier-Loustalot, M. F. 2003. Application of carbohydrate analysis to verify honey authenticity. *J. Chromatogr. A* 1021: 145–155.

Cui, S. W. 2005. *Food carbohydrates: Chemistry, physical properties and applications.* Boca Raton, CRC Press.

Davis, F., Terry, L. A., Chope, G. A. and Faul, C. F. J. 2007. Effect of extraction procedure on measured sugar concentrations in Onion (*Allium cepa* L.) bulbs. *J. Agric Food Chem.* 55: 4299–4306.

de la Fuente, E., Sanz, M. L., Martinez-Castro, I., Sanz, J. and Ruiz-Matute, A. I. 2007. Volatile and carbohydrate composition of rare unifloral honeys from Spain. *Food Chem.* 105: 84–93.

Dourado, F., Vasco, P., Barros, A., Mota, M., Coimbra, M. A. and Gama, F. M. 2003. Characterisation of Chilean hazelnut (*Gevuina avellana*) tissues: Light microscopy and cell wall polysaccharides. *J. Sci. Food Agric.* 83: 158–165.

Eremeeva, T. 2003. Size-Exclusion chromatography of enizmatically treated cellulose and related polysaccharides: a review. *J. Biochem. Biophys. Methods* 56: 253–264.

García-Cañas, V. and Cifuentes, A. 2008. Recent advances in the application of capillary electromigration methods for food analysis. *Electrophoresis* 29: 294–309.

Geisser, A., Hendrich, T., Boehm, G. and Stahl, B. 2005. Separation of lactose from human milk oligosaccharides with simulated moving bed chromatography. *J. Chromatogr. A* 1092:17–23.

Gemeiner, P., Mislovicova, D., Tkac, J., et al. 2009. Lectinomics II. A highway to biomedical/clinical diagnostics. *Biotechnol. Adv.* 27: 1–15.

Gibson, P. R., Newham, E., Barrett, J. S., Shepherd, S. J. and Muir, J. G. 2007. Fructose malabsorption and the bigger picture. *Aliment. Pharmaciol. Ther.* 25: 349–363.

Gniechwitz, D., Reichardt, N., Ralph, J., Blaut, M., Steinhart, H. and Bunzel, M. 2008. Isolation and characterisation of a coffee melanodin fraction. *J. Sci. Food Agric.* 88: 2153–2160.

Godshall, M. A., Roberts, E. J. and Miranda, X. M. 2001. Composition of the soluble, nondyalizable components in raw cane sugar. *J. Food Process. Preservation* 25: 323–335.

Hartmann, G., Piber, M. and Koehler, P. 2005. Isolation and chemical characterisation of water-extractable arabinoxylans from wheat and rye during breadmaking. *Eur. Food Res. Technol.* 221: 487–492.

Harvey, D. J. 2008. Analysis of carbohydrates and glycoconjugates by matrix-assisted laser desorption/ionization mass spectrometry: An update covering the period 2001–2002. *Mass Spectrom. Rev.* 27: 125–201.

Haska, L., Nyman, M. and Andersson, R. 2008. Distribution and characterisation of fructan in wheat milling fractions. *J. Cereal Sci.* 48: 768–774.

Haslam, S. M., North, S. J. and Dell, A. 2006. Mass spectrometric analysis of N and O-glycosylation of tissues and cells. *Current Opin. Struct. Biol.* 16: 584–591.

Hernandez, O., Ruiz-Matute, A. I., Olano, A., Moreno, F. J. and Sanz, M. L. 2009. Comparison of fractionation techniques to obtain prebiotic galactooligosaccharides. *Int. Dairy J.* 19: 531–536.

Herrero, M., García-Cañas, V., Simo, C. and Cifuentes, A. 2009. Recent advances in the application of capillary electromigration methods for food analysis and foodomics. *Electrophoresis* in press.

Hounsome, N., Hounsome, B., Tomos, D. and Edwards-Jones, G. 2008. Plant metabolites and nutritional quality of vegetables. *J. Food Sci.* 73: R48–R65.

Hsu, K. L. and Mahal, L. K. 2009. Sweet tasting chips: Microarray-based analysis of glycans. *Current Opin. Chem. Biol.* 13: 427–432.

Ibañez, E. and Cifuentes, A. 2003. New analytical techniques in food science. *Crit. Rev. Food Sci.* 41: 413–450.

Jager, A. V., Tonin, F. G. and Tavares, M. F. M. 2007. Comparative evaluation of extraction procedures and method validation for determination of carbohydrates in cereals and dairy products by capillary electrophoresis. *J. Sep. Sci.* 30: 586–594.

Kakita, H., Kamishima, H., Komiya, K. and Kato, Y. 2002. Simultaneous analysis of monosaccharides and oligosaccharides by high-performance liquid chromatography with post-column fluorescence derivatization. *J. Chromatogr. A* 961: 77–82.

Karakach, T. K., Huenupi, E. C., Soo, E. C., Walter, J. A., Afonso, L. O. B. 2009. H-1-NMR and mass spectrometric characterization of the metabolomic response of juvenile Altantic salmon (Salmo salar) to long-term handling stress. *Metabolomics* 5: 123–137.

Kamada, T., Nakajima, M., Nabetani, H. and Iwamoto, S. 2002. Pilot-scale study of the purification and concentration of non-digestible saccharides from Yacon Rootstock using membrane technology. *Food Sci. Technol. Res.* 8: 172–177.

Khandurina, J. and Guttman, A. 2005. High resolution capillary electrophoresis of oligosaccharide structural isomers. *Chromatographia* 62: S37–S41.

Kim, H. S., Yi, S. Y., Kim, J., et al. 2009a. Novel application of surface plasmon resonance biosensor chips for measurement of advanced glycation end products in serum of Zucker diabetic fatty rats. *Biosens. Bioelectron.* 25: 248–252.

Kim, Y. P., Park, S., Oh, E., Oh, Y. H. and Kim, H. S. 2009b. On-chip detection of protein glycosylation based on energy transfer between nanoparticles. *Biosens. Bioelectron.* 24: 1189–1194.

Kitahara, K. and Copeland. L. 2004. A simple method for fractionating debranched starch using a solid reversed-phase cartridge. *J. Cereal Sci.* 39: 91–98.

Klampfl, C. W. and Buchberger, W. 2001. Determination of carbohydrates by capillary electrophoresis with electrospray-mass spectrometric detection. *Electrophoresis* 22: 2737–2742.

Kolhed, M. and Larlberg, B. 2005. Capillary electrophoretic separation of sugars in fruit juices using on-line mid infrared Fourier transform detection. *Analyst* 130: 772–778.

Kuligowski, J., Quintas, C., Garrigues, S. and de la Guardia, M. 2008. On-line gradient liquid chromatography-Fourier transform infrared spectrometry determination of sugars in beverages using chemometric background correction. *Talanta* 77: 779–785.

Lamari, F. N., Kuhn, R. and Karamanos, N. K. 2003. Derivatization of carbohydrates for chromatographic, electrophoretic and mass spectrometric structure analysis. *J. Chromatogr. B* 793: 15–36.

Lastovickova, M., Chmelik, J. and Bobalova, J. 2009. The combination of simple MALDI matrices for the improvement of intact glycoproteins and glycans analysis. *Int. J. Mass Spectrom.* 281: 82–88.

Louw, S., Pereira, A. S., Lynen, F., Hanna-Brown, M. and Sandra, P. 2008. Serial coupling of reversed-phase and hydrophilic interaction liquid chromatography to broaden the elution window for the analysis of pharmaceutical compounds. *J. Chromatogr. A* 1208: 90–94.

Mabel, M. J., Sangeetha, P. T., Platel, K., Srinivasan, K. and Prapulla, S. G. 2008. Physicochemical characterization of fructooligosaccharides and evaluation of their suitability as a potential sweetener for diabetics. *Carbohydrate Res.* 343: 55–66.

Martínez-Montero C., Rodríguez Dodero, M. C., Guillén Sánchez, D. A. and Barroso, C. G. 2004. Analysis of low molecular weight carbohydrates in food and beverages: A review. *Chromatographia* 59: 15–30.

Martínez-Villaluenga, C., Cardelle-Cobas, A., Olano, A., Corzo, N., Villamiel, M. and Jimeno, M. L. 2008. Enzymatic synthesis and identification of two trisaccharides produced from lactulose by transgalactosylation. *J. Agric. Food Chem.* 56: 557–563.

Megherbi, M., Herbretau, B., Dessalces, G. and Grenier-Loustalot, M. F. 2008. Solid phase extraction of oligo- and polysaccharides; application to malodextrins and honey qualitative analysis. *J. Liq. Chromatogr. Rel. Technol.* 31: 1033–1046.

Megherbi, M., Herbreteau, B., Faure, R.and Salvador, A. 2009. Polysaccharides as a marker for detection of corn sugar syrup addition in honey. *J. Agric. Food Chem.* 57: 2105–2111.

Mendiola, J. A., Herrero, M., Cifuentes, A. and Ibáñez, E. 2007. Use of compressed fluids for sample preparation: Food applications. *J. Chromatogr. A* 1152: 234–246.

Momenbeik, F., Johns, C., Breadmore, M. C., Hilder, E. F., Macka, M. and Haddad, P. R. 2006. Sensitive determination of carbohydrates labelled with p-nitroaniline by capillary electrophoresis with photometric detection using a 406 nm light-emitting diode. *Electrophoresis* 27: 4039–4046.

Montañés, F., Fornari, T., Martin-Alvarez, P. J., Corzo, N., Olano, A. and Ibañez, E. 2006. Selective recovery of tagatose from mixtures with galactose by direct extraction with supercritical CO2 and different cosolvents. *J. Agric. Food Chem.* 54: 8340–8345.

Montañés, F., Fornari, T., Martin-Alvarez, P. J., et al. 2007. Selective fractionation of disaccharide mixtures by supercritical CO2 with ethanol as co-solvent. *J. Supercrit. Fluids* 41: 61–67.

Montilla, A., Moreno, F. J. and Olano, A. 2005. A reliable gas capillary chromatographic determination of lactulose in dairy samples. *Chromatographia*, 62: 311–314.

Montilla, A., van de Langemaat, J., Olano, A. and del Castillo, M. D. 2006a. Determination of oligosaccharides by conventional high resolution gas chromatography. *Chromatographia* 63: 453–458.

Montilla, A., Ruiz-Matute, A. I., Sanz, M. L., Martinez-Castro, I. and del Castillo, M. D. 2006b. Difructose anhydrides as quality markers of honey and coffee. *Food Res. Int.* 39: 801–806.

Morales, V. Olano, A. and Corzo, N. 2004. Ratio of maltose to maltulose and furosine as quality parameters for infant formula. *J. Agric. Food Chem.* 52: 6732–6736.

Morales, V., Corzo, N. and Sanz, M. L. 2008. HPAEC-PAD oligosaccharide analysis to detect adulterations of honey with sugar syrups. *Food Chem.* 107: 922–928.

Morales, V., Sanz, M. L., Olano, A. and Corzo, N. 2006. Rapid separation on activated charcoal of high oligosaccharides in honey. *Chromatographia* 64: 233–238.

Morales-Cid, G., Simonet, B. M., Cardenas, S. and Valvarcel, M. 2007. On-capillary sample cleanup method for the electrophoretic determination of carbohydrates in juice samples. *Electrophoresis* 28: 1557–1563.

Muir, J. G., Rose, R., Rosella, O. et al. 2009. Measurement of short-chain carbohydrates in common Australian vegetables and fruits by high-performance liquid chromatography (HPLC). *J. Agric. Food Chem.* 57: 554–565.

Muir, J. G., Shepherd, S. J., Rosella, O., Rose, R., Barrett, J. S. and Gibson, P. R. 2007. Fructan and free fructose content of common Australian vegetables and fruit. *J. Agric. Food Chem.* 55: 6619–6627.

Murphy, S. P. and Johnson, R. 2003. The scientific basis of recent US guidance on sugars intake. *Am. J. Clin. Nutr.* 78: 827S–833S.

Ninonuevo, M. R., Park, Y., Yin, H., et al. 2006. A strategy for annotating the human milk glycome. *J. Agric. Food Chem.* 54: 7471–7477.

Ninonuevo, M., An, H., Yin, H., et al. 2005. Nanoliquid chromatography-mass spectrometry of oligosaccharides employing graphitized carbon chromatography on microchip with a high-accuracy mass analyzer. *Electrophoresis* 26: 3641–3646.

Okatch, H., Torto, N. and Armateifio, J. 2003. Characterisation of legumes by enzymatic hydrolysis, microdialysis sampling, and micro-high-performance anion-exchange chromatography with electrospray ionisation mass spectrometry. *J. Chromatogr. A* 992:67–74.

Paredes, E., Maestre, S. E., Prats, S. and Todoli, J. L. 2006. Simultaneous determination of carbohydrates, carboxylic acids, alcohols and metals in foods by high-performance liquid chromatography inductively coupled plasma atomic emission spectrometry. *Anal. Chem.* 78: 6774–6782.

Park, Y. K and Yetley, E. A. 1993. Intakes and food sources of fructose in the United States. *Am. J. Clin. Nutr.* 58: 737S–747S.

Pierce, K. M., Hope, J. M., Hoggard, J. C. and Synovec, R. E. 2006. A principal component analysis based method to discover chemical differences in comprehensive two-dimensional gas chromatography with time-of-flight mass spectrometry (GC×GC-TOFMS) separations of metabolites in plant samples. *Talanta* 70: 797–804.

Rovio, S., Yli-Kauhaluoma, J. and Siren, H. 2007. Determination of neutral carbohydrates by CZE with direct UV detection. *Electrophoresis* 28: 3129–3135.

Ruiz-Matute, A. I., Sanz, M. L., Corzo, N., et al. 2007a. Purification of lactulose from mixtures with lactose using pressurized liquid extraction with ethanol-water at different temperatures. *J. Agric. Food Chem.* 55: 3346–3350.

Ruiz-Matute, A. I., Sanz, M. L. and Martinez-Castro, I. 2007b. Use of gas chromatography-mass spectrometry for identification of a new disaccharide in honey. *J. Chromatogr. A* 1157: 480–483.

Ruiz-Matute, A. I., Ramos, L., Martinez-Castro, I. and Sanz, M. L. 2008. Fractionation of honey carbohydrates using pressurized liquid extraction with activated charcoal. *J. Agric. Food Chem.* 56: 8309–8313.

Soga, T. 2002. "Analysis of carbohydrates in food and beverages by HPLC and CE." In: *Carbohydrate analysis by modern chromatography and electrophoresis*, edited by Z. El Rassi. Pages 483–503. Amsterdam, Elsevier Science.

Soria, A. C., Sanz, M. L. and Villamiel, M. 2009. Determination of minor carbohydrates in carrot (*Daucus carota L.*) by GC-MS. *Food Chem.* 114: 758–762.

Stikarovska, M. and Chmelik, J. 2004. Determination of neutral oligosaccharides in vegetables by matrix-assisted laser desorption/ionization mass spectrometry. *Anal. Chim. Acta,* 520: 47–55.

Suzuki, S., Kelly, J. F., Locke, S. J., Thibault, P. and Honda, S. 2003. "Derivatization of Carbohydrates." In *Capillary electrophoresis of carbohydrate,* edited by P. Thibault and S. Honda. New York, Humana Press.

Takahashi, N. and Kato, K. 2003. GALAXY (Glycoanalysis by the three axes of MS and chromatography): A web application that assists structural analyses of N-glycans. *Trends Glycosci. Glycotech.* 15: 235–251.

Takahashi, N., Nakagawa, H., Fujikawa, K., Kawamura, Y. and Tomiya, N. 1995. Three-dimensional elution mapping of pyridylaminated N-linked neutral and sialyl oligosaccharides. *Anal. Biochem.* 226: 139–146.

Terrab, A., Vega-Perez, J. M., Diez, M. J. and Heredia, F. J. 2001. Characterization of northwest Moroccan honeys by gas chromatographic-mass spectrometric analysis of their sugar components. *J. Sci. Food Agric.* 82: 179–185.

Torto, N., Hofte, A., Van Der Hoeven, R., et al. 1998. Microdialysis introduction high-performance anion-exchange chromatography/ionspray mass spectrometry for monitoring on-line desalted carbohydrate hydrolysates. *J. Mass Spectrom.* 33: 334–341.

Van de Lagemaat, J., Silvan, J. M., Moreno, F. J., Olano, A. and del Castillo, M. D. 2007. In vitro glycation and antigenicity of soy proteins. *Food Res. Int.* 40: 153–160.

Wei, Y. and Ding, M. Y. 2002. Ethanolamine as modifier for analysis of carbohydrates in foods by HPLC and evaporative light scattering detection. *J. Liq. Chromatogr. Rel. Technol.* 25: 1769–1778.

Weston, R. J. and Brocklebank, L. K. 1999. The oligosaccharide composition of some New Zealand honeys. *Food Chem.* 64: 33–37.

Xiaoli, X., Liyi, Y., Shuang, H., et al. 2008. Determination of oligosaccharide contents in 19 cultivars of chickpea (*Cicer arietinum L*) seeds by high performance liquid chromatography. *Food Chem.* 111: 215–219.

Yagi, H. and Kato, K. 2009. Multidimensional HPLC mapping method for the structural analysis of anionic N-glycans. *Trends Glycosci. Glycotech.* 21: 95–104.

Yagi, H., Takahashi, N., Yamaguchi, Y., et al. 2005. Development of structural analysis of sulfated N-glycans by multidimensional high performance liquid chromatography mapping methods. *Glycobiology* 15: 1051–1060.

Zaia, J. 2004. Mass spectrometry of oligosaccharides. *Mass Spectrom. Rev.* 23: 161–227.

Zamfir, A. D., Bindila, L., Lion, N., Allen, M., Girault, H. H. and Peter-Katalinic, J. 2005. Chip electrospray mass spectrometry for carbohydrate analysis. *Electrophoresis* 26: 3650–3657.

Zhai, C., Li, C., Qiang, W., Xei, J., Yu, X. and Ju, H. 2007. Amperometric detection of carbohydrates with a portable silicone/quartz capillary microchip by designed fracture sampling. *Anal. Chem.* 79: 9427–9432.

Zhang, L., Xu, J., Zhang, L., Zhang, W. and Zhang, Y. 2003. Determination of 1-phenyl-3-methyl-5-pyrazolone-labeled carbohydrates by liquid chromatography and micellar electrokinetic chromatography. *J. Chromatogr. B* 793: 159–165.

8 Determination and Speciation of Trace Elements in Foods

Stephen G. Capar and Piotr Szefer

CONTENTS

8.1 INTRODUCTION

Food is routinely analyzed for a variety of elements to assess possible nutritional or toxicological implications and to ensure compliance with government regulations or product quality[1–3] and to determine authenticity or geographic origin.[4] The level of a particular element varies greatly among the large variety of foods but is generally consistent for a particular food product. Element levels that may be considered safe or that fulfill a nutrient requirement may also be considered toxic at higher levels.[5–8] Many elements routinely monitored in food have not been identified as nutrients or have not been found to be toxic at the levels normally found in food. The terms listed in Table 8.1 are used to conveniently describe the ranges of element mass fractions of interest in food. Element levels may be evaluated with respect to a specific food or compiled for many foods and used to estimate dietary intakes. Duplicate diet or total diet studies are usually undertaken by national governments or large organizations to estimate intakes and assess the health of the population and safety of the food supply.[9] Total-diet studies entail the analysis of a large number of foods selected on the basis of consumption. Duplicate-diet studies are another method for studying element intakes. In such studies, conducted over a relatively short period, a selected population provides a replicate portion size of the foods they consumed that is analyzed for the analytes of interest. These studies are usually designed to assess specific exposure issues.[10,11] Limited surveys of food are also performed and usually focus on a specific food[12,13] or category (e.g., infant food, dried desserts) collected at retail stores.[14] The quality of the planning, collection, preparation, element analysis, and data evaluation determines the value of intake results from total-diet and duplicate-diet studies. Table 8.2 lists the elements most commonly monitored in foods, the usual purpose for their monitoring, and the usual range of interest. Many of these elements can be defined as "metals," but this term and the term "heavy metals" are discouraged since their meaning is not always understood.[15] Many countries routinely monitor the level of elements in their foods,[16–23] and compilations of element levels are available for some elements.[24,25] The techniques commonly used for determining the levels of elements in foods are listed in Table 8.3

8.2 PREPARATION OF ANALYTICAL SOLUTION

Horwitz[26] has presented terminology for sampling and analytical operations that are useful for describing a "sample" through the various stages from collection to

TABLE 8.1

Terms for Mass Fraction Ranges

Term (%)	Mass Fraction (%)	Mass Fraction
Major	0.1–100	1–1000 g/kg
Minor	0.001–0.1	10–1000 mg/kg
Trace	0.000001–0.001	0.01–10 mg/kg
Ultratrace	0.000001	10 µg/kg

TABLE 8.2

Elements Commonly Monitored in Food

Element	Primary Purpose	Range of Interest
Aluminum (Al)	Toxicity	Trace
Arsenic (As)	Toxicity	Trace/ultratrace
Boron (B)	Nutrition	Trace
Cadmium (Cd)	Toxicity	Trace/ultratrace
Calcium (Ca)	Nutrition	Major/minor
Chromium (Cr)	Nutrition/toxicity	Trace/ultratrace
Copper (Cu)	Nutrition	Minor/trace
Fluorine (F)	Nutrition/toxicity	Trace
Iodine (I)	Nutrition/toxicity	Trace
Iron (Fe)	Nutrition	Minor/trace
Lead (Pb)	Toxicity	Trace/ultratrace
Magnesium (Mg)	Nutrition	Minor/trace
Manganese (Mn)	Nutrition	Minor/trace
Mercury (Hg)	Toxicity	Trace/ultratrace
Molybdenum (Mo)	Nutrition	Trace/ultratrace
Nickel (Ni)	Toxicity	Trace/ultratrace
Phosphorus (P)	Nutrition	Major/minor
Potassium (K)	Nutrition	Major/minor
Selenium (Se)	Nutrition/toxicity	Trace/ultratrace
Sodium (Na)	Nutrition	Major/minor
Tin (Sn)	Toxicity	Minor/trace
Zinc (Zn)	Nutrition	Minor/trace

TABLE 8.3

Primary Determinative Techniques Used for Monitoring Elements in Food

Technique	Abbreviation
Flame atomic absorption spectrometry	FAAS
Electrothermal atomization atomic absorption spectrometry	ETA-AAS
Hydride generation atomic absorption spectrometry	HG-AAS
Cold vapor-atomic absorption spectrometry	CV-AAS
Inductively coupled plasma atomic emission spectrometry	ICP-AES
Inductively coupled plasma mass spectrometry	ICP-MS
Ion-selective electrode	ISE
Anodic stripping voltammetry	ASV
Colorimetry	—
Neutron activation analysis	NAA

analysis. The sample's composition must not be adversely affected by contamination, change in moisture, or decay, for example, during the collection, shipping, storage, and preparation for analysis. Therefore, the containers used at each stage must be known not to contaminate the sample with the element(s) of interest or absorb the element from the sample. Hoenig[27] has provided general guidance on preventing the problems during preparation of the analytical sample.

The edible portion of a food is analyzed for assessing nutritional or toxicological properties of an element during intake or exposure. The portion of the food analyzed should always be described to enable comparison with other reported findings. For packaged foods, the edible portion is usually the entire contents of the package, but may include special consumer practices such as removal of liquids. For raw products, consumer practices may vary, allowing a choice for the edible portion (e.g., optional peeling of apples or potatoes). Fish and marine products may require definition of the edible portion.[28] Whether the food is cooked or prepared as consumed (e.g., fat removed, condiments added) is also important for interpreting analytical results.[29] Washing raw food products with water, especially produce, is a practice normal to food preparation and should always be performed with water of a quality commensurate with the element(s) and the levels to be determined. Usually, deionized water or water of higher quality is used.

Most determinative techniques for elements in food require the destruction of the food's organic matter. This is usually accomplished by dry ashing (i.e., calcination) or wet digestion with oxidizing acids. An analytical solution is prepared from these digestions for introduction into the instrument used for measuring the element's concentration. A proficiency testing report that summarized results for a number of years indicated that current commonly used digestion procedures for food analysis of elements (Pb, Cd, Hg, As, Sn) are dry ashing, microwave (MW) heating of acids, open vessels heating of acids, and heating in pressure bombs.[30] Hoenig[27] has provided an excellent summary of these preparation steps for trace element analysis that are considered the most critical part of the analysis, and he maintains that errors associated with sample preparation are difficult to uncover.

8.2.1 HOMOGENIZATION

The laboratory sample is made into a homogeneous analytical sample from which replicate analytical portions will produce equivalent analytical results and represent the laboratory sample. Obtaining a homogeneous mixture becomes more difficult, as analytical procedures use smaller analytical portions and measure elements at lower levels. In addition, many food analyses require that a composite be made of a relatively large amount of food, such as 12 large containers of peanut butter or 12 watermelons. Homogenization procedures vary depending on the physical state of the food and may require cutting, mixing, blending, chopping, or crushing by hand or mechanical equipment to obtain an analytical sample from which the analytical portions may be taken. Rains[31] has reported many useful instructions for homogenizing a variety of foods. Removing water from the food by freeze-drying (lyophilization) provides a material that is more easily homogenized and is a practice amenable to some foods and elements; however, during freeze-drying, loss of some trace elements may occur.[32,33] Foods requiring the use of homogenization equipment have the potential of being contaminated by abrasion of the equipment's surface. Although care must be taken to avoid contamination from the equipment, reliable ways to routinely assess contamination are lacking. This source of contamination is not uncovered by laboratory quality control analyses such as the use of reference materials and method blanks. Equipment with food contact surfaces coated with plastic or other noncontaminating material should be used if available, and equipment that uses titanium blades instead of stainless steel should be used when contaminants from steel are of concern. Contamination from homogenization equipment has been documented. Using breakfast meals, Razagui and Barlow[34] demonstrated contamination of Fe, Zn, Cu, Pb, Cd, and Cr from blenders. They offered a means to reduce the potential of contamination by cleaning the equipment with a solution of 2% ethylenediaminetetraacetic acid (EDTA) and 2% citric acid. Cubadda et al.[35] studied the potential contamination of wheat by 15 elements (Al, As, Cd, Co, Cr, Cu, Fe, Mn, Mo, Ni, Pb, Se, Sn, V, and Zn) from homogenization equipment (porcelain and glass mortars and four commercial grinders). Depending on the equipment, they found at least a few among 10 elements (Al, Cd, Co, Cr, Cu, Fe, Mn, Mo, Ni, and Pb) were likely to contaminate the sample. Use of agate and stainless steel grinders on plant materials was compared, and the stainless steel grinder, on average, contributed higher levels of the elements studied (Pb, Cd, Ni, and Cr).[36] Cryogenic grinding is showing promise in providing homogeneous materials with low potential for contamination. Food that is generally difficult to homogenize with conventional equipment can be homogenized well using commercial cryogrinding equipment.[37–39] This equipment may alleviate some of the heterogeneity problems associated with analysis of relatively small analytical portions. Initially, custom cryogenic systems accommodated only a few grams of sample, whereas current commercial equipment can accommodate up to about 100g.

The analytical portion should be taken out of the equipment immediately after homogenization before liquids and solids separate. Homogenates must be stored in containers that do not contaminate the analytical sample. If analytical samples are frozen prior to aliquoting, then they should be thoroughly mixed before aliquoting the analytical portion. Repeated freezing and thawing may affect the integrity of the sample by changing the moisture level.

8.2.2 Dry Ashing

Dry ashing is a technique that has been used for many years to prepare foods for the determination of elements and is used in many standard methods.[40–42] The ashing is usually performed using a temperature-programmable muffle furnace, and an ashing aid may be used to assist in retaining certain elements. Mader, Száková, and Curdová[43] reviewed dry-ashing procedures and reported how best to perform dry ashing. They noted that dry ashing is usually performed at a maximum temperature of about 450°C–500°C and that the most frequently used ashing aid is magnesium nitrate. Dry-ashing procedures require drying the sample, sometimes in a convection oven, before ashing, and the ashing temperature is increased at a slow rate to avoid splattering or combustion. After initial ashing, the residue is dissolved in acid and may require reashing to obtain a carbon-free ash. Presence of insoluble residue at this stage usually indicates the presence of silica (especially plant materials), which retains the elements of interest and requires treatment by HF.[44] Ashing vessels made of quartz, porcelain, or (preferably) platinum are commonly used, and analytical portions are about 5–25 g. Alkaline dry ashing (or alkali fusion) is also used for preparing food samples for F and I determination, and sodium or potassium hydroxide is normally used as an ash aid. The advantages of dry ashing are that a large analytical portion can be used, which may minimize homogeneity difficulties. In addition, a relatively large number of samples can be processed with minimum operator attention over the few days that are needed to complete the ashing.

Volatilization of some elements of interest using dry ashing is a concern. Dry ashing is inappropriate for preparing food samples for Hg determination due to low volatility of Hg. For plant samples, Vassileva et al.[45] found that magnesium nitrate as an ashing aid suppresses As and Se volatilization during dry ashing and that the ashing aid is not necessary for determining As in terrestrial plants. Fecher and Ruhnke[46] demonstrated cross contamination of Pb and Cd during dry ashing with no ashing aid at 450°C. Koh et al.[47] studied the effect of dry ashing plant material on 22 elements and indicated that there was some loss of Al, Cr, and V and possible contamination of Na using porcelain crucibles. Using a dry-ashing procedure, Sun, Waters, and Mawhinney[48] analyzed seven food-related reference materials for 13 elements (Al, B, Ca, Cu, Fe, K, Mg, Mn, Na, P, S, Sr, and Zn) by inductively coupled plasma atomic emission spectrometry (ICP-AES) and observed loss of S and low recovery of Na and K.

8.2.3   WET DIGESTION

8.2.3.1   Convection Heating

Wet digestion of food has traditionally been performed with HNO_3–$HClO_4$, HNO_3–H_2SO_4, or HNO_3–$HClO_4$–H_2SO_4.[31,42,49] Various procedures for performing the digestion exist, but generally for the HNO_3–$HClO_4$–H_2SO_4, the acid mixture is added to the analytical portion (2 g dry or 5–10 g wet) in a borosilicate or quartz Kjedhal flask and allowed to digest at room temperature for a few hours or overnight. The flask is gently heated to boiling. As HNO_3 distills, $HClO_4$ begins to react with any remaining organic material. If charring occurs at this stage, the flask is removed from the heat, a few drops of HNO_3 are added, and heating is continued. This step may need to be repeated. $HClO_4$ is then distilled, (white fumes) and the digestion continued until the appearance of white fumes of sulfur trioxide. This digestion has also been used for preparing foods samples for I determination using a cold finger.[50] The large amounts of corrosive reagents used and the safety considerations of $HClO_4$ made this digestion procedure less favorable. In addition, high-pressure ashing devices are available that use convection heating with the ashing vessels under pressure and are able to achieve temperatures up to 320°C.[51]

Tinggi, Reilly, and Patterson[52] studied HNO_3–$HClO_4$–H_2SO_4 and HNO_3–H_2SO_4 (to fumes of H_2SO_4) and the addition of HF to HNO_3–H_2SO_4 for the analysis of various food samples for the determination of Mn and Cr by electrothermal atomization atomic absorption spectrometry (ETA-AAS). They found that both procedures produced good results and that HF was not needed. Schelenz and Zeiller[53] found that an HF pretreatment prior to a HNO_3–$HClO_4$ wet digestion was necessary for the analysis of Al in food by ICP-AES. Hoenig et al.[54] developed wet digestion procedures for plant reference materials using sand-bath heating and open-vessel MW digestion with HF and evaporation to dryness, and results were compared to that of a dry-ash procedure. Al, Ca, Cr, Cu, Fe, P, K, Na, Mg, Mn, and Zn were determined by ICP-AES; and Cd, Pb, Ni, Co, As, and Se were determined by ETA-AAS. Focused MW digestion (open-vessel) without HF resulted in low recovery of Al (60%), Fe (87%), Ca and Mg (90%), and Ni (80%). Hg, which was determined by cold vapor atomic absorption spectrometry (CV-AAS), was lost when using HF and when the sample was evaporated to dryness; however, use of a sand bath or MW H_2SO_4–H_2O_2 digestion produced satisfactory results for Hg. Kakulu, Osibanjo, and Ajayi[55] compared two wet digestion procedures (HNO_3–H_2SO_4 and HNO_3–H_2O_2) and a dry-ash procedure (muffle furnace 500°C) for analysis of fish and shellfish (~10 g diluted to 25 mL) for the determination of Pb, Cu, Ni, Zn, Mn, and Fe by flame atomic absorption spectrometry (FAAS). All three decomposition procedures appeared equivalent except for low results for Pb using HNO_3–H_2SO_4, which they suspected was due to precipitation of lead sulfate. Their results using HNO_3–H_2O_2 and FAAS were comparable to the determinations by ICP-AES. Using HNO_3–$HClO_4$ digestion, Sun, Waters, and Mawhinney[48] analyzed seven food-related reference materials for 12 elements (Al, Ca, Cu, Fe, K, Mg, Mn, Na, P, S, Sr, and Zn) by ICP-AES and observed low recovery of Al, Na, and K.

Digestion procedures for Hg must prevent the loss of Hg due to its volatility. Wet digestions are commonly used with HNO_3 alone or in combination with $HClO_4$ and/or H_2SO_4. Burguera and Burguera[56] evaluated 10 wet digestion procedures on reference materials and sausage fortified with inorganic Hg. Their results using CV-AAS indicated that to prevent Hg loss, one needed to use sealed vessels or a condenser and to perform the digestion at a low temperature. Four commonly used wet digestion procedures for Hg were studied by Adeloju, Dhindsa, and Tandon[57] on fish. Digestions were performed in a loosely stoppered flask at a maximum temperature of 90°C, and Hg was determined using CV-AAS. Spike recovery of inorganic Hg was good, but that of organic Hg (e.g., CH_3HgCl) was good only for the procedures using HNO_3–H_2SO_4 and HNO_3–H_2O_2. On the basis of additional results for reference materials, the authors recommended the use of HNO_3–H_2SO_4. Adair and Cobb[58] reported a relatively fast method to extract Hg from small analytical portions of bovine kidney (0.1–0.3 g) by digesting with small amounts of HNO_3–H_2SO_4 in vials with heating at 90 °C, treatment with H_2O_2 and determination of Hg using CV-AAS. They validated the method using the recovery of added inorganic and organic Hg to samples.

8.2.3.2 Microwave Heating

MW-assisted digestion[59] is increasingly used to replace traditional heating for wet digestion procedures.[60] Generally, MW-assisted digestion systems use either open or closed vessels, with the latter used to retain volatile elements and achieve higher temperatures.[61] The digestion procedure accommodates less sample mass than traditional procedures but requires less amount of reagents. Digestion time is much shorter than that of traditional procedures, but preparation of vessels takes more time. A standard method for food analysis using MW-assisted digestion is available,[62] and the U.S. Environmental Protection Agency (EPA) provides a comprehensive MW total digestion method that can be used for determining 26 elements.[63]

Lamble and Hill[64] presented an excellent review of many MW digestion procedures that have been reported for the determination of elements, which includes analysis of food matrices and issues related to accuracy. As with other digestion procedures, HF should be used in the digestion to recover Al,[65] although there are conflicting results.[64] Sun, Waters, and Mawhinney[48] found that MW-assisted digestion using HNO_3, HNO_3–H_2O_2, and HNO_3–H_2O_2–HF of seven food-related reference materials and determination of 13 elements (Al, B, Ca, Cu, Fe, K, Mg, Mn, Na, P, S, Sr, and Zn) produced good results from all three procedures for all the elements except for three materials, where good recovery of Al was consistent only with the digestion using HNO_3–H_2O_2–HF. Sahuquillo, Rubio, and Rauret[66] studied the determination of Cr in plants by wet digestion procedures using $HClO_4$ or H_2SO_4 with convection heating and MW digestion (opened focused). A loss of Cr was observed in some materials when using convection heating and $HClO_4$ but not with H_2SO_4. No loss of Cr with either acid mixture was observed when using MW digestion (opened focused). Gawalko et al.[67] studied closed- and open-vessel MW digestions with cereal samples and obtained good results for both techniques for Cd, Cu, Pb, and Se. Dolan and Capar[68] developed a MW-assisted digestion procedure that uses a single program to digest a variety of foods, which bases the analytical portion mass (0.41–9.5 g) on the food's energy content. Carrilho et al.[38] used a single program for MW-assisted digestion of animal and plant samples by drying (freeze-drying or oven drying) and cryogenic grinding the samples prior to digesting 250-mg plant material or 300-mg animal-derived material. Wasilewska et al.[69] studied the efficiency of oxidation in wet digestion using four techniques: high-pressure wet ashing, pressurized MW in closed quartz vessels, pressurized MW in closed Teflon® (E.I. Du Pont de Nemours & Company, Inc., Wilmington, Delaware) vessels, and pressurized MW in open vessels. Residual carbon, including some organoarsenic compounds, were not completely oxidized until attaining 300°C, a temperature that is not achieved by many wet digestion systems. Determining As from incompletely oxidized As compounds using hydride generation (HG) techniques would not produce accurate results. In addition, residual carbon content may also affect the accuracy of measurements by ICP-AES near the detection limits for As and Se. Analysis of other elements in a cod muscle reference material (As, Cd, Cu, Fe, Hg, Mn, and Zn) was not affected by the residual carbon content.

8.3 DETERMINATION BY ATOMIC ABSORPTION SPECTROMETRY

Atomic absorption spectrometry (AAS) is one of the older techniques still being used for determining elements in foods, although its utilization is decreasing relative to that of simultaneous multielement techniques. Background correction (usually with Zeeman effect, deuterium, or Smith-Hieftje devices) is required for most elements and is built into the operation of modern instruments. Atomic absorption techniques are used for food analysis in many laboratories for nutritional elements using FAAS or ETA-AAS; Cd and Pb using ETA-AAS; As and Se by HG-AAS or ETA-AAS; and Hg by CV-AAS. A proficiency testing report on food analysis for Pb, Cd, Hg, As, and Sn indicated that results from AAS-related techniques (FAAS and ETA-AAS) were not as good as those from inductively coupled plasma-related techniques (ICP-AES and inductively coupled plasma mass spectrometry [ICP-MS]). However, they caution that specialization of the staff who use the latter techniques may be the reason for the better performance and not the ease or difficulty of the techniques themselves.[30]

8.3.1 FLAME

FAAS[70] can determine many nutritional elements and some toxic elements at the levels of interest in food. Analytical solutions are aspirated into a flame that is usually air–acetylene or nitrous oxide–acetylene, depending on the element being determined. Analytical solutions are prepared by either wet digestion or dry ashing. Enrichment of the analytical solution by chelation-solvent extraction may be necessary to measure low levels of some elements in foods. FAAS instruments are relatively simple to operate, and if analysis of only a couple of elements is required, FAAS technique may be faster than more-sophisticated multielement techniques. Examples of standard methods that use FAAS are Cd[71] and Pb[72] in food; Cu, Fe, and Zn in food[40,62]; Sn in canned food[73]; Zn in food,[74] Zn,[75] Cu,[76] and Fe[77] in fruits and vegetables; Zn in milk,[78] Ca and Mg in cheese[79]; and Ca, Cu, Fe, Mg, Mn, K, Na, and Zn in infant formula.[41] Table 8.4 lists references to recent surveys on elements in food that used FAAS.

Rains[31] has provided methods for food analysis using FAAS for determination of Na, K, Mg, Ca, Al, V, Cr, Mn, Fe, Co, Ni, Cu, Zn, Mo, Cd, Sn, and As, and Se with a hydride generator. Miller-Ihli[88] has demonstrated the capability of FAAS for food

---

TABLE 8.4
Surveys of Food for Elements That Used Flame Atomic Absorption Spectrometry

Food	Elements	Reference
Total diet	Ca, Cu, Mg, Na, P, Zn	80
Total diet	Ca, Cu, Mg, Mn, Zn	81
Total diet, duplicate diet	Cd, Cu, Mn, Zn	22
Duplicate diet	Ca, Cu, Fe, Mg, Mn, K, Na, Zn	11
Fruits	Ca, Cu, Fe, Mg, Mn, K, Na, Zn	82
Nuts	Ca, Cu, Fe, Mg, Zn	83
Canned seafood	Fe, Zn	84
Canned tuna fish	Cd, Pb	85
Coffee	Ca, Cr, Fe, K, Mg, Mn, Ni, Sr, Zn	12
Coffee	Cd, Co, Cu, Mn, Ni, Pb, Zn	86
Rice	Na	87

analysis of elements related to food-labeling regulations (Ca, Cu, Cr, Fe, Mg, Mn, and Zn) by either dry-ash or wet digestion procedures. Leblebici and Volkan[89] dry-ashed 100 g of sugar and determined relevant levels of Cu, Fe, and Pb using FAAS. A collaborative study on Ca and Mg in food using MW digestion and FAAS was successfully completed.[90]

8.3.2   ELECTROTHERMAL ATOMIZATION

ETA (also commonly referred to as graphite furnace) AAS[70,91,92] can determine many nutritional and toxic elements at the levels of interest in food. A small aliquot of analytical solution is injected into a graphite tube (pyrolytically or nonpyrolytically coated depending on element), usually having a graphite platform, and is heated electrically in three stages to dry the solution, ash or pyrolyze the residue, and atomize the analyte. Usually, a chemical matrix modifier is added to the analytical solution for assisting in retaining the analyte while removing the matrix during the ashing step. Various matrix modifiers are used depending on the element being determined. Analytical solutions are prepared by either wet digestion or dry ashing, and MW wet digestion is becoming widely used.[60] In addition, this technique is used to analyze foods directly, for example, by using a slurry of the food. Examples of standard methods that use ETA-AAS are Cd and Pb in food[40,62]; Cd[93] and Pb[94] in fruits and vegetables; Sn in canned evaporated milk[95]; Cd in oils and fats,[96] Cd,[97] Pb,[98] P,[99] and Cu, Fe, and Ni[100] in animal and vegetable fats and oils; Cd[101] and Pb[102] in starch-based foods; and Pb in sugars and syrups, using a HNO_3 solubilization.[103] Table 8.5 lists references to recent surveys on elements in food that used ETA-AAS.

Allen, Siitonen, and Thompson[116] demonstrated the use of open-vessel MW digestion of edible oils for the determination of Cu, Pb, and Ni by ETA-AAS using external calibration and instrument conditions allowing simultaneous determination. A method for determining Cd and Pb using simultaneous AAS with ETA and closed-vessel MW digestion was developed for a variety of foods (primarily vegetables and fruits).[117] A matrix modifier composed of tungsten, palladium, and tartaric acid was found to be preferable for the determination of Pb in cokies.[118] Julshamn, Thorlacius, and Lea[119] performed a collaborative study of a closed-vessel MW digestion method for determining As in seafood by ETA-AAS. They concluded that the method is suitable for determining As at levels as low as 2.5 mg/kg dry weight. Fedorov, Ryabchuk, and Zverev[120] compared ETA-AAS and HG-AAS for the determination of As in food and concluded that HG-AAS produced more precise and accurate results. Rosa et al.[121] studied various matrix modifiers for determining Se in vegetables by ETA-AAS and found that palladium nitrate produced optimum results. Veillon and Patterson[122] reviewed the difficulties of determining Cr when ETA-AAS is used as the primary analytical technique. Miller-Ihli[108] has used a modified dry-ash procedure to analyze a variety of foods for the determination of Cr and found that for most foods, direct calibration could be used instead of the more time-consuming standard additions. Adachi et al.[123] developed a method for determining V in foods using chelation and extraction prior to the determination by ETA-AAS. Bermejo-Barrera et al.[124] developed an indirect method for determining I in milk samples. Iodide in the sample was

TABLE 8.5

Surveys of Food for Elements That Used Electrothermal Atomization Atomic Absorption Spectrometry

Food	Elements	Reference
Total daily diet	Cr	81
Total diet	Cd, Co, Pb, Ni	104
Total diet	Cd, Pb, Ni	105
Duplicate diet	Al, Cd, Co, Cr, Ni, Pb	9
Duplicate diet, fruit juices	Al	106
Food, prepared dishes	Cd, Cr, Ni, Pb	107
Vegetables, fruits, dairy products, grain products, meats, eggs, fats, beverages, condiments, snack foods	Cr	108
Meat, milk, vegetables, grains	As, Cd, Pb	109
Cereals, vegetables, fruits, meat, fish, eggs, beverages	Cd, Pb, Ni, Se	110
Grain products, meat, dairy products, vegetables, fruits	Cr, Mn	52
Seafood, cereals, vegetables, dairy products, olive oil	Cr	111
Meat, seafood, dairy products, cereal products, vegetables, fruits, fats and oils, nuts	Cr	112
Canned seafood	Al, Cr, Cu, Pb	84
Seafood, vegetables, olive oils, dairy products, stimulant drink and infusions, fruit juices, soft drinks	Al	113, 114
Infant cereals	Cd, Pb	115

oxidized to iodine and distilled using MW heating. Iodine was then reduced back to iodide and combined with Hg(II)-2,2′-dipyridyl, extracted into isobutyl methyl ketone, and then Hg was determined in the extract by ETA-AAS. The method produced good results for reference materials, recovery of added I, and was applied to various types of milk samples.

ETA-AAS has been used to directly analyze foods without wet digestion or dry ashing to enable faster sample preparation and minimize contamination and analyte loss. The problems with direct analysis include the need for a very homogeneous analytical sample, overcoming high background during atomization, and buildup of carbonaceous residues in the graphite tube. Liquid or pureed foods are most often analyzed by direct injection. Milk was analyzed for Al, Cr, Mn, and Mo by ETA-AAS using a suspension medium containing ethanol, HNO_3, and H_2O_2.[125] Cr was determined by direct injection in various dairy products using Triton X-100, treating the graphite tube with ammonium molybdate and using magnesium nitrate as a matrix modifier.[111] Milk slurries were analyzed for Al, Cd, Cr, Cu, Pb, Mn, Ni, Se, and Zn by shaking sample portions with blown zirconia spheres and Triton X-100, removing the spheres, adding silicone antifoaming emulsion, and using various matrix modifiers depending on the element.[126] Results from the direct injection analysis compared well with those from samples prepared by MW digestion. Slurries of various dairy products were prepared for the determination of Al by shaking samples with blown zirconia spheres and Triton X-100, removing the spheres, adding silicone antifoaming emulsion, treating the graphite tube with ammonium molybdate, and using magnesium nitrate as a matrix modifier.[113] Fats and oils are another type of food products for which direct injection is used with ETA-AAS. Direct injection methods include the determination of Cd using a palladium matrix modifier,[96] Pb using a lecithin solution in cyclohexane matrix modifier,[127] and Cu, Fe, and Ni using a niobium matrix modifier for determining Fe.[128] Butter was analyzed for Cu and Fe by direct injection by dissolving a sample portion in butylamine, water, and tetrahydrofuran.[129] Edible oils and fats were analyzed for Cd by simply heating the samples to a liquid state prior to injection.[130] Optimum performance was found by using a palladium chloride modifier and uncoated graphite tubes with a pyrolytic graphite-coated platform. Pureed baby food is another food product for which efforts have been made to use direct injection. Viñas et al.[131] did considerable work applying direct injection ETA-AAS to the analysis of baby foods for As and Cd; Pb and Se[132]; Co, Cu, Mn, and Ni[133]; and Al and Cr.[134] A food suspension was prepared with Triton X-100 (ethanol for Al), H_2O_2, HNO_3, silicone antifoam, and an appropriate matrix modifier. Hg was determined in baby foods in a suspension of Triton X-100, HNO_3, and silicone antifoam and using the matrix modifiers potassium permanganate and silver nitrate.[135] Results from the direct injection analyses compared well with those for the samples prepared by MW digestion. Extension of direct injection to samples other than liquid or pureed foods requires extensive homogenization to obtain a representative aliquot for injection. Cryogenic grinding was used for the determination of Cd and Pb in a variety of foods with different water, fiber, and fat content.[39] The foods were chopped, freeze-dried, and cryogenically ground, and then a portion was mixed with HNO_3 and Triton X-100 and sonicated before an aliquot was analyzed.

Fresh or frozen seafood samples were analyzed for Se by direct injection.[136] The samples were homogenized in a blender, dried, and ground with a mixer mill. The analytical portion was taken from the fraction of this mixture, which was 100 µm, and mixed by sonication with HNO_3 and Triton X-100 and then analyzed with a matrix modifier composed of palladium nitrate and magnesium nitrate. The samples were freeze-dried and ground in a ball mill, and slurry suspensions were prepared with ethanol, H_2O_2, HNO_3, and nickel nitrate. Direct injection for vegetables was evaluated for the determination of Cr, Co, and Ni.[137] The samples were cut, dried, and ground using a household grinder and mortar and a suspension was made with HNO_3 and H_2O_2 with mixing by sonication. Results for Co and Ni were generally accurate, but Cr recovery was slightly low (85%).

8.3.3 HYDRIDE GENERATION

Hydride generation-atomic absorption spectrometry (HG-AAS)[92,138,139] is commonly used to determine As and Se in foods. Foods must be rigorously digested to oxidize refractory organometallic compounds, especially organoarsenic compounds, to compare with the normally used calibration standards. Samples are usually digested using HNO_3–$HClO_4$–H_2SO_4 or dry ashing with magnesium nitrate and magnesium oxide, and the analytical solution is prepared with HCl. Digestion with only HNO_3 is an alternative, but a high temperature must be achieved for reliable measurement of As.[140] Warming of the analytical solution is needed to prereduce

Se from Se(VI) to Se(IV), and As is prereduced from As(V) to As(III), usually with sodium or potassium iodide prior to generation of their hydrides with sodium borohydride. The hydrides are transferred to a quartz tube or graphite furnace and heated to produce atomization for atomic absorption. HG-AAS is used in a standard method for the determination of As and Se in food.[141] Recent surveys that used HG-AAS for the determination of As and Se include analysis of total-[105] and duplicate-diet foods,[11] determination of As in total- and duplicate-diet foods,[22] and determination of Se in total-diet foods,[142] meat,[143] composite dishes,[144] seafood,[145] and soy milk and infant formula.[146]

For the determination of As in food, Fedorov, Ryabchuk, and Zverev[120] reported that HG-AAS produced more precise and accurate results than ETA-AAS. Mindak[147] has determined As and Se using closed-vessel MW digestion with HNO_3 followed by a dry-ash step to convert organoarsenic compounds into hydride-forming species for accurate measurement by HG-AAS. For the same purpose, Ringmann et al.[148] used HNO_3, sodium persulfate, and sodium fluoride as reagents for MW digestion and achieved good results on seafood reference materials. The required prereduction of Se(VI) to Se(IV), usually performed in a water bath, and elimination of interfering nitrogen oxides using amidosulfuric acid were combined by Schloske, Waldner, and Marx[149] into a single 10-minute MW heating after the initial digestion. Kabengera et al.[150] developed a relatively simple method for the determination of As in foods using HNO_3–H_2O_2 digestion and HG by FAAS. Tinggi[151] has developed HNO_3–$HClO_4$–H_2SO_4 digestion using a programmable heating block for determining Se in meat products by HG-AAS. The digestion did not require constant operator attention.

8.3.4 COLD VAPOR

There are many techniques for the determination of Hg,[152] but CV-AAS is still the technique being used primarily for the determination of Hg in food. Hg vapor is generated from the analytical solution using stannous chloride or sodium borohydride and swept into a cell for atomic absorption. Traditionally, the cell was maintained at room temperature, but heating of the cell has improved the reliability of the technique. The term "*hydride generation*" is sometimes incorrectly used to describe this technique when sodium borohydride is used to produce Hg vapor. Standard methods are available for the determination of Hg in fruits and vegetables[153] and in fish.[154] Recent surveys that used CV-AAS include analysis of total-diet foods,[22,105,155] various foods,[109,110] and seafood.[13,84,85,156,157]

Gas–liquid separation is critical to acceptable precision for determining Hg by CV-AAS. Corns et al.[158] demonstrated improvement in long-term stability of Hg measurements using a semipermeable membrane dryer tube to remove moisture from the gas stream. Dabeka, Bradley, and McKenzie[159] developed a method for detecting low levels of Hg in foods using a low-temperature digestion. Larsen et al.[110] applied high-pressure ashing and CV-AAS to analyze a variety of foods for Hg. Zenebon et al.[160] developed a digestion method for Hg in a variety of foods using H_2O_2–H_2SO_4 in open vessels and heating in a water bath at 80°C. The method is stated as being simple, rapid (overnight digestion), sensitive, and inexpensive. Julshamn and Brenna[161] performed a collaborative study of a closed-vessel MW digestion method for determining Hg in seafood by CV-AAS and obtained repro-ducibility relative standard deviations ranging from 7.7% to 16.6%. Tao, Willie, and Sturgeon[162] used tetramethylammonium hydroxide to analyze food-related reference materials for Hg by flow injection CV-AAS with online decomposition of organomercury using potassium permanganate. They stated that large batches of samples could be analyzed, and analyte loss and contamination were considerably reduced. Perring and Andrey[163] validated the CV-AAS method for Hg in food products using high-pressure ashing and MW-assisted digestion. They found stannous chloride to be a more robust reducing agent than sodium borohydride, and the two digestion techniques yielded comparable accuracy and precision.

8.4 DETERMINATION BY INDUCTIVELY COUPLED PLASMA ATOMIC EMISSION SPECTROMETRY

ICP-AES[164] is commonly used for analyzing food, particularly for elements at major and minor levels. The technique's multielement capability makes it useful for dietary studies. Instrument improvements including axially viewed plasmas[165,166] and charge transfer devices as two-dimensional detectors[167] have improved the utility of ICP-AES. The charge transfer devices make the required assessment of background emission and interfering analytes much easier to perform. Examples of standard methods that use ICP-AES are Ca, Cu, Fe, Mg, Mn, P, K, Na, and Zn in infant formula[49]; B, Ca, Cu, K, Mg, Mn, P, and Zn in plants[168]; and P in animal and vegetable fats and oils.[169] Table 8.6 lists references to recent surveys on elements in food that used ICP-AES.

Todolí et al.[175] reviewed the state of knowledge on matrix effects on ICP-AES caused by elements present at high concentrations. They concluded that there is still not a clear understanding of the mechanism responsible for the interferences and

TABLE 8.6
Surveys of Food for Elements That Used Inductively Coupled Plasma Atomic Emission Spectrometry

Food	Elements	Reference
Total diet study	Ca, Cu, Fe, Mg, Mn, K, P, Na, Zn	105
Composite dishes	Al, B, Ca, Cr, Cu, Fe, K, Mg, Mn, Mo, P, Na	144
Infant formula	Al, Ag, As, Ba, Be, Ca, Cd, Co, Cr, Cu, Fe, Hg, Mg, Mn, Mo, Na, Ni, Pb, Sb, Sn, Sr, Ti, Tl, U, V, Zn	170
Fruit, juice, juice products	Ag, Al, As, Ba, Be, Bi, Ca, Cd, Co, Cr, Cu, Fe, Ga, In, K, Li, Mg, Mn, Na, Ni, P, Pb, Rb, Se, Sr, Tl, V, Zn	171
Wild edible mushrooms	Al, Ca, K, Mg, Na, P, Si	172
Rice	Al, Ca, Cu, Fe, K, Mg, Mn, P, Zn	87
Orange juice	Al, B, Ca, Fe, Mg, P, K, Si, Na, Sr, Ti	173
Tea beverages	Al, Ba, Ca, Cu, Fe, K, Mg, Mn, Na, Sr, Zn	174

more efficient methods for overcoming this type of matrix effect are needed. They also stated that quantification by the method of standard additions was probably the best method for overcoming matrix effects in ICP-AES. Mochizuki, Hondo, and Ueda[176] evaluated potential interferences on 24 elements and also advocated using standard additions for quantification. Grotti, Leardi, and Frache[177] studied the interferences on ICP-AES caused by four mineral acids (HCl, HNO_3, H_2SO_4, $HClO_4$) and found that their combined effects were more complex than addition of their single effects. Canals, Gras, and Contreras[178] eliminated HNO_3 interference in ICP-AES using a heated vertical cyclonic spray chamber and a semipermeable membrane dryer tube.

Miller-Ihli[88] demonstrated the capability of ICP-AES for food analysis of Ca, Co, Cu, Cr, Fe, Mg, Mn, Ni, P, V, and Zn by either dry-ash or wet digestion procedures. Dolan and Capar[68] used ICP-AES to determine 20 elements in food using an analytical portion based on the food's energy content and a single MW-assisted digestion program. Carrilho et al.[38] also used ICP-AES to determine 13 elements in plant and animal samples by using a single program for MW-assisted digestion but first drying and grinding the samples prior to digesting a fixed mass. Lomer et al.[179] developed a method using ICP-AES to measure Ti in food for assessing the amount of the food additive titanium dioxide. Food samples were digested using H_2SO_4 heated to 250°C. Perring and Basic-Dvorzak[180] validated a method for determining Sn in canned food using ICP-AES. Both MW-assisted digestion and high-pressure ashing procedures were used and produced comparable results.

Hua, Kay, and Indyk[181] improved the performance of their slurry nebulization ICP-AES method[182] for determining Ca, Cu, Fe, K, Mg, Mn, Na, P, and Zn in infant formulas by using a tertiary amine: EDTA dispersant. The procedure enhanced overall slurry stability for a wide range of samples, improved precision, and improved recovery of Ca, Mg, Cu, Zn, and P.

8.5   DETERMINATION BY INDUCTIVELY COUPLED PLASMA MASS SPECTROMETRY

ICP-MS[183] has become an established multielement determinative technique since its first development in the early 1980s. The technique is capable of detecting elements at the trace and ultratrace levels with a large dynamic range. Compared with ICP-AES, there is a sensitivity improvement of about two to three orders of magnitude; however, the instrument is more costly and complex to operate, and there is potential of spectral interferences from molecular species. To suppress isobaric interferences using a quadrupole ICP-MS, collision and reaction cells[184] or a dynamic reaction cell[185] are now being used. A double-focusing magnetic-sector ICP-MS is still the ultimate means of overcoming these isobaric interferences, but the cost of the instrument is much higher than that of other ICP-MS instruments. Todolí and Mermet[186] provided an extensive review of the interferences in atomic spectrometry caused by acids and outlined a possible strategy to minimize the effects of acids on ICP-AES and ICP-MS. Table 8.7 lists references to recent surveys on elements in food that used ICP-MS.

TABLE 8.7

Surveys of Food for Elements That Used Inductively Coupled Plasma Mass Spectrometry

Food	Elements	Reference
Total diet	As, Cd, Pb, Hg, Se, Sn, Zn	23
Total diet	As, Cd, Pb	155
Total diet	Br, I	187
Duplicate diet	As, Cd, Cr, Pb	10
Duplicate meals	Al, As, Ca, Cd, Co, Cr, Cu, Fe, Hg, K, Li, Mg, Mn, Mo, Na, Ni, Pb, Sb, Se, Sn, Zn	188
Fish, food (from plants)	I	189
Infant formula, milk	Ba, Be, Bi, Cd, Co, Cs, Cu, La, Li, Mn, Mo, Pb, Rb, Sb, Sn, Sr, Tl, Zn	190
Dairy products	I	191
Cabbage, sprouts	Ag, Al, Au, B, Ba, Be, Bi, Cd, Ce, Co, Cr, Cs, Cu, Dy, Er, Eu, Fe, Ga, Gd, Hf, Hg, Ho, In, Ir, La, Li, Lu, Mg, Mn, Mo, Nb, Nd, Ni, P, Pb, Pd, Pr, Pt, Rb, Re, Rh, Ru, S, Sb, Sc, Se, Si, Sm, Sn, Sr, Ta, Tb, Te, Th, Ti, Tl, Tm, U, V, Y, Yb, Zn, Zr	192
Onions, peas	Ag, Al, Au, B, Ba, Be, Bi, Ca, Cd, Ce, Co, Cr, Cs, Cu, Dy, Er, Eu, Fe, Ga, Gd, Ge, Hf, Ho, In, Ir, K, La, Li, Lu, Mn, Mo, Na, Nb, Nd, P, Pb, Pr, Pt, Rb, Re, Ru, S, Sb, Sc, Se, Si, Sm, Sn, Sr, Tb, Te, Th, Ti, Tl, Tm, U, V, W, Y, Yb, Zn, Zr	193
Orange juice	Ag, Al, As, B, Ba, Bi, Ca, Cd, Ce, Co, Cr, Cs, Cu, Eu, Fe, Ga, Gd, I, Li, Mg, Mn, Mo, Na, Nd, Ni, Pb, Pr, Rb, Sb, Sc, Se, Si, Sm, Sn Sr, Th, Ti, Tl, V, W, Y, Zn, Zr	194
Orange juice	Ba, Co, Cu, Li, Lu, Mn, Mo, Ni, Rb, Sn, V, Zn	173
Potato	Ag, Al, Au, Ba, Bi Ca, Cd Co, Cr, Cs, Cu, Dy, Er, Fe, Ga, Gd, Ho, In, Ir, La, Lu, Mn, Mo, Nb, Nd, P, Pb, Pd, Pr, Pt, Rb, Re, Rh, Ru, Sb, Sc, Sm, Sn, Sr, Ta, Tb, Th, Ti, Tl. Tm, u, V, Y, Yb, Zn	195
Rice	As, Cd, Mo, Ni	87
Wild edible mushrooms	Ag, Rb, Cd, Hg, Pb, Cs, Sr, Tl, In, Bi, Th, U, Ce, Pr, Nd, Sm, Eu, Gd, Tb, Dy, Ho, Er, Tm, Yb, La, Lu, Zn, Fe, Cu, Mn, Ba	172

Crews[196] described the advantages and disadvantages of ICP-MS and provided examples of food analysis. Salomon, Jenne, and Hoenig[197] addressed analytical problems beyond those associated with routine analysis by ICP-MS. They discussed problems such as inadequate sample dilution, inappropriate calibrations, and insufficient representativity of procedure blanks. Baxter et al.[198] described the quality control criteria they used, based on many years of experience, using ICP-MS for the analysis of foods.

Baker, Bradshaw, and Miller-Ihli[199] analyzed food-related reference materials for 13 elements using a quadrupole ICP-MS. Samples were prepared using

$HNO_3-H_2O_2$ digestion in quartz tubes with a block heater set to a maximum of 100°C. A determined correction factor was used for the interference of Ca on Fe. Cubadda et al.[200] developed a routine method for quadrupole ICP-MS determination of 15 elements in food-related reference materials using MW-assisted digestion. Spectral interferences caused by C, Cl, and Ca were corrected by applying correction factors and a protocol for routine high-throughput analysis of food was provided. Martino, Sánchez, and Sanz-Medel[201] investigated the use of double-focusing magnetic-sector ICP-MS for the determination of 15 elements in milk products using MW-assisted digestion. The double-focusing magnetic-sector ICP-MS overcame many of the polyatomic interferences using quadrupole ICP-MS. Zhang et al.[202] compared ICP-MS and ETA-AAS for the determination of Cd and Pb in duplicate diets and found the ICP-MS analysis to be much faster and as accurate and precise as ETA-AAS, but Pb intakes based on the ICP-MS results were 10%–20% lesser. However, they concluded that ICP-MS method could be used as a routine analytical method for food analysis. Zbinden and Andrey[203] developed a routine method for food analysis using a high-pressure ashing device and ICP-MS for the determination of Al, As, Cd, Hg, Pb, and Se in food. They overcame the problem of residual carbon interference on As, Pb, and Se by adding isopropanol to the analytical solutions. Bhandari and Amarasiriwardena[204] developed an ICP-MS method for the determination of Pb and seven other elements in maple syrup using MW-assisted digestion. Residual carbon content was suitable for ICP-MS determination. D'Ilio et al.[205] studied the analysis of rice by double-focusing magnetic-sector ICP-MS for the determination of As and nine other elements and compared the results to those obtained by either quadrupole ICP-MS or ICP-AES. They discussed spectral interferences associated with ICP-MS and the effectiveness or need for using double-focusing magnetic-sector ICP-MS to accurately measure the elements.

Rädlinger and Heumann[206] developed a relatively fast method for the determination of I in food products using ICP-MS with quantification by isotope dilution. They studied the extraction of I by tetramethylammonium hydroxide and MW-assisted digestion with HNO_3-HClO_4. Both digestions were considered suitable for routine analysis although the HNO_3-HClO_4 digestion was faster and provided a more complete dissolution of the sample. Fecher, Goldmann, and Nagengast[207] developed a method for the determination of I in dietetic foods using a tetramethylammonium hydroxide extraction with heating up to 90°C. An interlaboratory trail demonstrated the good performance of the method for the product studied. The authors cautioned that soluble I can be extracted by the procedure, but results for matrices other than those studied must be verified. Julshamn, Dahl, and Eckhoff[208] developed an ICP-MS method for the determination of I in seafood using a MW-assisted HNO_3 digestion with ammonia as a stabilizer prior to determination. They discussed many of the problems associated with the determination of I by ICP-MS. Haldimann, Eastgate, and Zimmerli[209] developed an ICP-MS method for the determination of I in food using isotope dilution and miniature sample introduction systems to reduce the long washout time associated with I and to allow direct determination of I from digestions using only HNO_3. A number of food-related reference materials produced good results, and results for total-diet food samples were in agreement with those obtained by neutron activation analysis.

8.6 OTHER DETERMINATIVE TECHNIQUES

8.6.1 ELECTROCHEMICAL

Except for the determination of F by ion-selective electrode (ISE), routine determination of elements in food by electrochemical techniques has been used less often than spectrometric techniques. Other electrochemical techniques, such as stripping voltammetry, provide a viable alternative for determining many elements in food.[43,210] Standard methods using electrochemical techniques include the determination of I in infant formula[211] and F in plants[212] using ISE; Cd and Pb in food[213] using anodic stripping voltammetry; and Zn in fruits and vegetables using polarography.[214] Recent surveys that used electrochemical techniques include analysis of total-diet foods,[104] fish and staple foods,[215] and mechanically separated chicken[216] for the determination of F by ISE, coffee for Cd, Cu, and Pb by differential pulse anodic stripping voltammetry,[12] mushrooms for Hg by anodic stripping voltammetry,[217] and canned evaporated milk for Cu, Pb, and Zn by alternating current polarography.[218]

Malde, Bjorvatn, and Julshamn[219] developed a simple method for measuring F in food using alkali fusion with sodium hydroxide and determination by ISE. They studied the influence on the results of the ashing aid, sample mass, and analytical solution storage time. They validated the method by analysis of reference materials, seafood products, and tea and recovery of added F to samples.

Ratana-Ohpas et al.[220] used constant current stripping analysis and potentiometric stripping analysis for the determination of Sn in fruit juices. The analytical solution was prepared by adding HNO_3 and HCl to the juice and diluting with water. An analysis could be performed in about 10 minutes, and results from both the electrochemical techniques agreed satisfactorily with those obtained from FAAS. Qiong et al.[221] developed a method for the determination of Sn in canned fruit juices using single-sweep polarography in a solution of oxalic acid–methylene blue. Juices were digested with HNO_3, and results for Sn agree with those obtained using a spectrophotometric method. Guanghan et al.[222] developed a single-sweep polarographic procedure for the determination of F by forming a complex with La(III)-alizarin complexone. Samples of maize and rice were dry ashed, and then F was distilled from the ash solution. The results agreed with those obtained from ISE, and the recovery of added F was good. Yao, Chen, and Wei[223] developed a procedure for determining I in food using a quartz crystal microbalance. The free iodine was extracted into carbon tetrachloride and absorbed at the gold electrodes of the microbalance, where the I level was estimated by a decrease in the quartz crystal's oscillation frequency. Various food samples and a reference material were dry ashed with potassium hydroxide and zinc sulfate. Analytical results compared well with those obtained from a gas chromatographic method, and a small interlaboratory trial of the method produced good results. Lo Coco et al.[224] determined Cd and Pb in wheat by derivative potentiometric stripping analysis and prepared the samples using a dry-ash procedure with magnesium nitrate ash aid. Recoveries of added Cd and Pb were good, and the results compared well with those obtained from ICP-MS. Lambert and Turoczy[225] compared various digestion methods for the determination of Se in fish by cathodic stripping voltammetry. Of the digestion procedures evaluated, only a combination of wet and dry-ashing procedures produced reliable results. Inam and Somer[226] analyzed milk for the determination of Pb and Se by differential pulse stripping voltammetry using an HNO_3–$HClO_4$ digestion, with care to prevent charring and possible loss of Se during the digestion. Sancho et al.[227] determined Co and Ni in refined beet sugar by adsorptive cathodic stripping voltammetry as their dimethylglyoxime complexes. The samples were simply dissolved in water for analysis. Sancho et al. also applied stripping voltammetric methods for the determination of As and Cu,[228] Cd, Pb, and Zn,[229] and Hg[230] in refined beet sugar.

8.6.2 COLORIMETRY

Colorimetry is used for the routine determination of I in food using methods by either Moxon and Dixon,[231] based on the catalytic effect of iodide on the destruction of the thiocyanate ion by the nitrite ion and using an alkaline dry-ash sample preparation or Fischer, L'Abbé, and Giroux,[50] based on the iodide-catalyzed reduction of Ce(IV) by As(III) and using wet digestion sample preparation. Both methods require the use of automated analyzers. Perring, Basic-Dvorzak, and Andrey[232] modified the method by Moxon and Dixon for the analysis of inorganic I in fortified culinary products. They reduced the rate of the colorimetric reaction that allowed detection of I without an automated analyzer and preparation of the culinary products by dissolving in water. Recoveries were good, and the results by the method were comparable to those obtained by ICP-MS. Recent surveys that used colorimetry to determine I include analysis of total-diet foods[23] and dairy products.[233]

8.6.3 NEUTRON ACTIVATION ANALYSIS

Neutron activation analysis is a multielement technique that is used for the analysis of food.[234–236] The nuclear properties of the technique have different interferences than normally used atomic-based techniques and offer a means for an independent confirmation analysis. For neutron activation analysis, a source of neutrons is necessary, which requires access to a nuclear reactor or other source of neutrons. The sample is exposed to neutrons, which results in the formation of radioactive isotopes of the element, and the gamma rays emitted during the decay of these isotopes (or isotopes of decay products of the element) are measured by gamma-ray spectrometry for quantification. The most common neutron activation techniques applied to routine food analysis are instrumental neutron activation analysis (INAA), radiochemical neutron activation analysis (RNAA), neutron capture prompt gamma-ray activation analysis, and epithermal instrumental neutron activation analysis.[237] Table 8.8 lists references to recent surveys on elements in food that used nuclear techniques.

A formalized method for determining Na in food and other related materials by INAA was validated by an interlaboratory trial.[249] The method minimized the number of procedural options but allowed application to a variety of neutron activation facilities. Balaji et al.[250] used INAA with a single-comparator method to analyze cereals and pulses for 15 elements (Al, Ca, Cl, Co, Br, Fe, K, La, Mg, Mn, Na, Rb, Se, Sm, and Zn). The single-comparator method used gold as a comparator instead

TABLE 8.8
Surveys of Food for Elements That Used Neutron Activation Analysis

Food	Elements	Reference
Total diet	B, Cl, H, K, Na, S	238
Total diet	Ca, Co, Cr, Cs, Fe, I, K, Se, Sr, Th, Zn	239
Total diet	Al, As, Br, Ca, Cl, Co, Cr, Cs, Fe, I, K, Mg, Mn, Na, Rb, Sc, Se, Sm, Sr, Th, U, Zn	240
Total diet	I, Se	81
Duplicate diet	Hg	11
Duplicate diet	Br, Ca, Cl, Co, Cr, Cs, K, Fe, Mn, Mg, Mo, Na, Rb, Se, Zn	241
Duplicate diet	Se, Zn	242
Dried desserts	Sb, Cr, Co, Fe, Mn, K, Rb, Sc, Na, Zn	14
Foods	Al	243
Foods	Ca, Cl, Co, Cr, Cu, Fe, K, Mg, Mn, Mo, Na, Se, Sn, Zn	244
Salt	I	245
Coffee	Ba, Ce, Co, Cr, Cs, Eu, Fe, Hf, La, Rb, Sc, Zn	86
Coffee	Al, As, Ba, Ca, Ce, Co, Cr, Cs, Dy, Eu, Fe, Gd, Hf, K, La, Lu, Mg, Mn, Na, Rb, Sb, Sc, Se, Sm, Sr, Ta, Tb, Th, Ti, Tm, U, V, Yb, Zn	246
Cereals, oils, sweeteners, vegetables	Br, Ca, Cl, Co, Cu, I, K, Mg, Mn, Na, Rb, S, Ti, V	247
Dairy products	Co, Cr, Fe, Rb, Se, Zn	248

of a multielement standard for quantification of the elements. Kucera, Randa, and Soukal[251] studied three activation analysis procedures for determining I in food. Accuracy and precision estimates were developed using food-related reference materials, and the lowest I detection limit achieved was 1 μg/kg. Chao et al.[252] evaluated the detection of I using NAA under various irradiation and measurement conditions and various interfering levels of Na and Cl.

8.7 ELEMENT SPECIATION

8.7.1 INTRODUCTION

Investigations of trace elements species in biological matrices are very important from both environmental and food points of view. Speciation analysis is aimed at determining individual chemical forms of mineral components characterized by different bioavailability, mobility, and toxicity. Recently, speciation has become the main topic of interest to ensure food safety and to determine nutritional quality and is becoming an increasingly important tool in defining and guaranteeing food safety.[253] The knowledge of the total level of trace elements in food is very important for establishing dietary requirements; however, the concentration of the individual element species in food is also needed to evaluate the safety and nutritional quality of food. Although most regulations concerning trace elements in food are based on the total element concentration, only a few regulations address specific forms of chemical elements. However, organizations such as the EPA, the World Health Organization, European Environmental Commission, and the U.S. Food and Drug Administration seem to recognize the great importance of element speciation[254] and are just starting to recommend and/or regulate speciation analysis for As, Hg, or Sn.[255] The importance of trace element speciation in nutritional sciences has been overviewed.[256]

The determination of various oxidation states of so-called inherently unstable trace elements (e.g., Cr, Fe, As, Se) or inherently stable organometallic species (e.g., tributyltin [TBT], methylmercury [CH_3Hg], arsenobetaine [AB], monomethylarsonic acid [MMA], methylead) are of great interest for environmentalists and food scientists and constitute an analytical challenge. Because the species are mostly unstable and occur in different matrices at extremely low quantities (e.g., µg/kg), sensitive and selective methods should be applied to the quantitative isolation and determination of given element species in biological materials including food products.[257] The most popular chromatographic methods, such as gas chromatography (GC) and high-performance liquid chromatography (HPLC), are coupled to the powerful detectors—for example, AAS, mass spectrometry (MS), flame photometric detector (FPD), ICP-AES, ICP-MS, and atomic fluorescence spectrometry (AFS). The use of ICP-MS coupled to various separation techniques—such as liquid chromatography (LC), GC, supercritical fluid chromatography, (SFC), and capillary electrophoresis (CE)—for elemental speciation has recently been paid a great attention.[255,258–263] In the past decade, electrospray ionization mass spectrometry (ESI-MS) has been used in elemental speciation (e.g., ionspray-MS/MS) for the quantification of TBT in a certified reference material (CRM).[264] Hyphenated speciation techniques are characterized by adequate selectivity and sensitivity for specific forms of chemical elements in drinking water and solid samples including food.[253,265–272] Hyphenated techniques have been widely used for the analytical quantification of species of essential and nonessential elements—for example, Fe, Zn, Cu, Co, V, Pt, Ag, Ca,[273–279] As,[148,253,274,275,280–287] Sb,[288–290] Mn,[291] Cr,[253,274,292–297] Mo,[292] Hg,[253,298] Se,[253,299,300] Cd,[274,275,301–303] Pb,[274,304–310] Ni,[274,275,311] Tl,[312] Sn,[255,313,314] Te,[315] S, N, P, Cl, I,[254,274,316,317] and Al.[274,318]

Very labile element species may occur on time scales significantly shorter than those typically used for off-line analysis. Therefore, exclusively fast and noninvasive techniques should be used (e.g., ISEs or real-time spectroscopic methods).[319]

Individual steps for speciation analysis are as follows: sampling, conservation and storage, subsampling, extraction or digestion, derivatization, separation, detection, calibration, calculation of results, and reporting.[298,319,320]

Application of isotope dilution MS in elemental speciation is discussed by Huo, Kingston, and Larget.[321] According to Encinar,[322] isotope dilution analysis will likely be applied in determining the nutritional and clinical applications of different element species used as additives in the treatment of patients and in supplemented diets.

The need for appropriate CRMs for speciation analysis has been justified by Quevauviller[323] and Emons.[324] CMRs are an important and often necessary component of quality assurance and control (QA/QC) systems for speciation analysis (e.g., in food control and environmental biomonitoring). A typical sample matrix of mussel tissue and fish muscle used in marine monitoring might also be appropriate for use as human diet material.[324]

8.7.2 Arsenic

Arsenic, as a nonessential metalloid, occurs in inorganic and organic forms exhibiting significant differences in its toxicity. Inorganic As species are more toxic than organic As species. Based on the toxicity, As compounds can be arranged as follows: As(III), As(V), MMA, dimethylarsinic acid (DMA), trimethylarsine oxide (TMAO). Organic As compounds, such as AB and arsenocholine (AC), occurring mainly in seafood (e.g., edible molluscs, shrimp, lobster, algae), are considered nontoxic.[253,255] However, transformations of As in marine food chains are still being studied, and therefore, food safety assessment requires data not only on total As content but also on the As species in a specific food.[253] For instance, in fish products, the total concentration of As is relatively high, and a great percentage is the nontoxic arsenobetaine. Consequently, consumers do not risk consuming excessive amounts of As with fish. On the other hand, the lack-of-toxicity data for organoarsenic species such as the arsenosugars does not allow risk assessment for these compounds. Therefore, simple and validated analytical methods and legislation of As based on its species are needed for the proper evaluation of the quality of fish and fish products.[325]

Many analytical procedures have been reported for the separation and determination of As species in biological matrices including food products. A number of these methods are based on gas or liquid chromatographic techniques coupled with spectroscopic or electrochemical detection,[253,326] while AAS, ICP-AES, and ICP-MS were the most popular and powerful detectors.[260,275] These specific element detectors have been widely coupled with GC or liquid chromatography (HPLC) for the separation and quantitative detection of As species.[275] ICP-MS coupled with HPLC or CE has recently become more frequently used detection technique for the determination of specific forms of As. Several species of As, namely As(III), As(V), MMA, DMA, AB, TMAO, AC, and tetramethylarsonium ion (TMI) have been determined by techniques based on coupled anion- and cation-exchange HPLC-ICP-MS. For instance, Wrobel et al.[327] determined As species in fish tissues using HPLC-ICP-MS. This rapid and sensitive technique has also been used for the speciation analysis of As in edible scallops including quality assurance analyses of reference materials (e.g., dogfish muscle, bovine liver, and oyster tissue).[328] A gradient anion-exchange HPLC-ICP-MS method has been developed for the separation and determination of As species in lobster.[329] Differences between recoveries of the various As species have resulted from differences in parameters such as the sample handling, extraction technique, and/or conditions of the extraction applied.

The HG technique has been widely used in the determination of chemical forms of As in biological matrices. One of the major routes for As entering humans is by the consumption of fish and seafood.[254] Speciation of As in manufactured seafood products has been studied by the ion-exchange HPLC-HG-AAS method.[255] Chatterjee[330] applied MW-assisted extraction followed by HPLC-ICP-MS for the quantification of As species in oyster tissue, that is, phosphate-arsenosugar, AB, DMA, inorganic As, methylarsonic acid, AC, TMAO, and TMI. Multidimensional liquid chromatography with parallel ICP-MS and electrospray MS/MS detection has been used for determining As species in Laminaria algae and oysters.[331,332] This analytical strategy based on ICP-MS and ES MS/MS seems to be the most comprehensive of the published research concerning the characterization of marine biological material and relevant food carried out by LC coupled to dual ICP and ES MS/MS detection. This procedure allows the establishment of a complete map of arsenic species in a natural sample. The purification technique SE-AE was used for arsenosugars A, C, D, As(V), and DMA; SE-AE-RP was used for arsenosugars B and G.[332]

An HPLC-HG-AFS method for the direct determination of As(III), DMA, MMA, and As(V) in beers has been developed by Coelho et al.[333] An HPLC-MW-HG-AAS coupled method has been described for the above-mentioned species of As and also for AB and AC separation and determination in fish samples.[334] Interlaboratory studies have been organized by the Standards, Measurements, and Testing Programme in order to certify the concentrations of As species in food-related samples (e.g., fish and mussel tissue).[335–337] The analytical methods applied in the certification study were as follows: HPLC-UV-HG-ICP-OES, GC-CT-HG-QFAAS, HPLC-QFAAS (DMA), HPLC-ICP-MS, and HPLC-ICP-OES (AB). The overview by Gong et al.[338] indicates that methods based on HPLC separation with detection by ICP-MS, HG-AAS, and ES-MSD appeared the most useful for As speciation in biological matrices including foodstuffs. These hyphenated techniques for extracting As species from solid samples are required for obtaining reliable information on speciation of As. According to Van Dael,[253] current speciation techniques are sufficiently selective and sensitive for As speciation in food and drinking water.

8.7.3 Mercury

Organomercury compounds are characterized by very high toxicity. Significant bioaccumulation of methylmercury (CH_3Hg) in seafood has resulted in a serious food safety problem as fish have the ability to concentrate CH_3Hg in their muscle by a factor of 105–107. This can lead to dangerously elevated levels of Hg in seafood, even in regions with typical aquatic Hg levels. The U.S. Food and Drug Administration established a guideline for CH_3Hg in seafood at a level of 1 mg/kg.[253] In consequence, the need for routine species analysis of Hg by control laboratories has recently stimulated developing different analytical methods mostly based on coupling of reversed-phase HPLC to CV-AAS or CV-ICP-MS.[255] The usual applications of GC-ICP-MS for biological samples concern the determination of CH_3Hg.[339] Direct GC with electron capture detection (GC-ECD) has been applied by Mizuishi et al.[340] for the analysis of CH_3Hg in fish samples. The authors used HBr-methanol-treated capillary columns; the detection limits were about 5 μg/L on the packed column and 5 μg/L on the capillary. Pereiro and Diaz[341] described the current status of a rapid and low-cost GC-MW-induced plasma atomic emission detector (MIP-AED) specifically useful for routine determination of Hg, as well as Sn and Pb species, in environmental control laboratories. Multicapillary GC coupled with an ICP-MS detector has been reviewed by Ackley, Sutton, and Caruso.[259] The use of HPLC with cold vapor generation (CV) coupled with AAS allows for the detection of Hg species in concentration range between 0.1 and 1 ng Hg (II); as for CH_3Hg and C_2H_5Hg, the detection is dependent on the system. However, generation of a CV from organic compounds of Hg requires an extra step for conversion to Hg(II), usually online using, for example, UV irradiation.[342]

Application of MIP-AES as an element-specific detector for speciation analysis has been reported by Heltai et al.[343] Headspace analysis using GC-MIP-AES has been used for the determination of CH_3Hg in fish tissue.[344] Headspace and solvent extraction techniques with GC-AFS have been applied in the certification of CH_3Hg in reference biological materials such as tuna fish and mussel *Mytilus edulis*.[345] The accuracy of both these techniques appeared to be satisfactory. GC with a MW-induced plasma atomic emission spectrometer (GC-MIP-AES) has also been applied for the determination of CH_3Hg in fish tissue.[346] The detection limit was estimated to be 0.1 mg/kg of CH_3Hg (as Hg). A better detection limit was obtained by Grinberg et al.,[347] who used solid-phase microextraction combined with tandem GC and furnace atomization plasma emission spectrometry (SPME-GC-FAPES) for the quantification of CH_3Hg and inorganic Hg in fish samples; the detection limits amounted to 1.5 and 0.7 µg/kg. An electrothermal vaporization with ICP-MS (ETV-ICP-MS) permitted the quantitative speciation of Hg without sample pretreatment and avoiding the risk of species interconversion (CH_3Hg and Hg(II) into elemental Hg).[348] Such transformation was observed during derivatization preceding GC-ICP-MS. ETV-ICP-MS was recommended for the Hg speciation in fish samples; the limits of detection were 2 and 6 µg/kg for CH_3Hg and inorganic Hg, respectively.

Multicapillary GC has become the preferred sample introduction technique for the time-resolved introduction of elemental species to selective detectors owing to its high resolution and high sensitivity. This technique has been successfully applied by Rosenkranz and Bettmer[349] for rapid separation of Hg species in fish and seafood (lobster). A novel automatic system based on the line coupling of HPLC separation, postcolumn MW digestion, and cold vapor atomic fluorescence spectrometry detection has been recommended by Liang et al.[350] for the determination of four Hg species: mercury chloride, methylmercury chloride (MMC), ethylmercury chloride, and phenylmercury chloride. The method has been successfully applied to determine MMC in seafood.

A critical review of the different strategies for Hg speciation investigations in environmental and seafood-related samples has been carried out by Sanchez Uria and Sanz-Medel.[351] As can be concluded from an overview by Van Dael,[253] both hyphenated speciation techniques (e.g., CV-AAS, CV-AFS, FT-IR, and MIP-AES, or furnace atomization plasma emission spectrometry) and less modern speciation methods offer adequate selectivity and sensitivity for Hg speciation in food and are sufficiently sensitive to comply with food safety assessment.

8.7.4 SELENIUM

Because the toxic properties of Se are dependent on its speciation, this has stimulated the development of analytical procedures for the determination of Se compounds in biological and nutritional samples.[255] The range between safe and toxic doses of Se is narrow, and the bioavailability and toxicity of Se are dependent on the Se species in the diet. The absorption of Se has been established to be higher from organic compounds of Se; hence, the knowledge of Se speciation in foods is important for a better understanding of utilization of this metalloid. From the overview by Van Dael,[253] speciation of Se and its distribution in foods have been reported for selected food items—for example, soybean, wheat, enriched garlic, onion, broccoli, Se-rich yeast, cooked cod, and milk (human, cow, goat, and sheep).

Many methods based on GC coupled with AAS, AFS, AES, or ICP-MS have been used for the quantification of Se species in different sample matrices. The coupling of pervaporation–atomic fluorescence (PV-AFS) has been proposed by Moreno, Pérez-Conde, and Cámara[357] for the identification of these species and pervaporation–gas chromatography–atomic fluorescence (PV-GC-AFS) for their individual quantification. The recommended PV-GC-AFS method (complementary to the previously used PV-AFS) was applied to the analysis of volatile Se species in garlic, allowing their direct determination within the concentration range of 0.5–1 mg/kg (as Se). The PV technique appears to be an excellent tool to test the presence of volatile Se species and an alternative to headspace for their extraction from a solid matrix such as garlic without any further pretreatment and introduction into a capillary GC.[352]

An extraction procedure for five Se species based on enzymatic treatments (protease XIV, lipase VII, and protease VIII) of biological material has been performed using HPLC-MAD-HG-AFS.[285] Enzymatic digestion was an easy step to release the Se species from the samples analyzed; however, the extracts must be submitted to a cleanup procedure to be suitable for HPLC-AFS analysis. Moreover, recoveries of Se species were low, resulting possibly from the incomplete extraction of these compounds from the complex matrix of shellfish.[285]

Anion-exchange HPLC-ICP-MS has been used to determine Se species extracted from cooked fish (cod) by enzyme digestion at pH 6.8.[255] Kannamkumarath et al.[353] used hyphenated techniques (e.g., HPLC-ICP-MS) for the determination of Se distribution and speciation in different types of nut after enzymatic hydrolysis (proteinase K). Separation of proteins was accomplished by size-exclusion chromatography (SEC) and the results obtained showed that about 12% of total Se was weakly bound to proteins. Hydrolysis of proteins with methanesulfonic acid and an improved HPLC-ICP-MS technique have been applied[354] to the determination of selenomethionine (Se-M) in yeast and nuts. Better cleavage of Se-M was attained using this technique compared to the enzymatic hydrolysis. Modern trends in the speciation of Se by the hyphenated techniques have been evaluated by Uden.[355] From this overview, the reversed-phase ion-pair HPLC and ion chromatography ICP-MS are powerful techniques for the separation and determination of Se species in selenium-enriched yeast, garlic, and mushroom, and selenized yeast and algal extract.

Ion-pair reversed-phase HPLC-ICP-MS has been used for speciation analysis of Se-Cys, Se-M, and Me_3Se^+ in biological samples (e.g., yeast).[255] According to the latter authors, reversed-phase chiral HPLC coupled to an MW-HG-ICP-MS detector has been applied for determining Se species (e.g., D,L-selenomethionine) in enriched yeast. A flow injection system used in conjunction with HG-AAS, and MW-aided prereduction of Se(VI) to Se(IV) with HCl:HBr has been developed for quantifying these two inorganic species.[356] This sensitive method with good linearity, reproducibility, and high sampling frequency has been applied to the determination of the Se species in orange juice samples. Van Dael reported[253] that HPLC-ICP-MS is an effective technique for the identification of five Se species (i.e., Se(IV), Se (VI), Se-M, Se-Cys, and Met-Se-Cys) and other (several) species in the Se-enriched garlic, onion, and broccoli.

A review by Guerin, Astruc, and Astruc[357] describes the most commonly applied speciation analyses that were or could be used to determine Se and As species in different matrices including foods such as fish, mussels, vegetables, mushrooms, and mineral water. Recently, CE has made much progress in speciation analysis of Se in environmental and biological samples—for example, in nutritional supplements and human milk.[358] A three-step sample preparation method based on the proteolytic enzymatic process for speciation analysis of Se-enriched edible mushroom *Agaricus bisporus* connected with the HPLC HHPN (hydraulic high-pressure nebulizer)-AFS system has been applied.[359] Improved Se extraction by sequential enzymatic processes for its speciation in samples of this species of mushroom has been developed by Dernovics, Stefánka, and Fodor.[360] According to Van Dael,[253] current speciation techniques offer good selectivity and sensitivity for the determination of Se species in drinking water and food.

8.7.5 TIN

Sn does not constitute any serious problems in foods except an occasional elevated total level in some canned products, usually acidic foods.[180,361] Speciation of organotin compounds, which have been used for many years as (for example) antifouling paint agents, is recently of great interest owing to their high toxicity to fish, crustaceans, and molluscs. TBT is highly bioaccumulated in fish and other edible marine organisms.[362] According to an overview by Belfroid, Purperhart, and Ariese,[363] tolerable average residue levels (TARL) for TBT in seafood are exceeded in the case of one or even more samples in 9 of the 22 countries (i.e., Canada, France, Italy, Japan, Korea, Poland, Taiwan, Thailand, and the United States). However, for most countries, there is no available information on organotin levels in seafood. This overview advocates adoption of maximum residue limits (MRLs) for TBT in seafood; hence, there is a need to monitor TBT levels more regularly.[363] In order to determine organotin species, many analytical methods couple GC separation to spectroscopic detection. For obtaining volatile compounds, derivatization procedures such as HG[364] and alkylation by Grignard reagents[365,366] or by sodium tetraethylborate[364] are successfully applied to determination of organotin compounds. The HPLC methods based on specific interactions such as reversed-phase, ion-pair, and ion-exchange micellar have been coupled to the detectors ICP-AES and ICP-MS for organotin speciation.[255] A rapid HPLC-HG-ICP-AES method has been applied to determine the concentration of organotins compounds in molluscs.[255] GC-FPD has been used for determination of butyltins, mainly TBT, DBT, and MBT in marine molluscs,[366,367] fish,[367,368] and fish-eating birds.[368] Determination of butyltins (except MBT) in marine molluscs has also been achieved using an ion-trap detector (ITD) coupled to GC.[364] A GC coupled to a mass selective detector with multiple selected monitoring (GC/MS-SIM) has been successfully adopted by Nemanic et al.[369] for speciation analysis of pentylated organotins in mussel and water. The method was characterized by satisfactory sensitivity; the limit of detection was in the µg/kg (as Sn) range and the RSD lower than 10% for the mussel analyzed.

8.7.6 OTHER CHEMICAL ELEMENTS

A better understanding of the distribution of various species of trace elements in foods and their subsequent behavior in the gut should be very helpful in making decisions concerning dietary requirements and related legislation.[361] These requirements apply to all analyses, including the determination of both the total trace element concentrations and concentrations of their chemical forms.[361] Important topics concern the migration of Ni and Cr from food contact materials, although there is a lack of information on their species in normal diet.[325]

Simultaneous speciation analysis of multiple chemical elements such as As, Se, Sb, and Te in fish extracts has been performed by online coupling of anion-exchange HPLC with ICP-MS.[315] SEC with elemental specific detection is frequently applied to the analysis of trace element species in protein-rich matrices (e.g., extracts of animal and plant tissues). Therefore, the technique can successfully be used in the analysis of specific food matrices. For instance, Cu and Zn species have been analyzed in legume samples (common white bean, chickpea and lentil seeds, and defatted soybean flour) by online hyphenation of SEC and ICP-MS.[278] The main advantage of the SEC/ICP-MS is the simplicity of application. The differences between isotope dilution (ID) and external calibration (EC) methods were the same order of magnitude as the precision and therefore did not play a significant role in practical use.[278] SEC coupled with ICP-MS has also been used to investigate the speciation of Cu, Cd, Zn, Se, As, and Ca and Fe, Zn, Cu, Ag, Cd, Sn, and Pb in tissue extracts of fish (largemouth bass) and mussel, respectively.[276,303] The use of this multielement speciation method should explain any interactions between multiple groups of trace elements within the various protein fractions of biological matrices such as fish muscle or mussel soft tissue.

Another example of successful application of SEC was a speciation study of Fe in milk and infant formulas whey by SEC-SEC-HPLC ETA-AAS.[370] The limit of detection was 1.4 µg/L; Fe was bound principally to the proteins attributable to its molecular weight of 3 and 76 kDa in breast milk. A systematic study concerning Cd species in two contaminated vegetable foodstuffs has been carried out by gel permeation chromatography (GPC) with ET-AAS.[302] The most important Cd-binding form in cytosols of all the plants analyzed was found to be the high-molecular-weight Cd species (HMW-Cd-SP, 150–700 kDa).

Online determination of Sb(III) and Sb(V) in liver tissue has been performed by flow-injection HG-AAS.[289] Methods of Cr analysis with discussion of unique techniques of sampling, storage, handling, and separation for its speciation have been discussed in detail.[294] An overview of the application with developing trends of CE to simultaneous separation and determination of various species of inorganic elements has been presented by Timerbaev.[371] According to Crews,[301] speciation analyses of foods must have multi-disciplinary character and should not be carried out in isolation. Because the analytical methods available have become more sophisticated and very expensive, further developments will be possible in the case of collaboration between researchers, institutes and even between countries.[301]

REFERENCES

1. Murphy, S.P., Dietary reference intakes for the U.S. and Canada: update on implications for nutrient databases, *J. Food Compos. Anal.*, 15, 411, 2002.
2. Berg, T. and Licht, D., International legislation on trace elements as contaminants in food: a review, *Food Addit. Contam.*, 19, 916, 2002.
3. Berg, T. and Larsen, E.H., Speciation and legislation—where are we today and what do we need for tomorrow? *Fresenius J. Anal. Chem.*, 363, 431, 1999.
4. Crews, H.M., "Trace element analysis for food authenticity studies." In *Analytical Methods Food Authentication*, Ashurst, P.R., and Dennis, J.J., Eds., Blackie, London, 1998.
5. Standing Committee on the Scientific Evaluation of Dietary Reference Intakes, *Dietary Reference Intakes for Calcium, Phosphorus, Magnesium, Vitamin D and Fluoride*, National Academy Press, Washington, D.C., 1997, http://www.nap.edu/books/0309063507/html/index.html (accessed on 24 April 2003).
6. Panel on Dietary Antioxidants and Related Compounds, *Dietary Reference Intakes for Vitamin C, Vitamin E, Selenium and Carotenoids*, National Academy Press, Washington, D.C., 2000, http://www.nap.edu/books/0309069351/html (accessed on 24 April 2003).
7. Panel on Micronutrients, *Dietary Reference Intakes for Vitamin A, Vitamin K, Arsenic, Boron, Chromium, Copper, Iodine, Iron, Manganese, Molybdenum, Nickel, Silicon, Vanadium, and Zinc*, National Academy Press, Washington, D.C., 2001 http://www.nap.edu/books/0309072794/html, (accessed on 24 April 2003).
8. The European Commission, 2003, Tolerable upper intake levels for vitamins and minerals, http://ec.europa.eu/food/fs/sc/scf/out80_en.print.html, (accessed on 15 July 2011).
9. Petersen, B.J., and Barraj, L.M., Assessing the intake of contaminants and nutrients: an overview of methods, *J. Food Compos. Anal.*, 9, 243, 1996.
10. Scanlon, K.A., et al., A longitudinal investigation of solid-food based dietary exposure to selected elements, *J. Exposure Anal. Environ. Epidemiol.*, 9, 485, 1999.
11. Jorhem, L., Becker, W., and Slorach, S., Intake of 17 elements by Swedish women, determined by a 24-h duplicate portion study, *J. Food Compos. Anal.*, 11, 32, 1998.
12. Suseela, B., et al., Daily intake of trace metals through coffee consumption in India, *Food Addit. Contam.*, 18, 115, 2001.
13. Storelli, M.M., Stuffler, R.G., and Marcotrigiano, G.O., Total and methylmercury residues in tuna-fish from the Mediterranean Sea, *Food Addit. Contam.*, 19, 715, 2002.
14. Kyritsis, A., Kanias, G.D., and Tzia, C., Nutritional values of trace elements in dried desserts, *J. Radioanal. Nucl. Chem.*, 217, 209, 1997.
15. Duffus, J.H., "Heavy metal"—a meaningless term? (IUPAC Technical Report), *Pure Appl. Chem.*, 74, 793, 2002.
16. U.S. Food and Drug Administration, 2008, Total Diet Study, http://www.fda.gov/Food/FoodSafety/FoodContaminantsAdulteration/TotalDietStudy/default.htm, (accessed on 15 July 2011).
17. United Kingdom Food Standards Agency, Food surveys, 2011, http://www.food.gov.uk/science/surveillance/, (accessed on 15 July 2011).
18. Ministry of Agriculture, Fisheries, *Lead, Arsenic and Other Metals in Food—Food Surveillance Paper No. 52*, The Stationery Office, London, 1998.

19. Ministry of Agriculture, Fisheries, *Cadmium, Mercury and Other Metals in Food—Food Surveillance Paper No. 53*, The Stationery Office, London, 1998.
20. Food Standards Australia New Zealand, 2003, The 20th Australian Total Diet Survey—a total diet survey of pesticide residues and contaminants, http://www.foodstandards.gov.au/scienceandeducation/publications/20thaustraliantotaldietsurveyjanuary2003/20thaustraliantotaldietsurveyfullreport/, (accessed on 15 July 2011).
21. National Institute of Public Health in Prague, Environment and health monitoring—Project No. IV: reported alimentary diseases and total diet study in the Czech Republic, April 18, 2003, http://www.chpr.szu.cz/monitor/monitor.html (accessed on 24 April 2003).
22. Tsuda, T., et al., Market basket and duplicate portion estimation of dietary intakes of cadmium, mercury, arsenic, copper, manganese, and zinc by Japanese adults, *J. AOAC Int.*, 78, 1363, 1995.
23. New Zealand Ministry of Health, 2000, 1997/1998 New Zealand total diet survey, http://www.moh.govt.nz/moh.nsf/pagesmh/975, (accessed on 15 July 2011).
24. Souci, S.W., Fachmann, W., and Kraut, H., *Food Composition and Nutrition Tables*, 6th ed., CRC Press, Boca Raton, 2000.
25. U.S. Department of Agriculture, 2011, USDA National Nutrient Database for Standard Reference, Release 23, http://www.ars.usda.gov/nutrientdata, (accessed on 15 July 2011).
26. Horwitz, W., Nomenclature for sampling in analytical chemistry (recommendations 1990), *Pure Appl. Chem.*, 62, 1193, 1990.
27. Hoenig M., Preparation steps in environmental trace element analysis—facts and traps, *Talanta*, 54, 1021, 2001.
28. Official Methods of Analysis of AOAC International, 17th Ed., Rev 1, Official Method 937.07, *Fish and Marine Products—Treatment and Preparation of Sample—Procedure*, AOAC International, Gaithersburg, MD, 2002.
29. Torelm, I., et al., Variations in major nutrients and minerals due to interindividual preparation of dishes from recipes, *J. Food Compos. Anal.*, 10, 14, 1997.
30. Rose, M., et al., A review of analytical methods for lead, cadmium, mercury, arsenic and tin determination used in proficiency testing, *J. Anal. At. Spectrom.*, 16, 1101, 2001.
31. Rains, T.C., "Application of atomic absorption spectrometry to the analysis of foods." In *Atomic Absorption Spectrometry; Theory, Design and Applications*, Haswell, S.J., Ed., Elsevier, Amsterdam, 1991.
32. Fourie, H.O., and Peisach, M., Loss of trace elements during dehydration of marine zoological material, *Analyst*, 102, 193, 1977.
33. Horvat, M., and Byrne, A.R., Preliminary study of the effects of some physical parameters on the stability of methylmercury in biological samples, *Analyst*, 117, 665, 1992.
34. Razagui, I.B., and Barlow, P.J., A chemical clean-up procedure to reduce trace metal contamination from laboratory blenders, *Food Chem.*, 44, 309, 1992.
35. Cubadda, F., et al., Influence of laboratory homogenization procedures on trace element content of food samples: an ICP-MS study on soft and durum wheat, *Food Addit. Contam.*, 18, 778, 2001.
36. Stringari, G., et al., Influence of two grinding methods on the uncertainty of determinations of heavy metals in atomic absorption spectrometry/electrothermal atomisation of plant samples, *Accredit. Qual. Assur.*, 3, 122, 1998.
37. Gouveia, S.T., et al., Homogenization of breakfast cereals using cryogenic grinding, *J. Food Engineer.*, 51, 59, 2002.
38. Carrilho, E.N.V.M., et al., Microwave-assisted acid decomposition of animal- and plant-derived samples for element analysis, *J. Agric. Food Chem.*, 50, 4164, 2002.
39. Santos, D., et al., Determination of Cd and Pb in food slurries by GFAAS using cryogenic grinding for sample preparation, *Anal. Bioanal. Chem.*, 373, 183, 2002.

40. Official Methods of Analysis of AOAC International, 17th Ed., Rev 1, Official Method 999.11. *Determination of Lead, Cadmium, Copper, Iron, and Zinc in Foods—Atomic Absorption Spectrophotometry after Dry Ashing—NMLK–AOAC Method*, AOAC International, Gaithersburg, MD, 2002.

41. Official Methods of Analysis of AOAC International, 17th Ed., Rev 1, Official Method 985.35. *Minerals in Infant Formula, Enteral Products, and Pet Foods—Atomic Absorption Spectrophotometric Method*, AOAC International, Gaithersburg, MD, 2002.

42. Skurikhin, I.M., Methods of analysis for toxic elements in foods. Part IV. General method of ashing for the determination of toxic elements, *J. AOAC Int.*, 76, 257, 1993.

43. Mader, P., Száková, J., and Curdová, E., Combination of classical dry ashing with stripping voltammetry in trace element analysis of biological materials: a review of literature published after 1978, *Talanta*, 43, 521, 1996.

44. Hoenig, M., Critical discussion of trace element analysis of plant matrices, *Sci. Total Environ.*, 176, 85, 1995.

45. Vassileva, E., et al., Revisitation of mineralization modes for arsenic and selenium determinations in environmental samples, *Talanta*, 54, 187, 2001.

46. Fecher, P. and Ruhnke, G., Cross contamination of lead and cadmium during dry ashing of food samples, *Anal. Bioanal. Chem.*, 373, 787, 2002.

47. Koh, S., et al., Losses of elements in plant samples under the dry ashing process, *J. Radioanal. Nucl. Chem.*, 239, 591, 1999.

48. Sun, D.-S., Waters, J.K., and Mawhinney, T.P., Determination of thirteen common elements in food samples by inductively coupled plasma atomic emission spectrometry: comparison of five digestion methods, *J. AOAC Int.*, 83, 1218, 2000.

49. Official Methods of Analysis of AOAC International, 17th Ed., Rev 1, Official Method 984.27. *Calcium, Copper, Iron, Magnesium, Manganese, Phosphorus, Potassium, Sodium, and Zinc in Infant Formula—Inductively Coupled Plasma Emission Spectroscopic Method*, AOAC International, Gaithersburg, MD, 2002.

50. Fischer, P.W.F., L'Abbé, M.R., and Giroux, A., Colorimetric determination of total iodine in foods by iodide-catalyzed reduction of Ce^{14}, *J. Assoc. Off. Anal. Chem.*, 69, 687, 1986.

51. White, R.T., Kettisch, P., and Kainrath, P., The high pressure asher: a high-performance sample decomposition system as an alternative to microwave-assisted digestion, *At. Spectrosc.*, 19, 187, 1998.

52. Tinggi, U., Reilly, C., and Patterson, C., Determination of manganese and chromium in foods by atomic absorption spectrometry after wet digestion, *Food Chem.*, 60, 123, 1997.

53. Schelenz, R. and Zeiller, E., Influence of digestion methods on the determination of total Al in food samples by ICP-ES, *Fresenius J. Anal. Chem.*, 345, 68, 1993.

54. Hoenig, M., et al., Critical discussion on the need for an efficient mineralization procedure for the analysis of plant material by atomic spectrometric methods, *Anal. Chim. Acta*, 358, 85, 1998.

55. Kakulu, S.E., Osibanjo, O., and Ajayi, S.O., Comparison of digestion methods for trace metal determination in fish, *Intern. J. Environ. Anal. Chem.*, 30, 209, 1987.

56. Burguera, J.L., and Burguera, M., Evaluation of ten digestion methods for use prior to mercury determination, *J. Food Compos. Anal.*, 1, 159, 1988.

57. Adeloju, S.B., Dhindsa, H.S., and Tandon, R.K., Evaluation of some wet decomposition methods for mercury determination in biological and environmental material by cold vapour atomic absorption spectroscopy, *Anal. Chim. Acta*, 285, 359, 1994.

58. Adair, B.M. and Cobb, G.P., Improved preparation of small biological samples for mercury analysis using cold vapor atomic absorption spectroscopy, *Chemosphere*, 38, 2951, 1999.

59. Kingston, H.M.S. and Haswell, S.J., *Microwave-Enhanced Chemistry*, American Chemical Society, Washington, D.C., 1997.

60. Chakraborty, R., et al., Literature study of microwave-assisted digestion using electrothermal atomic absorption spectrometry, *Fresenius J. Anal. Chem.*, 355, 99, 1996.

61. Kingston, H.M.S. and Walter, P.J., "The Art and Science of Microwave Sample Preparations for Trace and Ultratrace Elemental Analysis." In *Inductively Coupled Plasma Mass Spectrometry*, Montaser, A., Ed., Wiley-VCH, New York, 1998.

62. Official Methods of Analysis of AOAC International, 17th Ed., Rev 1, Official Method 999.10. *Lead, Cadmium, Zinc, Copper, and Iron in Foods—Atomic Absorption Spectrophotometry after Microwave Digestion*, AOAC International, Gaithersburg, MD, 2002.

63. Environmental Protection Agency, SW-846 EPA Method 3052 rev. 0, Microwave assisted acid digestion of siliceous and organically based matrices, *Test Methods for Evaluating Solid Waste*, 3rd ed., 3rd update, U.S. EPA, Washington, D.C., 1996, SW846 On-Line: http://www.epa.gov/epawaste/hazard/testmethods/sw846/online/3_series.htm, (accessed on 15 July 2011).

64. Lamble, K.L. and Hill, S.J., Microwave digestion procedures for environmental matrices, *Analyst*, 123, 103R, 1998.

65. Sun, D.-H., Waters, J.K., and Mawhinney, T.P., Microwave digestion with HNO_3–H_2O_2–HF for the determination of total aluminum in seafood and meat by inductively coupled plasma atomic emission spectrometry, *J. Agric. Food Chem.*, 45, 2115, 1997.

66. Sahuquillo, A., Rubio, R., and Rauret, G., Classical wet ashing versus microwave-assisted attacks for the determination of chromium in plants, *Analyst*, 124, 1, 1999.

67. Gawalko, E. J., et al., Comparison of closed-vessel and focused open-vessel microwave dissolution for determination of cadmium, copper, lead, and selenium in wheat, wheat products, corn bran, and rice flour by transverse-heated graphite furnace atomic absorption spectrometry, *J. AOAC Int.*, 80, 379, 1997.

68. Dolan, S.P. and Capar, S.G., Multi-element analysis of food by microwave digestion and inductively coupled plasma-atomic emission spectrometry, *J. Food Compos. Anal.*, 15, 593, 2002.

69. Wasilewska, M., et al., Efficiency of oxidation in wet digestion procedures and influence from the residual organic carbon content on selected techniques for determination of trace elements, *J. Anal. At. Spectrom.*, 17, 1121, 2002.

70. Anderson, K.A., *Analytical Techniques for Inorganic Contaminants*, Chapter 2, AOAC International, Gaithersburg, 1999.

71. Official Methods of Analysis of AOAC International, 17th Ed., Rev 1, Official Method 973.34. *Cadmium in Food—Atomic Absorption Spectrophotometric Method*, AOAC International, Gaithersburg, MD, 2002.

72. Official Methods of Analysis of AOAC International, 17th Ed., Rev 1, Official Method 972.25. *Lead in Food—Atomic Absorption Spectrophotometric Method*, AOAC International, Gaithersburg, MD, 2002.

73. Official Methods of Analysis of AOAC International, 17th Ed., Rev 1, Official Method 985.16. *Tin in Canned Foods—Atomic Absorption Spectrophotometric Method*, AOAC International, Gaithersburg, MD, 2002.

74. Official Methods of Analysis of AOAC International, 17th Ed., Rev 1, Official Method 969.32. *Zinc in Food—Atomic Absorption Spectrophotometric Method*, AOAC International, Gaithersburg, MD, 2002.

75. International Organization for Standardization, ISO 6636–2, *Fruits, Vegetables and Derived Products—Determination of Zinc Content—Part 2: Atomic Absorption Spectrometric Method,* International Organization for Standardization, 1981

76. International Organization for Standardization, ISO 7952, *Fruits, Vegetables and Derived Products—Determination of Copper Content—Method Using Flame Atomic Absorption Spectrometry*, International Organization for Standardization, 1994.

77. International Organization for Standardization, ISO 9526, *Fruits, Vegetables and Derived Products—Determination of Iron Content by Flame Atomic Absorption Spectrometry*, International Organization for Standardization, 1990.

78. International Organization for Standardization, ISO 11813, *Milk and Milk Products—Determination of Zinc Content—Flame Atomic Absorption Spectrometric Method*, International Organization for Standardization, 1998.

79. Official Methods of Analysis of AOAC International, 17th Ed., Rev 1, Official Method 991.25. *Calcium, Magnesium, and Phosphorus in Cheese—Atomic Absorption Spectrophotometric and Colorimetric Method*, AOAC International, Gaithersburg, MD, 2002.

80. Lombardi-Boccia, G., et al., Content of some trace elements and minerals in the Italian total-diet, *J. Food Compos. Anal.*, 13, 525, 2000.

81. Pokorn, D., et al., Elemental composition (Ca, Mg, Mn, Cu, Cr, Zn, Se, and I) of daily diet samples from some old people's homes in Slovenia, *J. Food Compos. Anal.*, 11, 47, 1998.

82. Miller-Ihli, N.J., Atomic absorption and atomic emission spectrometry for the determination of the trace element content of selected fruits consumed in the United States, *J. Food Compos. Anal.*, 9, 301, 1996.

83. Plessi, M., et al., Dietary fiber and some elements in nuts and wheat brans, *J. Food Compos. Anal.*, 12, 91, 1999.

84. Tahán, J.E., et al., Concentration of total Al, Cr, Cu, Fe, Hg, Na, Pb, and Zn in commercial canned seafood determined by atomic spectrometric means after mineralization by microwave heating, *J. Agric. Food Chem.*, 43, 910, 1995.

85. Voegborlo, R.B., El-Methnani, A.M., and Abedin, M.Z., Mercury, cadmium and lead content of canned tuna fish, *Food Chem.*, 67, 341, 1999.

86. El-Dine, N.W., et al., Analysis of some different kinds of coffee using atomic and nuclear techniques, *Nucl. Sci. J.*, 38, 126, 2001.

87. Phuong, T.D., et al., Elemental content of Vietnamese rice. Part 1. Sampling, analysis and comparison with previous studies, *Analyst*, 124, 553, 1999.

88. Miller-Ihli, N.J., Trace element determinations in foods and biological samples using inductively coupled plasma atomic emission spectrometry and flame atomic absorption spectrometry, *J. Agric. Food Chem.*, 44, 2675, 1996.

89. Leblebici, J. and Volkan, M., Sample preparation for arsenic, copper, iron, and lead determination in sugar, *J. Agric. Food Chem.*, 46, 173, 1998.

90. Julshamn, K., Maage, A., and Wallin, H.C., Determination of magnesium and calcium in foods by atomic absorption spectrometry after microwave digestion: NMKL collaborative study, *J. AOAC Int.*, 81, 1202, 1998.

91. Jackson, K. W., Ed., *Electrothermal Atomization for Analytical Atomic Spectrometry*, Wiley, Chichester, 1999.

92. Tsalev, D. L., Vapor generation or electrothermal atomic absorption spectrometry?—both! *Spectrochim. Acta. B*, 55, 917, 2000.

93. International Organization for Standardization, ISO 6561, *Fruits, Vegetables and Derived Products—Determination of Cadmium Content—Flameless Atomic Absorption Spectrometric Method*, International Organization for Standardization, 1983.

94. International Organization for Standardization, ISO 6633, *Fruits, Vegetables and Derived Products—Determination of Lead Content—Flameless Atomic Absorption Spectrometric Method*, International Organization for Standardization, 1984.

95. International Organization for Standardization, ISO 14377, *Canned Evaporated Milk—Determination of Tin Content—Method Using Graphite Furnace Atomic Absorption Spectrometry*, International Organization for Standardization, 2002.

96. Lacoste, F., Van Dalen, G., and Dysseler, P., The determination of cadmium in oils and fats by direct graphite furnace atomic absorption spectrometry (technical report), *Pure Appl. Chem.*, 71, 361, 1999.

97. International Organization for Standardization, ISO 15774, *Animal and Vegetable Fats and Oils—Determination of Cadmium Content by Direct Graphite Furnace Atomic Absorption Spectrometry*, International Organization for Standardization, 2000.

98. International Organization for Standardization, ISO 12193, *Animal and Vegetable Fats and Oils—Determination of Lead Content—Graphite Furnace Atomic Absorption Method*, International Organization for Standardization, 1994.

99. International Organization for Standardization, ISO 10540–2, *Animal and Vegetable Fats and Oils—Determination of Phosphorus Content—Part 2: Method Using Graphite Furnace Atomic Absorption Spectrometry*, International Organization for Standardization, 2003.

100. International Organization for Standardization, ISO 8294, *Animal and Vegetable Fats and Oils—Determination of Copper, Iron and Nickel Contents—Graphite Furnace Atomic Absorption Method*, International Organization for Standardization, 1994.

101. International Organization for Standardization, ISO 11212–4, *Starch and Derived Products—Heavy Metals Content—Part 4: Determination of Cadmium Content by Atomic Absorption Spectrometry with Electrothermal Atomization*, International Organization for Standardization, 1997.

102. International Organization for Standardization, ISO 11212–3, *Starch and Derived Products—Heavy Metals Content—Part 3: Determination of Lead Content by Atomic Absorption Spectrometry with Electrothermal Atomization*, International Organization for Standardization, 1997.

103. Official Methods of Analysis of AOAC International, 17th Ed., Rev 1, Official Method 997.15. *Lead in Sugars and Syrups—Graphite Furnace Atomic Absorption Method*, AOAC International, Gaithersburg, MD, 2002.

104. Dabeka, R.W. and McKenzie, A.D., Survey of lead, cadmium, fluoride, nickel, and cobalt in food composites and estimation of dietary intakes of these elements by Canadians in 1986–1988, *J. AOAC Int.*, 78, 897, 1995.

105. Capar, S.G. and Cunningham, W.C., Element and radionuclide concentrations in food: FDA total diet study 1991–1996, *J. AOAC Int.*, 83, 157, 2000.

106. Tripathi, R.M., et al., Daily intake of aluminium by adult population of Mumbai, India, *Sci. Total Environ.*, 299, 73, 2002.

107. Alberti-Fidanza, A., Burini, G., and Perriello, G., Trace elements in foods and meals consumed by students attending the faculty cafeteria, *Sci. Total Environ.*, 287, 133, 2002.

108. Miller-Ihli, N. J., Graphite furnace atomic absorption spectrometry for the determination of the chromium content of selected U.S. foods, *J. Food Compos. Anal.*, 9, 290, 1996.

109. Malmauret, L., et al., Contaminants in organic and conventional foodstuffs in France, *Food Addit. Contam.*, 19, 524, 2002.

110. Larsen, E.H., et al., Monitoring the content and intake of trace elements from food in Denmark, *Food Addit. Contam.*, 19, 33, 2002.

111. Lendinez, E., et al., Chromium in basic foods of the Spanish diet: Seafood, cereals, vegetables, olive oils and dairy products, *Sci. Total Environ.*, 278, 183, 2001.

112. Bratakos, M.S., Lazos, E.S., and Bratakos, S.M., Chromium content of selected Greek foods, *Sci. Total Environ.*, 290, 47, 2002.

113. López, F.F., et al., Aluminum content in foods and beverages consumed in the Spanish diet, *J. Food Sci.*, 65, 206, 2000.

114. López, F.F., et al., Aluminium content of drinking waters, fruit juices and soft drinks: contribution to dietary intake, *Sci. Total Environ.*, 292, 205, 2002.

115. de Togores, M.R., Farré, R., and Frigola, A.M., Cadmium and lead in infant cereals—electrothermal-atomic absorption spectroscopic determination, *Sci. Total Environ.*, 234, 197, 1999.

116. Allen, L.B., Siitonen, P., and Thompson, H.C., Determination of copper, lead, and nickel in edible oils by plasma and furnace atomic spectroscopies, *J. Am. Oil Chem. Soc.*, 75, 477, 1998.

117. Correia, P.R.M., Oliveira, E., and Oliveira, P.V., Simultaneous determination of Cd and Pb in foodstuffs by electrothermal atomic absorption spectrometry, *Anal. Chim. Acta*, 405, 205, 2000.

118. Acar, O., Kiliç, Z., and Türker, A.R., Determination of lead in cookies by electrothermal atomic absorption spectrometry with various chemical modifiers, *Food Chem.*, 71, 117, 2000.

119. Julshamn, K., Thorlacius, A., and Lea, P., Determination of arsenic in seafood by electrothermal atomic absorption spectrometry after microwave digestion: NMKL collaborative study, *J. AOAC Int.*, 83, 1423, 2000.

120. Fedorov, P.N., Ryabchuk, G.N., and Zverev, A.V., Comparison of hydride generation and graphite furnace atomic absorption spectrometry for the determination of arsenic in food, *Spectrochim. Acta, Part B*, 52, 1517, 1997.

121. Rosa, C.R., et al., Effect of modifiers on thermal behaviour of Se in acid digestates and slurries of vegetables by graphite furnace atomic absorption spectrometry, *Food Chem.*, 79, 517, 2002.

122. Veillon, C., and Patterson, K.Y., Analytical issues in nutritional chromium research, *J. Trace Elem. Exp. Med.*, 12, 99, 1999.

123. Adachi, A., et al., Determination of vanadium in foods by atomic absorption spectrophotometry, *Anal. Lett.*, 32, 2327, 1999.

124. Bermejo-Barrera, P., et al., Microwave-assisted distillation of iodine for the indirect atomic absorption spectrometric determination of iodide in milk samples, *J. Anal. At. Spectrom.*, 16, 382, 2001.

125. Viñas, P., et al., Electrothermal atomic absorption spectrometric determination of molybdenum, aluminium, chromium and manganese in milk, *Anal. Chim. Acta*, 356, 267, 1997.

126. García, E. M., et al., Trace element determination in different milk slurries, *J. Dairy Res.*, 66, 569, 1999.

127. Official Methods of Analysis of AOAC International, 17th Ed., Rev 1, Official Method 994.02. *Lead in Edible Oils and Fats—Direct Graphite Furnace Atomic Absorption Spectrophotometric Method*, AOAC International, Gaithersburg, MD, 2002.

128. Official Methods of Analysis of AOAC International, 17th Ed., Rev 1, Official Method 990.05. *Copper, Iron, and Nickel in Edible Oils and Fats—Direct Graphite Furnace Atomic Absorption Spectrophotometric Method*, AOAC International, Gaithersburg, MD, 2002.

129. Lelièvre, C., et al., A rapid method for the direct determination of copper and iron in butter by GFAAS, *At. Spectrosc.*, 21, 23, 2000.

130. van Dalen, G., Determination of cadmium in edible oils and fats by direct electrothermal atomic absorption spectrometry, *J. Anal. At. Spectrom.*, 11, 1087, 1996.

131. Viñas, P., Pardo-Martínez, M., and Hernández-Córdoba, M., Slurry atomization for the determination of arsenic in baby foods using electrothermal atomic absorption spectrometry and deuterium background correction, *J. Anal. At. Spectrom.*, 14, 1215, 1999.

132. Viñas, P., Pardo-Martínez, M., and Hernández-Córdoba, M., Rapid determination of selenium, lead and cadmium in baby food samples using electrothermal atomic absorption spectrometry and slurry atomization, *Anal. Chim. Acta*, 412, 121, 2000.

133. Viñas, P., Pardo-Martínez, M., and Hernández-Córdoba, M., Determination of copper, cobalt, nickel, and manganese in baby food slurries using electrothermal atomic absorption spectrometry, *J. Agric. Food Chem.*, 48, 5789, 2000.

134. Viñas, P., Pardo-Martínez, M., and Hernández-Córdoba, M., Determination of aluminium and chromium in slurried baby food samples by electrothermal atomic absorption spectrometry, *J. AOAC Int.*, 84, 1187, 2001.

135. Viñas, P., et al., Determination of mercury in baby food and seafood samples using electrothermal atomic absorption spectrometry and slurry atomization, *J. Anal. At. Spectrom.*, 16, 633, 2001.

136. Méndez, H., et al., Determination of selenium in marine biological tissues by transverse heated electrothermal atomic absorption spectrometry with longitudinal zeeman background correction and automated ultrasonic slurry sampling, *J. AOAC Int.*, 84, 1921, 2001.

137. Carlosena, A., Gallego, M., and Valcárcel, M., Evaluation of various sample preparation procedures for the determination of chromium, cobalt and nickel in vegetables, *J. Anal. At. Spectrom.*, 12, 479, 1997.

138. Dedina, J. and Tsalev, D.L., *Hydride Generation Atomic Absorption Spectrometry*, Wiley, Chichester, 1995.

139. Yan, X.-P. and Ni, Z.-M., Vapor generation atomic absorption spectrometry, *Anal. Chim. Acta*, 291, 89, 1994.

140. Fecher, P., and Ruhnke, G., Determination of arsenic and selenium in foodstuffs—methods and errors, *At. Spectrosc.*, 19, 204, 1998.

141. Official Methods of Analysis of AOAC International, 17th Ed., Rev 1, Official Method 986.15. *Arsenic, Cadmium, Lead, Selenium, and Zinc in Human and Pet Foods—Multielement Method*, AOAC International, Gaithersburg, MD, 2002.

142. Hussein, L. and Bruggeman, J., Selenium analysis of selected Egyptian foods and estimated daily intakes among a population group, *Food Chem.*, 65, 527, 1999.

143. Díaz-Alarcón, J.P., et al., Determination of selenium in meat products by hydride generation atomic absorption spectrometry–selenium levels in meat, organ meats, and sausages in Spain, *J. Agric. Food Chem.*, 44, 1494, 1996.

144. Sawaya, W.N., et al., Nutritional profile of Kuwaiti composite dishes: Minerals and vitamins, *J. Food Compos. Anal.*, 11, 70, 1998.

145. Plessi, M., Bertelli, D., and Monzani, A., Mercury and selenium content in selected seafood, *J. Food Compos. Anal.*, 14, 461, 2001.

146. Foster, L.H. and Sumar, S., Selenium concentrations in soya based milks and infant formulae available in the United Kingdom, *Food Chem.*, 56, 93, 1996.

147. Mindak, W.R. and Dolan, S.P., Determination of arsenic and selenium in food using a microwave digestion-dry ash preparation and flow injection hydride generation atomic absorption spectrometry, *J. Food Compos. Anal.*, 12, 111, 1999.

148. Ringmann, S., et al., Microwave-assisted digestion of organoarsenic compounds for the determination of total arsenic in aqueous, biological, and sediment samples using flow injection hydride generation electrothermal atomic absorption spectrometry, *Anal. Chim. Acta*, 452, 207, 2002.

149. Schloske, L., Waldner, H., and Marx, F., Optimization of sample pre-treatment in the HG-AAS selenium analysis, *Anal. Bioanal. Chem.*, 372, 700, 2002.

150. Kabengera, C., et al., Optimization and validation of arsenic determination in foods by hydride generation flame atomic absorption spectrometry, *J. AOAC Int.*, 85, 122, 2002.

151. Tinggi, U., Determination of selenium in meat products by hydride generation atomic absorption spectrophotometry, *J. AOAC Int.*, 82, 364, 1999.

152. Clevenger, W.L., Smith, B.W., and Winefordner, J.D., Trace determination of mercury: a review, *Crit. Rev. Anal. Chem.*, 27, 1, 1997.

153. International Organization for Standardization (1984) ISO 6637, Fruits, Vegetables and Derived Products—Determination of Mercury Content—Flameless Atomic Absorption Method.

154. Official Methods of Analysis of AOAC International, 17th Ed., Rev 1, Official Method 977.15. *Mercury in Fish—Alternative Flameless Atomic Absorption Spectrophotometric Method*, AOAC International, Gaithersburg, MD, 2002.

155. Llobet, J. M., et al., Concentrations of arsenic, cadmium, mercury, and lead in common foods and estimated daily intake by children, adolescents, adults, and seniors of Catalonia, Spain, *J. Agric. Food Chem.*, 51, 838, 2003.

156. Storelli, M.M. and Marcotrigiano, G.O., Total mercury levels in muscle tissue of swordfish (*xiphias gladius*) and bluefin tuna (*thunnus thynnus*) from the Mediterranean Sea (Italy), *J. Food Prot.*, 64, 1058, 2001.

157. Love, J.L., Rush, G.M., and McGrath, H., Total mercury and methylmercury levels in some New Zealand commercial marine fish species, *Food Addit. Contam.*, 20, 37, 2003.

158. Corns, W.T., et al., Effects of moisture on the cold vapour determination of mercury and its removal by use of membrane dryer tubes, *Analyst*, 117, 717, 1992.

159. Dabeka, R.W., Bradley, P., and McKenzie, A.D., Routine, high-sensitivity, cold vapor atomic absorption spectrometric determination of total mercury in foods after low-temperature digestion, *J. AOAC Int.*, 85, 1136, 2002.

160. Zenebon, O., et al., Rapid food decomposition by H_2O_2-H_2SO_4 for determination of total mercury by flow injection cold vapor atomic absorption spectrometry, *J. AOAC Int.*, 85, 149, 2002.

161. Julshamn, K., and Brenna, J., Determination of mercury in seafood by flow injection-cold vapor atomic absorption spectrometry after microwave digestion: NMKL interlaboratory study, *J. AOAC Int.*, 85, 626, 2002.

162. Tao, G., Willie, S.N., and Sturgeon, R.E., Determination of total mercury in biological tissues by flow injection cold vapor generation atomic absorption spectrometry following tetramethylammonium hydroxide digestion, *Analyst*, 123, 1215, 1998.

163. Perring, L., and Andrey, D., Optimization and validation of total mercury determination in food products by cold vapor AAS: comparison of digestion methods and with ICP-MS analysis, *At. Spectrosc.*, 22, 371, 2001.

164. Montaser, A., and Golightly, D.W., *Inductively Coupled Plasmas in Analytical Atomic Spectrometry*, 2nd edn., VCH, New York, 1992.

165. Brenner, I. B., and Zander, A.T., Axially and radially viewed inductively coupled plasmas—a critical review, *Spectrochim. Acta, Part B*, 55, 1195, 2000.

166. Silva, F.V., et al., Evaluation of inductively coupled plasma optical emission spectrometers with axially and radially viewed configurations, *Spectrochim. Acta Part B*, 57, 1905, 2002.

167. Broekaert, J.A.C., State-of-the-art and trends of development in analytical atomic spectrometry with inductively coupled plasmas as radiation and ion sources, *Spectrochim. Acta, Part B*, 55, 739, 2000.

168. Official Methods of Analysis of AOAC International, 17th Ed., Rev 1, Official Method 985.01. *Metals and Other Elements in Plants and Pet Foods—Inductively Coupled Plasma Emission Spectroscopic Method*, AOAC International, Gaithersburg, MD, 2002.

169. International Organization for Standardization, ISO 10540–3, *Animal and Vegetable Fats and Oils—Determination of Phosphorus Content—Part 3: Method Using Inductively Coupled Plasma (ICP) Optical Emission Spectroscopy*, International Organization for Standardization, 2002.

170. Ikem, A., et al., Levels of 26 elements in infant formula from USA, UK, and Nigeria by microwave digestion and ICP-OES, *Food Chem.*, 77, 439, 2002.

171. Barnes, K.W., Trace metal determination in fruit juice, and juice products using an axially viewed plasma, *At. Spectrosc.*, 18, 84, 1997.

172. Falandysz, J., et al., ICP/MS and ICP/AES elemental analysis (38 elements) of edible mushrooms growing in Poland, *Food Addit. Contam.*, 18, 503, 2001.

173. Simpkins, W.A., et al., Trace elements in Australian orange juice and other products, *Food Chem.*, 71, 423, 2000.
174. Fernández, P.L., et al., Multi-element analysis of tea beverages by inductively coupled plasma atomic emission spectrometry, *Food Chem.*, 76, 483, 2002.
175. Todolí, J.L., et al., Elemental matrix effects in ICP-AES, *J. Anal. At. Spectrom.*, 17, 142, 2002.
176. Mochizuki, M., Hondo, R., and Ueda, F., Simultaneous analysis for multiple heavy metals in contaminated biological samples, *Biol. Trace Elem. Res.*, 87, 211, 2002.
177. Grotti, M., Leardi, R., and Frache, R., Combined effects of inorganic acids in inductively coupled plasma optical emission spectrometry, *Spectrochim. Acta Part B*, 57, 1915, 2002.
178. Canals, A., Gras, L., and Contreras, H., Elimination of nitric acid interference in ICP-AES by using a cyclonic spray chamber/nafion membrane-based desolvation system, *J. Anal. At. Spectrom.*, 17, 219, 2002.
179. Lomer, M.C.E., et al., Determination of titanium dioxide in foods using inductively coupled plasma optical emission spectrometry, *Analyst*, 125, 2339, 2000.
180. Perring, L. and Basic-Dvorzak, M., Determination of total tin in canned food using inductively coupled plasma atomic emission spectroscopy, *Anal. Bioanal. Chem.*, 374, 235, 2002.
181. Hua, K.M., Kay, M., and Indyk, H.E., Nutritional element analysis in infant formulas by direct dispersion and inductively coupled plasma-optical emission spectrometry, *Food Chem.*, 68, 463, 2000.
182. McKinstry, P.J., Indyk, H.E., and Kim, N.D., The determination of major and minor elements in milk and infant formula by slurry nebulization and inductively coupled plasma-optical emission spectrometry, *Food Chem.*, 65, 245, 1999.
183. Montaser, A., *Inductively Coupled Plasma Mass Spectrometry*, Wiley-VCH, New York, 1998.
184. Tanner, S.D., Baranov, V.I., and Bandura, D.R., Reaction cells and collision cells for ICP-MS: a tutorial review, *Spectrochim. Acta, Part B*, 57, 1361, 2002.
185. Tanner, S.D., Baranov, V.I., and Vollkopf, U., A dynamic reaction cell for inductively coupled plasma mass spectrometry (ICP-DRC-MS) part III. optimization and analytical performance, *J. Anal. At. Spectrom.*, 15, 1261, 2000.
186. Todolí, J.-L. and Mermet, J.-M., Acid interferences in atomic spectrometry: analyte signal effects and subsequent reduction, *Spectrochim. Acta, Part B*, 54, 895, 1999.
187. Rose, M., et al., Bromine and iodine in 1997 UK total diet study samples, *J. Environ. Monit.*, 3, 361, 2001.
188. Noël, L., Leblanc, J.-C., and Guérin, T., Determination of several elements in duplicate meals from catering establishments using closed vessel microwave digestion with inductively coupled plasma mass spectrometry detection: estimation of daily dietary intake, *Food Addit. Contam.*, 20, 44, 2003.
189. Eckhoff, K.M. and Maage, A., Iodine content in fish and other food products from East Africa analyzed by ICP-MS, *J. Food Compos. Anal.*, 10, 270, 1997.
190. Krachler, M., Rossipal, E., and Irgolic, K.J., Trace elements in formulas based on cow and soy milk and in Austrian cow milk determined by inductively coupled plasma mass spectrometry, *Biol. Trace Elem. Res.*, 65, 53, 1998.
191. Larsen, E.H., Knuthsen, P., and Hansen, M., Seasonal and regional variations of iodine in Danish dairy products determined by inductively coupled plasma mass spectrometry, *J. Anal. At. Spectrom.*, 14, 41, 1999.
192. Bibak, A., et al., Concentrations of 63 elements in cabbage and sprouts in Denmark, *Commun. Soil Sci. Plant Anal.*, 30, 2409, 1999.
193. Gundersen, V., et al., Comparative investigation of concentrations of major and trace elements in organic and conventional Danish agricultural crops. 1. Onions (*Allium cepa* hysam) and Peas (*Pisum sativum* ping pong), *J. Agric. Food Chem.*, 48, 6094, 2000.

194. Martin, G.J., et al., Optimization of analytical methods for origin assessment of orange juices. II. ICP-MS determination of trace and ultra-trace elements, *Analusis*, 25, 7, 1997.

195. Bibak, A., et al., Concentrations of 50 major and trace elements in Danish agricultural crops measured by inductively coupled plasma mass spectrometry. Potato (*Solanum tuberosum* folva), *J. Agric. Food Chem.*, 47, 2678, 1999.

196. Crews, H.M., "Inductively coupled plasma-mass spectrometry (ICP-MS) for the analysis of trace element contaminants in foods." In *Progress in Food Contaminant Analysis*, Gilbert, J., Ed., Blackie, London, 1996.

197. Salomon, S., Jenne, V., and Hoenig, M., Practical aspects of routine trace element environmental analysis by inductively coupled plasma-mass spectrometry, *Talanta*, 57, 157, 2002.

198. Baxter, M.J., et al., Quality control in the multi-element analysis of foods using ICP-MS, in *Plasma Source Mass Spectrometry: Developments and Applications*, Holland, G. and Tanner, S.D., Eds., The Royal Society of Chemistry, London, 1997.

199. Baker, S.A., Bradshaw, D.K., and Miller-Ihli, N.J., Trace element determinations in food and biological samples using ICP-MS, *At. Spectrosc.*, 20, 167, 1999.

200. Cubadda, F., et al., Multielement analysis of food and agricultural matrixes by inductively coupled plasma-mass spectrometry, *J. AOAC Int.*, 85, 113, 2002.

201. Martino, F.A. R., Sánchez, M. L. F., and Sanz-Medel, A., The potential of double focusing-ICP-MS for studying elemental distribution patterns in whole milk, skimmed milk and milk whey of different milks, *Anal. Chim. Acta*, 442, 191, 2001.

202. Zhang, Z.-W., et al., Determination of lead and cadmium in food and blood by inductively coupled plasma mass spectrometry: a comparison with graphite furnace atomic absorption spectrometry, *Sci. Total Environ.*, 205, 179, 1997.

203. Zbinden, P., and Andrey, D., Determination of trace element contaminants in food matrixes using a robust, routine analytical method for ICP-MS, *At. Spectrosc.*, 19, 214, 1998.

204. Bhandari, S.A., and Amarasiriwardena, D., Closed-vessel microwave acid digestion of commercial maple syrup for the determination of lead and seven other trace elements by inductively coupled plasma-mass spectrometry, *Microchem. J.*, 64, 73, 2000.

205. D'Ilio, S., et al., Arsenic content of various types of rice as determined by plasma-based techniques, *Microchem. J.*, 73, 195, 2002.

206. Rädlinger, G., and Heumann, K.G., Iodine determination in food samples using inductively coupled plasma isotope dilution mass spectrometry, *Anal. Chem.*, 70, 2221, 1998.

207. Fecher, P.A., Goldmann, I., and Nagengast, A., Determination of iodine in food samples by inductively coupled plasma mass spectrometry after alkaline extraction, *J. Anal. At. Spectrom.*, 13, 977, 1998.

208. Julshamn, K., Dahl, L., and Eckhoff, K., Determination of iodine in seafood by inductively coupled plasma/mass spectrometry, *J. AOAC Int.*, 84, 1976, 2001.

209. Haldimann, M., Eastgate, A., and Zimmerli, B., Improved measurement of iodine in food samples using inductively coupled plasma isotope dilution mass spectrometry, *Analyst*, 125, 1977, 2000.

210. Brainina, Kh.Z., Malakhova, N.A., and Stojko, N.Yu., Stripping voltammetry in environmental and food analysis, *Fresenius J. Anal. Chem.*, 368, 307, 2000.

211. Official Methods of Analysis of AOAC International, 17th Ed., Rev 1, Official Method 992.24. *Iodine in Ready-To-Feed Milk-Based Infant Formula—Ion-Selective Electrode Method*, AOAC International, Gaithersburg, MD, 2002.

212. Official Methods of Analysis of AOAC International, 17th Ed., Rev 1, Official Method 975.04. *Fluoride in Plants—Potentiometric Method*, AOAC International, Gaithersburg, MD, 2002.

213. Official Methods of Analysis of AOAC International, 17th Ed., Rev 1, Official Method 982.23. *Cadmium and Lead in Food—Anodic Stripping Voltammetric Method*, AOAC International, Gaithersburg, MD, 2002.

214. International Organization for Standardization, ISO 6636–1, *Fruits, Vegetables and Derived Products—Determination of Zinc Content—Part 1: Polarographic Method,* International Organization for Standardization, 1986.
215. Malde, M.K., et al., Fluoride content in selected food items from five areas in East Africa, *J. Food Compos. Anal.,* 10, 233, 1997.
216. Fein, N.J. and Cerklewski, F.L., Fluoride content of foods made with mechanically separated chicken, *J. Agric. Food Chem.,* 49, 4284, 2001.
217. Alonso, J., et al., Accumulation of mercury in edible macrofungi: influence of some factors, *Arch. Environ. Contam. Toxicol.,* 38, 158, 2000.
218. Ramonaityte, D.T., Copper, zinc, tin and lead in canned evaporated milk, produced in Lithuania: the initial content and its change at storage, *Food Addit. Contam.,* 18, 31, 2001.
219. Malde, M.K., Bjorvatn, K., and Julshamn, K., Determination of fluoride in food by the use of alkali fusion and fluoride ion-selective electrode, *Food Chem.,* 73, 373, 2001.
220. Ratana-Ohpas, R., et al., Determination of tin in canned fruit juices by stripping potentiometry, *Anal. Chim. Acta,* 333, 115, 1996.
221. Qiong, L., et al., Determination of trace tin in foods by single-sweep polarography, *Food Chem.,* 64, 129, 1999.
222. Guanghan, L., et al., Polarographic determination of trace fluoride in foods, *Food Chem.,* 66, 519, 1999.
223. Yao, S.-Z., Chen, P., and Wei, W.-Z., A quartz crystal microbalance method for the determination of iodine in foodstuffs, *Food Chem.,* 67, 311, 1999.
224. Lo Coco, F., et al., Determination of lead (II) and cadmium (II) in hard and soft wheat by derivative potentiometric stripping analysis, *Anal. Chim. Acta,* 409, 93, 2000.
225. Lambert, D.F., and Turoczy, N.J., Comparison of digestion methods for the determination of selenium in fish tissue by cathodic stripping voltammetry, *Anal. Chim. Acta,* 408, 97, 2000.
226. Inam, R. and Somer, G., A direct method for the determination of selenium and lead in cow's milk by differential pulse stripping voltammetry, *Food Chem.,* 69, 345, 2000.
227. Sancho, D., et al., Determination of nickel and cobalt in refined beet sugar by adsorptive cathodic stripping voltammetry without sample pretreatment, *Food Chem.,* 71, 139, 2000.
228. Sancho, D., et al., Determination of copper and arsenic in refined beet sugar by stripping voltammetry without sample pretreatment, *Analyst,* 123, 743, 1998.
229. Sancho, D., et al., Determination of zinc, cadmium and lead in untreated sugar samples by anodic stripping voltammetry, *Analyst,* 122, 727, 1997.
230. Sancho, D., et al., Determination of mercury in refined beet sugar by anodic stripping voltammetry without sample pretreatment, *Food Chem.,* 74, 527, 2001.
231. Moxon, R.E.D., and Dixon, E.J., Semi-automatic method for the determination of total iodine in food, *Analyst,* 105, 344, 1980.
232. Perring, L., Basic-Dvorzak, M., and Andrey, D., Colorimetric determination of inorganic iodine in fortified culinary products, *Analyst,* 126, 985, 2001.
233. Cressey, P.J., Iodine content of New Zealand dairy products, *J. Food Compos. Anal.,* 16, 25, 2003.
234. Dermelj, M., et al., Applicability of neutron activation analysis (NAA) in quantitative determination of some essential and toxic trace elements in food articles, *Z. Lebensm. Unters. Forsch.,* 202, 447, 1996.
235. Contis, E.T., Use of nuclear techniques for the measurement of trace elements in food, *J. Radioanal. Nucl. Chem.,* 243, 53, 2000.
236. Anderson, D.L., et al., Nuclear methods for food analysis at the U. S. Food and Drug Administration, *J. Radioanal. Nucl. Chem.,* 249, 29, 2001.
237. Alfassi, Z.B., Instrumental neutron activation analysis (INAA), in *Determination of Trace Elements,* Alfassi, Z.B., Ed., VCH, New York, 1994, chap. 7.

238. Anderson, D.L., Cunningham, W.C., and Lindstrom, T.R., Concentrations and intakes of H, B, S, K, Na, Cl, and NaCl in foods, *J. Food Compos. Anal.*, 7, 59, 1994.
239. Dang, H.S., Jaiswal, D.D., and Nair, S., Daily dietary intake of trace elements of radiological and nutritional importance by the adult Indian population, *J. Radioanal. Nucl. Chem.*, 249, 95, 2001.
240. Cho, S.Y., et al., Daily dietary intake of elements of nutritional and radiological importance by adult Koreans, *J. Radioanal. Nucl. Chem.*, 249, 39, 2001.
241. Maihara, V.A., et al., Determination of mineral constituents in duplicate portion diets of two university student groups by instrumental neutron activation, *J. Radioanal. Nucl. Chem.*, 249, 21, 2001.
242. Aras, N. K., et al., Dietary intake of zinc and selenium in Turkey, *J. Radioanal. Nucl. Chem.*, 249, 33, 2001.
243. Soliman, K., and Zikovsky, L., Concentration of Al in food sold in Montreal, Canada, and its daily dietary intake, *J. Radioanal. Nucl. Chem.*, 242, 807, 1999.
244. Waheed, S., et al., Instrumental neutron activation analysis of 23 individual food articles from a high altitude region, *J. Radioanal. Nucl. Chem.*, 254, 597, 2002.
245. Nyarko, B.J.B., et al., Epiboron instrumental neutron activation analysis for the determination of iodine in various salt samples, *J. Radioanal. Nucl. Chem.*, 251, 281, 2002.
246. Vega-Carrillo, H. R., Iskander, F.Y., and Manzanares-Acuña, E., Elemental content in ground and soluble/instant coffee, *J. Radioanal. Nucl. Chem.*, 252, 75, 2002.
247. Soliman, K., and Zikovsky, L., Determination of Br, Ca, Cl, Co, Cu, I, K, Mg, Mn, Na, Rb, S, Ti and V in cereals, oils, sweeteners and vegetables sold in Canada by neutron activation analysis, *J. Food Compos. Anal.*, 12, 85, 1999.
248. Gambelli, L., et al., Minerals and trace elements in some Italian dairy products, *J. Food Compos. Anal.*, 12, 27, 1999.
249. Cunningham, W.C., Anderson, D.L., and Capar, S.G., Determination of sodium in biological materials by instrumental neutron activation analysis, *J. AOAC Int.*, 80, 871, 1997.
250. Balaji, T., et al., Multielement analysis in cereals and pulses by k_0 instrumental neutron activation analysis, *Sci. Total Environ.*, 253, 75, 2000.
251. Kucera, J., Randa, Z., and Soukal, L., A comparison of three activation analysis methods for iodine determination in foodstuffs, *J. Radioanal. Nucl. Chem.*, 249, 61, 2001.
252. Chao, J.H., et al., Evaluation of minimum detectable amounts of iodine in biological samples by neutron activation analysis, *J. Radioanal. Nucl. Chem.*, 254, 577, 2002.
253. Van Dael, P., "Trace element speciation in food: A tool to assure food safety and nutritional quality," in *Trace Element Speciation for Environment, Food and Health*, Ebdon, L., et al., Eds., RSC, Cambridge, 2001, 233.
254. Sutton, K.L. and Heitkemper, D.T., "Speciation analysis of biological, clinical and nutritional samples using plasma spectrometry," in *Elemental Speciation. New Approaches for Trace Element Analysis*, Caruso, J.A., Sutton, K.L., and Ackley, K.L., Eds., Elsevier Science B.V., Amsterdam, 2000, 501.
255. González, E.B., and Sanz-Medel, A., "Liquid chromatographic techniques for trace element speciation analysis," in *Elemental Speciation. New Approaches for Trace Element Analysis*, Caruso, J.A., Sutton, K.L., and Ackley, K.L., Eds., Elsevier Science B.V., Amsterdam, 2000, 81.
256. Fairweather-Tait, S.J., The importance of trace element speciation in nutritional sciences, *Fresenius J. Anal. Chem.*, 363, 536, 1999.
257. Buldini, P.L., Cavalli, S., and Trifirò, A., State-of-the-art ion chromatographic determination of inorganic ions in food. *J. Chromatogr. A*, 789, 529, 1997.
258. Zoorob, G.K., McKiernan, J.W., and Caruso, J.A., ICP-MS for elemental speciation studies, *Mikrochim. Acta*, 128, 145, 1998.

259. Ackley, K.L., Sutton, K.L., and Caruso, J.A., "The use of ICP-MS as a detector for elemental speciation studies," in *Elemental Speciation, New Approaches for Trace Element Analysis*, Caruso, J.A., Sutton, K.L., and Ackley, K.L., Eds., Elsevier Science B.V., Amsterdam, 2000, 249.

260. Ebdon, L., and Fisher, A., "The use of ICP-AES as a detector for elemental speciation studies," in *Elemental Speciation. New Approaches for Trace Element Analysis*, Caruso, J.A., Sutton, K.L., and Ackley, K.L., Eds., Elsevier Science B.V., Amsterdam, 2000, 227.

261. Evans E.H., and O'Connor, "Plasma sources as alternatives to the atmospheric pressure ICP for speciation studies," in *Elemental Speciation. New Approaches for Trace Element Analysis*, Caruso, J.A., Sutton, K.L., and Ackley, K.L., Eds., Elsevier Science B.V., Amsterdam, 2000, 315.

262. Olesik, J.W., "Capillary electrophoresis for elemental speciation studies," in *Elemental Speciation. New Approaches for Trace Element Analysis*, Caruso, J.A., Sutton, K.L., and Ackley, K.L., Eds., Elsevier Science B.V., Amsterdam, 2000, 151.

263. Uden, P.C., "Gas chromatographic and supercritical fluid chromatographic techniques for elemental speciation," in *Elemental Speciation. New Approaches for Trace Element Analysis*, Caruso, J.A., Sutton, K.L., and Ackley, K.L., Eds., Elsevier Science B.V., Amsterdam, 2000, 123.

264. Barnett, D.A., Handy R., and Horlick G., "Electrospray Ionization Mass Spectrometry," in *Elemental Speciation. New Approaches for Trace Element Analysis*, Caruso, J.A., Sutton, K.L., and Ackley, K.L., Eds., Elsevier Science B.V., Amsterdam, 2000, 383.

265. Quevauviller, P., Improvement of quality control of speciation analysis using hyphenated techniques. A decade of progress within the European Community, *J. Chromatogr. A.*, 750, 25, 1996.

266. Ellis, L.A. and Roberts, D.J., Chromatographic and hyphenated methods for elemental speciation analysis in environmental media, *J. Chromatogr. A*, 774, 3, 1997.

267. Lobinski, R., Elemental speciation and coupled techniques, *Appl. Spectrosc.*, 51, 260, 1997.

268. Nemanic, T.M., et al., Organotin compounds in the marine environment of the Bay of Piran, northern Adriatic Sea, *J. Environ. Monit.*, 4, 426, 2002.

269. Ritsema, R., and Donard, O.F.X., Organometallic compound determination in the environment by hyphenated techniques, in: *Sample Handling and Trace Analysis Pollutants: Techniques, Applications and Quality Assurance*, Barceló, D., Ed., Elsevier Science B.V., Amsterdam, 1999, 1003.

270. Stewart, I.I., Electrospray mass spectrometry: a tool for elemental speciation, *Spectrochim. Acta Part B*, 54, 1649, 1999.

271. Szpunar, J., Bouyssiere, B., and Lobinski, R. "Sample preparation techniques for elemental speciation studies," in *Elemental Speciation. New Approaches for Trace Element Analysis*, Caruso, J.A., Sutton, K.L., and Ackley, K.L., Eds., Elsevier Science B.V., Amsterdam, 2000, 7.

272. Adams, F., Ceulemans, M., and Slaets, S., GC hyphenations for speciation analysis of organometal compounds, *LC GC Europe*, 14, 548, 2001.

273. Barefoot, R.R., Distribution and speciation of platinum group elements in environmental matrices, *Trends Anal. Chem.*, 18, 702, 1999.

274. Das, A.K., de la Guardia, M., and Cervera, M.L., Literature survey of on-line elemental speciation in aqueous solutions, *Talanta*, 55, 1, 2001.

275. Muñoz-Olivas, R. and Cámara C., Speciation related to human health, in: *Trace Element Speciation for Environment, Food and Health*, Ebdon, L., et al., Eds., RSC, Cambridge, 2001, 331.

276. Ferrarello, C.N., Fernández de la Campa, M.R., and Sanz-Medel, A., Multielement trace-element speciation in metal-biomolecules by chromatography coupled with ICP-MS, *Anal. Bioanal. Chem.*, 373, 412, 2002.

277. Kamnev, A.A., et al., Trace cobalt speciation in bacteria and at enzymic active sites using emission Mössbauer spectroscopy, *Anal. Bioanal. Chem.*, 372, 431, 2002.

278. Mestek, O., et al., Quantification of copper and zinc species fractions in legume seeds extracts by SEC/ICP-MS: validation and uncertainty estimation, *Talanta*, 57, 1133, 2002.

279. Richarz, A.-N. and Brätter, P., Speciation analysis of trace elements in the brain of individuals with Alzheimer's disease with special emphasis on metallothioneins, *Anal. Bioanal. Chem.*, 372, 412, 2002.

280. Albert J., Rubio, R., and Rauret, G., Arsenic speciation in marine biological materials by LC-UV-HG-ICP/OES, *Fresenius J. Anal. Chem.*, 351, 415, 1995.

281. Albert, J., Rubio, R., and Rauret, G., Extraction method for arsenic speciation in marine organisms, *Fresenius J. Anal. Chem.*, 351, 420, 1995.

282. Larsen, E.H., and Berg, T., "Trace element speciation and international food legislation— A Codex Alimentarius position paper on arsenic as a contaminant," in *Trace Element Speciation for Environment*, Food and Health, Ebdon, L., et al., Eds., RSC, Cambridge, 2001, 251.

283. Larsen, E.H., et al., Arsenic speciation in shrimp and mussel from the mid-Atlantic hydrothermal vents, *Mar. Chem.*, 57, 341, 1997.

284. Gómez-Ariza, J. L., et al., A comparison between ICP-MS and AFS detection for arsenic speciation in environmental samples, *Talanta*, 51, 257, 2000.

285. Gómez-Ariza, J.L., et al., Pretreatment procedure for selenium speciation in shellfish using high-performance liquid chromatography-microwave-assisted digestion-hydride generation atomic fluorescence spectrometry, *Appl. Organometal. Chem.*, 16, 265, 2002.

286. Vilanó M. and Rubio, R., Determination of arsenic species in oyster tissue by microwave-assisted extraction and liquid chromatography-atomic fluorescence detection, *Appl. Organometal. Chem.*, 15, 658, 2001.

287. Chen, Z.L., Lin, J.-M., and Naidu, R., Separation of arsenic species by capillary electrophoresis with sample-stacking techniques, *Anal. Bioanal. Chem.*, 375, 679, 2003.

288. Smichowski, P., Madrid, Y., and Cámara, C., Analytical methods for antimony speciation in waters at trace and ultratrace levels. A review. *Fresenius J. Anal. Chem.*, 360, 623, 1998.

289. Petit de Peña, Y., et al., On line determination of antimony (III) and antimony (V) in liver tissue and whole blood by flow injection—hydride generation—atomic absorption spectrometry, *Talanta*, 55, 743, 2001.

290. Miekeley, N., Mortari, S.R., and Schubach, A.O., Monitoring of total antimony and its species by ICP-MS and on-line ion chromatography in biological samples from patients treated for leishmaniasis, *Anal. Bioanal. Chem.*, 375, 666, 2002.

291. Chau, Y.K., Yang, F., and Brown, M., Determination of methylcyclopentadienyl-manganese tricarbonyl (MMT) in gasoline and environmental samples by gas chromatography with helium microwave plasma atomic emission detection, *Appl. Organometallic Chem.*, 11, 31, 1997.

292. Yao, W., and Byrne, R.H., Determination of trace chromium (VI) and molybdenum (VI) in natural and bottled mineral waters using long pathlength absorbance spectroscopy (LPAS), *Talanta*, 48, 277, 1999.

293. Cámara, C., Cornelis, R., and Quevauviller, P., Assessment of methods currently used for the determination of Cr and Se species in solutions, *Trends Anal. Chem.*, 19, 189, 2000.

294. Kotas, J. and Stasicka, Z., Chromium occurrence in the environment and methods of its speciation, *Environ. Pollut.*, 107, 263, 2000.

295. Marqués, M. J., et al., Chromium speciation in liquid matrices: a survey of the literature, *Fresenius J. Anal. Chem.*, 367, 601, 2000.

296. Darrie, G., "The importance of chromium in occupational health," in *Trace Element Speciation for Environment, Food and Health*, Ebdon, L., et al., Eds., RSC, Cambridge, 2001, 315.

297. Chai, Z., et al., Overview of the methodology of nuclear analytical techniques for speciation studies of trace elements in the biological and environmental sciences, *Anal. Bioanal. Chem.*, 372, 407, 2002.

298. Quevauviller, P. and Morabito, R., Evaluation of extraction recoveries for organometallic determinations in environmental matrices, *Trends Anal. Chem.*, 19, 86, 2000.

299. Olivas, R.M., et al., Analytical techniques applied to the speciation of selenium in environmental matrices, *Anal. Chim. Acta*, 286, 357, 1994.

300. Lobinski, R., et al., Species-selective determination of selenium compounds in biological materials, *Pure Appl. Chem.*, 72, 447, 2000.

301. Crews, H.M., Speciation of trace elements in foods, with special reference to cadmium and selenium: is it necessary? *Spectrochim. Acta Part B*, 53, 213, 1998.

302. Günther, K., and Kastenholz, G.J.B., Characterization of high molecular weight cadmium species in contaminated vegetable food, *Fresenius J. Anal. Chem.*, 368, 281, 2000.

303. Jackson, B.P., et al., Trace element speciation in largemouth bass (*Micropterus salmoides*) from a fly ash settling basin by liquid chromatography-ICP-MS, *Anal. Bioanal. Chem.*, 374, 203, 2002.

304. Pyrzyñska, K., Organolead speciation in environmental samples: a review, *Mikrochim. Acta*, 122, 279, 1996.

305. Szpunar-Lobiñska, J., et al., Separation techniques in speciation analysis for organometallic species, *Fresenius J. Anal. Chem.*, 351, 351, 1995.

306. Salih, B., Speciation of inorganic and organolead compounds by gas chromatography-atomic absorption spectrometry and the determination of lead species after pre-concentration onto diphenylthiocarbazone-anchored polymeric microbeads, *Spectrochim. Acta Part B*, 55, 1117, 2000.

307. Crnoja, M., et al., Determination of Sn- and Pb-organic compounds by solid-phase microextraction-gas chromatography-atomic emission detection (SPME-GC-AED) after in situ propylation with sodium tetrapropylborate, *J. Anal. Atom. Spectrom.*, 16, 1160, 2001.

308. Wasik, A., and Namieñnik, J., Speciation of organometallic compounds of tin, lead, and mercury, *Polish J. Environ. Studies*, 10, 405, 2001.

309. Forsyth, D.S. and Taylor, J., Detection of organotin, organomercury, and organolead compounds with a pulsed discharge detector (PDD), *Anal. Bioanal. Chem.*, 375, 133, 2002.

310. Rodriguez Pereiro, I. and Carro Diaz, A., Speciation of mercury, tin, and lead compounds by gas chromatography with microwave-induced plasma and atomic-emission detection (GC-MIP-AED), *Anal. Bioanal. Chem.*, 372, 74, 2002.

311. Williams, S.P., "Occupational health and speciation using nickel and nickel compounds as an example," in *Trace Element Speciation for Environment, Food and Health*, Ebdon, L., et al., Eds., RSC, Cambridge, 2001, 297.

312. Schedlbauer, O.F. and Heumann, K.G., Development of an Isotope Dilution Mass Spectrometric Method for dimethylthallium speciation and first evidence of its existence in the ocean, *Anal. Chem.*, 71, 5459, 1999.

313. White, S., et al., Speciation of organo-tin compounds using liquid chromatography—atmospheric pressure ionisation mass spectrometry and liquid chromatography—inductively coupled plasma mass spectrometry as complementary techniques, *J. Chromatogr. A*, 794, 211, 1998.

314. Elgethun, K., Neumann, C., and Blake, P., Butyltins in shellfish, finfish, water and sediment from the Coos Bay estuary (Oregon, USA), *Chemosphere*, 41, 953, 2000.

315. Lindemann, T., et al., Stability studies of arsenic, selenium, antimony and tellurium species in water, urine, fish and soil extracts using HPLC/ICP-MS, *Fresenius J. Anal. Chem.*, 368, 214, 2000.

316. De Carvalho, L. M. and Schwedt, G., Sulfur speciation by capillary zone electrophoresis: conditions for sulfite stabilization and determination in the presence of sulfate, thiosulfate and peroxodisulfate, *Fresenius J. Anal. Chem.*, 368, 208, 2000.

317. Unger-Heumann, M., Rapid tests—a convenient tool for sample screening with regard to element speciation, in: *Trace Element Speciation for Environment, Food and Health*, Ebdon, L., et al., Eds., RSC, Cambridge, 2001, 211.

318. Bi, S.-P., et al., Analytical methodologies for aluminium speciation in environmental and biological samples—a review, *Fresenius J. Anal. Chem.*, 370, 984, 2001.

319. Rosenberg, E., and Ariese, F., "Quality control in speciation analysis," in *Trace Element Speciation for Environment, Food and Health*, Ebdon, L., et al., Eds., RSC, Cambridge, 2001, 17.

320. Gómez-Ariza, J. L., et al., "Sample treatment and storage in speciation analysis," in *Trace Element Speciation for Environment, Food and Health*, Ebdon, L., et al., Eds., RSC, Cambridge, 2001, 51.

321. Huo, D., "Skip" Kingston, H.M., and Larget, B., "Application of isotope dilution in elemental speciation: speciated isotope dilution mass spectrometry (SIDMS)," in *Elemental Speciation. New Approaches for Trace Element Analysis*, Caruso, J.A., Sutton, K.L., and Ackley, K.L., Eds., Elsevier Science B.V., Amsterdam, 2000, 277.

322. Encinar, J.R., Isotope dilution analysis for speciation, *Anal. Bioanal. Chem.*, 375, 41, 2003.

323. Quevauviller, Ph., "Certified reference materials: a tool for quality control of elemental speciation analysis" in *Elemental Speciation. New Approaches for Trace Element Analysis*, Caruso, J.A., Sutton, K.L., and Ackley, K.L., Eds., Elsevier Science B.V., Amsterdam, 2000, 531.

324. Emons, H., Challenges from speciation analysis for the development of biological reference materials, *Fresenius J. Anal. Chem.*, 370, 115, 2001.

325. Cornelis, R., et al., The EU network on trace element speciation in full swing, *Trends Anal. Chem.*, 19, 210, 2000.

326. Jain, C.K., and Ali, I., Arsenic: occurrence, toxicity and speciation techniques, *Wat. Res.*, 34, 4304, 2000.

327. Wrobel, K., et al., Determination of As(III), As(V), monomethylarsonic acid, dimethylarsinic acid and arsenobetaine by HPLC-ICP-MS: analysis of reference materials, fish tissues and urine, *Talanta*, 58, 899, 2002.

328. Lai, V. W-M., Cullen, W.R., and Ray, S., Arsenic speciation in scallops, *Mar. Chem.*, 66, 81, 1999.

329. Brisbin, J.A., B'Hymer, C., and Caruso, J.A., A gradient anion exchange chromatographic method for the speciation of arsenic in lobster tissue extracts, *Talanta*, 58, 133, 2002.

330. Chatterjee, A., Determination of total cationic and total anionic arsenic species in oyster tissue using microwave-assisted extraction followed by HPLC-ICP-MS, *Talanta*, 51, 303, 2000.

331. McSheehy, S., et al., Investigations of arsenic speciation in oyster test reference material by multidimensional HPLC-ICP-MS and electrospray tandem mass spectrometry (ES-MS-MS), *Analyst*, 126, 1055, 2001.

332. McSheehy, S., et al., Multidimensional liquid chromatography with parallel ICP-MS and electrospray MS-MS detection as a tool for the characterization of arsenic species in algae, *Anal. Bioanal. Chem.*, 372, 457, 2002.

333. Coelho, N.M.M., et al., High performance liquid chromatography—atomic fluorescence spectrometric determination of arsenic species in beer samples, *Anal. Chimica Acta*, 482, 73, 2003.

334. Villa-Lojo, M.C., et al., Coupled high performance liquid chromatography-microwave digestion-hydride generation-atomic absorption spectrometry for inorganic and organic speciation in fish tissue, *Talanta*, 57, 741, 2002.

335. Lagarde, F., et al., Improvement scheme for the determination of arsenic species in mussel and fish tissues, *Fresenius J. Anal. Chem.*, 363, 5, 1999.

336. Lagarde, F., et al., Preparation of pure calibrants (arsenobetaine and arsenocholine) for arsenic speciation studies and certification of an arsenobetaine solution—CRM 626, *Fresenius J. Anal. Chem.*, 363, 12, 1999.

337. Lagarde, F., et al., Certification of total arsenic, dimethylarsinic acid and arsenobetaine contents in a tuna fish powder (BCR-CRM 627), *Fresenius J. Anal. Chem.*, 363, 18, 1999c.

338. Gong, Z., et al., Arsenic speciation analysis, *Talanta*, 58, 77, 2002.

339. Bouyssiere, B., Szpunar, J., and Lobinski, R., Gas chromatography with inductively coupled plasma mass spectrometric detection in speciation analysis, *Spectrochim. Acta Part B*, 57, 805, 2002.

340. Mizuishi, K., et al., Direct GC determination of methylmercury chloride on HBr-methanol-treated capillary columns. *Chromatographia*, 44, 386, 1997.

341. Pereiro, I.R., and Diaz, A.C., Speciation of mercury, tin, and lead compounds by gas chromatography with microwave-induced plasma and atomic-emission detection (GC-MIP-AED), *Anal. Bioanal. Chem.*, 372, 74, 2002.

342. Harrington, Ch.F., The speciation of mercury and organomercury compounds by using high-performance liquid chromatography, *Trends Anal. Chem.*, 19, 167, 2000.

343. Heltai, Gy., et al., Application of MIP-AES as element specific detector for speciation analysis, *Fresenius J. Anal. Chem.*, 363, 487, 1999.

344. Donard, O.F.X., et al., "Trends in speciation analysis for routine and new environmental issues," in *Elemental Speciation. New Approaches for Trace Element Analysis*, Caruso, J.A., Sutton, K.L., and Ackley, K.L., Eds., Elsevier Science B. V., Amsterdam, 2000, 451.

345. Baeyens, W., et al., Investigation of headspace and solvent extraction methods for the determination of dimethyl- and monomethylmercury in environmental matrices, *Chemosphere*, 39, 1107, 1999.

346. Palmieri, H.E.L., and Leonel, L.V., Determination of methylmercury in fish tissue by gas chromatography with microwave-induced plasma atomic emission spectrometry after derivatization with sodium tetraphenylborate, *Fresenius J. Anal. Chem.*, 363, 466, 2000.

347. Grinberg, P., et al., Solid phase microextraction capillary gas chromatography combined with furnace atomization plasma emission spectrometry for speciation of mercury in fish tissue, *Spectrochim. Acta Part A*, 58, 427, 2003.

348. Vanhaecke, F., Resano, M., and Moens, L., Electrothermal vaporisation ICP-mass spectrometry (ETV-ICP-MS) for the determination and speciation of trace elements in solid samples—A review of real-life applications from the author's lab, *Anal. Bioanal. Chem.*, 375, 133, 2002.

349. Rosenkrantz, B., and Bettmer, J., Rapid separation of elemental species by multicapillary GC, *Anal. Bioanal. Chem.*, 373, 461, 2002.

350. Liang, L.-N., et al., Speciation analysis of mercury in seafood by using high-performance liquid chromatography on-line coupled with cold-vapor atomic fluorescence spectrometry via a post column microwave digestion, *Anal. Chim. Acta*, 477, 131, 2003.

351. Sanchez Uria, J. E., and Sanz-Medel, A., Inorganic and methylmercury speciation in environmental samples, *Talanta*, 47, 509, 1998.

352. Moreno, M. E., Pérez-Conde, C., and Cámara C., The effect of the presence of volatile organoselenium compounds on the determination of inorganic selenium by hydride generation, *Anal. Bioanal. Chem.*, 375, 666, 2003.

353. Kannamkumarath, S. S., et al., HPLC-ICP-MS determination of selenium distribution and speciation in different types of nut, *Anal. Bioanal. Chem.*, 373, 454, 2002.

354. Wrobel, K., et al., Hydrolysis of protein with methanesulfonic acid for improved HPLC-ICP-MS determination of seleno-methionine in yeast and nuts, *Anal. Bioanal. Chem.*, 375, 133, 2003.

355. Uden, P. C., Modern trends in the speciation of selenium by hyphenated techniques, *Anal. Bioanal. Chem.*, 373, 422, 2002.

356. Gallignani, M., et al., Sequential determination of Se(IV) and Se(VI) by flow injection-hydride generation-atomic absorption spectrometry with HCl/HBr microwave aided pre-reduction of Se(VI) to Se(IV), *Talanta*, 52, 1015, 2000.

357. Guerin, T., Astruc, A., and Astruc, M., Speciation of arsenic and selenium compounds by HPLC hyphenated to specific detectors: a review of the main separation techniques, *Talanta*, 50, 1, 1999.

358. Pyrzyñska, K., Analysis of selenium species by capillary electrophoresis, *Talanta*, 55, 657, 2001.

359. Stefánka, Zs., et al., Comparison of sample preparation methods based on proteolytic enzymatic processes for Se-speciation of edible mushroom (*Agaricus bisporus*) samples, *Talanta*, 55, 437, 2001.

360. Dernovics, M., Stefánka, Zs, and Fodor, P., Improving selenium extraction by sequential enzymatic processes for Se-speciation of selenium-enriched *Agaricus bisporus*, *Anal. Bioanal. Chem.*, 372, 473, 2002.

361. Crews, H. M., The importance of trace element speciation in food issues, in: *Trace Element Speciation for Environment, Food and Health*, Ebdon, L., et al., Eds., RSC, Cambridge, 2001, 223.

362. Hoch, M., Organotin compounds in the environment—an overview, *Appl. Geochem.*, 16, 719, 2001.

363. Belfroid, A. C., Purperhart, M., and Ariese, F., Organotin levels in seafood, *Mar. Pollut. Bull.*, 40, 226, 2000.

364. St-Jean, S. D., et al., Butyltin conentrations in sediments and blue mussels (*Mytilus edulis*) of the southern Gulf of St. Lawrence, Canada, *Environ. Technology*, 20, 181, 1999.

365. Minganti, V., Capelli, R., and De Pellegrini, R., Evaluation of different derivatization methods for the multielement detection of Hg, Pb and Sn compounds by gas chromatography-microwave induced plasma-atomic emission spectrometry in environmental samples, *Anal. Bioanal. Chem.*, 351, 471, 1995.

366. Sudaryanto, A., et al., Occurrence of butyltin compounds in mussels from Indonesian coastal waters and some Asian countries, *Water Sci. Technol.*, 42, 71, 2000.

367. Albalat, A., et al., Assessment of organotin pollution along the Polish coast (Baltic Sea) by using mussels and fish as sentinel organisms, *Chemosphere*, 47, 165, 2002.

368. Kannan, K., and Falandysz, J., Butyltin residues in sediment, fish, fish-eating birds, harbour porpoise and human tissues from the Polish coast of the Baltic Sea, *Mar. Pollut. Bull.*, 34, 203, 1997.

369. Nemanic, T. M., et al., Organotin compounds in the marine environment of the Bay of Piran, northern Adriatic Sea, *J. Environ. Monit.*, 4, 426, 2002.

370. Bermejo, P., et al., Speciation of iron in breast milk and infant formulas whey by size exclusion chromatography-high performance liquid chromatography and electrothermal atomic absorption spectrometry, *Talanta*, 50, 1211, 2000.

371. Timerbaev, A. R., Element speciation analysis by capillary electrophoresis, *Talanta*, 52, 573, 2000.

9 Analysis of Vitamins for Health, Pharmaceutical, and Food Sciences

Semih Ötleş and Yildiz Karaibrahimoglu

CONTENTS

9.1   INTRODUCTION

Vitamins are organic compounds that promote and regulate essential biochemical reactions within the human body, which is generally unable to synthesize them—they have to be obtained from food in trace amounts for growth, health, and reproduction, and lack of these compounds in the diet results in overt symptoms of deficiency. Vitamins themselves do not provide energy but help convert fat and carbohydrates into energy and assist in forming bone and tissue.[1] The vitamins include vitamin D, vitamin E, vitamin A, and vitamin K—fat-soluble vitamins—and thiamin (vitamin B_1), riboflavin (vitamin B_2), pantothenic acid, folate (folic acid), vitamin B_{12} (cyanocobalamin), biotin, vitamin B_6, niacin, and vitamin C (ascorbic acid)—water-soluble vitamins.[2]

As more is learned about vitamins, it is becoming apparent that the differences between these two groups (water-soluble and fat-soluble vitamins) concern more than just solubility. Although the modes of action of the fat-soluble vitamins are not nearly as clearly understood as those of the water-soluble vitamins, it is evident that the coenzyme activities of the former are different from that of water-soluble vitamins. Vitamins are required in the diet in only tiny amounts, in contrast to the energy components of the diet. The energy components of the diet are sugars, starches, fats, and oils, and these occur in relatively large amounts in the diet. For many years, the National Research Council of the U.S. National Academy of Sciences has taken responsibility for establishing guidelines on quantity of the various nutrients that should be consumed by men and women at various ages. These are called "RDAs" (recommended dietary allowances and often referred to as recommended daily allowances). RDAs are defined as the levels of intake of essential nutrients that, on the basis of scientific knowledge, are judged by the Food and Nutrition Board to be adequate to meet the known nutrient needs of practically all healthy persons. RDAs are categorized by age, weight, height, sex, pregnancy, and lactation. The RDA of fat-soluble vitamins and water-soluble vitamins is shown in Tables 9.1 and 9.2, respectively, which is a standard for the daily amounts of vitamins needed for a healthy person.[3] Unfortunately, the amounts the experts came up with give us only the bare minimum required to ward off deficiency diseases such as beriberi, rickets, scurvy, and night blindness. They do not indicate the amounts needed to maintain maximum health, instead of borderline health.

The proper balance of vitamins is also important for the proper functioning of all vitamins. Scientific research has proved that an excess of an isolated vitamin can produce the same symptoms as a deficiency. The water-soluble vitamins are not stored in the body in significant amounts and need to be replaced daily by food or supplement to maintain adequate levels. These vitamins can be rapidly depleted in conditions interfering with intake or absorption. However, the fat-soluble vitamins are better stored in the body, and if not excreted, toxic levels could occur. In Table 9.3, the structures, functions in body, good dietary sources, and deficiency signs of fat- and water-soluble vitamins are summarized.[4–10] Many vitamins function together to regulate several processes within the body. A lack of vitamins or a diet that does not provide

TABLE 9.1
RDA of Fat-Soluble Vitamins[a]

| Age | Energy(kcal) | Protein (g) | Vitamin A | | Vitamin D | | Vitamin E | | Vitamin K(μg) |
			(IU)	(RE) (μg)	(IU)	(μg)	(IU)	(TE) mg	
				Children					
4–6	1800	30/24	2500	500	400	5	9	7	2/20
7–10	2400/2000	36/28	3300	500	400	5	10	7	2/30
				Males					
15–18	3000	54/59	5000	1000	400	5	15	10	2/65
19–24	3000/2900	54/58	5000	1000	400	5	15	10	2/70
25–50	2700	56/63	5000	1000	—	5	15	10	2/80
501	2400	56/63	5000	1000	—	10	15	10	2/80
				Females					
15–18	2100	48/44	4000	800	400	5	12	8	2/55
19–24	2100	46/46	4000	800	400	5	12	8	2/60
25–50	2000	46/50	4000	800	—	5	12	8	2/65
501	1800	46/50	4000	800	—	10	12	8	2/65

[a] The first figure refers to the old RDA listing; the second figure refers to the newer DRI listing.

TABLE 9.2
RDA of Water-Soluble Vitamins[a]

Age	Ascorbic Acid (mg)	Folacin/ Folate (mcg)	Niacin (mg)	Riboflavin (mg)	Thiamine (mg)	Vitamin B_6 (mg)	Vitamin B_{12} (µg)
			Children				
4–6	40/45	200/75	12	1.1	0.9	0.9/1.1	1.5/1.0
7–10	40/45	300/100	16/13	1.2	1.2/1.0	1.2	2.0/1.4
			Males				
15–18	45/60	400/200	20	1.8	1.5	2.0	3.0/2.0
19–24	45/60	400/200	20/19	1.8/1.7	1.5	2.0	3.0/2.0
25–50	45/60	400/200	18/19	1.6/1.7	1.4/1.5	2.0	3.0/2.0
501	45/60	400/200	16/15	1.5/1.4	1.2	2.0	3.0/2.0
			Females				
15–18	45/60	400/180	14/15	1.4/1.3	1.1	2.0/1.5	3.0/2.0
19–24	45/60	400/180	14/15	1.4/1.3	1.1	2.0/1.6	3.0/2.0
25–50	45/60	400/180	13/15	1.2/1.3	1.0/1.1	2.0/1.6	3.0/2.0
501	45/60	400/180	12/13	1.1/1.2	1.0	2.0/1.6	3.0/2.0

[a] The first figure refers to the old RDA listing; the second figure refers to the newer DRI listing.

adequate amounts of certain vitamins can upset the body's internal balance or block one or more metabolic reactions. Vitamin B_1 deficiency results in beriberi, a disease resulting in atrophy, weakness of the legs, nerve damage, and heart failure. Vitamin C deficiency results in scurvy, a disease that involves bleeding. Specific diseases uniquely associated with deficiencies of vitamin B_6, riboflavin, or pantothenic acid have not been found in humans, though persons who have been starving or consuming poor diets for several months might be expected to be deficient in most of the nutrients, including vitamins B_2, B_6, and pantothenic acid. Niacin deficiency causes pellagra, which involves skin rashes and scabs, diarrhea, and mental depression. Vitamin B_{12} deficiency may occur with the failure to consume meat and milk or other dairy products. Vitamin B_{12} deficiency causes megaloblastic anemia and, if severe enough, can result in irreversible nerve damage. Mild or moderate folate deficiency is common throughout the world and can result from the failure to eat green, leafy vegetables and fruits and fruit juices. Folate deficiency causes megaloblastic anemia, which is characterized by the presence of large abnormal cells called megaloblasts in the circulating blood. The symptoms of megaloblastic anemia are tiredness and weakness. Vitamin A deficiency is common throughout the poorer parts of the world and causes night blindness. Severe vitamin A deficiency causes xerophthalmia, a disease that if left untreated results in total blindness. Vitamin D deficiency causes rickets, a disease that leads to weakening of the bones. Vitamin E deficiency occurs rarely and causes nerve damage. Vitamin K deficiency causes spontaneous bleeding (Table 9.3).[4–10]

Although a great deal is known about vitamins, investigations continue to determine their chemical structures, properties, and functions.

TABLE 9.3
Properties of Vitamins (and Related Nutrients)

Vitamin	Structure	Functions in Body	Good Dietary Sources	Deficiency Signs
Water-Soluble Vitamins				
B$_1$ (thiamin)	[chemical structure]	Used in energy metabolism (the conversion of protein, carbohydrates, and fat into energy). Supports normal appetite and needed for brain and nervous system function. A role in detoxification and heart functions.	Bread, brewer's yeast, blackstrap molasses, brown rice, cereals, cereal products, egg yolk, peas, potatoes, milk, and meat.	Beriberi, a disease resulting in atrophy, weakness of the legs, nerve damage, and heart failure.
B$_2$ (riboflavin)	[chemical structure]	Essential for cellular energy metabolism. Supports hormone production, neurotransmitter function, healthy eyes and skin, and the production of red blood cells.	Brewer's yeast, milk, meat, cereals, eggs, yogurt, blackstrap molasses, brown rice, egg yolk, peas.	May result in itching and burning eyes; cracks and sores in the mouth and lips; bloodshot eyes; purplish tongue; dermatitis; retarded growth; digestive disturbances; trembling; sluggishness; oily skin.
B$_3$ (niacin/ niacinamide)	[chemical structure]	Used in the release of energy from carbohydrates. Health of skin, nervous system, digestive system. Essential for normal growth.	Milk, brewer's yeast, eggs, meat, fish, peanuts, potatoes, poultry, rice, and all protein-containing foods.	Pellagra, gastrointestinal disturbance, nervousness, headaches, fatigue, mental depression, vague aches and pains, irritability, loss of appetite, insomnia, skin disorders, muscular weakness, indigestion, bad breath, canker sores.

Vitamin	Structure	Function	Sources	Deficiency symptoms
B_5 (pantothenic acid)		Used in the release of energy from carbohydrates. Part of a coenzyme vital for energy release. Important for the health of the adrenal gland. Formation of antibodies.	Liver, brewer's yeast, legumes, salmon, chicken, porridge, meat, nuts, cereals, eggs, dairy products.	Painful and burning feet, skin abnormalities, retarded growth, dizzy spells, digestive disturbances, vomiting, restlessness, stomach stress, muscle cramps.
B_6 (pyridoxine)		Important in protein synthesis and the manufacture of hormones, red blood cells, and enzymes. Vital for maintaining a healthy nervous system, skin, muscles, and blood and crucial for a healthy immune system. Involved with conversion of the omega 6 essential fatty acids, which play a role in hormone health.	Liver, wholegrain foods, brewer's yeast, meat/poultry, peanuts, walnuts, bananas, fish, broccoli, potatoes, soybeans, milk, peanuts.	Nervousness, insomnia, skin eruptions, loss of muscular control, anemia, mouth disorders, muscular weakness, dermatitis, arm and leg cramps, loss of hair, slow learning, and water retention.
B_{12}		Involved with red blood cell formation and health of nervous system. Absorption requires intrinsic factor, which is secreted by the stomach. Aids in the replication of the genetic code within each cell and plays a role in the processing of carbohydrates, protein, and fats in the body.	Meat, pork, liver, fish, eggs, dairy products, also fortified soya products, yeast extract, miso, bananas, kelp, peanuts.	Pernicious anemia, poor appetite, growth failure in children, tiredness, brain damage, nervousness, neuritis, degeneration of spinal cord, depression, lack of balance.

continued

TABLE 9.3 (continued)
Properties of Vitamins (and Related Nutrients)

Vitamin	Structure	Functions in Body	Good Dietary Sources	Deficiency Signs
Folic acid (or folate)		Regulates cell division and the transfer of inherited traits from one cell to another. Supports the health of gums, red blood cells, skin, gastrointestinal tract, and immune system. Reduces risk of neural tube birth defects. Heart health—helps regulate blood levels of homocysteine.	Wheat germ, nuts, green leafy vegetables (broccoli, Brussels sprouts), peas, chickpeas, brewer's yeast, brown rice, oranges, bananas, liver, fish, eggs.	Gastrointestinal disorders, anemia, vitamin B_{12} deficiency, premature gray hair.
Biotin		Part of coenzyme used in energy metabolism and fat synthesis. Supports health of skin, hair, and mucous membranes.	Organ meats (kidney, liver), eggs, yeast extract, cereals, beer, some fruit, and vegetables.	Extreme exhaustion, drowsiness, muscle pain, loss of appetite, depression, grayish skin color.

	Structure	Function	Sources	Deficiency
Choline and inositol		Lipotropic factors help to prevent fat accumulations in the liver. Choline is involved in formation of the brain chemical acetylcholine. Inositol is involved in nerve transmission and the regulation of certain enzymes.	Liver, meat, nuts, brewer's yeast, pulses, bread. Components of lecithin.	Cirrhosis and fatty degeneration of the liver, hardening of the arteries, heart problems, high blood pressure, hemorrhaging of kidneys due to choline deficiency; high blood cholesterol, constipation, eczema, hair loss due to inositol deficiency.
PABA		Involved in B vitamin metabolism as an enzyme cofactor, amino acid metabolism, and health of red blood cells.	Liver, yogurt, green leafy vegetables, eggs, wheat germ, and molasses.	Extreme fatigue, eczema, irritability, depression, nervousness, constipation, headaches, digestive disorders, hair turning gray prematurely.
C		An important antioxidant that helps protect cells against damage caused by free radicals. Immune system (antibodies and white blood cells), strengthening resistance to infection. Collagen formation. Iron absorption. Formation of corticosteroid hormones in the adrenal gland. Mild antihistamine. Plays a role in healthy gums, skin, and vision.	Fresh fruit and vegetables, especially bell peppers (capsicum), black currants, Brussels sprouts, fresh oranges, strawberries, melons, asparagus, rosehips, broccoli, and watercress.	Scurvy, soft and bleeding gums, swollen or painful joints, slow-healing wounds and fractures, bruising, nosebleeds, tooth decay, loss of appetite, muscular weakness, skin hemorrhages, capillary weakness, anemia, impaired digestion.

continued

TABLE 9.3 (continued)

Properties of Vitamins (and Related Nutrients)

Vitamin	Structure	Functions in Body	Good Dietary Sources	Deficiency Signs
Bioflavonoids	(chemical structure: flavonoid ring system with rings labeled A, B, C; positions numbered 1, 2, 3, 4, 5, 6, 7, 8; O and O atoms shown)	A complex closely associated with vitamin C. There are many compounds in this group, including rutin and quercetin Bioflavonoids can increase the absorption of vitamin C. They are protective antioxidants; some have anti-inflammatory properties.	Citrus fruits, black currants, fruits, buckwheat.	Night blindness; increased susceptibility to infections; rough, dry, scaly skin; loss of smell and appetite; frequent fatigue; lack of tearing; defective teeth, and retarded growth of gums.
		Fat-Soluble Vitamins		
A (retinol)	(chemical structure with OH group)	Involved in normal eyesight, immune system response, embryonic development. Mucous membranes. Skin health. Development of bones and teeth.	Milk, cream, cheese, butter, eggs, liver, cod liver oil, mackerel, herring.	Night blindness; increased susceptibility to infections; rough, dry, scaly skin; loss of smell and appetite; frequent fatigue; lack of tearing; defective teeth, and retarded growth of gums.
Beta carotene	(chemical structure)	Precursor of vitamin A; converted to vitamin A only when body needs it. Functions as for vitamin A, but also a significant antioxidant. One of the carotenoid compounds—important antioxidant plant pigments.	Carrots, spinach, watercress, sweet potato, broccoli, apricots, tomatoes. (Beta carotene provides orange pigment in fruit and vegetables.)	Night blindness; increased susceptibility to infections; rough, dry, scaly skin; loss of smell and appetite; frequent fatigue; unable to tear; defective teeth, and retarded growth of gums.

	Structure	Function	Sources	Deficiency
D		Converted from cholesterol by action of sunlight on the skin. Required for bone and teeth strength. For correct calcium absorption as it is changed to a calcium-controlling hormone in the body. For healthy immune function.	Milk, cheese, butter, margarines, eggs, liver, fish, cod liver oil.	Osteoporosis (brittle bone disease), rickets (in children), and osteomalacia (bone thinning in adults).
E		Fat-soluble antioxidant. Protects the fatty parts of cell walls; prevents oxidation of polyunsaturated fatty acids. Reduces oxygen requirement of muscles. Prevents degeneration of nerves and muscles.	Wheat germ oil, other vegetable/plant oils, green leafy vegetables, whole grains.	A rupture of red blood cells, loss of reproductive powers, lack of sexual vitality, abnormal fat deposits in muscles, degenerative changes in the heart and other muscles; dry skin.
K		Required for the synthesis of blood-clotting proteins. Involved in bone formation and the regulation of blood calcium levels.	Manufactured by colon bacteria. Meat, liver, green leafy vegetables, soya margarines, tea. Occurs widely in other foods.	Spontaneous bleeding.

9.2 METHOD DEVELOPMENTS IN VITAMIN ANALYSIS

The qualitative and quantitative analysis of vitamins in foods, whether inherently present or added during food manufacture, plasma, serum, and pharmaceuticals has become important to the food industry, to assess the nutritional quality of foodstuff, and to the medical and pharmaceutical industries, for protection or control of human health. Increasing interest in good eating habits in humans and also in animals has meant greater awareness of the vital role that vitamins play in growth and health. In addition, the presence of fruits and vegetables in the daily diet and the consumption of vitamin-supplemented and preserved foods have substantially increased. Vitamins may be leached during the storage and processing of foods, with a possible loss through chemical reactions so it is extremely important to have available preparations to replace the possible lack of vitamins in daily diet. Hence, multivitamin pharmaceuticals are widely employed. These facts, together with the introduction of food-labeling regulations, lead to a need for very powerful analytical separation techniques for the quality control of these complex preparations. This has stimulated research on accurate and efficient analytical methods to quantify vitamins. However, the instability and the complexity of the matrices in which they are usually analyzed remain a problem.[11,12]

Although the literature is replete with descriptions of different methods for the analysis of such diversified products, the search for better methods continues. Attempts to quantify fat- and water-soluble vitamins in samples have resulted in a large number of methods. Data on amount of vitamin given in the literature are based on indirect determination by curative bioassays done earlier and recalculated in an uncertain way. The standard and official analytical methods, which are tedious, sometimes nonspecific, and time consuming, involve pretreatment of the samples through complex chemical, physical, or biological reactions to eliminate the interferences commonly found, followed by individual methods for each different vitamin. These methods include spectrophotometric, polarographic, fluorimetric, enzymatic, and microbiological procedures. In addition to colorimetric methods, there are numerous nonspectrophotometric methods, and their number and kind are increasing rapidly. These include titrimetry, voltammetry, fluorometry, potentiometry, radioimmunoassay (RIA), enzyme-linked immunosorbent assay, kinetic-based chemiluminescence (CL), flow injection analyses, and chromatography. Individual vitamins can be chromatographed isocratically and in certain combinations of two or three vitamins. The simultaneous chromatography of more complicated mixtures may require a gradient elution program. Determinations can be carried out, among other methods, by ion-exchange chromatography (IEC), normal-phase chromatography, ion-pairing chromatography , or reversed-phase chromatography, being the most common method. Most of the published methods involve the use of complex buffered mobile phases. Several bonded and stationary phases or column-packing materials are developed, and several detection methods can be applied, such as ultraviolet (UV)-visible absorbance with a single or variable wavelength or photodiode array, fluorimetric, or electrochemical.[12,13]

High-performance capillary electrophoresis is a relatively new separation technique that has several advantages over reversed-phase-high-pressure liquid chromatography (RP-HPLC). The advantages are higher efficiency, higher resolution, and method simplification in addition to the availability of various techniques developed for charged and uncharged analytes in free-zone capillary electrophoresis and micellar electro-kinetic capillary chromatography.[14] Vitamins were determined in different kinds of pharmaceutical formulations, such as tablets, injections, syrup, and gelatin capsules, using micellar electrokinetic chromatography or capillary zone electrophoresis.[15–17]

The use of mass spectrometry (MS) is gaining ground and can be combined with several other separation techniques, such as gas liquid chromatography, HPLC, and capillary electrophoresis (CE). The coupled technique LC–MS enables the separation of nonvolatile thermally labile substances for introduction into the mass spectrometer for reliable identification. The assay of vitamins is important in food analysis because of their important biological activity in humans, and few reports dealing with the LC–MS analysis of fat-soluble and water-soluble vitamins have been published recently. Nevertheless, the few studies on the LC–MS analysis of these natural compounds have revealed that this coupled technique has considerable potential in the characterization and determination of vitamins.[18]

9.3 PROCESSES FOR SAMPLE PREPARATION

Vitamins, which usually occur in trace amounts, are difficult to be analyzed in foods, serum, and plasma, which contain interfering substances. Several successful methods have been developed for extraction, cleanup, and quantification of vitamins from a wide variety of food matrices, serum, and plasma.[19–21] Because most research for method development are not carried out under the conditions associated with the particular problems of the individual analyst, it is common that analysts will be required to do some method development or modification to solve their current problems.[22] Therefore, care is required in each new application to ensure correct preparation and separation.[11]

The sample preparation process for the analysis of vitamins has a direct impact on accuracy, precision, and quantification limits and is often the rate-determining step for many analytical methods. Aqueous samples must be processed to isolate and concentrate organic analytes from the sample matrix and provide a suitable sample extract for instrumental analysis. Although the significance of sample preparation is often overlooked, it is arguably the most important step in the analytical process.[23] Bioassay and microbiological methods can detect even the tiny amounts of vitamins in serum, plasma, and food samples and are sufficiently specific to not require much sample cleanup, but these methods are time consuming and of limited precision. Fluorometric and spectrophotometric methods are considered most desirable because of their sensitivity, rapidity, and specificity, but they require additional pretreatments to separate water- and fat-soluble vitamins from interfering substances. Therefore, various chromatographic procedures (column and paper chromatography, thin-layer chromatography) have been used for the purification of extracts.[24]

The extraction protocols suggested for the determination of vitamins in the free form are more varied. Some authors have recommended chemical methods. Others prefer an acid phosphatase or a takadiastase, sometimes combined with a β-glucosidase treatment. In some cases, enzymatic treatments were preceded by hydrolysis of mineral acids. The first step in the analysis is extraction from the food sample, which is done with a combination of either strong acid or alkali or solvent with heating. This is necessary because measurement techniques cannot be used for solid food substances. The next step is enzyme treatment to release any of the vitamins bound to proteins and other components in the food matrix.[17-27]

Analytical chemists continue to search for sample preparation procedures that are faster, easier, safer, and less expensive to provide accurate and precise data with reasonable quantitation limits. Solid-phase extraction (SPE) techniques have been developed to replace many traditional liquid–liquid extraction methods for the determination of vitamins in aqueous samples.[23] SPE is used as a method for fractionating the vitamins into two fractions (water-soluble and fat-soluble), in case both of them exist in the same sample. The vitamins are then measured using chemical, microbiological, or HPLC techniques. Sample cleanup procedures are often used prior to HPLC analysis to remove interfering compounds. The use of supercritical fluids in vitamin analysis provides an interesting alternative to the use of organic solvents. The main advantages of using supercritical fluids over conventional organic solvents are the minimal consumption of organic solvents, exclusion of oxygen, and reduction of heat. Modern supercritical fluid extraction offers shorter extraction times, potentially higher selectivity, and increased sample throughput (due to available automated instruments) compared to conventional solvent extraction techniques. The chromatography analogue supercritical fluid chromatography permits the separation of compounds of widely different polarities and molecular masses and eliminates the need for derivatization of vitamins.[14,27,28]

9.4 WATER-SOLUBLE VITAMINS

9.4.1 Vitamin C

The determination of L-ascorbic acid and L-dehydroascorbic acid (its oxidized form) in serum, plasma, pharmaceuticals, and foods has received considerable attention, and reliable assays have been available in many situations for decades. In the FLAIR report, the reasons for variability are summarized as follows:

Vitamin C is easily oxidized; therefore, plasma samples must be immediately acidified and frozen; a variety of assay procedures are available, which differ in fundamental principle; a quality control material does not exist with certified L-ascorbic acid levels for use as an in-house method validation tool; methods that measure only the reduced or oxidized forms of vitamin C are still in use; some assay methods are subject to positive or negative interferences by other components in the plasma.[21]

At present, vitamin C is determined separately using widely different techniques including colorimetry, titrimetry, UV spectrophotometry (2,6-dichlorophenolindophenol, 2,4-dinitrophenylhydrazine), and CL (Cu(II)-luminol, Ce(IV)-rhodamine 6G, Fe(II)-luminol-O_2, $KMnO_4$-luminol, H_2O_2-hemin-luminol, and H_2O_2-luminol-peroxidase CL), as well as kinetic, electrochemical (polarography, solid electrode voltammetry, amperometry), fluorometric (*ortho*-phenylenediamine), chromatographic (HPLC-UV absorbance, fluorescence, diode array, electrochemical detection, gas chromatography (GC)-flame ionization detector, electron capture detector, Nitrogen phosphorus detector; paper chromatography, thin-layer chromatography), and other spectroscopic methods (Fourier transform infrared photoacoustic and Fourier transform Raman spectroscopy).[29-33]

A variety of analytical procedures exist for detecting vitamin C, but no procedure is entirely satisfactory due to the lack of specificity and problems arising from numerous interfering substances contained in most foods.[4,34] Among these, many methods have been investigated to suit to the analysis of biological fluids such as urine and blood samples. The use of urinary levels of ascorbic acid has limited value, but the measurement of vitamin C content in blood is of importance because it reflects the body's vitamin C status and therefore requires methods for its analysis that can provide precise and accurate results without compromising their sensitivity and selectivity.[35] Application of GC to determine vitamin C has been on a limited scale, and only a few studies that deal with its application to food have appeared.[36] The use of Fourier transform Raman spectroscopy in food and pharmaceuticals is gaining acceptance because of its excellent precision, good signal-to-noise ratio, and minimal data acquisition time.[37]

Nowadays, the most common mode of analysis of vitamin C is HPLC, which overcomes the main drawbacks of chemical methods in the presence of interfering substances, because it is more accurate, selective, and sensitive—and it is the only technique that can distinguish between the L- and D-isomers of vitamin C.[26,30] There is a wide range of HPLC techniques such as RP-HPLC, normal-phase-HPLC, IEC, paired-ion with different columns (C_{18}, NH_2), and detections (UV, fluorescence, coulochemical). Vitamin C assays have begun to use CE, and GC/MS is also proving to be a powerful alternative for certain specific tasks and method validation.[26,30,34,38–42]

9.4.2 VITAMIN B GROUP

The first assays developed for the vitamin B group were animal assays, followed by microbiological methods.[22] Vitamin B_6 (*Saccharomyces carlsbergensis as the test organism*), folic acid (*Lactobacillus casei*), vitamin B_{12} (*Lactobacillus leichmannii*), pantothenic acid (*Lactobacillus plantarum*), and biotin (*Allescheria boydii*) are usually assayed microbiologically. The literature contains numerous references to determine vitamin B group by a wide variety of techniques, including microbiological methods, classical gravimetric and titration methods, spectrophotometry, spectrofluorimetry, CL, selective membrane electrodes, potentiometry, capillary electrophoresis, chromatography (thin-layer chromatography, gas chromatography, HPLC), and hyphenated techniques (HPLC-MS, GC-MS, etc.).[36,43–57]

However, many of these methods suffer from lack of selectivity, are complicated and tedious, and require the use of either expensive instrumentation or dangerous reagents. Among them, gravimetric method, titration, and many electrochemical analysis methods are suitable for the determination of some of vitamin B group in high levels. There were small applications for the determination of vitamin B group by using the thin-layer chromatography, gas chromatography, and capillary electrophoresis. At present, the most common methods used are high-performance liquid chromatography, fluorimetry, and spectrophotometry. Fluorimetry is particularly useful in pharmaceutical analysis, and the lack of fluorescence in many compounds has led to the development of reagents that aid the formation of fluorescent derivatives. The most widely used method for the determination of thiamine is the so-called thiochrome method, which involves the oxidation of thiamine in alkaline solution and extraction of thiochrome formed from the aqueous phase into an organic phase, which is then measured fluorimetrically. Riboflavin is usually assayed fluorimetrically by measuring the characteristic yellowish green fluorescence. Niacin is determined by first hydrolyzing the samples with sulfuric acid to liberate nicotinic acid from combined forms. The pyridine ring of the nicotinic acid is opened with cyanogen bromide, and the fission product is coupled with sulfanilic acid to yield a yellow dye at 470 nm. The high-performance liquid chromatography used to determine vitamin B group, and their isomers has high sensitivity, good selectivity, and the ability of simultaneous multicomponent determination. Therefore, HPLC methods have been widely used to study this group and have been applied to the separation and determination of the complex samples, such as pharmaceuticals, blood, serum, and foods.[4,21,30,58–87]

9.5 FAT-SOLUBLE VITAMINS

The fat-soluble vitamins (vitamins A, D, E, and K) are of great importance in nutrition. They consist of different classes of chemical compounds, including retinoids, olefins, D_2, D_3 and its precursors, tocopherols, tocotrienols, menadiones, and menaquinones, which show distinct bioactivity according to International Union of Pure and Applied Chemistry. These compounds are linked to other compounds present in cellular membranes of foods and feeds by physical forces, chemical forces, or covalent linkage. These forces may affect bioavailability.[88]

Previously, the accepted method to determine fat-soluble vitamins was bioassay. True biological activities of fat-soluble vitamins are determined by their ability to prevent or reverse specific deficiency symptoms of each fat-soluble vitamin *in vivo*. Biochemical and histological methods are common, but there is no microbiological method for the determination of fat-soluble vitamins. Some success was claimed for chemical or physical methods applied to high-potency mixtures. The concentrations of fat-soluble vitamins in most food, blood, and serum samples are low, and these vitamins are usually found in the presence of larger amounts of other substances (e.g., sterols, triglycerides, phospholipids) that may cause interference. The advent of HPLC, GC, CE, and hyphenated techniques (e.g., HPLC-MS, GC-MS) has revolutionized the analysis of fat-soluble vitamins. It is now generally accepted that the activity of vitamin A in foods and biological samples is best determined by chromatographic separation of the carotenoids, followed by summation of the activities present in the various geometric and stereoisomers. The early analytical procedures, such as the Carr-Price reaction with antimony trichloride, gave results that were too high, and they were insensitive to *cis/trans* isomerization. In recent years, HPLC has overtaken GC as the method of choice for the analysis of vitamin E isomers because of the greater flexibility and convenience it offers in handling different sample matrices. For routine analysis of vitamin E in foods, the first step is to liberate the vitamin from the sample by saponification, but this procedure is not needed when extracting vitamin E from biological fluids. This step is followed by extraction and evaporation, and the residue is redissolved in a solvent that is compatible with the mobile phase prior to injecting it onto the column. Because of the low concentration of vitamin K in body fluids and the lack of a specific and sensitive physical property, chemical derivatization process or RIA, its quantitation required the development of highly efficient HPLC assays, with at least two sequential separation stages, to achieve an adequate purity and a high enough concentration for accurate quantitation by fluorimetry or electrochemical detection. Excellent methods based on competitive protein-binding assays and radioreceptor assay using [³H] metabolites have been available for the past two decades for the analysis of vitamin D in serum. These methods rely on extensive prepurification of extracts by LC or SPC. HPLC techniques for the analysis of fat-soluble vitamins have been the methods most often applied over the past 10 years even though SFC has been developed and improved.[4,21,24,30,34,88–117]

REFERENCES

1. Ötleş, S., Hisil, Y., and Kavas, A., Synergistic effects of vitamins. *J. Facul. Eng. B,* 6, (2), 147, 1988.
2. Colakoglu, M., and Ötleş, S., Factors affecting vitamin degradation and preventing methods of vitamins. *J. Facul. Eng. B,* 3, (2), 71–84 (1985).
3. www.anyvitamins.com/rda.htm
4. Fennema, O.R., *Food Chemistry,* 2nd Edition, Marcel Dekker, New York, 1985, 477.
5. Krause, M. V. and Mahan, L. K., *Food, Nutrition & Diet Therapy,* 7th Edition, W.B. Saunders Company, Philadelphia, 1984, 99.
6. Machlin, L.J., *Handbook of Vitamins,* Marcel Dekker, New York, 1984, 1.
7. Ötleş, S. and Colakoglu, M., Reiche Quellen von Vitamine. *J. Facul. Eng. ,* 5, (2), 119, 1987.
8. Ötleş, S. and Pire, R., Diet and cancer. *Hi-Tech,* 25, 132, 1999.
9. www.realhealth.co.uk/vitamins.htm
10. www.roche-vitamins.com
11. Hisil, Y. and Ötleş, S., Methods of vitamin analysis in foods and related problems. *J. Facul. Eng.,* 7, (1), 69, 1989.
12. Moreno, P. and Salvado, V., Determination of eight water- and fat-soluble vitamins in multi-vitamin pharmaceutical formulations by high-performance liquid chromatography, *J. Chrom.,* 870, 207, 2000.
13. Pire, R. and Ötleş, S., Chromatographic techniques and their importance in food industry. *Food Sci. Tech.,* 4, (2), 34, 1999.
14. Buskov, S., Moller, P., Sorensen, H., Sorensen, J. C., and Sorensen, S., Determination of vitamins in food based on supercritical fluid extraction prior to micellar electrokinetic capillary chromatographic analyses of individual vitamins, *J. Chrom.,* 802, 233, 1998.
15. Flotsing, L., Fillet, M., Bechet, L., Hubert, P., and Crommen, J., Determination of six water-soluble vitamins in a pharmaceutical formulation by CE, *J. Pharmaceut. Biomed. Anal.,* 15, 1113, 1997.
16. Jenkins, M. and Guerin, M., Capillary electrophoresis as a clinical tool, *J. Chrom. B,* 682, 123, 1996.
17. Schwiewe, J., Mrestani, Y., and Neubert, R., Application and optimization of CZE in vitamin analysis, *J. Chrom.,* 717, 255, 1995.
18. Careri, M., Bianchi, F., and Corradini, C., Recent advances in the application of mass spectrometry in food-related analysis, *J. Chrom.,* 970, 3, 2002.
19. Hagg, M., Effect of various commercially available enzymes in the LC determination with external standardization of thiamine and riboflavin in foods. *J. AOAC Int.,* 77, (3), 681, 1994.

20. Ötleş, S., Vergleichende Vitamin-B$_1$-und B$_2$-Bestimmungen mit verschiedenen Enzympraparaten in Lebensmitteln (Comparisons of vitamins B1 and B2 analysis by using of different enzymes in foods). *Zeitschrift für Lebensmittel-Untersuchung und-Forschung*, 193, 347, 1991.

21. Eitenmiller, R.R. and Landen, W., *Vitamin Analysis for Health and Food Sciences*, CRC Press LLC, New York, 1998, 1.

22. Augustin, J., Klein, B., Becker, D., and Venugopal, P., *Methods of Vitamin Assay*, John Wiley & Sons, New York, 1985, 1.

23. Beney, P., Breuer, G., Jacobs, G., Larabee-Zierath, D., Mollenhauer, P., Norton, K., and Wichman, M., Review, evaluation and application of solid phase extraction methods, *Hotline*, 35, (6), 1, 1996.

24. Ötleş, S., High performance liquid chromatography of water-soluble and fat-soluble vitamins. *CAST (Chromatography and Separation Technology)*, 10, (6), 20, 1999.

25. Gauch, R., Leuenberger, U., and Muller, U., Bestimmung der wasserloslicher Vitamine in Milch durch HPLC, *Z Lebensm Unters Forsch*, 195, 312, 1992.

26. Rizzolo, A. and Posello, S., Chromatographic determination of vitamins in foods, *J. Chrom.*, 624, 103, 1992.

27. Turner, C., King, J., and Mathiasson, L., Supercritical fluid extraction and chromatography for fat-soluble vitamin analysis, *J. Chrom.*, 936, 215, 2001.

28. Berg, H., Turner, C., Dahlberg, L., and Mathiasson, L., Determination of food constituents based on SFE: Applications to vitamins A and E in meat and milk, *J. Biochem. Biophys. Methods*, 43, 391, 2000.

29. Anonymous, Analytical Procedures for the determinations of vitamins in multivitamin preparations, Roche, 1974, 1.

30. Bates, C. J., *Vitamins: Fat and Water Soluble: Analysis, Encyclopedia of Analytical Chemistry*, John Wiley & Sons Ltd, Chichester, 2000, 1.

31. Bognar, A., Bestimmung von vitamin C in lebensmitteln mittels HPLC, *Deutsche Lebensmittel-Rundschau*, 84, (3), 73, 1988.

32. Bognar, A. and Daood, H., Simple in-line postcolumn oxidation and derivatization for the simultaneous analysis of AA and DHA in foods, *J. Chromatogr. Sci.*, 38, 162, 2000.

33. Ötleş, S., Comparative determination of ascorbic acid in bass (Morone lebrax) liver by HPLC and DNPH methods *int. J. food sci. nutr.*, 46, 229, 1995.

34. Ötleş, S., Applications of HPLC and GC to vitamin and lipid analysis. *LC.GC Europe*, 13, (12), 905, 2000.

35. Arya, S. and Jain, P., Non-spectrophotometric methods for the determination of Vitamin C, *Analytica Chimica Acta*, 417, 1, 2000.

36. Velisek, J. and Davidek, J., GLC of vitamins in foods, *J.micronutr. anal.*, 2, 25, 1986.

37. Yang, H. and Irudayaraj, J., Rapid determination of vitamin C by NIR, MIR and FT-Raman techniques, *J. Pharm. Pharmacol.*, 54, 1, 2002.

38. Antonelli, M., D'Ascenzo, G., Lagana, A., and Pusceddu, P., Food analyses: a new calorimetric method for ascorbic acid (vitamin C) determination, *Talanta*, 1, 2002.

39. Augustine, J., Beck, C., and Marousek, G., Quantitative determination of ascorbic acid in potatoes and potato products by HPLC, *J. Food Sci.*, 46, (1), 312, 1981.

40. Kissinger, P. and Pachla, L., Determinations of AA and DHA using LC with UV and EC detection, *Food Tech.*, 11, 108, 1987.

41. Nicholson, I., Macrae, R., and Richardson, D., Comparative assessment of HPLC methods for the determination of ascorbic acid and thiamin in foods, *Analyst*, 109, 267, 1984.

42. Vanderslice, J. and Higgs, D., HPLC analysis with FD of vitamin C in food samples, *J. Chromatogr. Sci.*, 22, 485, 1984.

43. Agostini, T. and Godoy, H., Simultaneous determination of nicotinamide, nicotinic acid, riboflavin, thiamin, and pyridoxine in enriched Brazilian foods by HPLC, *J. High Resol. Chromatogr.*, 20, 245, 1997.
44. Altria, K., Kelly, M., and Clark, B., Current applications in the analysis of pharmaceuticals by capillary electrophoresis, *Trends Anal. Chem.*, 17, (4), 204, 1998.
45. Augustin, Y., Simultaneous determination of thiamine and riboflavin in foods by LC, *J. Assoc. Off. Anal. Chem.*, 67, (5), 1012, 1984.
46. Bognar, A., Bestimmung von Riboflavin und Thiamin in Lebensmitteln mit Hilfe der HPLC, *Deutsche Lebensmittel-Rundschau*, 77, (12), 431, 1981.
47. Finglas, P. and Faulks, R., The HPLC analysis of thiamin and riboflavin in potatoes, *Food Chem.*, 15, 37, 1984.
48. Gauch, R., Leuenberger, U., and Muller, U., Die Bestimmung von Folsaure in Lebensmitteln mit HPLC, *Mitt. Gebiete Lebensm. Hyg.*, 84, 295, 1993.
49. Hasselman, C., Franck, D., Grimm, P., Diop, A., and Soules, C., HPLC analysis of thiamin and riboflavin in dietetic foods, *J.micronutr. anal.*, 5, 269, 1989.
50. Hisil, Y. and Ötleş, S. Identification of water soluble vitamins by high pressure liquid chromatography (H.P.L.C.), *J. Facul. Eng.*, 8, (1), 37, 1990.
51. Knifel, W. and Sommer, R., Bestimmung von Thiamin in Milch und Milch-Produkten mittels HPLC, *Nutrition*, 10, (7), 459, 1986.
52. Ötleş, S. and Hisil, Y., HPLC analysis of water soluble vitamins in eggs, *Ital. J. Food Sci.*, 1, 69, 1993.
53. Polesello, A. and Rizzolo, A., Application of HPLC to the determination of water soluble vitamins in foods, IVTPA, Venezia, 1993.
54. Rizzolo, A., Baldo, C. and Polesello, A., Application of HPLC to the analysis of niacin and biotin in Italian almond cultivars, *J. Chrom.*, 553, 187, 1991.
55. Toma, R. and Tabekhia, M., HPLC analysis of B-vitamins in rice and rice products, *J. Food. Sci.*, 44, (1), 263, 1979.
56. Williams, R., Baker, D., and Schmit, E., Analysis of water soluble vitamins by HPLC, *J. Chromatogr. Sci.*, 11, 618, 1973.
57. Woollard, D., New ion-pair reagent for the HPLC separation of B-group vitamins in pharmaceuticals, *J. Chrom.*, 301, 470, 1984.
58. Aboul-Kasim, E., Anodic adsorptive voltammetric determination of the vitamin B_1 (thiamine), *J. Pharmaceut. Biomed. Anal.*, 22, 1047, 2000.
59. Argoudelis, C., Simple high-performance liquid chromatographic method for the determination of all seven vitamin B_6-related compounds, *J. Chrom.*, 790, 83, 1997.
60. Barrales, O., Fernandez, D., and Diaz, M., A selective optosensor for UV spectrophotometric determination of thiamine in the presence of other vitamins B, *Analytica Chimica Acta*, 376, 227, 1998.
61. Berg, H., Scheik, F., Finglas, P., and Gortz, I., Third EU MAT intercomparison for the determination of vitamins B_1, B_2, B_6 in food. *Food Chem.*, 57, 1, 101, 1996.
62. Breithaupt, D., Determination of folic acid by ion-pair RP-HPLC in vitamin-fortified fruit juices after solid-phase extraction, *Food Chem.*, 74, 521, 2001.
63. Cheong, W., Kang, G., Lee, W., and Yoo, J., Rapid determination of water-soluble B group vitamins in urine by gradient LC-MS with a disposable home-made microcolumn, *J. Liq. Chrom. & Rel. Technol.*, 2, (5), 1967, 2002.
64. Cho, C., Ko, J., and Cheong, W., Simultaneous determination of water-soluble vitamins excreted in human urine after eating an overdose of vitamin pills by a HPLC method coupled with a solid phase extraction, *Talanta*, 51, 799, 2000.
65. Fernandez, R., Sales, M., Reis, B., Zagatto, E., Araujo, A., and Montonegro, N., Multi-task flow system for potentiometric analysis: its application to the determination of vitamin B_6 in pharmaceuticals, *J. Pharmaceut. Biomed. Anal.*, 25, 713, 2001.

66. Finglas, P., Berg, H., and Gortz, I., Improvements in the determination of vitamins in foods, *Food Chem.*, 57, (1), 91, 1996.
67. Gregory, J. and Feldstein, D., Determination of vitamin B–6 in foods and other biological materials by PI-HPLC, *J. Agric. Food Chem.*, 33, 359, 1985.
68. Gregory, J. and Kirk, J., Improved chromatographic separation and fluorometric determination of vitamin B–6 compounds in foods. *J. Food Sci.*, 42, 1073, 1977.
69. Gregory, J. and Kirk, J., Determination of urinary 4-pyridoxic acid using HPLC, *Am. J. Clin. Nutr.*, 32, 879, 1979.
70. Gregory, J., Sartain, D., and Day, B., Fluorometric determination of folacin in biological materials using HPLC, *Nutrition*, 341, 1984.
71. Hisil, Y. and Ötleş, S., Analysis of water soluble vitamins by reserve-phase liquid chromatography with ion-pair reagents. *J. Facul. Eng.*, 10, (1), 57, 1992.
72. Hu, Q., Zhou, T., Zhang, L., Li, H., and Fang, Y., Separation and determination of three water-soluble vitamins in pharmaceutical preparations and food by micellar electrokinetic chromatography with amperometric electrochemical detection, *Analytica Chimica Acta*, 437, 123, 2001.
73. Ivanovic, D., Popovic, A., Radulovic, D., and Medenica, M., Reversed-phase ion-pair HPLC determination of some water-soluble vitamins in pharmaceuticals, *J. Pharmaceut. Biomed. Anal.*, 18, 999, 1999.
74. Kneifel, W., Fluorimetrische HPLC-Bestimmung von Riboflavin in Milch (Fluorimetric HPLC method of riboflavine in milk), *Deutsche Molkerei Zeitung*, 9, 212, 1986.
75. Li, H., Chen, F., and Jiang, Y., Determination of vitamin B_{12} in multivitamin tablets and fermentation medium by high-performance liquid chromatography with fluorescence detection, *J. Chrom.*, 891, 243, 2000.
76. Liu, S., Zhang, Z., Liu, Q., Luo, H., and Zheng, W., Spectrophotometric determination of vitamin B_1 in a pharmaceutical formulation using triphenylmethane acid dyes, *J. Pharmaceut. Biomed. Anal.*, 30, 685, 2002.
77. Medina, A., Cordova, M., and Diaz, A., Flow injection-solid phase spectrofluorimetric determination of pyridoxine in presence of group B-vitamins, *Fresenius J Anal Chem.*, 363, 265, 1999.
78. Moreno, P. and Salvado, V., Determination of eight water- and fat-soluble vitamins in multi-vitamin pharmaceutical formulations by high-performance liquid chromatography, *J. Chrom.*, 870, 207, 2000.
79. Ndaw, S., Bergaentzle, M., Aoude-Werner, D., and Hasselmann, C., Extraction procedures for the liquid chromatographic determination of thiamin, riboflavin and vitamin B_6 in foodstuffs, *Food Chem.*, 71, 129, 2000.
80. Saidi, B. and Warthesen, J., Influence of pH and light on the kinetics of vitamin B_6 degradation, *J. Agr. Food. Chem.*, 31, 876, 1983.
81. Sharpless, K., Margolis, S., and Thomas, J., Determination of vitamins in food-matrix Standard Reference Materials, *J. Chrom.*, 881, 171, 2000.
82. Siddique, I. and Pitre, K., Voltammetric determination of vitamins in a pharmaceutical formulation, *J. Pharmaceut. Biomed. Anal.*, 26, 1009, 2001.
83. Steijns, L, Braams, K., Luiting, H., and Weide, J., Evaluation of nonisotopic binding assays for measuring vitamin B_{12} and folate in serum, *Clinica Chimica Acta*, 248, 135, 1996.
84. Stokes, P. and Webb, K., Analysis of some folate monoglutamates by high-performance liquid chromatography–mass spectrometry. *J. Chrom.*, 864, 59, 1999.
85. Vanderslice, J., Brownlee, S., and Cortissoz, M., LC determination of vitamin B–6 in foods, *J. Assoc. Off. Anal. Chem.*, 67, 999, 1984.
86. Wehling, R. and Wetzel, D., Simultaneous determination of pyridoxine, riboflavin and thiamin in fortified cereal products by HPLC, *J. Agr. Food. Chem.*, 32, 1326, 1984.

87. Wongyai, S., Determination of vitamin B_{12} in multivitamin tablets by multimode HPLC, *J. Chrom.*, 870, 217, 2000.
88. Paixao, J. and Campos, J., Determination of fat-soluble vitamins by RP-HPLC coupled with UV detection,, *J. Liq. Chrom. & Rel. Technol.*, 26, (4), 641, 2003.
89. Amin, A., Colorimetric determination of tocopheryl acetate (vitamin E) in pure form and in multi-vitamin capsules, *Eur. J. Pharm. Biopharm.*, 51, 267, 2001.
90. Atli, Y., Ötleş, S., and Akcicek, E., An important nutrient for health: vitamin E, *Food Tech.*, 3, (11), 66, 1998.
91. Bramley, P., Elmadfa, I., Kafatos, A., Kelly, F., Manios, Y., Roxborough, H., Schuh, W., Sheehy, P., and Wagner, K., Vitamin E, *J. Sci. Food Agr.*, 80, 913, 2000.
92. Brinkmann, E., Dehne, L., Oei, H., Tiebach, R., and Baltes, W., Separation of geometrical retinol isomers in food samples by using NB-HPLC, *J. Chrom.*, 693, 271, 1995.
93. Careri, M., Lugari, M., Mangia, A., Manini, P., and Spagnoli, S., Identification of vitamins A, D and E by PB-HPLC, *Fresenius J. Anal. Chem.*, 351, 768, 1995.
94. Delgado-Zamarreno, M., Gonzalez-Maza, I., Sanchez-Perez, A., and Carabias-Martinez, R., Separation and simultaneous determination of water-soluble and fat-soluble vitamins by electrokinetic capillary chromatography, *J. Chrom.*, 953, 257, 2002.
95. Dolnikowski, G., Sun, Z., Grusak, M., Peterson, J., and Booth, S., HPLC and GC/MS determination of deuterated vitamin K (phylloquinone) in human serum after ingestion of deuterium-labeled broccoli, *J. Nutr. Biochem.*, 13, 168, 2002.
96. Escriva, A., Esteve, M., Farre, R., and Frigola, A., Determination of liposoluble vitamins in cooked meals, milk and milk products by liquid chromatography, *J. Chrom.*, 947, 313, 2002.
97. Genestar, C. and Grases, F., Determination of vitamin A in pharmaceutical preparations by HPLC with DA detection, *Chromatographia*, 40, (3), 143, 1995.
98. Gomis, D., Fernandez, M., and Gutierrez Alvarez, D., Simultaneous determination of fat-soluble vitamins and provitamins in milk by microcolumn liquid chromatography, *J. Chrom.*, 891, 109, 2000.
99. Iwase, H., Determination of vitamin D in emulsified nutritional supplements by solid-phase extraction and column-switching high-performance liquid chromatography with UV detection, *J. Chrom.*, 881, 189, 2000.
100. Iwase, H., Simultaneous sample preparation for high-performance liquid chromatographic determination of Vitamin A and carotene in emulsified nutritional supplements after solid-phase extraction, *Analytica Chimica Acta*, 463, 21, 2002.
101. Jakob, E. and Elmadfa, I., Rapid and simple HPLC analysis of vitamin K in food, tissues, and blood, *Food Chem.*, 68, 219, 2000.
102. Ötleş, S., Detection and analysis of vitamin K by different methods, *J. Oil Soap Cosmetics*, 47, (5), 240, 1998.
103. Ötleş, S., High pressure liquid chromatographic analysis of water and fat-soluble vitamins, *J. Oil Soap Cosmetics*, 49, (1), 20, 2000.
104. Ötleş, S. and Atli, Y., Analysis of carotenoids in tomato paste by HPLC and OCC, *Am. Lab.*, 32, (15), 22, 2000.
105. Ötleş, S. and Hisil, Y., Simultaneous determination of vitamin E in eggs by high-pressure liquid chromatography, *J. Facul. Eng.*, 8, (1–2), 53, 1990.
106. Ötleş, S. and Hisil, Y., Analysis of vitamin A in eggs by high pressure liquid chromatography, *Die Nahrung/Food*, 35, (4), 391, 1991.
107. Ötleş, S. and Hisil, Y., The determination of vitamin D_3 by high pressure liquid chromatography (HPLC), *Food*, 16, (6), 377, 1994.
108. Qian, H. and Sheng, M., Simultaneous determination of fat-soluble vitamins A, D and E and pro-vitamin D2 in animal feeds by one-step extraction and high-performance liquid chromatography analysis, *J. Chrom.*, 825, 127, 1998.

109. Quiros, A., Lopez, J., and Lozano, J., Determination of carotenoids and liposoluble vitamins in sea urchin (Paracentrotus lividus) by high performance liquid chromatography, *Eur Food Res Technol*, 212, 687, 2001.
110. Parrish, D., Determination of vitamin D in foods, *Crit. Rev. Food Sci. Nutr.*, 11, 29, 1979.
111. Perez-Ruiz, T., Martinez-Lozano, C., Tomas, V., and Martin, J., Flow-injection fluorimetric determination of vitamin K1 based on a photochemical reaction, *Talanta*, 50, 49, 1999.
112. Rizzo, P., Dionisi-Vici, C., D'Ippoliti, M., Fina, F., Sabetta, G., and Federici, G., A simple and rapid HPLC method for simultaneous determination of plasma 7-dehydrocholesterol and vitamin E: its application in Smith–Lemli–Opitz patients, *Clinica Chimica Acta*, 291, 97, 2000.
113. Sforzini, A., Bersani, G., Stancari, A., Grossi, G., Bonoli, A., and Ceschel, G., Analysis of all-in-one parenteral nutrition admixtures by liquid chromatography and laser diffraction: study of stability, *J. Pharmaceut. Biomed. Anal.*, 24, 1099, 2001.
114. Strobel, M., Heinrich, F., and Biesalski, H., Improved method for rapid determination of vitamin A in small samples of breast milk by high-performance liquid chromatography, *J. Chrom.*, 898, 179, 2000.
115. Ternes, W., Kramer, P., Menzel, R., and Zeilfelder, K., Verteilung von fettloslichen Vitaminen (A, E, D2, D3) und Carotenoiden in den Lipiden von Granula und Plasma des Eigelbs (Separations of fat soluble vitamins (A, E, D2) and carotenoids in the lipids and plasma of eggyolk), *Archiv fur Geflugelkunde*, 5, 261, 1995.
116. Turner, C., Persson, M., Mathiasson, L., Adlercreutz, P., and King, J., Lipase-catalyzed reactions in organic and supercritical solvents: application to fat-soluble vitamin determination in milk powder and infant formula, *Enzym. Microb. Tech.* 29, 111, 2001.
117. Zenkevich, I., Kosman, V., Makarov, V., and Dadali, Y., Quantitative refractometric–spectrophotometric analysis of oil solutions of vitamins A and E, *Russ. J. Appl. Chem.*, 74, (6), 1034, 2001.

10 Analysis of Carotenoids and Chlorophylls in Foods

Jae Hwan Lee and Steven J. Schwartz

CONTENTS

10.1 INTRODUCTION

Pigments greatly influence the sensory characteristics and consumer acceptance of foods. Among categories of natural pigments, carotenoids and chlorophylls are two representative lipid-soluble colorants that are associated with each other in photosynthetic tissues during maturation and senescence of many fruits and vegetables.[1] Carotenoids are a large group of yellow-to-red colored compounds with over 700 different chemical structures.[1,2] Carotenoids can be found in many animals, including birds, insects, and marine animals, and in plants, such as vegetables, flowers, and fruits. Chlorophylls are green pigments ubiquitously found in photosynthetic tissues of plants and photosynthetic organisms.[3] Animals, including humans, must consume carotenoids and chlorophylls through daily diets because of their lack of ability to synthesize these pigments.[4] Some carotenoids and chlorophyll derivatives have beneficial effects on human health. Naturally occurring pigments from plants are generally considered safe because of their presence in edible plant materials and are exempted from toxicological testing. Carotenoids containing retinoid structure (β-ionone ring), such as α- and β-carotenes, serve as precursors of provitamin A. Carotenoids can act as good singlet oxygen quenchers and free radical scavengers due to the presence of many double bonds in their structures.[5] Also, some carotenoids have been reported to possess the ability of enhancing immune response, to be involved in cell communication, and to decrease the risks of chronic degenerative diseases and age-related macular degeneration.[6,7] Consumption of lycopene in tomatoes and tomato products and β-carotene is associated with potential health benefits including decrease in the risk of cervical, colon, prostate, rectal, stomach, and other types of cancers and diseases.[8] Geometric isomers of carotenoids may have different biological activities in certain situations. For example, ingested lycopene is predominately (~95%) in the all-*trans* form, while *cis*-isomers of lycopene represent approximately 50% of the total lycopene in blood and up to 80% in prostate tissues.[8,9] Provitamin A activities among *cis/trans* carotenoid isomers are different. Carotenoids with *cis* form have less provitamin A activity and antioxidant capacity than carotenoids with *trans* form.[10,11] Chlorophyll derivatives are also known to be associated with anticarcinogenic and antimutagenic activities.[12,13]

Because of the important roles of carotenoids and chlorophylls as nutrients, pigments, and health-beneficial nutraceuticals, analytical procedures have been developed to achieve efficient, sensitive, and reproducible data on the types, concentrations, and geometric isomers of carotenoids and chlorophylls in food samples through refinement of instrumentation and methodology. In addition, many studies have tried to detect the contents of both the carotenoids and chlorophylls in vegetables and fruits using the same extraction method and analytic tools owing to their lipophilic characters.[14–16] This chapter focuses on recently developed techniques and reported literature for the analysis of carotenoids and chlorophylls.

10.2 CAROTENOIDS

10.2.1 STRUCTURE AND EXTRACTION METHODS

10.2.1.1 Chemical Structure and Properties

Carotenoids are a group of tetraterpenoids. The basic carotenoid structural backbone comprises isoprenoid units formed either by head-to-tail or by tail-to-tail biosynthesis. There are primarily two classes of carotenoids: carotenes and xanthophylls. Carotenes are hydrocarbon carotenoids, and xanthophylls contain oxygen in the form of hydroxyl, methoxyl, carboxyl, keto, or epoxy groups. The structures of carotenoids are acyclic, monocyclic, or bicyclic. For example, lycopene is acyclic, γ-carotene is monocyclic, and α- and β-carotenes are bicyclic carotenoids. Double bonds in carotenoids are in conjugated forms, and usually the all-*trans* forms of carotenoids are found in plant tissues (Figure 10.1).[4,17]

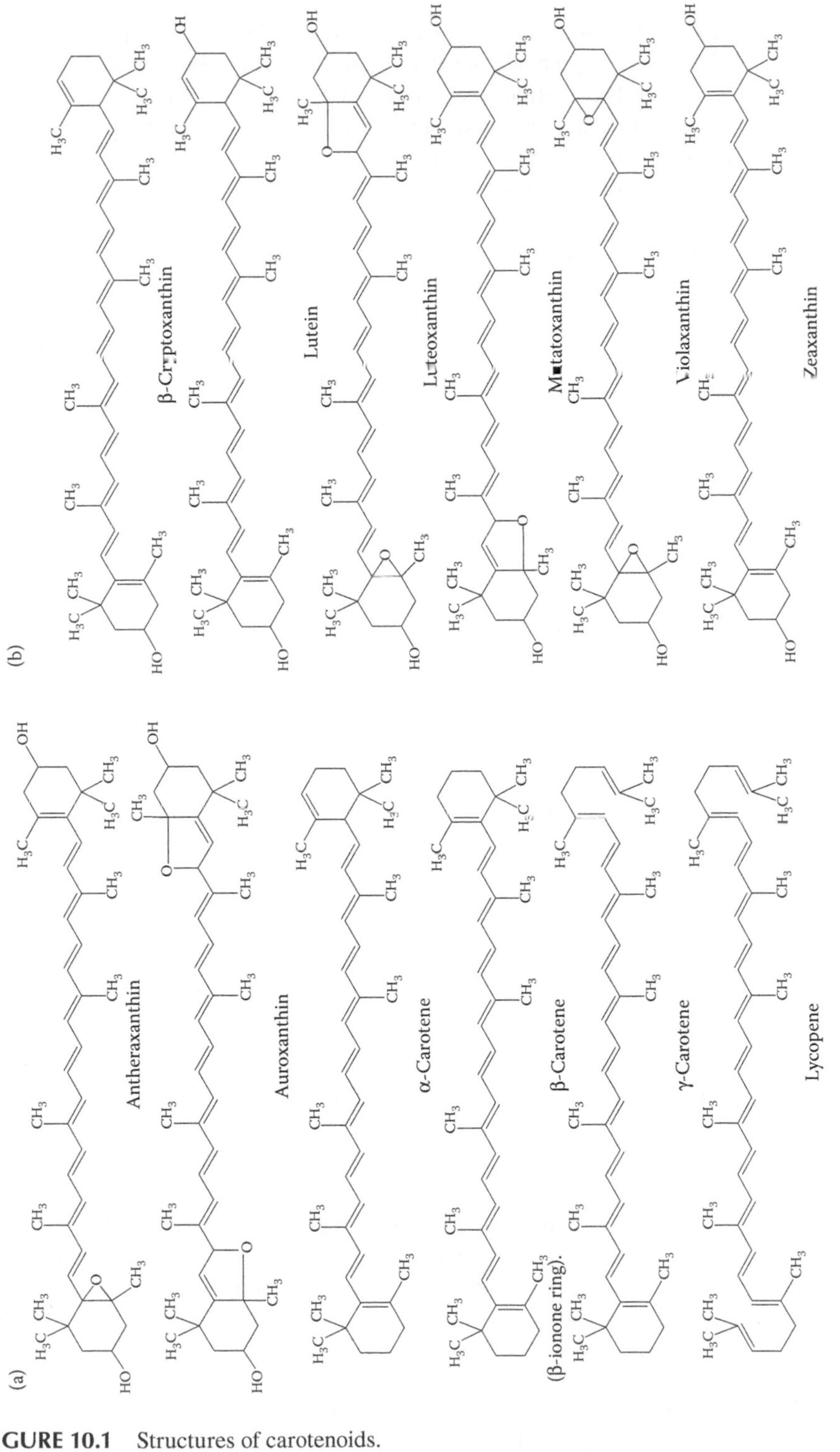

FIGURE 10.1 Structures of carotenoids.

Carotenoids in the food matrix are relatively stable during typical thermal processing.[18] However, carotenoids in solution are sensitive to heat, acid, oxygen, or light exposure.[4] Several precautions such as carrying out experiments under dim light, evaporation by a rotary evaporator under nitrogen gas flow, storage in the dark under nitrogen or argon at −20°C, and use of antioxidants such as butylated hydroxyanisole (BHA), pyrogallol, or ascorbic acid are necessary while handling carotenoids.[19,20]

The inherent instability and presence of diverse isomers make it difficult to obtain some commercially obtainable standard carotenoid compounds; therefore, quantitative and qualitative analyses of carotenoids are challenging. AOAC official methods of analysis recommend organic solvent extraction of carotenoids from food matrix, followed by open-column chromatographic separation and spectrophotometric determination.[21] Many current studies on carotenoids in foods use high-performance liquid chromatography (HPLC) coupled with a photodiode array (PDA) detector as a separation technique. Newer methods employ electrochemical detectors to enhance sensitivity and mass spectrometry to validate the identity of carotenoids.

10.2.1.2 Extraction, Separation, and Purification Methods

10.2.1.2.1 Extraction and Saponification

Because of the wide variety and complex chemical structures of carotenoids, selecting one reference method to extract all carotenoids in foods is almost impossible. Liquid–liquid extraction and supercritical fluid extraction (SFE) have been applied to extract mixtures of carotenoids in the food matrix. Liquid–liquid extraction uses a mixture of organic solvents to penetrate the tissue matrix, precipitate other compounds, and extract total carotenoids.[22] All classes of carotenoids are lipophilic compounds and are soluble in oils and organic solvents. However, the presence of moisture in foods requires solvents to be miscible with water to allow the penetration of the solvent into the food matrix.[23] Generally, water-miscible solvents including acetone, ethanol, methanol, or mixtures of these polar solvents and other more nonpolar solvents such as hexane, tetrahydrofuran (THF), diethyl ether, and methylene chloride have been used in many studies to extract carotenoids in foods. Mixtures of organic solvents used vary depending on the food matrix and the polarity of target carotenoids. The use of a mixture of hexane and acetone (1:1, v/v) or hexane alone was suggested for the extraction of carotenoids in dehydrated plant materials.[22] Methanol (MeOH) alone or a mixture of MeOH with other more nonpolar solvents is also widely used: MeOH and THF mixture (1:1, v/v) for the extraction of carotenoids in various vegetables and fruits,[24,25] a mixture of chloroform and MeOH (2:1, v/v) with 0.01% BHA in fat-cured crude sausage,[26] a mixture of dichloromethane and MeOH (2:1, v/v) in orange juice and tomato products,[27] and diethyl ether extraction followed by MeOH extraction for xanthophylls in red peppers and paprika.[28] De Sio et al.[27] reported the recovery of dichloromethane:MeOH (2:1, v/v) extraction for lutein, α-carotene, β-carotene, and lycopene as 99.1%, 99.3%, 98.7%, and 101.5%, respectively. A more complex mixture of organic solvents such as MeOH:ethyl acetate:light petroleum ether (1:1:1, v/v/v) was used for xanthophyll extraction in fruits and vegetables.[29] For the extraction of carotenoids in orange juice, the mixture of hexane:acetone:ethanol (50:25:25, v/v/v) was recommended.[30] THF containing 0.01% BHT was used for the extraction of carotenoids in green leafy vegetables, figs, and olive oil.[31]

Taungbodhitham et al.[32] compared the extraction efficiency of acetone:hexane (4:6, v/v), ethanol:hexane (4:3, v/v), chloroform:MeOH (2:1, v/v), dichloromethane:MeOH (2:1, v/v), hexane:isopropanol (3:2, v/v), and acetone:petroleum ether (1:1, v/v) mixtures for lycopene and β-carotene in canned tomato juices. Mixtures of acetone:hexane (4:6, v/v) and ethanol:hexane (4:3, v/v) showed the highest lycopene and β-carotene extraction efficiency. Lin and Chen[33] compared the extraction efficiency of carotenoids in tomato juice using five organic solvent mixtures and concluded that the mixture of ethanol (EtOH):hexane (4:3, v/v) showed better extraction efficiency than acetone:hexane (3:5, v/v), EtOH:acetone:hexane (2:1:3, v/v/v), ethyl acetate:hexane (1:1, v/v), and ethyl acetate (100%).

Addition of sodium, magnesium, or calcium carbonate is recommended to neutralize plant tissues containing acids.[34] Liquid–liquid extraction should be repeated two or three more times to increase the extraction efficiency of carotenoids. Remaining water in the organic layer should be removed through the addition of anhydrous Na_2SO_4.

SFE using carbon dioxide (CO_2) has been developed to extract carotenoids from various sources, including carrot,[35] vitamin supplements,[36] and byproducts of tomato processing.[37] SFE is a rapid, selective, and simple extraction technique. Compared with traditional solvent extraction methods, SFE using nontoxic CO_2 has reduced potential for oxidation, can be easily automated, and may produce higher yields, particularly for samples with high-fat foods. In addition, CO_2 has a relatively low critical temperature (31°C); therefore, it can extract thermally labile compounds efficiently. However, the low polarity of CO_2 makes the extraction of polar carotenoids difficult and requires the addition of polarity modifiers to CO_2. Gómez-Prieto et al.[38] reported that CO_2 of fluid density 0.90 g/mL and without a modifier was good for extracting all-*trans*-lycopene from tomato. Polarity modifiers including chloroform, EtOH, and MeOH have been tested to improve the extraction efficiency of carotenoids using CO_2.

Enzyme hydrolysis and solid-phase extraction have been applied to enhance the recovery efficiency of organic solvent extraction of carotenoids from food matrix. Enzymes such as glucosidase, cellulase, hemicellulase, and pectinase can help increasing the extraction yield of lutein in marigold flowers.[39,40]

Matrix solid-phase dispersion (MSPD) method was used instead of liquid–liquid extraction or solid-phase extraction for the extraction of lutein and zeaxanthin.[41] MSPD extraction is rapid (preparation time is less than 30 minutes) and can reduce the oxidation and isomerization of carotenoids as the carotenoids are protected within the matrix until the elution step.

Xanthophylls, carotenoids with hydroxyl groups, are predominantly present in the ester form linked with fatty acids in fruits and vegetables such as lutein, zeaxanthin, and β-cryptoxanthin.[17] The number of hydroxyl groups in the xanthophylls influences the degree of esterification. For example, carotenoids with a monohydroxyl group such as β-cryptoxanthin can exist either freely or as monoesters, while those with dihydroxyl groups such as lutein or zeaxanthin may be found either freely or as mono- or diesters. Therefore, through saponification, much simple results of HPLC chromatograms can be obtained. Carotenoids in high fat- and oil-containing foods such as butter and milk are in the lipid matrix. Saponification with aqueous, ethanolic, or methanolic KOH has been used to remove lipids and chlorophylls and to change the ester forms of carotenoids into free forms. According to the AOAC method, saponification for carotenoids was conducted at room temperature for 16 hours or at 56°C for 20 minutes with KOH.[21] High reaction temperature may give rise to unexpected isomerization. Depending on the characteristics of the extraction medium, methanolic or ethanolic KOH can be selected. If hexane is used as the extraction solvent for carotenoids, ethanolic KOH is preferred because methanol is immiscible with hexane. Reaction time for saponification can vary depending on the sample matrix. In the case of orange juice, Fisher and Rouseff[42] suggested 1 hour as the optimum time, while 0.5 hour of saponification time may be sufficient.[43]

However, the formation of *cis* carotenoids by high heat treatment, formation of artifacts, and loss of xanthophylls and xanthophyll esters may occur during saponification.[1,44,45] Recently, chromatographic methods for the simultaneous determination of free and ester forms of carotenoids in fruit samples have been developed, and the need for saponification is diminishing.[29,45,46]

10.2.1.2.2 Separation and Purification

Carotenoids present in organic solvent extracts can be separated using open-column chromatography, as recommended by the AOAC official method,[21] or by thin-layer chromatography (TLC). The isolated carotenoid fractions are eluted by mixtures of organic solvents with different polarities and further purified either by crystallization under nitrogen or by a preparative HPLC method. Development of stationary phases for TLC, including octyl-, octadecyl, cyano-, dio-, and aminopropyl silica, is useful for separation, purification, and isolation of specific carotenoids.[20,47]

10.2.2 Chromatographic Techniques

10.2.2.1 Chromatographic Detectors

The conjugated double bonds of carotenoids can absorb light energy between 400 and 500 nm, and these spectral characteristics are often used for the identification of carotenoids and the measurement of purity of carotenoids.[48] As a chromatographic detector, PDA has replaced the fixed wavelength UV–Vis detector. PDA can record the entire spectral range from 190 to 800 nm and provide information on peak identification and purity criteria of carotenoids.[20]

A coulometric electrochemical array detector was introduced to analyze carotenoids in vegetable oils,[49] raw or processed carrot,[50] and raw tomato.[51] Electrochemical detection is based on the oxidation and reduction properties of the compound of interest.[50] Coulometric electrochemical array detectors generate current voltage curves that can be used to identify and differentiate geometric isomers of carotenoids.[50] The detection limits of electrochemical array detectors for *trans* β-carotene were 10 fmol; that is, these detectors are approximately 100–1000 times more sensitive than UV–Vis detectors.[50]

Luterotti et al.[52] used thermal lens spectrometry to detect *trans* and *cis* β-carotene in vegetable oils. Thermal lens spectrometry is one of the laser photothermal detection methods, which is based on indirect absorbance measurement through thermal lens effects.[34,52] The detection limit of thermal lens spectrometry for carotenoids is approximately 100 times more sensitive than that of the conventional UV–Vis detection method.[53]

Strohschein et al.[54] elucidated the structure of β-carotene isomers and carotenoid isomers using HPLC coupled with nuclear magnetic resonance (NMR) spectrometry. D_2O was used as a mobile phase for HPLC-NMR to minimize the effects of solvents in the NMR spectra. Tiziani et al.[55] extracted carotenoids from tomato juice using $CDCl_3$ as solvent and analyzed them using a high-resolution multidimensional NMR spectrometer with a cryogenic probe. Rapid identification of lycopene isomers was possible by using one-dimensional NMR and minimizing further isomerization of (*all-E*)-lycopene, as compared to HPLC data.[55]

Mass spectrometry coupled with HPLC can elucidate the structural information of carotenoids. Classical ionization techniques such as electron impact and chemical ionization produce abundant fragmentation, but molecular ions are not always observed from carotenoid molecules. Soft ionization methods, including electrospray ionization (ESI), atmospheric pressure chemical ionization (APCI), fast atom bombardment (FAB), particle-beam electron-capture negative ion chemical ionization (PB-EC-NCI), and matrix-assisted laser desorption/ionization (MALDI), have been used to determine the molecular weight of carotenoids. More information on structural isomers can be obtained from fragmentation pattern provided by post source-decay coupled with MALDI or by collision-induced dissociation with ESI, APCI, or FAB. ESI and APCI are widely used ionization techniques for MS because of their comparable sensitivity at the low picamolar level and easy operational procedures. Although ESI forms primarily molecular ions with little fragmentation, APCI produces molecular ions and/or protonated or deprotonated molecules depending upon the mobile phase conditions, especially for xanthophylls.[56]

HPLC-ESI-MS has been used for the analysis of carotenoid composition of commercial marigold flower extract,[57] for β-carotene, and a mixture of xanthophylls, including β-cryptoxanthin, lutein, zeaxanthin, canthaxanthin, and astaxanthin.[58] Breithaupt et al.[59] used HPLC-APCI-MS to characterize lutein monoesters from marigold flowers and lutein diesters from cape gooseberry, kiwan, and pumpkin. Breithaupt and Bamedi[60] analyzed free and ester forms of carotenoids in eight potato cultivars using HPLC-APCI-MS. In this work, organic solvent extracts were purified through silica gel open-column chromatography and lipase hydrolysis of residual triacylglycerides for the direct injection of samples to HPLC-APCI-MS. Dachtler et al.[41] analyzed stereoisomers of β-carotene, lutein, and zeaxanthin in spinach by HPLC-APCI-MS coupled with NMR. HPLC-APCI-MS can separate and identify the different types of carotenoids but cannot differentiate the stereoisomers of carotenoids because of the same fragmentation pattern of each carotenoid. NMR can help identifying stereoisomers of carotenoids. van Breemen et al.[61] used FAB tandem mass spectrometry (MS/MS) for the analysis of 17 different carotenoids. MALDI time-of-flight post source-decay MS has been used to identify free and fatty acid esters of carotenoids from tangerine and orange concentrates,[62] and carotenoids and carotenol fatty acid esters.[63] Tian et al.[64] characterized six lutein esters from marigold extracts using HPLC-APCI-MS. Detailed structural information on lutein esters was obtained using in-source collisionally induced dissociation method.

10.2.2.2 Normal-Phase HPLC

Normal-phase (NP) HPLC with polar stationary phases can provide a more complete separation of polar carotenoids and their geometric isomers. Silica-based nitrile-bonded columns[65,66] and alumina columns have been used as stationary phases for carotenoid analysis. Humphries and Khachik[66] analyzed dietary lutein, zeaxanthin, and their geometric isomers in fruits, vegetables, wheat, and pasta products using a 5-μm silica-based nitrile-bonded column (25 cm × 4.6 mm ID). All sample extracts were subjected to saponification. Depending on the samples, different mixtures of organic solvents and gradients were used as mobile phases. Khachik et al.[67] separated lutein from marigold flower, kale, and human plasma using 5-μm silica-based nitrile-bonded columns (25 cm × 4.6 mm ID) and characterized lutein with ^{1}H and ^{13}C NMR and UV–Vis spectrophotometry. Hadden et al.[57] used a Zorbax SIL 5-μm column (25 cm × 4.6 mm ID) and a β-Cyclobond 5-μm column (25 cm × 4.6 mm ID), respectively, for the carotenoid analysis of marigold flower extracts. A mixture of hexane and ethyl acetate (75:25, v/v for Zorbax SIL column and 87:13, v/v for β-Cyclobond column) was used as the mobile phase at a flow rate of 2.0 mL/minute.

10.2.2.3 Reversed-Phase HPLC

Reversed-phase (RP) HPLC is widely used to separate nonpolar carotenes and some polar xanthophylls in foods. Breithaupt et al.[59] differentiated lutein monoesters and diesters from marigold flower, cape gooseberry, kiwano, and pumpkin using a 5-μm C_{30} column (25 cm × 4.6 mm ID) and a mixture of MeOH:MTBE:water (81:15:4, v/v/v or 6:90:4, v/v/v) as the mobile phase. The detection limit of RP HPLC-APCI-MS was estimated at 0.5 μg/mL for lutein diesters using lutein dimyristate as a reference compound. De Sio et al.[68] analyzed carotenes and xanthophylls from non-concentrated orange juice, tomato puree, tomato paste, and other vegetables using a 5-μm YMC C_{30} column (25 cm × 4.6 mm ID) and a linear gradient of solvent mixture A (MeOH:water, 95:5, v/v, containing 0.1% BHT and 0.05% triethyl amine) and solvent mixture B (methylene chloride containing 0.1% BHT and 0.05% triethyl amine) as eluent. In this work, 0.1% BHT was added to both the extraction solvent and a mobile-phase eluent to prevent oxidation of carotenoids during extraction and HPLC separation. Lee et al.[69] analyzed antheraxanthin, violaxanthin, cryptoxanthin, lutein, and carotenes from Earlygold sweet orange using a 3-μm YMC C_{30} carotenoid column (25 cm × 4.6 mm ID). A mixture of acetonitrile:MeOH (75:25, v/v), 100% MTBE, and water was used as a mobile phase with gradient. Acetonitrile:MeOH (75:25, v/v) containing 0.05% triethyl amine, instead of 100% MeOH, significantly increased the resolution and band spacing of coeluted carotenoids.

Chromatographic analysis of *cis/trans* geometric carotenoid isomers is necessary because of the differences in provitamin A activity.[11] NP HPLC methods with silica-based nitrile-bonded columns or calcium hydroxide stationary columns have been successfully used to separate xanthophyll and carotene isomers (Table 10.1).[66,70] In RP HPLC, C_{18} stationary phases are suitable for the separation of the different carotenoids but are not very efficient for the separation of *trans* and all other *cis* geometric isomers of carotenoids.[41,71,72] However, some studies have used a polymeric synthesized C_{18} column for the separation of *trans/cis* β-carotene isomers (Table 10.1).[73] A C_{30} column,

TABLE 10.1
HPLC Methodologies for the Geometric Isomer Analysis of Carotenoids in Foods

Carotenoids	Food Samples	Column	Mobile Phase	References
Lutein and zeaxanthin	Fruits and vegetables[a]	Silica-based nitrile-bonded	Hexane: dichromethane: MeOH:DIPEA	66
Carotenes (α, β, γ, ζ) and phytofluene	Carrot, baby food, and fresh tomato	Calcium hydroxide column	Hexane alone or hexane with modifiers[b]	70
β-carotene	Vegetable oils[c]	Vydac 218TP54, a 5-μm polymeric C_{18}	MeOH:THF (90:10, v/v)	73
Lutein and zeaxanthin	Spinach	ProntoSIL C_{30}	Acetone:H_2O (86:14, v/v)	41
Lycopene and α- and β-carotenes	Tomato products	3-μm polymeric C_{30}	40%–50% MTBE in MeOH	71
Lycopene and α- and β-carotenes	Vegetable oils[d]	5-μm polymeric YMC C_{30}	MeOH:MTBE: $CH_3CO_2NH_4$:H_2O (88:5:5:2, v/v/v/v) MeOH: MTBE: $CH_3CO_2NH_4$ (20:78:2, v/v/v)	49
Lycopene and β-carotene	Tomato	3-μm polymeric C_{30}	MTBE:MeOH: EtOAc (4:5:1, v/v/v)	75
Lycopene	Raw tomato	3-μm polymeric C_{30}	MeOH:MTBE: $CH_3CO_2NH_4$ (95:3:2, v/v/v or 25:73:2, v/v/v)	50
α- and β-carotenes and β-cryptoxanthin	Fruits and vegetables[e]	5-μm polymeric C_{30}	MeOH:MTBE (89:11, v/v)	76
α- and β-carotenes and lutein	Carrots	5-μm polymeric C_{30}	MeOH:MTBE (3:97 to 38:62, v/v)	77
α- and β-carotenes, lutein, and zeaxanthin	Carrots, tomato paste, Chinese wolfberries, and Betatene	3-μm polymeric C_{30}	MeOH:MTBE	78

MeOH: methanol, MTBE: methyl *tert*-butyl ether, $CH_3CO_2NH_4$: ammonium acetate, H_2O: water, THF: tetrahydrofuran, EtOAc: ethanol acetate, DIPEA: *N,N,*-diisopropylethylamine.

[a] Kale, collard greens, broccoli, spinach, green beans, fresh parsley, pumpkin, butternut, orange, papaya, mango, nectarine, lettuce, canned sweet corn, peas, lima bean, lasagne, egg noodles, wheat, and freekeh.

[b] Phytofluene:100% hexane, α-carotene: 0.1% p-methylanisole in hexane, β-carotene: 2.0% p-methylanisole or 2.0% acetone in hexane, γ-carotene: 2.45% benzyl ether in hexane, ζ-carotene: 1.4% benzyl ether in hexane.

[c] Linseed oil, wheat germ oil enriched with vitamin E, olive oil, safflower oil, sesame oil, and almond oil.

[d] Germ oil, soybean oil, virgin olive oil, and virgin olive oil with spinach leaves.

[e] Broccoli, cantaloupe, carrots, coolards, orange juice, peaches, spinach, sweet potatoes, tomatoes, and tomato-based vegetarian vegetable soup.

which has intermediate pore diameter of 200 Å and a polymerically bonded 30-carbon alkyl chain, was developed to prepare higher carbon-content silica and therefore stronger retention ability for the carotenoids.[72,74] Emenhiser et al.[72] showed the capacity of a polymeric synthesized C_{30} column for geometric-isomer separation of asymmetrical carotenoids (lutein, β-cryptoxanthin, and α-carotene) and symmetrical carotenoids (β-carotene, zeaxanthin, and lycopene). These columns can separate carotenoids of a wide range of polarity and the geometric isomers of carotenoids with a gradient mobile phase. Successful separation of geometric isomers of carotenes (lycopene, α-carotene, and β-carotene) and xanthophylls (lutein, zeaxanthin, and β-cryptoxanthin) in foods such as tomato products, vegetable oils, spinach, carrots, and marigold using a polymeric C_{30} column has been reported[18,41,49,51,75–78] (Table 10.1 and Figure 10.2).

10.2.3 CRITICAL PARAMETERS FOR CAROTENOID ANALYSIS

Many factors, including sample handling, sample preparation, extraction methods, and HPLC conditions, should be carefully considered for HPLC carotenoid analysis in foods to obtain reproducible, sensitive, and accurate results.[22,48]

The water content in the sample is one of the important considerations. Low-moisture samples simplify the extraction procedure, while complete dehydration or excessive water content in samples is not recommended.[22,48]

The choice of organic solvents for carotenoid extraction depends on the polarity of the pigments: nonpolar solvents such as hexane are a good choice for nonpolar or esterified forms of carotenoids, while ethanol, MeOH, and acetone are useful for extraction of polar carotenoids. If the characteristics of polarity of the pigment are unknown, an acetone/hexane (1:1, v/v) mixture is suitable.[22]

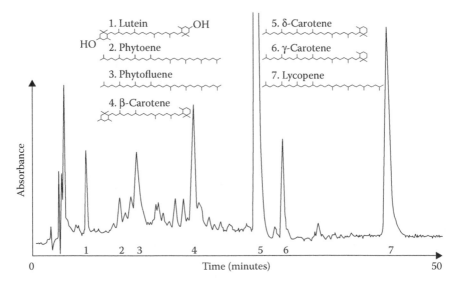

FIGURE 10.2 Typical HPLC chromatogram of carotenoids in tomato products. (Adapted from Nguyen, M., Francis, D., and Schwartz, S.J., Thermal isomerisation susceptibility of carotenoids in different tomato varieties, *J. Sci. Food Agric.*, 81, 910, 2001.)

Depending on the relative content of lipid and the presence of chlorophylls, saponification may be necessary. Samples containing xanthophyll esters and low carotenoid content, such as spinach and eggs, need saponification, while samples that do not contain xanthophyll esters (or high lipid) and low carotenoid content, such as carrot, do not need a saponification step.[74]

Stationary phase, particle size, particle shape, pore diameter, surface coverage, and monomeric or polymeric synthesis are important parameters in column selection. Generally, reversed-phase columns have been used for nonpolar carotenoids and normal-phase column for polar carotenoids. Particles of smaller sizes (3 µm) can be used to prepare short-length columns, which have higher efficiency, require less solvent consumption, and enhance the resolution of carotenoid peaks.[69,72] Spherical and uniform particle shape can enhance reproducibility.[70] Pore diameter influences the carbon load of the column and retention of carotenoids. Smaller pore diameters provide greater surface areas, higher carbon load, and better carotenoid retention. However, very small pore size may limit the passage of large carotenoid molecules.[80] Stationary phase synthesis methods, monomeric or polymeric, can affect the column performance. Monomeric synthesis, which involves the binding of carbon chains to the silica surface using monochlorosilanes, produces a more reproducible stationary phase but generally does not resolve geometric isomers. Polymeric synthesis, which involves the binding of carbon chains to the silica surface using trichlorosilanes with a small amount of water, is useful to separate geometric isomers.[79] To increase the carbon load in polymeric synthesized column, a longer alkyl chain such as C_{30} was developed.

Subambient temperature condition during HPLC analysis can increase the resolution of carotenoid separation because of a more rigid bonded phase, resulting in greater selectivity to molecular shape and size.[81,82] Modifications in mobile phases do not change the selectivity of carotenoid separation dramatically. Primary solvents should be weak organic solvents with low viscosity, for example, acetonitrile and MeOH. Modifiers such as chloroform, dichloromethane, THF, ethylacetate, and acetone can be added to mobile phases to enhance solubility, retention ability, and selectivity. Gradient elution mobile phase profiles have advantages that they can separate a wide range of analytes, shorten the time for analysis, and elute strongly retained carotenoids. However, there are some disadvantages such as the need for column reequilibration after each analysis and greater variability between solvents in terms of retention time, peak height, and area measurement. Gradient elution is not recommended if isocratic condition can achieve the same performance. Losses of carotenoids from nonspecific adsorption and oxidation have been reported.[83] Adding ammonium acetate to the mobile phase or using biocompatible frit instead of stainless steel frit can minimize the losses of carotenoids.[83]

10.3 CHLOROPHYLLS

10.3.1 STRUCTURE AND EXTRACTION METHODS

10.3.1.1 Chemical Structure and Properties

Chlorophylls are naturally occurring green pigments found in photosynthetic plants and vegetables and the most abundant among biologically produced pigments. The structure of chlorophyll is based on a porphyrin ring complexed with magnesium, which is a fully unsaturated and conjugated tetrapyrrole macrocyclic structure linked with methylene bridges. Several chlorophylls, including chlorophyll *a*, *b*, *c*, *d*, and *e*, are found in plants and photosynthetic organisms.[19] Chlorophylls *a* (blue-green) and *b* (yellow-green) are predominate in higher green plants at the ratio of 3:1 (Figure 10.3), while chlorophylls *c*, *d*, and *e* are found in photosynthetic algae.[3,4,19] All naturally occurring chlorophylls have a propionic acid residue (C-7 position), generally linked with an alcohol side chain, phytol. Phytol is a 20-carbon mono-unsaturated isoprenoid alcohol. Magnesium in chlorophylls can be removed easily by acids, and chlorophylls *a* and *b* form pheophytins *a* and *b* (olive-brown color), respectively. Hydrolysis of the phytol side chain forms the water-soluble chlorophyl-lide derivative. In addition, chlorophyllase can catalyze the cleavage of phytol chains from chlorophylls and pheophytins (Figure 10.3d). Porphyrins with carbomethoxy groups at C10 and hydrogen at C7 and C8 are needed for the enzymatic reaction.[4] Removal of both the phytol group and magnesium can generate pheophorbide, which is also water-soluble. The C10 carbomethoxy groups (CO_2CH_3) of pheophytins and pheophorbide can be removed by thermal processing, and as a result, pyropheophy-tin and pyropheophorbide are formed, respectively (Figure 10.3). Mild processing conditions such as freezing and drying and mild heating such as blanching generate C10 epimers of the native chlorophyll structures, which have no detrimental effects on the loss of color.[4]

Heating process inverts C10 carbomethoxy groups of chlorophylls into isomers. For example, chlorophylls *a* and *b* become chlorophylls *a'* and *b'*. Chlorophylls *a'* and *b'* and pheophytins are less polar than parent chlorophylls and more strongly absorbed in reversed-phase HPLC column. Chlorophyll *b* is more heat stable than chlorophyll *a* because of the electron withdrawal effects of C3 formyl group.[4]

The stability of chlorophylls is influenced by pH. In basic pH medium (9.0), chlo-rophyll is very stable to heating process, while in an acidic pH (3.0), it is unstable and becomes pheophytin when magnesium ion is removed. Further heating induces removal of C10 carbomethoxy groups from pheophytins, thereby forming pyropheo-phytins, which are olive-green. The severity of heat treatment can be indicated by the amount of pyropheophytins.

Chlorophylls in alcohol solution can be oxidized to form allomerization prod-ucts. The allomerized products are 10-hydroxychlorophylls and 10-methoxylactones, which are blue-green.[4]

Chlorophylls and their derivatives are extremely sensitive to acid, base, oxygen, heat, and light due to the presence of many conjugated double bonds. The surround-ing carotenoids and other lipids protect chlorophylls in plants from light irradiation. Once this protection is removed, the presence of oxygen and light irradiation acceler-ate the photodegradation of chlorophylls to form colorless products including methyl ethyl maleimide, glycerol, and organic acids. Opening the tetrapyrrole ring and frag-mentation into lower molecular weight compounds are believed to be the pathways for the photodegradation of chlorophylls through reaction with singlet oxygen and hydroxyl radicals. When exposed to light, chlorophylls and similar porphyrins can produce singlet oxygen as photosensitizers. Therefore, chlorophylls should be stored in the dark at reduced temperature.

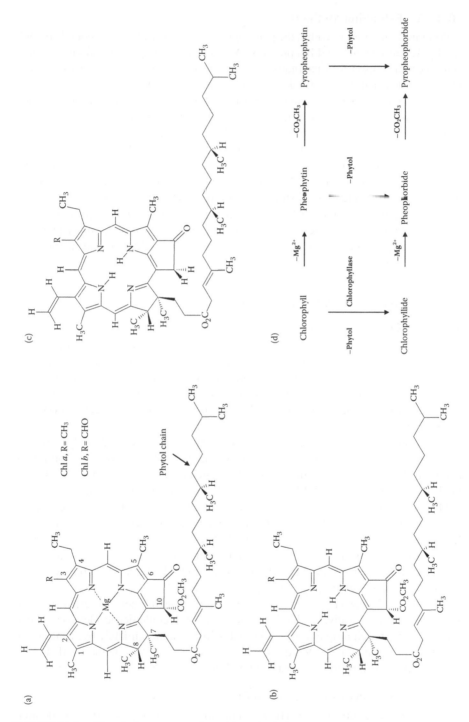

FIGURE 10.3 Chemical structures of (a) chlorophyll, (b) pheophytin, and (c) pyropheophytin and (d) their reactions.

10.3.1.2 Extraction Methods

Chlorophylls are present within the plant matrix with many other compounds including lipids, carotenoids, and lipoproteins. Acetone and diethyl ether are organic solvents generally used for quantitative extraction of chlorophyll derivatives in fresh plant samples and freeze-dried samples, respectively. Water-soluble chlorophyll derivatives can be extracted using cold MeOH followed by centrifugation.[84] Simon and Helliwell[85] compared the efficiency of solvents and extraction methods for chlorophyll a from freshwater green algae. In this work, MeOH was reported to be a better solvent than acetone, and probe sonication a better method than bath sonication, tissue grinding, or maceration by mortar and pestle. $CaCO_3$ saturated dimethyl sulfoxide was used to extract chlorophylls in French bean.[86] Vagi et al.[87] used supercritical carbon dioxide for the extraction of chlorophylls from marjoram.

The quantitative determination of total chlorophyll content especially chlorophyll a and b in organic extracts can be done by spectrophotometric assay. Chromatographic techniques are appropriate for both qualitative and quantitative analysis of chlorophylls.[19]

10.3.2 Chromatographic Techniques

10.3.2.1 Chromatographic Detectors

Conjugated double bonds in chlorophylls and their derivatives absorb light energy (blue, near 428 and 453 nm, and red, near 642 and 661 nm, spectral ranges) and generate fluorescent light, which allows identifying and quantifying the major pigment compounds in sample extracts. Identification of chlorophylls and their derivatives can be made more efficient by using a PDA than by using a fixed-wavelength UV–Vis detector.[88]

Mass spectrometry using electron impact ionization can provide only pyrolytic fragments due to the high mass, low volatility, and thermal instability of chlorophylls and their derivatives. Soft ionization methods such as laser desorption,[89] MALDI,[90] field desorption,[91] FAB,[92,93] ESI,[94] and APCI[95] have been used to identify chlorophylls and their derivatives. Operation at normal atmospheric pressure makes APCI and ESI compatible with HPLC. APCI methods that can detect as low as nanogram levels of chlorophylls are approximately 1000 times more sensitive than thermospray ionization. ESI methods can detect approximately 2 pmol of these compounds. Tandem mass spectrometry (MS/MS) with FAB and APCI ionization has also been reported to analyze chlorophylls and their derivatives.[95,96]

Puspitasari-Nienaber et al.[49] used a coulometric electrochemical array detector for chlorophyll derivatives. The detection limit of a coulometric electrochemical array detector for chlorophylls was 0.5 pmol, which is approximately five times more sensitive than that of a UV–Vis detector. The advantage of this method is that chlorophylls in oils can be directly analyzed without saponification.

10.3.2.2 High-Performance Liquid Chromatography

The widespread application of HPLC techniques on chlorophyll analysis demonstrates the technique's effectiveness, accuracy, and reliability.[88] RP HPLC methods are used most often for chlorophyll analysis, while a few applications have been reported using NP HPLC method. NP HPLC was used to separate metallo-chlorophyll complexes[97] and allomerization products of chlorophylls.[96] Canjura et al.[97] analyzed chlorophylls and Zn-chlorophyll complexes in thermally processed green peas with a normal-phase silica column (25 cm × 4.6 mm ID) and a gradient solvent system of hexane/isopropanol. Jie et al.[96] used a Spherisorb S5 W column (25 cm × 4.6 mm ID) as NP stationary phase and a mixture of 2-propanol:MeOH (1:1, v/v) in hexane as the mobile phase.

RP HPLC using a C_{18} column has been used extensively for the separation of predominant natural chlorophylls a and b, nonpolar derivatives such as pheophytins and pyropheophytins, polar derivatives such as chlorophyllides and pheophorbides, and metalloporphyrin derivatives such as copper and zinc pheophytins. Mobile phases of RP HPLC differ depending on the target compounds and sample matrix. Schwartz and coworkers[88,98–100] developed RP HPLC methodologies using a C_{18} column and methanol-based mobile phases with water and ethyl acetate as mobile phase modifiers. Ferruzzi[101] separated chlorophylls and their derivatives in fresh, heat- and acid-treated, and $ZnCl_2$-treated spinach puree using a Vydac 201TP54 C_{18} (150 mm × 4.6 mm ID) reversed-phase column. The mobile phase used was a gradient elution of solvent A (MeOH:water—75:25, v/v) and solvent B (ethyl acetate) (Figure 10.4). Teung and Chen[93] analyzed pyrochlorophylls and their derivatives in heat-treated spinach leaves using a 5-μm C_{18} column (25 cm × 4.6 mm ID). In this work, a total of 10 pigments, including chlorophyll b and its epimer, chlorophyll a and its epimer, pyrochlorophyll a and b, pyropheophytin a and b, and pheophytin a and its epimer, were detected with a mixture of acetonitrile:MeOH:chloroform:hexane (75:12.5:7.5:7.5, v/v/v/v) as a mobile phase. Gauthier-Jaques et al.[95] monitored chlorophyll degradation compounds in spinach powder and dried beans using a Lichrospher C_{18} column (25 cm × 4.6 mm ID). Mixtures of 1 M ammonium acetate:MeOH and acetone:MeOH were used as mobile phases with a linear gradient. Gauthier-Jaques et al.[95] also compared chlorophylls and their derivatives from conventionally canned beans and Veri-Green-processed beans, which has zinc-complexed chlorophylls instead of magnesium-complexed chlorophylls to enhance the thermal stability. Chromatographic and mass spectrometric results of zinc pheophytins were parallel to those of their counterparts with magnesium. Mangos and Berger[102] analyzed chlorophylls and their degradation products in spinach using a Nucleosil ODS (25 cm × 4.6 mm ID). Mixtures of MeOH, ammonium acetate buffer, or acetone were mobile phases, and ammonium acetate buffer enhanced the retention behavior of polar chlorophyll derivatives such as chlorophyllides a and b and pyrochlorophyllides a and b. Borrmann et al.[103] analyzed chlorophyll, greenish chlorophyll derivatives, and colorless chlorophyll metabolites in soybean seeds using RP HPLC—a diode array detector. A Shim-pack column (VPODS, 250 × 4.6 mm ID) with a Thermoquest precolumn (11 × 4.6 mm) was the stationary phase, and a mixture of methanol, 1-M ammonium acetate, and acetone was used as the mobile phase with gradient condition.

Simultaneous analysis of carotenoids and chlorophylls in fruits and vegetables was also reported.[46,49] Elution order of pigments in green pepper using C_{30} 5-μm column was neoxanthin, violaxanthin, chlorophyll b, lutein, zeaxanthin, chlorophyll

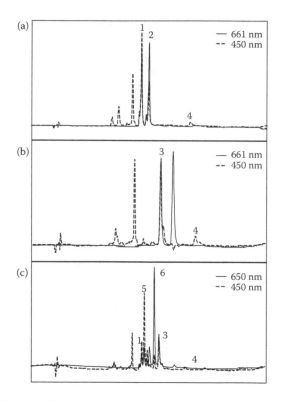

FIGURE 10.4 Typical HPLC chromatogram of chlorophyll derivatives in (a) fresh, (b) heat- and acid-treated, and (c) ZnCl2-treated spinach puree. Peak identifications: 1 = chlorophyll *b*; 2 = chlorophyll *a*; 3 = pheophytin *b*; 4= pheophytin *a*; 5 = Zn-pheophytin *b*; 6 = Zn-pheophytin *a*. (Adapted from Ferruzzi, M., Digestive stability, human intestinal cell uptake, and bioactivity of dietary chlorophyll derivatives. Ph.D. thesis. The Ohio State University, Columbus, 2001.)

a, and hydrocarbon carotenes. Puspitasari-Nienaber et al.[49] simultaneously detected chlorophylls and carotenoids in olive oil using C_{30} YMC column (25 cm × 4.6 mm ID). Lutein was eluted first in olive oil from a spinach extract and was followed by chlorophyll *b*, chlorophyll *a*, pheophytin *a*, and β-carotene.

10.3.3 CRITICAL PARAMETERS FOR CHLOROPHYLL ANALYSIS

Because of the inherent instability of chlorophylls and their derivatives, care should be taken during sample preparation, extraction, storage, and HPLC analysis. Most plant materials can be extracted directly without any pretreatment for chlorophyll analysis. Samples containing extremely high water content should be freeze-dried before extraction because of the effect of water content on the changes in absorption coefficients and the absorption maxima of pigments.[84] Small amount of MgO or $MgCO_3$ should be added during extraction to neutralize plant acids that cause the formation of pheophytin *a* from chlorophyll *a*.

Polar solvents such as acetone, ethanol, and methanol are useful for chlorophyll extraction in water-containing plant materials, but the solvent extracts should be transferred to hydrophobic organic solvents such as diethyl ether to prevent formation of allomeric chlorophylls in alcohol solutions and formation of crystals of pigments during refrigerated storage.[84] Because of the instability of chlorophylls in the presence of oxygen and light, extracted pigments should be placed in amber vials and analyzed immediately. If storage is necessary prior to analysis, extracts can be dried under nitrogen gas and kept at subzero temperature under a nitrogen or argon atmosphere without light exposure.[74]

For spectrophotometric analysis, the absorption maxima of chlorophylls and their derivatives strongly depend on the type of organic solvents used for solubilization. Generally, increasing the solvent polarity shifts the absorption maxima and absorption spectra to longer wavelengths. Therefore, standard chlorophyll compounds should be tested to determine the absorption maxima and absorption spectra in the particular spectrophotometric condition.[84]

HPLC methodologies change depending on the polarity and chemical structures of chlorophylls and their derivatives. Major developments in HPLC methods for foods are focused on the analysis of chlorophylls and their thermally induced derivatives such as pheophytins and pyropheophytins. If a C_{18} reversed-phase column is used as the stationary phase, MeOH-based mobile phases with modifiers can be used with different gradients depending on the target analytes, such as chlorophylls and their nonpolar derivatives, polar chlorophyll derivatives, or metalloporphyrins.[14,88,98] Advanced knowledge on the type and concentrations of chlorophylls in the test sample allows better judgment in choosing appropriate protocols, sample dilution volumes, and injection volumes. The HPLC analytical column should be cleaned and recalibrated with appropriate initial gradient condition between analyses, at the end of each day, which will help obtaining accurate and reproducible results and increasing column life.[88]

REFERENCES

1. Schwartz, S.J., Pigment analysis, in *Food analysis: Introduction to chemical analysis of foods*, Nielsen, S., Ed., Aspen Publishers, Gaithersburg, MD, 1998, 261.
2. Britton, G., Carotenoids, in *Natural Food Colorants*, Hendry, G.A.F. and Houghton, J.D., Eds., AVI, New York, 1992, 141.
3. Hendry, G.A.F., Chlorophylls, in *Natural Food Colorants*, Lauro, G.J. and Francis, F.J., Eds., Marcel Dekker, New York, 2000, 228.
4. von Elbe, J. and Schwartz, S.J., Colorants, in *Food Chemistry*, Fennema, O., Ed., Marcel Dekker, New York, 1996, 651.
5. Foote, C., Photosensitized oxidation and singlet oxygen: consequences in biological systems, in *Free Radicals in Biology*, Pryor, W.A., Ed., Academic Press, New York, 1976, 85.
6. Hughes, D.A., Dietary carotenoids and human immune function, *Nutrition*, 17, 823, 2001.
7. Parisi, V., Tedeschi, M., Gallinaro, G., Varano, M., Saviano, S., Piermarocchi, S., CARMIS Study. Carotenoids and antioxidants in age-related maculopathy Italian study: multifocal electroretinogram modifications after 1 year, *Ophthalmology*, 115, 324, 2008.
8. Hadley, C. et al., Tomatoes, lycopene, and prostate cancer: progress and promise, *Exp. Biol. Med.*, 227, 869, 2002.

9. Clinton, S.K. et al., cis-trans Lycopene isomers, carotenoids, and retinol in the human prostate, *Cancer Epidemiology, Biomarkers & Prevention*, 5, 823, 1996.

10. Schieber, A., Carle, R., Occurrence of carotenoid *cis*-isomers in food: Technological, analytical, and nutritional implications, *Trends Food Sci. Technol.*, 16, 416, 2005

11. Sweeney, J.P. and Marsh, A.C., Liver storage of vitamin A in rats fed carotene stereoisomers, *J. Nutr.*, 103, 20, 1973.

12. Dashwood, R.H., Chlorophylls as anticarcinogens (review), *Int. J. Oncology*, 10, 721, 1997.

13. Dashwood, R. et al., Chemopreventive properties of chlorophylls towards aflatoxin B1: a review of the antimutagenicity and anticarcinogenicity data in rainbow trout, *Mutation Research*, 399, 245, 1998.

14. Schoefs, B., Chlorophyll and carotenoid analysis in food products. Properties of the pigments and methods of analysis, *Trends Food Sci. Technol.*, 13, 361,2002.

15. Gandul-Rojas, B., Roca-L. Cepero, M., Mı́nguez-Mosquera, M.I., Chlorophyll and carotenoid patterns in olive fruits, *Olea europaea* Cv. Arbequina, *J. Agric. Food Chem.* 47, 2207, 1999.

16. McGhie, T.K., Ainge, G.D., Color in fruit of the genus *Actinidia*: carotenoid and chlorophyll compositions, *J. Agric. Food Chem. 50*, 117, 2002.

17. deMan, J., *Principles of Food Chemistry*. Aspen, Gaithersburg, Maryland, 1999, 239.

18. Nguyen, M.L. and Schwartz, S.J., Lycopene stability during food processing, *Pro. Soc. Exp. Bio. Med.*, 218, 101, 1998.

19. Ferruzzi, M.G. and Schwartz, S.J., Overview of chlorophylls in foods, in *Current protocols in Food Analytical Chemistry*, Schwartz, S.J., Ed., John Wiley & Sons Inc, New York, 2001, F4.1.1.

20. Oliver, J. and Palou, A., Chromatographic determination of carotenoids in foods, *J. Chomatogr. A.*, 881, 543, 2000.

21. AOAC international, *Official Methods of Analysis of AOAC international*, 16th ed., Arlington, VA, 2000, 45.1.03.

22. Rodriguez, G., Extraction, isolation, and purification of carotenoids, in *Current Protocols in Food Analytical Chemistry*, Schwartz, S.J., Ed., John Wiley & Sons Inc, New York, 2001, F2.1.1.

23. O'neil, C.A. and Schwartz, S.J., Chromatographic analysis of cis/trans carotenoid isomers, *J. Chromatogr.*, 624, 235, 1992.

24. Hart, D. and Scott, K., Development and evaluation of an HPLC method for the analysis of carotenoids in foods, and the measurement of the carotenoids center of vegetables and fruits commonly consumed in the UK, *Food Chem.*, 54, 101, 1995.

25. Konings, E. and Roomans, H., Evaluation and validation of an LC method for the analysis of carotenoids in vegetables and fruit, *Food Chem.*, 59, 599, 1997.

26. Oliver, J., Palou, A., and Pons, A., Semi-quantification of carotenoids by high-performance liquid chromatography: saponification-induced losses in fatty foods, *J. Chromatogr. A.*, 829, 393, 1998.

27. De Sio, F. et al., A chromatographic procedure for the determination of carotenoids and chlorophylls in vegetable products, *Acta Alimentaria*, 30, 395, 2001.

28. Weissenber, M. et al., Isocratic non-aqueous reversed-phase high performance liquid chromatographic separation of capsanthin and capsorubin in red peppers (Capsicum annuum L.), paprika and oleoresin, *J. Chromatogr. A*, 757, 89, 1997.

29. Breithaupt, D. and Bamedi, A., Carotenoid esters in vegetables and fruits: a screening with emphasis on beta-cryptoxanthin esters, *J. Agric. Food Chem.*, 49, 2064, 2001.

30. Meléndez-Martínez, A.J., Vicario, I.M., Heredia, F.J., Review: Analysis of carotenoids in orange juice, *J. Food Composit. Analysis,* 20, 638, 2007.

31. Su, Q. et al., Identification and quantitation of major carotenoids in selected components of the Mediterranean diet: green leafy vegetables, figs and olive oil, *Eur. J. Clin. Nutr.*, 56, 1149, 2002.
32. Taungbodhitham, A.K. et al., Evaluation of extraction method for the analysis of carotenoids in fruits and vegetables, *Food Chem.*, 63, 577, 1998.
33. Lin, C.H., Chen, B.H., Determination of carotenoids in tomato juice by liquid chromatography, *J. Chromatogr. A.*, 1012, 103, 2003.
34. Su, Q., Rowley, K.G., and Balazs, N.D.H., Carotenoids: separation methods applicable to biological samples, *J. Chromatogr. B.*, 781, 393, 2002.
35. Barth, M.M. et al., Determination of optimum conditions for supercritical fluid extraction of carotenoids from carrot (Daucus carota L.) Tissue, *J. Agric. Food Chem.*, 43, 2876, 1995.
36. Burrt, B.J. et al., Supercritical fluid extraction and reversed-phase liquid chromatography methods for vitamin A and b-carotene. Heterogeneous distribution of vitamin A in the liver, *J. Chromatogr. A*, 762, 201, 1997.
37. Rozzi, N.L. et al., Supercritical fluid extraction of lycopene from tomato processing byproducts, *J. Agric. Food Chem.*, 50, 2638, 2002.
38. Gómez-Prieto, M.S., Caja, M.M., Herraiz, M., Santa-María G.,Supercritical fluid extraction of *all-trans*-lycopene from tomato, *J. Agric. Food Chem.*, 51, 3, 2003.
39. Delgado-Vargas, F. and Paredes-Lopez, O., Enzymic treatment to enhance carotenoid content in dehydrated marigold flower meal, *Plant Foods Hum. Nutr.*, 50, 163, 1997.
40. Delgado-Vargas, F.P. and Paredes-Lopez. O., Effects of enzymic treatment on carotenoid extraction from marigold flowers (Tagetes erecta), *Food Chem.*, 58, 255, 1996.
41. Dachtler, M. et al., Combined HPLC-MS and HPLC-NMR on-line coupling for the separation and determination of lutein and zeaxanthin stereoisomers in spinach and in retina, *Anal. Chem.*, 73, 667, 2001.
42. Fisher, J.F., Rouseff, R.L., Solid-phase extraction and HPLC determination of beta.-cryptoxanthin and alpha- and beta-carotene in orange juice, *J. Agric. Food Chem.*, 34, 985, 1986.
43. Cortés, C., Esteve, M.J., Frígola, A., Torregrosa, F. Identification and quantification of carotenoids including geometrical isomers in fruit and vegetable juices by liquid chromatography with ultraviolet–diode array detection, *J. Agric. Food Chem.*, 52, 2203, 2004.
44. Kimura, M., Rodriguez-Amaya, D.B., and Godoy, H.T., Assessment of the saponification step in the quantitative determination of carotenoids and provitamins A, *Food Chem.*, 35, 187, 1990.
45. Granado, F. et al., A fast, reliable and low-cost saponification protocol for analysis of carotenoids in vegetables, *J. Food Comp. Anal.*, 14, 479, 2001.
46. Burns, J., Fraser, P.D., and Bramley, P.M., Identification and quantification of carotenoids, tocopherols and chlorophylls in commonly consumed fruits and vegetables, *Phytochemistry*, 62, 939, 2003.
47. Cserhati, T. and Forgacs, E., Review, liquid chromatographic separation of terpenoid pigments in foods and food products, *J. Chromatogr. A.*, 936, 119, 2001.
48. Scott, K., Detection and Measurement of Carotenoids by UV/VIS Spectrophotometry, in *Current Protocols in Food Analytical Chemistry*, Schwartz, S.J., Ed., John Wiley & Sons Inc., New York, 2001, F2.2.1.
49. Puspitasari-Nienaber, N.L., Ferruzzi, M.G., and Schwartz, S.J., Simultaneous detection of tocopherols, carotenoids, and chlorophylls in vegetable oils by direct injection C30 RP-HPLC with coulometric electrochemical array detection, *J. Am. Oil Chem. Soc.*, 79, 633, 2002.

50. Ferruzzi, M. et al., Carotenoid determination in biological microsamples using liquid chromatography with a coulometric electrochemical array detector, *Anal. Biochem.*, 256, 74, 1998.

51. Ferruzzi, M.G. et al., Analysis of lycopene geometrical isomers in biological microsamples by liquid chromatography with coulometric array detection, *J. Chromatogr. B.*, 760, 289, 2001.

52. Luterotti, S., Franko, M., and Bicanic, D., Fast quality screening of vegetable oils by HPLC-thermal lens spectrometric detection, *J. Am. Oil Chem. Soc.*, 79, 1027, 2002.

53. Franko, M., Van De Bovenkamp, P., and Bicanic, D., Determination of trans-beta-carotene and other carotenoids in blood plasma using high-performance liquid chromatography and thermal lens detection, *J. Chromatogr. B.*, 718, 47, 1998.

54. Strohschein, S. et al., Structure elucidation of b-carotene isomers by HPLC-NMR coupling using a C30-bound phase, *Fresenius' J. Anal. Chem.*, 357, 498, 1997.

55. Tiziani, S., Schwartz, S.J., Vodovotz, Y. Profiling of carotenoids in tomato juice by one- and two-dimensional NMR. *J. Agric. Food Chem.*, 54, 6094, 2006.

56. van Breemen, R.B., Mass spectrometry of Carotenoids, in *Current Protocols in Food Analytical Chemistry*, Schwartz, S.J., Ed., John Wiley & Sons Inc., New York, 2001, F2.4.1.

57. Hadden, W.L. et al., Carotenoid composition of marigold (Tagetes erecta) flower extract used as nutritional supplement, *J. Agric. Food Chem.*, 47, 4189, 1999.

58. Careri, M. et al., Use of eluent modifiers for liquid chromatography particle-beam electron-capture negative-ion mass spectrometry of carotenoids, *Rapid Communications in Mass Spectrometry*, 13, 118, 1999.

59. Breithaupt, D.E., Wirt, U., and Bamedi, A., Differentiation between lutein monoester regioisomers and detection of lutein diesters from marigold flowers (Tagetes erecta l.) and several fruits by liquid chromatography-mass spectrometry, *J. Agric. Food Chem.*, 50, 66, 2002.

60. Breithaupt, D.E. and Bamedi, A., Carotenoids and carotenoid esters in potatoes (Solanum tuberosum L.): new insights into an ancient vegetable, *J. Agric. Food Chem.*, 50, 7175, 2002.

61. van Breemen, R.B., Schmitz, H.H., and Schwartz, S.J., Fast atom bombardment tandem mass spectrometry of carotenoids, *J. Agric. Food Chem.*, 43, 384, 1995.

62. Wingerath, T.S. et al., Fruit juice carotenol fatty acid esters and carotenoids as identified by matrix-assisted laser desorption ionization (MALDI) mass spectrometry, *J. Agric. Food Chem.*, 44, 2006, 1996.

63. Kaufmann, R.W. et al., Analysis of carotenoids and carotenol fatty acid esters by matrix-assisted laser desorption ionization (MALDI) and MALDI -post-source-decay mass spectrometry, *Anal. Biochem.*, 238, 117, 1996.

64. Tian, Q., Duncan, C., and Schwartz, S.J., Atmospheric pressure chemical ionization mass spectrometry and in-source fragmentation of lutein esters, *Journal of Mass Spectrometry*, 38, 990, 2003.

65. Khachik, F. et al., Identification, quantification, and relative concentrations of carotenoids and their metabolites in human milk and serum, *Anal. Chem.*, 69, 1873, 1997.

66. Humphries, J.M. and Khachik, F., Distribution of lutein, zeaxanthin, and related geometrical isomers in fruit, vegetables, wheat, and pasta products, *J. Agric. Food Chem.*, 51, 1322, 2003.

67. Khachik, F., Steck, A., and Pfander, H., Isolation and structural elucidation of (13Z,13'Z,3R,3'R,6'R)-lutein from marigold flowers, kale, and human plasma, *J. Agric. Food Chem.*, 47, 455, 1999.

68. De Sio, F., Servillo, L. Loiuduce, R., Laratta, B., Castaldo, D., A chromatographic procedure for the determination of carotenoids and chlorophylls in vegetable products. *Acta Allimentaria*, 30, 395, 2001.

69. Lee, H.S., Castle, W.S., and Coates, G.A., High-performance liquid chromatography for the characterization of carotenoids in the new sweet orange (Earlygold) grown in Florida, USA, *J. Chromatogr. A.*, 913, 371, 2001.
70. Schmitz, H.H., Emenhiser, C., and Schwartz, S.J., HPLC separation of geometric carotene isomers using a calcium hydroxide stationary phase, *J. Agric. Food Chem.*, 43, 1212, 1995.
71. Nguyen, M., Francis, D., and Schwartz, S.J., Thermal isomerisation susceptibility of carotenoids in different tomato varieties, *J. Sci. Food Agric.*, 81, 910, 2001.
72. Emenhiser, C., Sander, L.C., and Schwartz, S.J., Capability of polymeric C_{30} stationary phase to resolve cis-trans carotenoid isomers in reversed-phase liquid chromatography, *J. Chromatogr. A.*, 707, 1995.
73. Luterotti, S. et al., Ultrasensitive assays of trans- and cis-b-carotenes in vegetable oils by high performance liquid chromatography-thermal lens detection, *Analytica Chimica Acta*, 460, 193, 2002.
74. Craft, N., Chromatographic Techniques for carotenoid separation, in *Current Protocols in Food Analytical Chemistry*, Schwartz, S.J., Ed., John Wiley & Sons Inc., New York, 2001, F2.3.1.
75. Ishida, B.K. et al., A modified method for simple, rapid HPLC analysis of lycopene isomers, *Acta Horticulturae*, 542, 235, 2001.
76. Lessin, W.J. et al., Quantification of cis-trans isomers of provitamin a carotenoids in fresh and processed fruits and vegetables, *J. Agric. Food Chem.*, 45, 3728, 1997.
77. Emenhiser, C. et al., Separation of geometrical carotenoid isomers in biological extracts using a polymeric C30 column in reversed-phase liquid chromatography, *J. Agric. Food Chem.*, 44, 3887, 1996.
78. Boehm, V. et al., Trolox equivalent antioxidant capacity of different geometrical isomers of a-carotene, b-carotene, lycopene, and zeaxanthin, *J. Agric. Food Chem.*, 50, 221, 2002.
79. Craft, N.E., Carotenoid reversed-phase high-performance liquid chromatography methods: Reference compendium, *Methods Enzymol.*, 213, 185, 1992.
80. Craft, N.E. and Soares, J.H. Jr., Relative solubility, stability, and absorptivity of lutein and b-carotene in organic solvents, *J. Agric. Food Chem.*, 40, 431, 1992.
81. Bohm, V., Use of column temperature to optimize carotenoid isomer separation by C30 high performance liquid chromatography, *J. Separ. Sci.*, 24, 955, 2001.
82. Craft, N.E., Wise, S.A., and Soares, J.H. Jr., Optimization of an isocratic high-performance liquid chromatographic separation of carotenoids, *J. Chromatogr.*, 589, 171, 1992.
83. Epler, K.S. et al., Evaluation of reversed-phase liquid chromatographic columns for recovery and selectivity of selected carotenoids, *J. Chromatogr.*, 595, 1992.
84. Lichtenthaler, K.B. and Buschmann, C., Extraction of photosyntheic tissues: chlorophylls and carotenoids, in *Current Protocols in Food Analytical Chemistry*, Schwartz, S.J., Ed., John Wiley & Sons Inc., New York, 2001, F 4.2.1.
85. Simon, D. and Helliwell, S., Extraction and quantification of chlorophyll a from freshwater green algae, *Water Research*, 32, 2220, 1998.
86. Prasad, V.K.V., Singh, C.P., and Singh, I.S., Chlorophyll extraction from french bean (*Phaseolus vulgaris* L.) leaves using dimethyl sulfoxide: a simple method, *Indian Biologist*, 25, 58, 1993.
87. Vagi, E. et al., Recovery of pigments from Origanum majorana L. by extraction with supercritical carbon dioxide, *J. Agric. Food Chem.*, 50, 2297, 2002.
88. Ferruzzi, M. and Schwartz, S.J., Chromatographic separation of chlorophylls, in *Current protocols in Food Analytical Chemistry*, Schwartz, S.J., Ed., John Wiley & Sons Inc., New York, 2001, F4.4.1.

89. Tabet, J. et al., Time-resolved desortion. III. The metastable decomposition of chlorophyll a and some derivatives, *Int. J. Mass Spectrom. Ion Processes*, 65, 105, 1985.

90. Liu, S. et al., Investigation of a series of synthetic cationic porphyrins using matrix-assisted laser desorption/ionization time-of-flight mass spectrometry, *Rapid Communications in Mass Spectrometry*, 13, 2034, 1999.

91. Daugherty, R.D. et al., Hydration behavior of chorophyll a: a field desorption mass spectral study, *J. Am. Chem. Soc*, 102, 416, 1980.

92. van Breemen, R.B., Canjura, F.L., and Schwartz, S.J., Identification of chlorophyll derivatives by mass spectrometry, *J. Agric. Food Chem.*, 39, 1452, 1991.

93. Teng, S.S. and Chen, B.H., Formation of pyrochlorophylls and their derivatives in spinach leaves during heating, *Food Chem.*, 65, 367, 1999.

94. Zissis, K.D., Dunkerley, S., and Brereton, R.G., Chemometric techniques for exploring complex chromatograms: Application of diode array detection high performance liquid chromatography electrospray ionization mass spectrometry to chlorophyll a allomers., *Analyst*, 124, 971, 1999.

95. Gauthier-Jaques, A.B. et al., Improved method to track chlorophyll degradation, *J. Agric. Food Chem.*, 49, 1117, 2001.

96. Jie, C.W., Walker, J.S., and Keely, B.J., Atmospheric pressure chemical ionization normal phase liquid chromatography mass spectrometry and tandem mass spectrometry of chlorophyll a allomers, *Rapid Communications in Mass Spectrometry*, 16, 473, 2002.

97. Canjura, F.L., Watkins, R.H., and Schwartz, S.J., Color improvement and metallo-chlorophyll complexes in continuous flow aseptically processed peas, *J. Food Sci.*, 64, 987, 1999.

98. Schwartz, S.J., Woo, S.L., and von Elbe, J.H., High-performance liquid chromatography of chlorophylls and their derivatives in fresh and processed spinach, *J. Agric. Food Chem.*, 29, 533, 1981.

99. Canjura, F.L. and Schwartz, S.J., Separation of chlorophyll compounds and their polar derivatives by high-performance liquid chromatography, *J. Agric. Food Chem.*, 39, 1102, 1991.

100. Schwartz, S.J. and Lorenzo, T.V., Chlorophyll stability during continuous aseptic processing and storage, *J. Food Sci.*, 56, 1059, 1991.

101. Ferruzzi, M., Digestive stability, human intestinal cell uptake, and bioactivity of dietary chlorophyll derivatives. Ph.D. thesis, The Ohio State University, Columbus, 2001.

102. Mangos, T.B. and Berger, R.G., Determination of major chlorophyll degradation products, *Zeitschrift fuer Lebensmittel-Untersuchung und -Forschung*, 204, 345, 1997.

103. Borrmann, D., Castelhano de Andrade, J., and Lanfer-Marques, U.M. Chlorophyll degradation and formation of colorless chlorophyll derivatives during soybean (*Glycine max* L. Merill) seed maturation. *J. Agric. Food Chem.*, 57, 2030, 2009.

11 Analysis of Polyphenols in Foods

Fereidoon Shahidi and Marian Naczk

CONTENTS

11.1 INTRODUCTION

Phenolic compounds in food and nutraceuticals originate from plant sources (van Sumere 1989; Shahidi 2000, 2002; Shahidi and Naczk 2004). Their occurrence in animal tissues and nonplant materials is generally due to the ingestion of plant foods. Plants and foods contain a large variety of phenolic derivatives, including simple phenols, phenylpropanoids, benzoic acid derivatives, flavonoids, stilbenes, tannins, lignans, and lignins. These rather varied substances are essential for the growth and reproduction of plants and also act as antifeedants and antipathogens (Butler 1992). The contribution of phenolics to the pigmentation of plant foods is also well recognized. In addition, phenolics function as antibiotics, natural pesticides, signaling substances for establishing symbiosis with rhizobia, attractants for pollinators, protective agents against ultraviolet (UV) light, insulating materials to make cell walls impermeable to gas and water, and structural materials to give plants stability.

Many properties of plant products are associated with the presence, type, and content of their phenolic compounds. The astringency of foods, the beneficial health effects of certain phenolics or their potential antinutritional properties when present in large quantities, is significant to both producers and consumers of foods. Anthocyanins are found widely in foods, especially in fruits, and also in floral tissues (Harborne and Williams 2001). They may also be used as nutraceuticals in the dried and powdered form from certain fruit or fruit by-product sources. Anthocyanins are responsible for red, blue, violet, and purple colors of most plant species and their fruits and other products. For example, after their extraction, colorants produced from the skin of grapes can be used by the food industry (Francis 1993).

11.2 BIOSYNTHESIS AND CLASSIFICATION

Phenylalanine ammonia lyase catalyzes the release of ammonia from phenylalanine and leads to the formation of a carbon–carbon double bond, yielding *trans*-cinnamic acid. In some plants and grasses, tyrosine is converted into 4-hydroxycinnamic acid via the action of tyrosine ammonia lyase (Figure 11.1). Introduction of an hydroxyl group into the *para* position of the phenyl ring of cinnamic acid proceeds via catalysis by monooxygenase utilizing cytochrome P_{450} as the oxygen-binding site. The *p*-coumaric acid so formed may be hydroxylated further in positions 3 and 5 by hydroxylases and possibly methylated via *O*-methyl transferase with *S*-adenosylmethionine as methyl donor; this leads to the formation of caffeic, ferulic, and sinapic acids (Figure 11.2). All these compounds possess a phenyl ring (C6) and a C3 side chain, thus are collectively termed phenylpropanoids, which serve as precursors for the synthesis of lignins and many other compounds.

Benzoic acid derivatives are produced via the loss of a two-carbon moiety from phenylpropanoids. Salicylic acid is a benzoic acid derivative that acts as a signaling substance (Raskin 1992). After infection or UV irradiation, many plants increase their salicylic acid content, which may induce the biosynthesis of defense substances. Aspirin, the acetyl ester of salicylic acid, was first isolated from the bark of the willow tree. Similar to phenylpropanoid series, hydroxylation and possible methylation of hydroxybenzoic acid lead to the formation of dihydroxybenzoic acid (protocatechuic

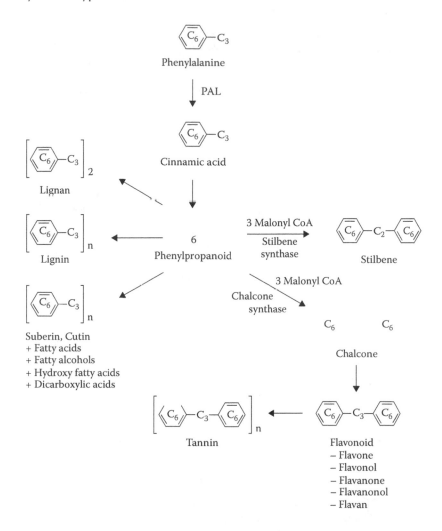

FIGURE 11.1 Production of phenylpropanoids, flavonoids, tannins, and other phenolics from phenylalanine. PAL denotes phenylalanine ammonia lyase.

acid), vanillic acid, syringic acid, and gallic acid (Figure 11.3). Hydroxybenzoic acids are commonly present in the bound form in foods and are components of complex structures, such as lignins and hydrolyzable tannins. They are also found in the form of organic acids and as sugar derivatives (Schuster and Herrmann 1985). Also, they are present in the free form (Mosel and Herrmann 1974a,b; Schmidtlein and Herrmann 1975; Sohr and Herrmann 1975a,b).

Conventionally, phenylpropanoids (e.g., cinnamic acid family) and benzoic acid derivatives are collectively termed "*phenolic acids*" in the food science literature. However, it should be noted that this nomenclature is not necessarily correct from a chemical and structural viewpoint. Nonetheless, we refer to these compounds as phenolic acids in the remainder of this chapter.

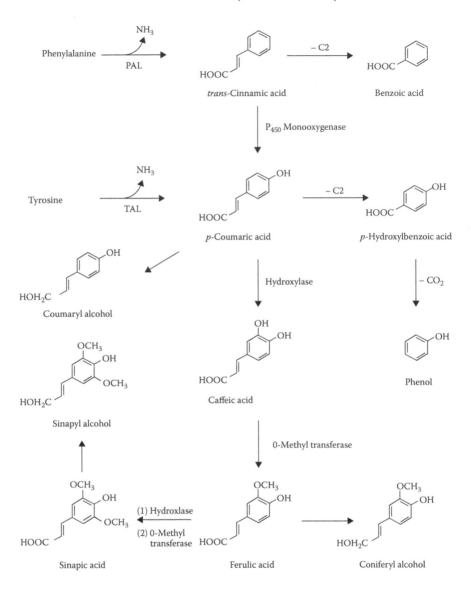

FIGURE 11.2 Formation of phenylpropanoids of cinnamic acid and benzoic acid derivatives and related compounds from phenylalanine and tyrosine. PAL denotes phenylalanine ammonia lyase and TAL denotes tyrosine ammonia lyase.

Decarboxylation of benzoic acid and phenylpropanoids derivatives leads to the formation of simple phenols (Figure 11.2). Thermal degradation of lignin or microbial transformation may also produce simple phenols in foods (Maga 1978); thus, vinyl-substituted phenols may be produced by decarboxylation of hydroxycinnamic acids (Pyysalo et al. 1977). Exposure of 4-vinylguaiacol to oxygen leads to the formation of vanillin (Fiddler et al. 1967). A number of simple phenols—namely phenol, o-cresol, 4-ethylphenol, guaiacol, 4-vinylguaiacol, and eugenol—are found

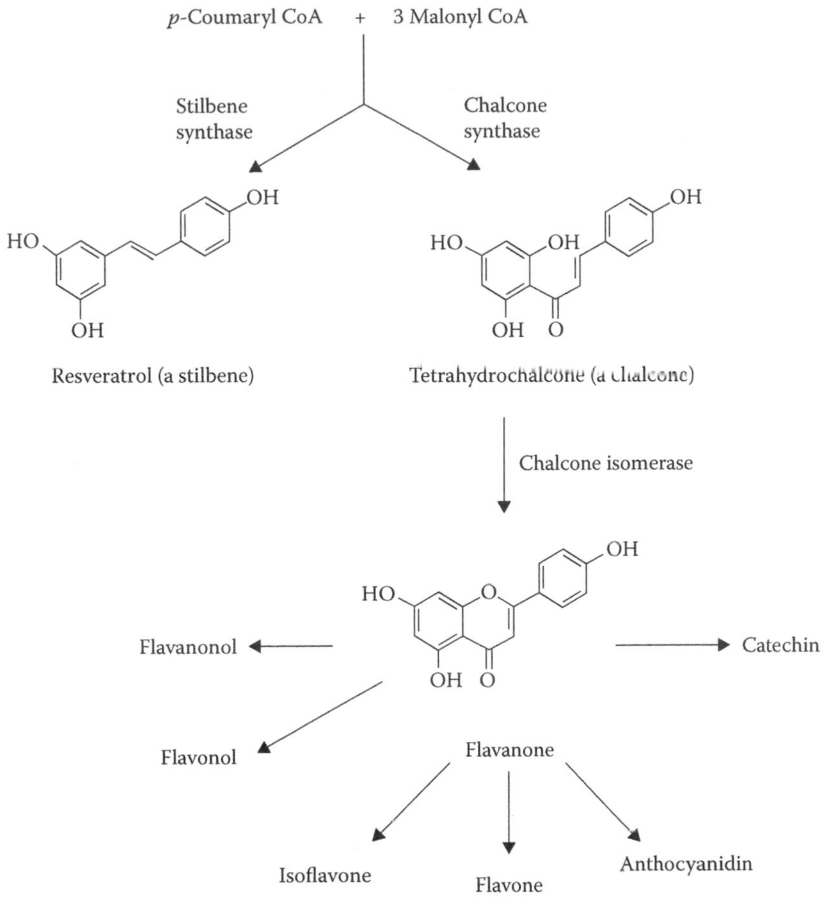

FIGURE 11.3 Production of flavonoids and stilbenes from phenylpropanoids (coumaryl CoA) and malonyl CoA.

in foods of plant origin. Reduction products of phenylpropanoids also yield phenolic alcohols, such as sinapyl alcohol, coniferyl alcohol, and coumaryl alcohol, among others.

Coumarins are lactones of *cis-O*-hydroxycinnamic acid derivatives present in certain foods of plant origin and exist in the free form or as glycosides. The most important of these are simple coumarins, furanocoumarins (also called psoralens), and pyranocoumarins, which are present in both free and glycosidic forms in foods. This group of compounds also acts as phytoalexins in response to disease infection. It is worth noting that most studies on coumarins have been carried out on Umbelliferae and Rutaceae families (Murray et al. 1982). Presence of psoralens in different parts of fruits and vegetables of these families has been confined, as high levels are found in parsnip roots (Murray et al. 1982).

Flavonoids, including flavones, isoflavones, and anthocyanidins, are formed by condensation of a phenylpropane (C6–C3) compound via participation of three

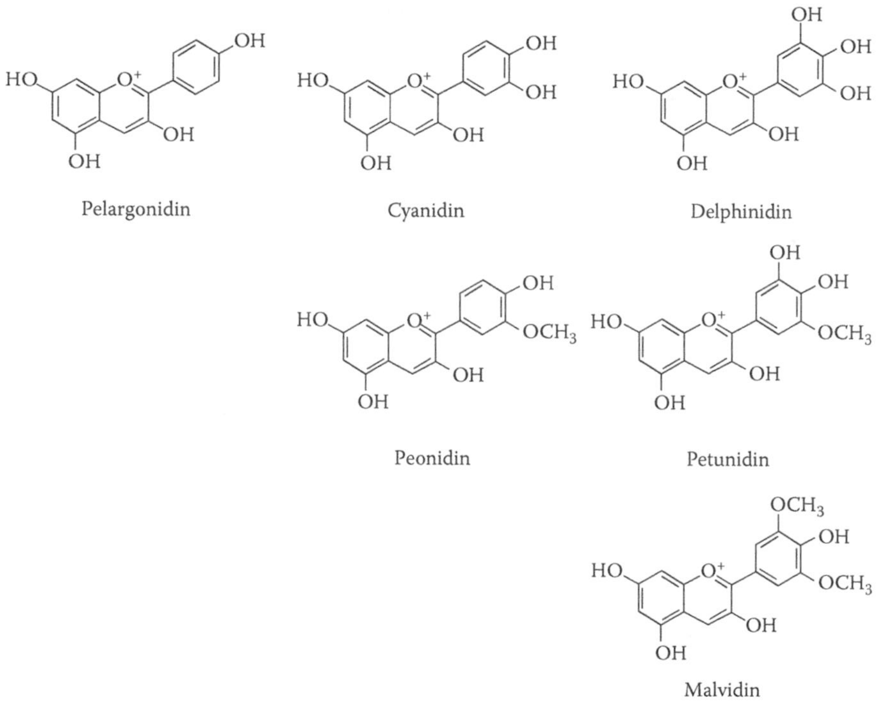

Pelargonidin   Cyanidin   Delphinidin

Peonidin   Petunidin

Malvidin

FIGURE 11.4   Chemical structures of anthocyanidins.

molecules of malonyl coenzyme A, which leads to the formation of chalcones that subsequently cyclize under acidic conditions (Figure 11.3). Thus, flavonoids have the basic skeleton of diphenylpropane (C_6–C_3–C_6) with different oxidation level of the central pyran ring. This also applies to stilbenes, but in this case, after introduction of the second phenyl moiety, one carbon atom of the phenylpropane is split off. Stilbenes are potent fungicides in plants (e.g., viniferin in grapevine). In the case of flavonoids and isoflavonoids, depending on the substitution and unsaturation patterns, flavones, flavonones, flavonols, and flavanonols as well as flavan-3-ols and related compounds may be formed.

Among flavonoids, anthocyanins and catechins, known collectively as flavans because of lack of the carbonyl group in the 3-position, are important; flavan-3-ols and flavan-3,4-diols belong to this category. Anthocyanins are glycosidically bound anthocyanidins (Figure 11.4) present in many flowers and fruits. Chalcones and flavones are yellow; anthocyanins are water-soluble pigments responsible for bright red, blue, and violet color of fruits and other foods (Mazza and Miniati 1994). Thus, the bright red skin of radishes, the red skin of potatoes, and the dark skin of eggplants are due to the presence of anthocyanins. Blackberries, red and black raspberries, blueberries, cherries, currants, concord and other red grapes, pomegranates, ripe gooseberries, cranberries, and plums contain anthocyanins. Almost 200 different anthocyanins have been identified in plants. The anthocyanin pelargonin contains pelargonidin as its chromophore. Introduction of one or more hydroxyl groups in

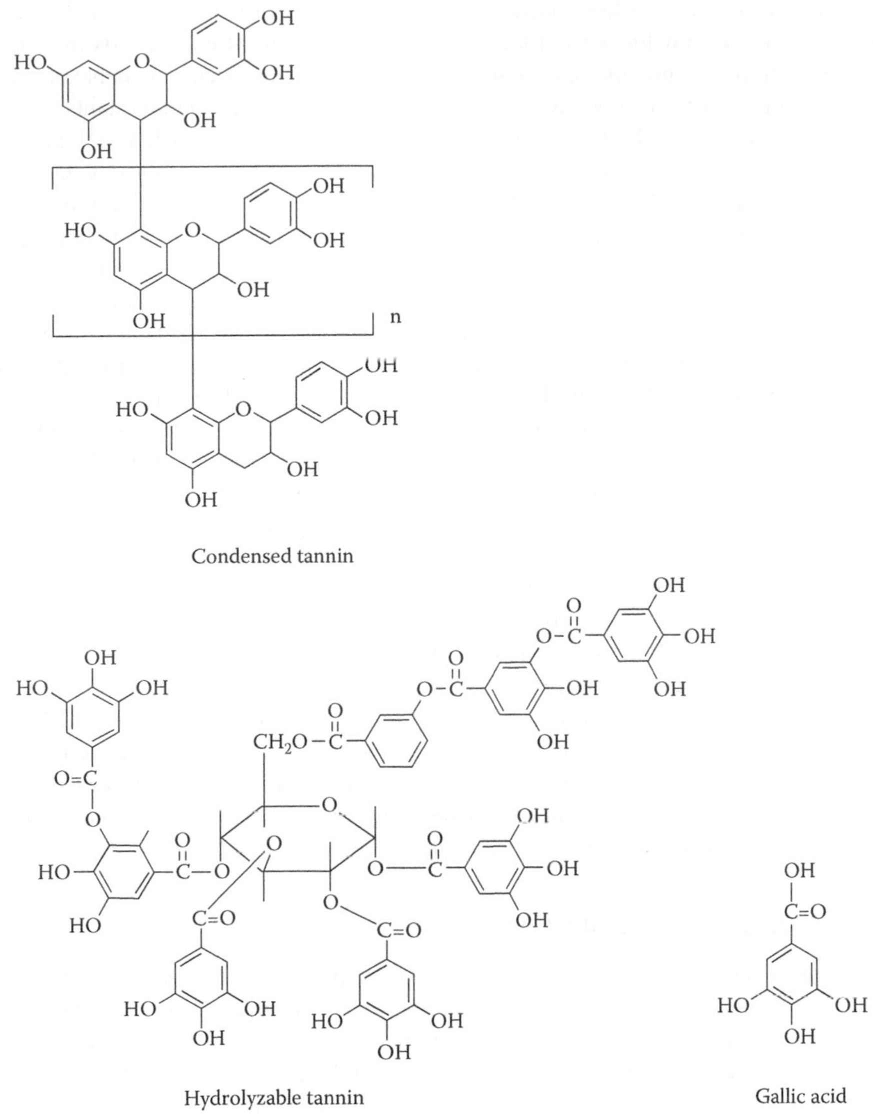

Condensed tannin

Hydrolyzable tannin   Gallic acid

FIGURE 11.5 Chemical structures of tannins.

3′ and 5′ positions of the phenyl residue and their successive methylation yields five additional anthocyanidins that differ in color. Most anthocyanins occur as monogly-cosides and diglycosides of pelargonidin, cyanidin, peonidin, delphinidin, petunidin, and malvidin. Meanwhile, anthocyanins assume different colors when subjected to pH variation in solutions. In addition, catechins and epicatechins, found in different plants, and in high amounts in green tea leaves, are similar in their structures to anthocyanidins, but are colorless.

Tannin is a collective term used to describe a variety of plant polyphenols used in the tanning of raw hides to produce leather. This group of compounds includes both oligomeric and polymeric constituents. Tannins are polyphenols capable of precipitating proteins from aqueous solutions and are widely distributed in plants. They are found in especially high amounts in the bark of certain trees such as oak and in galls. Tannins also form complexes with certain types of polysaccharides, nucleic acids, and alkaloids (Ozawa et al. 1987). Depending on their structures, tannins are defined as hydrolyzable or condensed, the latter term also known as proanthocyanidin (Figure 11.5). The condensed tannins are oligomers and polymers of flavonoids, specifically flavan-3-ols, whereas hydrolyzable tannins are glycosylated gallic acids (Ferreira and Li 2000; Khanbabaee and Van Ree 2001). Gallic acids are linked to sugar molecules. The molecular weights of hydrolyzable tannins range from 500 to 2800 Da or more (Haddock et al. 1982). Based on their hydrolysis products, hydrolyzable tannins include gallotannins and ellagitannins. Condensed tannins include dimers, oligomers, and polymers of flavan-3-ols. Approximately, 50 procyanidins ranging from dimers to hexamers have been identified (Hemingway 1989). Based on the hydroxylation pattern of A- and B-rings, proanthocyanidins can be divided into procyanidins, propelargonidins, and prodelphinidins. The consecutive units of condensed tannins are linked through the interflavonoid bond between C-4 and C-8 or C-6 (Hemingway 1989). The molecular weights of proanthocyanidins isolated from fruits range between 2000 and 4000 Da (Macheix et al. 1990).

The phenolic groups of tannins bind very tightly with the –NH groups of peptides and proteins and thus prevent their hydrolysis and digestion in the stomach, making tannins antinutritional in nature. In the tanning process, tannins bind to the collagen of the animal hide and thus produce leather that can withstand attack by the degrading microorganisms. Because they inactivate aggressive enzymes, tannins also protect plants against attack by microorganisms.

11.3 ANALYSIS OF PHENOLICS

There are a number of reviews on the analysis of polyphenolics (Antolovich et al. 2000; Deshpande et al. 1986; Hagerman et al. 1997; Jackman et al. 1987; Makkar 1989; Porter 1989; Scalbert et al. 1989; Scalbert 1992; Tempel 1982). The assays used for the analysis of phenolics and polyphenolics can be classified either as those that determine total phenolics content or as those that quantify a specific group or class of compounds.

Quantification of phenolic compounds in plant materials is influenced by their chemical nature, the extraction method employed, sample particle size, storage time and conditions, assay method, selection of standards, and presence of interfering substances, such as waxes, fats, terpenes, and chlorophylls. Solubility of polyphenolics is governed by the type of solvent (polarity) used, degree of polymerization, interaction of phenolics with other food constituents, and formation of insoluble complexes. Methanol, ethanol, acetone, water, ethyl acetate, propanol, dimethylformamide, and their combinations are commonly used for the extraction of phenolics (Antolovich et al. 2000), and there is no satisfactory extraction procedure that is suitable for extraction of all phenolics or a specific class of phenolic substances in plant materials. Moreover, it is difficult to find a specific and suitable standard for quantification of phenolics due to the complexity of plant phenolic substances and existing differences in the reactivity of phenols toward reagents used for their quantification.

11.4 EXTRACTION PROCEDURES

The chemical nature of plant phenolics varies from simple to very highly polymerized substances that include varying quantities of phenolic acids, phenylpropanoids, anthocyanins, and tannins, as well as possible interactions of phenolics with carbohydrates, proteins, and other plant components. Some high-molecular-weight phenolics and their complexes may be insoluble. Therefore, the phenolic extracts of plant materials always include a mixture of different classes that are soluble in the solvent system used. Additional steps are required for the removal of unwanted phenolics and nonphenolic substances.

Krygier et al. (1982) extracted free and esterified phenolic acids from oilseeds using a mixture of methanol–acetone–water (7:7:6, v/v/v) at room temperature. Following this, the free phenolics were extracted with diethyl ether from the extract, and the extract was then treated with 4 M NaOH under nitrogen to liberate esterified phenolic acids. The hydrolyzate was acidified, and the liberated phenolic acids were extracted with diethyl ether. The leftover sample after exhaustive extraction with a mixture of methanol/acetone/water was treated with 4 M NaOH under nitrogen to liberate insoluble-bound phenolic acids. Other solvents, such as ethanol and acetone, have also been used for the extraction of phenolics with different proportions of water (Antolovich et al. 2000; Kozlowska et al. 1983; Naczk et al. 1992a).

Anthocyanins are usually extracted from plant material with an acidified organic solvent, most commonly methanol. This solvent system destroys the cell membranes, simultaneously dissolving the anthocyanins and stabilizing them. However, according to Moore et al. (1982a,b), the acid may change the native form of anthocyanins by breaking down their complexes with metals and copigments. The acidic extracts of anthocyanins are first concentrated under vacuum and then extracted, if necessary, with petroleum ether, ethyl acetate, or diethyl ether to remove lipids and unwanted polyphenols (Fuleki and Francis 1968a,b). The extract can also be partially purified using ion-exchange resins as described by Fuleki and Francis (1968c). Concentration of acidic extracts of anthocyanins before purification may, however, result in losses of labile acyl and sugar residues (Adams 1972). To avoid this detrimental effect, several researchers have proposed to reduce the acid contact with the pigments (Du and Francis 1975), use neutral organic solvents or boiling water (Adams 1972), or use weak organic acids such as formic acid or acetic acid for acidification of the solvents used for the extraction of pigments (Antolovich et al. 2000; Moore et al. 1982 a,b). Anthocyanins may also be recovered from the extract using solid-phase extraction (SPE) or solid-phase microextraction on C_{18} cartridges. Hong and Wrolstad (1990) fractionated anthocyanins by elution of the extract through a SPE cartridge with an alkaline borate solution. Only anthocyanins with o-dihydroxy groups (cyanidin, delphinidin, and petunidin) were preferentially eluted from the SPE cartridge with borate solution because of the formation of a hydrophilic borate–anthocyanin complex. Wang and Sporns (1999) also isolated anthocyanins from fruit juices and wine using a SPE cartridge but used a methanol–formic acid–water (70:2:28, v/v/v) solvent system for their elution.

Several solvent systems have been used, including absolute methanol, ethanol, acidified methanol, acetone, water, and their combination for the extraction of condensed tannins. For example, 1% HCl in methanol was used for the extraction of tannins from sorghum (Price et al. 1978) and dry beans (Deshpande and Cheryan 1985). Acetone/water (70:30, v/v) served best for the extraction of tannins from rapeseed/canola (Table 11.1) (Leung et al. 1979; Naczk and Shahidi 1991; Shahidi and Naczk 1989), beach pea (*Lathyrus maritimus* L.) (Chavan et al. 2001; Shahidi et al. 2001), and blueberries (Prior et al. 2001). However, acetone/water (60:40, v/v) solution was used for extraction of tannins from cider apple (Guyot et al. 2001) and grape skins (Labarbe et al. 1999).

The extraction period also affects the recovery of polyphenolics from plant materials. Extraction periods varying from 1 minute (Price and Butler 1977) to 24 hours (Burns 1971; Maxson and Rooney 1972) have been reported. Longer extraction times increase the chance of oxidation of phenolics unless reducing agents are added to the solvent system (Khanna et al. 1968). However, Naczk and Shahidi (1991) and Naczk et al. (1992a) found that a two-stage extraction with 70% (v/v) acetone, 1 minute each, at 10,000 rpm using a Polytron homogenizer, was sufficient for the extraction of tannins from commercial canola meals. Further extraction (up to six stages) only marginally enhanced the yield of extraction of other phenolic compounds (Table 11.2). Deshpande (1985) demonstrated that the optimum extraction time required for dry bean phenolics was 50–60 minutes. Tannins of cloves and allspice may be extracted with boiling water for 2 hours (AOAC 1980).

The recovery of polyphenols from food products is also influenced by the ratio of sample-to-solvent (R) (Table 11.3). Naczk and Shahidi (1991) and Naczk et al. (1992a) found that changing R from 1:5 to 1:10 increased the extraction of condensed

TABLE 11.1
Effect of HCl Addition and/or Boiling on the Recovery of Total Phenolics and Tannins from Canola Meal

Solvent System	Total Phenolics	Tannins
70% methanol	847.0	241.3
1% HCl in methanol	892.1	73.0
1% HCl in 70% methanol	1079.8	225.9
1% HCl in methanol 1 4 min of boiling	1053.3	112.5
70% acetone	805.8	321.3
1% HCl in 70% acetone	1010.7	216.9
1% HCl in 70% acetone 1 4 min of boiling	1051.6	338.7

Source: Naczk, M. and Shahidi, F. Critical evaluation of quantification methods of rapeseed tannins. Presented at 8th International Rapeseed Congress, Saskatoon, Canada, July 9–11, 1991.

Total phenolics are expressed as mg sinapic acid equivalents per 100 g meal; tannin content is expressed as mg catechin equivalents per 100 g meal.

TABLE 11.2
Effect of the Number of Extraction Steps on the Recovery of Total Phenolics and Tannins

Number of Extraction Steps	70% Acetone		70% Methanol	
	Total Phenolics	Tannins	Total Phenolics	Tannins
1	720.0	268.1	837.4	127.7
2	805.8	321.3	847.0	241.3
4	972.1	328.0	1025.7	243.8
6	1075.0	331.3	1081.0	235.5

Source: Naczk, M. and Shahidi, F. Critical evaluation of quantification methods of rapeseed tannins. Presented at 8th International Rapeseed Congress, Saskatoon, Canada, July 9–11, 1991.

Total phenolics are expressed as mg sinapic acid equivalents per 100 g meal; tannin content is expressed as mg catechin equivalents per 100 g meal.

TABLE 11.3
Effect of the Canola Meal to Solvent Ratio (R) on the Recovery of Total Phenolics and Tannins

R	70% Acetone		70% Methanol	
	Total Phenolics	Tannins	Total Phenolics	Tannins
1:5	773.5	257.3	844.3	188.3
1:10	805.8	321.3	874.0	241.3
1:20	948.0	324.6	886.7	275.4

Source: Naczk, M. and Shahidi, F. Critical evaluation of quantification methods of rapeseed tannins. Presented at 8th International Rapeseed Congress, Saskatoon, Canada, July 9–11, 1991.

Total phenolics are expressed as mg sinapic acid equivalents per 100 g meal; tannin content is expressed as mg catechin equivalents per 100 g meal.

tannins from commercial canola meals from 257.3 to 321.3 mg per 100 g of meal and total phenolics from 773.5 to 805.8 per 100 g of meal when using 70% acetone.

Deshpande and Cheryan (1985) demonstrated that the yield of tannin recovery from dry beans was strongly influenced by variations in the sample particle size. They found that the amount of vanillin assayable tannins decreased by about 25%–49% as the minimum size was reduced from 820 to 250 μm (Table 11.4).

11.5 SPECTROPHOTOMETRIC ASSAYS

A number of spectrophotometric methods for quantification of phenolic compounds in plant materials have been developed. These assays are based on different

TABLE 11.4

Effects of Particle Size on Tannin Analysis of Dry Beans Using 0.5% Vanillin Assay (mg Catechin Equivalents per 100 g Beans, on Dry Basis)

Minimum Particle Size (μm)	Bean Varieties			
	Pinto	Small Red	Viva Pink	Black Beauty
820	337	304	169	154
420	269	262	111	172
350	245	250	105	213
250	223	227	87	258

Source: Deshpande, S. S. and Cheryan, M., *J. Food Sci.,* 50, 905–910, 1985.

principles and are used to determine various structural groups present in phenolic compounds.

11.5.1 DETERMINATION OF TOTAL PHENOLICS

11.5.1.1 Folin–Denis Assay

The Folin–Denis assay is the most widely used procedure for quantification of total phenolics in plant materials (AOAC 1980). Reduction of phosphomolybdic–phosphotungstic acid (Folin–Denis reagent) to a blue-colored complex in an alkaline solution occurs in the presence of phenolic compounds (Folin and Denis 1912). Swain and Hillis (1959) modified the Folin–Denis method for routine analysis of a large number of samples.

11.5.1.2 Folin–Ciocalteu Assay

The Folin–Ciocalteu assay is also used for determination of the total content of phenolics in plant material (Brune et al. 1991; Deshpande and Cheryan 1987; Earp et al. 1981; Hoff and Singleton 1977; Maxson and Rooney 1972). The Folin–Ciocalteu reagent is not specific and detects all phenolic groups found in extracts, including those found in the extractable proteins. A disadvantage of this assay is the interference of reducing substances such as ascorbic acid with the determinations.

11.5.2 DETERMINATION OF CONDENSED TANNINS AND RELATED PHENOLICS

11.5.2.1 Vanillin Assay

The vanillin method is widely used for the quantification of proanthocyanins (condensed tannins) in plant materials (Goldstein and Swain 1963a) and grains (Burns 1971). The vanillin test is specific for flavan-3-ols, dihydrochalcones, and proanthocyanins, which have a single bond at the 2-, 3-position and possess free *meta*-hydroxy groups on the B-ring (Gupta and Haslam 1980; Sarkar and Howarth 1976). Catechin, a monomeric flavan-3-ol, is often used as a standard in the vanillin assay. According to Price et al. (1978) and Gupta and Haslam (1980), this may lead to overestimation

of tannin contents. The vanillin assay in methanol is more sensitive toward polymeric tannins than monomeric flavan-3-ols. The vanillin assay is generally recognized as a useful method for the detection and quantification of condensed tannins
in plant materials due to its simplicity, sensitivity, and specificity. The method can
be used for quantifying condensed tannins in the range of 5–500 µMg with precision
and accuracy of greater than 1 µMg when the optimum concentrations of reactants
and solvents are selected (Broadhurst and Jones 1978).

The possibility of interference with dihydrochalcones and anthocyanins
(Table 11.5) (Sarkar and Howarth 1976) as well as ascorbic acid or ascorbate and
chlorophylls (Broadhurst and Jones 1978; Sun et al. 1998a) has been reported.
Anthocyanins, at usual concentrations present in plant extracts, do not react with
vanillin (Broadhurst and Jones 1978; Sun et al. 1998a). However, their presence in
phenolic extract may lead to overestimation of tannins because maximum absorption
of tannin-vanillin adducts coincides with that of anthocyanins. However, this interference can be eliminated by the use of an appropriate blank (Broadhurst and Jones
1978). Ascorbic acid or ascorbate interferes with the vanillin assay in the presence
of sulfuric acid. This may be due to the oxidative nature of sulfuric acid (Broadhurst
and Jones 1978). Sun et al. (1998b) described a simple procedure for the removal of
ascorbic acid and ascorbates from wine and grape juice by fractionation with C_{18}
Sep-Pak cartridges. Chlorophylls also absorb in the region that coincides with the
maximum absorption displayed by tannin-vanillin pigment. Therefore, presence of
chlorophylls in plant phenolic extracts may lead to overestimation of tannins. This
interference may be eliminated by hexane extraction of chlorophylls from the aqueous phenolic solution (Sun et al. 1998a).

The vanillin method is based on the condensation of the vanillin reagent with
proanthocyanins in acidic solutions. Protonated vanillin, a weak electrophilic radical, reacts with the flavonoid ring at the 6- or 8- position. The intermediate product
of this reaction dehydrates readily to give a light pink to deep red cherry–colored

---

TABLE 11.5
Response of Some Flavonoid and Chromone Compounds to Vanillin Test

Class	Compound	Reaction with Vanillin
Flavonols	Catechin (5,7,3',4'-tetrahydroxyflavane-ol)	111
Dihydrochalcones	Phoretin (4,2',4',6'-tetrahydroxydihydrochalcone)	111
Chalcones	Butein (3,4,2',4'-tetrahydroxychalcone)	—
Flavonones	Naringenin (5,7,4'-trihydroxyflavavones)	—
Flavones	7-Hydroxyflavone	—
Flavononols	Dihydroxyquercitin	1
Flavonols	Kaempferol (5,7,4'-trihydroxyflavonol) Quercitin (5,7,3',4'-tetrahydroxyflavanol)	—
Chromones	Eugenin (2-methyl-5-hydroxy-7-methoxychromone)	—

Source: Sarkar, S. K. and Howarth, R. E., *J. Agric. Food Chem.*, 24, 317–320, 1976.
111 = very positive; 1 = positive; and 2 = negative.

product (Figure 11.6). The reaction of vanillin with phenolics containing *meta*-oriented di- and trihydroxy groups on the benzene rings is approximately stoichiometric (Goldstein and Swain 1963b; Price et al. 1978). The vanillin reaction is affected by the acidic nature and concentration of the substrate, reaction time, temperature, and vanillin concentration as well as water content (Sun et al. 1998a). A higher sensitivity of vanillin assay is experienced when sulfuric acid is used rather than HCl (Scalbert et al. 1989; Sun et al. 1998a). Improvement in carbonium ion stability may be responsible for this increase in sensitivity (Scalbert 1992). However, high acid concentrations should be avoided as they may promote self-reaction of vanillin and decomposition of proanthocyanidins (Broadhurst and Jones 1978; Beart et al. 1985; Scalbert 1992).

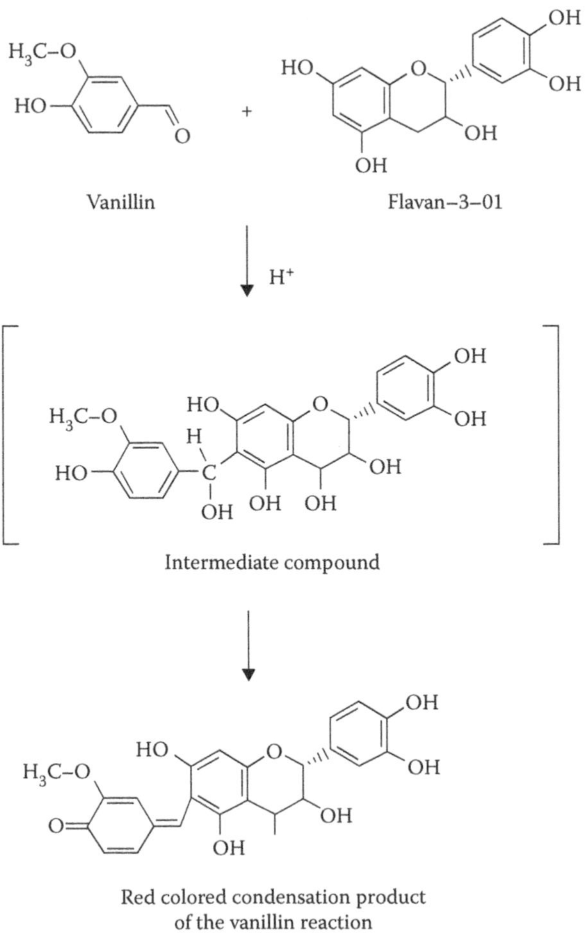

FIGURE 11.6 Condensation reactions of vanillin with leucocyanidin (Adapted from Salunkhe, D. K., Chavan, J. K., and Kadam, S. S., *Dietary Tannins: Consequences and Remedies,* CRC Press, Boca Raton, FL, 1989.)

According to Sun et al. (1998a), vanillin concentration used in the vanillin assay should be between 10 and 12 g/L. At higher vanillin concentrations, self-reaction of vanillin may lead to the formation of colored products (Broadhurst and Jones 1978). Furthermore, the color stability of vanillin-tannin adducts may increase when light is excluded and the temperature of the reaction is controlled (Broadhurst and Jones 1978). Thus, to obtain accurate and reproducible results, the vanillin assay has to be conducted at a fixed temperature (Gupta and Haslam 1980; Price et al. 1978; Sun et al. 1998a). The solvent used in the vanillin assay may also affect the reaction kinetics. In methanol, the time course of the vanillin reaction with catechin is different from that of tannins. However, use of glacial acetic acid in place of methanol affords similar kinetics for the monomers and polymers. This, according to Butler et al. (1982), suggests that in glacial acetic acid, vanillin reacts only with the terminal groups, and thus, the concentration of oligomeric molecules rather than the total content of flavan-3-ols is measured. The molar extinction coefficients of vanillin monomer and vanillin oligomer adducts are shown in Table 11.6. Butler et al. (1982) proposed that vanillin assay in glacial acetic acid may be used for estimating the degree of polymerization of condensed tannins.

11.5.2.2 DMCA Assay

The formation of a green chromophore between catechin and 4-[dimethylamino]-cinnamaldehyde (DMCA) was first reported by Thies and Fisher (1971). Subsequently, McMurrough and McDowell (1978), Delcour and de Verebeke (1985), and Treutter (1989) demonstrated that DMCA did not react with a wide range of flavonoids, including dihydrochalcones, flavanones, and flavononols and phenolic acids. Weak responses were detected for resorcinol, orcinol, naphthoresorcinol, and phloretin. Treutter (1989) also reported that DMCA reacted with indoles and terpenes. Later,

---

TABLE 11.6

UV Absorption Patterns of Various Phenolic Compounds

Compound	$e_{500} \times 10^{23}$ in Methanol	$e_{500} \times 10^{23}$ in Acetic Acid
Monomers		
Epicatechin	0.43	9.8
Catechin	0.30	8.5
Dimers		
B-2 (epicatechin)$_2$	2.52	11.2
B-3 (catechin)$_2$	1.82	12.3
Trimers		
B-9 (epicatechin)$_3$	4.16	12.2

Source: Butler, L. G., Price, M.L. and Brotherton, J. E., *J. Agric. Food Chem.*, 30, 1087–1089, 1982..

Li et al. (1996) reexamined the use of the DMCA assay for the determination of proanthocyanidins in plants. According to these authors, on a molar basis, the sensitivity of DMCA toward proanthocyanidins was four times greater than that of indole and 30,000 times greater than that of thymol.

The DMCA reagent, compared to the vanillin reaction carried out in methanol, reacts only with the terminal groups of condensed tannins. The presence of methanol, acetone, ethyl acetate, and dimethylformamide does not have any detrimental effect on the rate and intensity of the color development. However, the DMCA reagent is sensitive to both monomeric and polymeric units (McMurrough and McDowell 1978), and this may lead to overestimation of condensed tannins content.

11.5.2.3 Proanthocyanidin Assay

The proanthocyanidin assay is carried out in a solution of butanol–concentrated HCl (95:5, v/v). In the presence of hot HCl–butanol solution, proanthocyanidins (condensed tannins) are converted to anthocyanidins through autoxidation of carbocations formed by cleavage of interflavonoid bonds (Porter et al. 1986). The yield of the reaction depends on the concentration of HCl and water, temperature, reaction time, presence of transition metals, and degree of proanthocyanidin polymerization (Porter et al. 1986; Scalbert et al. 1989). Govindarajan and Mathew (1965) reported that water content of up to 20% did not affect the yield of anthocyanidins. Later, Porter et al. (1986) demonstrated that the maximum yield of anthocyanidins was obtained at a 6% (v/v) water content in the solvent. Both lower and higher levels of water had a detrimental effect on the formation of anthocyanidins. The presence of transition metals enhanced both the reproducibility and yield of conversion of proanthocyanidin to anthocyanidins. Fe^{+2} and Fe^{+3} were the most effective catalysts of anthocyanidins formation. The maximum yield of anthocyanidin was achieved at a Fe^{+3} concentration of 7.3×10^{24} M (Porter et al. 1986). Scalbert et al. (1989) reported that the reproducibility of the method was significantly improved if the reaction temperature was strictly controlled.

11.5.3 DETERMINATION OF INSOLUBLE CONDENSED TANNINS

Insoluble condensed tannins were first identified in plants by Bate-Smith (1973a). Subsequently, Terrill et al. (1992) fractionated condensed tannins into tannins soluble in organic solvents, tannins soluble in solution of sodium dodecyl sulfate (SDS) solution, and insoluble tannins. The procedure proposed by these authors included the extraction of soluble tannins with a mixture of acetone/water/diethyl ether (4.7:2.0:3.3, v/v/v), followed by the extraction of SDS-soluble tannins with a boiling aqueous solution of SDS containing 2-mercaptoethanol. The insoluble tannins were then determined directly on the remaining residue after extraction of SDS-soluble tannins using the proanthocyanidin assay. The residue was subjected to a one-step treatment with butanol/HCl. The one-step treatment with butanol/HCl was also used by Makkar et al. (1997), Degen et al. (1995), and Makkar and Singh (1991) for determination of insoluble condensed tannins in forages. Later, Harinder et al. (1999) reported incomplete recovery of insoluble condensed tannins after a one-step treatment of sample with butanol/HCl. These authors used solid-state ^{13}C NMR as a probe for the measurement of bound condensed tannins. Naczk et al. (2000) investigated a multistep treatment of canola hulls with butanol/HCl and reported that a six-step extraction of hulls, free of soluble tannins, with butanol/HCl, removed most of the insoluble condensed tannins as anthocyanidins.

11.5.4 Determination of Hydrolyzable Tannins

Various approaches have been used for screening of a large number of plant samples for their hydrolyzable tannins. Of these, the most widely used method is based on the reaction between potassium iodate and hydrolyzable tannins. This reaction was first described by Haslam (1965) and later used by Bate-Smith (1977) for the development of an analytical assay for estimating hydrolyzable tannins in plant materials. The original assay included mixing of the sample and potassium iodate, cooling the mixture in an ice-water bath for 40 minutes (Haslam 1965) or 90 minutes, (Bate-Smith 1977) and measuring the absorbance at 550 nm. Later, Willis and Allen (1998) demonstrated that chilling the samples after addition of potassium iodate was unnecessary. These authors recommended carrying out the reaction at 25°C for a time unique for each type of plant material. Several limitations have been reported for the original assay, including variability in the reaction time required for different tannins and for maximum color development, variability in spectral properties of chromophores formed from different hydrolyzable tannins, formation of yellow oxidation products by nontannin phenolics, the possibility of precipitate formation, and difficulties in reproducibility of the results (Inoue and Hagerman 1988; Hartzfeld et al. 2002; Waterman and Mole 1994). Hartzfeld et al. (2002) modified this assay by including a methanolysis step followed by oxidation with potassium iodate. The assay is based on gallic acid, which is a common structural component found in both gallotannins and ellagitannins. Methyl gallate, formed by methanolysis of hydrolyzable tannins in the presence of strong acids, reacts with potassium iodate to produce a red chromophore with a maximum absorbance at 525 nm. The oxidation reaction is carried out at 30°C for 50 minutes at pH 5.5. The detection limit of the method is 1.5 μg of methyl gallate. Methanol and acetone stabilize the chromophore, while the presence of water accelerates degradation of the resultant pigment. According to Hartzfeld et al. (2002), the modified potassium iodate method provides a good estimate for gallotannins but underestimates the content of ellagitannins.

Other analytical assays proposed for the quantification of hydrolyzable tannins in plant materials include the rhodanine (Inoue and Hagerman 1988) and sodium nitrate (Wilson and Hagerman 1990) methods. The rhodanine assay may be used for estimation of gallotannins and is based on determination of gallic acid in a sample subjected to acid hydrolysis under anaerobic conditions. On the other hand, the sodium nitrate assay is developed for quantitative determination of ellagic acid in sample hydrolyzates, but this assay requires large quantities of pyridine as a solvent.

11.5.4.1 Determination of Sinapine

Sinapine, a choline ester of sinapic acid (Figure 11.7), is found in seeds of *Brassica napus, Brassica campestris,* and *Crambe abyssinica* (Austin and Wolff 1968; Clandinin 1961). Methods for estimation of sinapine are based on the formation of

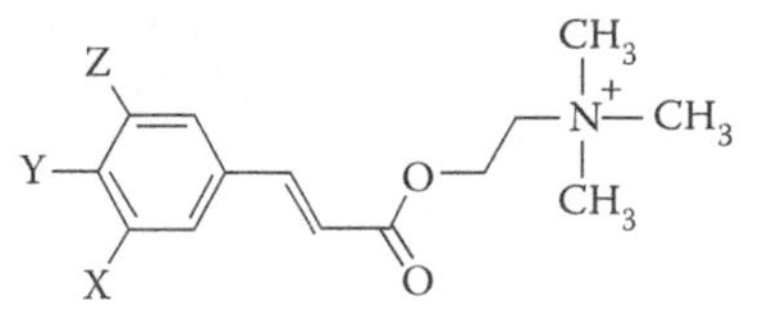

Sinapines	X	Y	Z
p-Coumaroylcholine	H	OH	H
Feruloylcholine	H	OH	OCH_3
Isoferuloylcholine	H	OCH_3	OH
Sinapine	OCH_3	OH	OCH_3
Sinapine glucoside	OCH_3	O–Glu	OCH_3

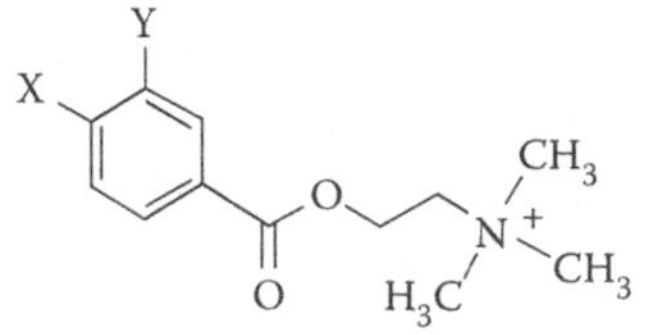

Sinapines	X	Y
4–Hydroxybenzoylcholine	OH	H
Hesperalin	OCH_3	OCH_3

FIGURE 11.7 Chemical structures of sinapines

a water-insoluble complex between Reinecke salt and a quaternary nitrogen base (Austin and Wolff 1968; Tzagoloff 1968) and formation of a colored complex between sinapine and titanium chloride (Ismail and Eskin 1979; Eskin 1980).

11.5.4.2 Determination of Chlorogenic Acid

A number of UV spectrophotometric methods have been used for determination of chlorogenic acid (Figure 11.8) in foodstuff (AOAC 1980; Corse et al. 1962; Merrit and Proctor 1959; Moores et al. 1948). A colorimetric method for quantification of chlorogenic acid in green coffee beans was proposed by Clifford and Wright (1976). Metaperiodate reagent was found to provide a rapid and simple means for measuring the total chlorogenic acid. However, successive use of molybdate and metaperiodate reagents enabled the estimation of total content of feruloylquinic acids and accurate measurement of caffeoylquinic acids and total chlorogenic acids. Molybdate reagent is specific for *ortho*-dihydroxyphenols, whereas metaperiodate reagent reacts with *ortho*- and *para*-dihydroxyphenols and their monomethyl ethers to produce a yellow-orange color (Clifford and Wright 1973).

11.5.4.3 Determination of Anthocyanins

Quantification of anthocyanins (Figure 11.4) takes advantage of their characteristic behavior in acidic media. Anthocyanins in acidic media exist in equilibrium between the colored oxonium ion and the colorless pseudobase form. The oxonium

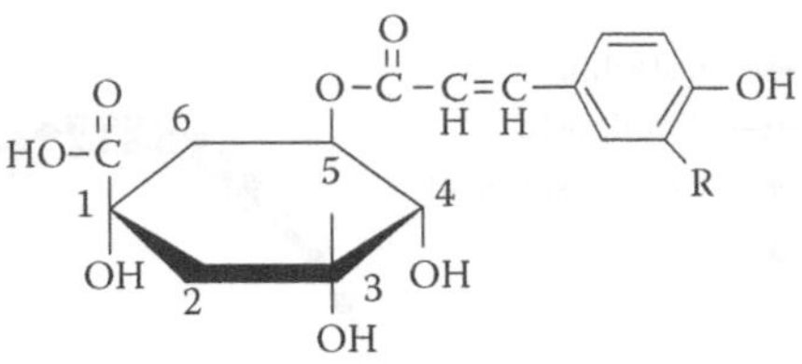

R = H   5–*p*–Coumaroylquinic Acid

R = OH   5–Caffeoylquinic Acid

R = OCH$_3$ 5–Feruloylquinic Acid

FIGURE 11.8   Chemical structures of chlorogenic acids

form comprises about 15% of the equilibrium mixture at pH 3.9 and 100% at pH 1, respectively (Timberlake and Bridle 1967). The analytical procedure for quantification of anthocyanins was first developed by Sondheimer and Kertesz (1948). This procedure was later modified by Swain and Hillis (1959) who suggested to express the concentration of pigments in terms of the change in the absorbance at 1 max between pH 3.5 and 1.0. Fuleki and Francis (1968b) suggested extending the pH difference between 4.5 and 1.0. They proposed to determine the optical density of the two samples at 515 nm for aliquots buffered to the above pH values. The total content of anthocyanins may also be estimated from the absorption of the total extracts at 520 nm using an average extinction coefficient (Moskowitz and Hrazdina 1981). However, Little (1977) reported that pH manipulation, both downward and upward, offered no advantage for products such as wine with a natural pH of around 3.4. The use of transreflectometric thin-layer measurements at two pH levels, natural (about 3.5) and pH 1.0, was suggested to estimate the content of anthocyanins.

11.5.5 DETERMINATION OF IRON-BINDING PHENOLIC GROUPS

Brune et al. (1991) developed a spectrophotometric method for determination and evaluation of iron-binding phenolic groups in foods. This method is based on the findings of Mejbaum-Katzenellenbogen and Kudrewicz-Hubicka (1966), who observed that Fe-galloyl and Fe-catechol complexes were colored differently. Ferric ions reacted with phenolic compounds containing three adjacent hydroxyl groups (galloyl groups) to form a blue-colored complex; phenolics having two adjacent hydroxyl groups (catechol groups) formed a green-colored complex. The absorption maximum for the iron-tannin complex was around 580 nm, and that for complexes of iron with catechin, pyrocatechol, caffeic acid, and chlorogenic acid was around 680 nm (Figure 11.9). The iron-binding reagent did not lend itself for determination of monohydroxy phenolics and dihydroxy phenolics with a *meta*- or *para*-hydroxylation pattern.

The above procedure involves a 16-hour extraction of phenolic compounds in the dark from dry or freeze-dried food samples with 50% dimethylformamide solution in 0.1 M acetate buffer (pH 4.4) at room temperature. To the filtered extract, ferric ammonium sulfate (FAS) reagent was added (the FAS reagent consisted of 1 M HCl

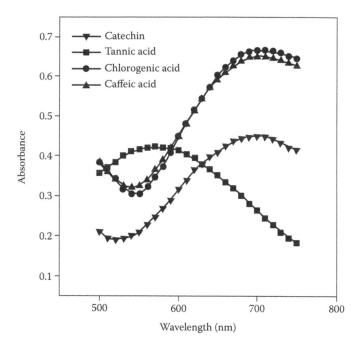

FIGURE 11.9   Absorbance of complexes of equimolar concentrations of sealed phenolics with iron.

solution containing 5% FAS and 1% gum arabic). The resulting color was read spectrophotometrically against the blank at both 578 and 680 nm. These wavelengths correspond to the absorption maxima of iron-galloyl (blue color) and iron-catechol (green color) complexes, respectively. The absorbances of the blank were also read and subtracted from those of the food sample at the two given wavelengths. Following this, the contents of catechol groups (expressed as catechin equivalents) and galloyl groups (expressed as tannic acid equivalents) were calculated from four standard curves obtained using tannic acid and catechin at the two specified wavelengths.

11.5.6   UV Spectrophotometric Assay

A number of approaches have been used to develop a simple and satisfactory UV spectrophotometric assay. Simple phenolics have absorption maxima between 220 and 280 nm (Table 11.7) (Owades et al. 1958a,b), but the absorption is affected by the nature of solvent and pH. Moreover, the possibility of interference by UV-absorbing substances such as proteins, nucleic acids, and amino acids should be considered. Therefore, development of a satisfactory UV assay is a rather cumbersome and difficult task. In addition, suitability of the UV assay depends on the material to be analyzed. Phenolics in tea and beer (Owades et al. 1958a,b), tea, coffee, and other beverages (Hoff and Singleton 1977), cereals, legumes (Davis 1982), and oilseeds

TABLE 11.7

UV Absorption Patterns of Various Phenolic Compounds[a]

Class of Compounds	UV Absorption, l_{max}
Benzoic acids	270–280
Hydroxycinnamic acids	290–300,[b] 305–330
Anthocyanin pigments	270–280, 315–325[c]
Flavonols	250–270, 330[b], 350–380
Flavan-3-ols	270–280
Coumarins	220–230, 310–350
Flavones	250–270, 330–350
Flavonones, flavononols	270–295, 300–330[b]
Chalcones	220–270, 300–320, 340–390
Aurones	240–270
Isoflavones	245–270, 300–350

Source: Macheix, J.-J., Fleuriet, A., and Billot, J., *Fruit Phenolics,* CRC Press Inc., Boca Raton, FL, 1990.

[a] Solvent was methanol, except for anthocyanin pigments for which methanol containing HCl 0.01% was used.

[b] Shoulder.

[c] In the case of acylation by hydroxycinnamic acids.

such as canola meals (Blair and Reichert 1984; Naczk et al. 1992b; Shahidi and Naczk 1990) have been estimated by this method. Determination of the absorbance at 325 nm has also been proposed for estimating the content of hydroxycinnamic acid derivatives (chlorogenic acid) in pears (Billot et al. 1978) and coffee (Moores et al. 1948).

Both UV and visible spectroscopic techniques are often used for identification of isolated phenolic compounds, particularly flavonoids (Mabry et al. 1970). The spectra of phenolics in methanol after addition of shift reagents are recorded. For example, the addition of sodium hydroxide or HCl is helpful in determining the phenolic and carboxylic acid groups present in the molecules (Jurd 1957). On the other hand, the absorption spectra of total phenolic extracts can be used to identify the presence of groups of predominant phenolic compounds (Macheix et al. 1990).

11.5.7 CHEMOMETRICS METHODS

Traditional spectroscopic assays may lead to overestimation of polyphenol contents of crude extracts from plant materials due to the overlapping of spectral responses. This problem can be overcome by using a chemometric technique to analyze the spectra such as partial least squares (PLS) or principal component analysis (PCA) (Kramer 1998; van der Voort 1992). Chemometric technique uses information (such as a spectrum) and chemical indices (such as concentration of a component) and establishes a mathematical relationship between them. It assumes that the chemical index (concentration) is correct and attributes weighings of the spectral information accordingly. The setting up of the model, correlating the information with a chemical index, is known as calibration (Beebe and Kowalski 1987).

Monedero et al. (1999) developed a chemometric technique for controlling the content of phenolic aldehydes and acids during production of wine subjected to accelerated aging. The wine was aged by the addition of oak wood extracts obtained by maceration of charred oak shavings. Charring time and/or the interactions between charring temperature and time were essential factors to control the content of 10 of the 11 phenolic compounds studied. Edelman et al. (2001) developed a rapid method for discrimination of Austrian red wines based on mid-infrared spectroscopy of phenolic extracts of wine. The wine samples were cleaned up by SPE before collection of spectra using Fourier transform infrared (FTIR) spectrophotometry. These authors also reported that the use of UV-visible spectroscopy was limited to the authentication of the Burgundy species Pinot Noir. Subsequently, Brenna and Pagliarini (2001) used a multivariate analysis for establishing a correlation between the polyphenolic composition and antioxidant power of red wines.

Briandet et al. (1996) applied PCA to differentiate between *arabica* and *robusta* instant coffees on the basis of their FTIR spectra. Spectra used in this study were obtained using the diffuse reflection infrared Fourier transform and attenuated total reflection sampling techniques. According to these authors, the discrimination between species of coffee was based on different contents of chlorogenic acid and caffeine. Later, Downey et al. (1997) successfully applied factorial discriminant analysis and PLS to develop a mathematical model for varietal authentication of lyophilized samples of coffee based on near- and mid-infrared spectra.

Schulz et al. (1999) used a near-infrared reflectance (NIR) spectroscopic method for prediction of polyphenols in green tea leaves. The contents of gallic acid and catechins were determined using reversed-phase-high-performance liquid chromatographic (HPLC) method. The PLS method was used to calibrate NIR spectra with the contents of gallic acid and catechins in tea. These models predicted the contents of catechins and gallic acid with good accuracy. On the other hand, Mangas et al. (1999) used a linear analysis procedure for discrimination of bitter and nonbitter cider apple varieties. The most discriminant variables were chlorogenic acid, phloretin 2-xyloglucoside, and an unidentified phenolic acid derivative with a maximum absorption at 316.7 nm. Using this model, 91.3% and 85.7% correct classification was obtained for internal and external evaluation of the model, respectively.

11.5.8 Nuclear Magnetic Resonance Spectroscopy

Various nuclear magnetic resonance (NMR) spectroscopic techniques have been used for structure elucidation of complex phenolics isolated from foods (Porter 1989). These include ^{1}H NMR and ^{13}C NMR, two-dimensional homonuclear (2D ^{1}H-^{1}H) correlated NMR spectroscopy (COSY), heteronuclear chemical shift correlation NMR (C-H HECTOR), totally correlated NMR spectroscopy (TOCSY), nuclear Overhauser effect in the laboratory frame spectroscopy (NOESY), and rotating frame of reference spectroscopy (ROESY) (Bax 1985; Bax and Grzesiek 1996; Belton et al. 1995; Derome 1987; Ferreira and Brandt 1989; Kolodziej 1992; Hemingway et al. 1992; Newman and Porter 1992). Combinations of high-resolution spectroscopic techniques with novel mathematical treatments of data provide a greater insight into structure elucidation of mixtures of compounds without their prior separation into individual components (Gerothanassis et al. 1998; Limiroli et al. 1996; Sacchi et al. 1996).

11.6 PROTEIN PRECIPITATION ASSAYS

Evaluation of biological activity of polyphenolics is based on their ability to form insoluble complexes with proteins. Many methods have been developed for quantification of biological activity of phenolics. Most of these methods are based on the ability of phenolics to precipitate proteins such as gelatin, bovine serum albumin (BSA), and hemoglobin and different enzymes (Table 11.8). Protein precipitation methods are highly correlated with the biological value of high–tannin content foods and feeds (Hahn et al. 1984).

11.6.1 BOVINE SERUM ALBUMIN ASSAYS

Several methods have been published that use BSA for determining the ability of polyphenols to precipitate proteins (Asquith and Butler 1985; Hagerman and Butler 1978; Hoff and Singleton 1977; Makkar et al. 1987, 1988; Marks et al. 1987; Naczk et al. 2001a; Silber et al. 1998; Verzele et al. 1986). Complete precipitation of a protein–polyphenolic complex depends on the structure and quantity of polyphenols and may take from 15 minutes to 24 hours (Hagerman and Robbins 1987; Makkar 1989). Therefore, the precipitation time should be standardized. In addition, the formation of a protein–polyphenol complex is affected by the presence of small amounts of residual organic solvents such as acetone used for the extraction of polyphenols (Dawra et al. 1988). Other factors that may influence the assay include the concentration and nature of proteins used in the test mixture and pH and ionic strength of the reaction mixture (Makkar 1989).

Some methods determine only the quantity of precipitable phenolics (Hagerman and Butler 1978); others may also determine the precipitation capacity of phenolics (Figure 11.10) (Asquith and Butler 1985; Dawra et al. 1988; Hagerman and Butler 1980; Makkar et al. 1987, 1988; Marks et al. 1987). Naczk et al. (2001a) proposed to express protein-precipitating potential of crude tannin extracts as a slope of lines (titration curves) reflecting the amount of tannin–protein complexes precipitated as a function of the amount of tannins added to the reaction mixture. The titration curves were obtained using both the protein precipitation assay (Hagerman and Butler 1978) and dye-labeled protein assay (Asquith and Butler 1985). The determination of slope values is based on the statistical analysis (linear regression) of experimental data involving the precipitation of tannin–protein complexes at a minimum of four different tannin levels. Determination of the slopes under standardized conditions (type and concentration of protein, pH, and temperature) afforded meaningful differences in the slope values for the crude tannin extracts from various canola and rapeseed hulls (Naczk et al. 2001a) as well as condensed tannins of beach pea, evening primrose, and faba bean (Naczk et al. 2001b).

TABLE 11.8
Some Methods for Determination of Tannins by Protein Precipitation

Tannins Measured	Remarks	References
Hemoglobin precipitation	Requires fresh blood, interference by plant pigments, saponins, etc.	Bate-Smith (1973b), Schultz et al. (1981)
AOAC	Accuracy 10%, not suitable for sorghum grains, suffers from disadvantages associated with FD, FC, PB methods	AOAC (1980)
b-Glucosidase inhibition	Relationship between enzyme activity and insoluble complex formation not known, cumbersome, expensive	Goldstein and Swain (1965)
Immobilized protein	Time-consuming, expensive, difficult to handle large number of samples at a time	Hoff and Singleton (1977)
Dye-labeled BSA precipitation	Simple but rather insensitive, preferentially forms soluble complexes	Asquith and Butler (1985)
Radial diffusion	Insensitive to acetone, simple but involves an element of subjectivity	Hagerman (1987)
BSA precipitation	Simple method, measures phenolics in the complex, nonspecific binding of phenols to the complex can introduce an error, not suitable for comparing tannins from different sources	Hagerman and Butler (1978)
	Indirect method, protein estimations by Lowry or Bradford assays, takes longer time, less sensitive	Martin and Martin (1982, 1983)
	Indirect method, protein determination by Kjeldahl method, less sensitive	Amory and Schubert (1987)
	Indirect method, protein determination by HPLC, more sensitive compared with above indirect methods	Verzele *et al.* (1986)
	Ninhydrin method used to measure protein in complex after its alkaline hydrolysis, takes only 20 min for the hydrolysis, assay a bit messy because of ninhydrin reagent	Makkar et al. (1987b)
	Same as above in the method of Makkar et al. (1987b) except that hydrolysis of the complex is under acidic condition and takes longer (22 h)	Marks et al. (1987)
	Measures both tannins and proteins	Makkar et al. (1988b)
	Highly sensitive, requires microquantity of tannins, allows quantitation of both hydrolyzable and condensed tannins in a plant extract	Dawra et al. (1988)

Source: Makkar, H. P. S., *J. Agric. Food Chem.,* 37, 1197–1202, 1989.

Asquith and Butler (1985) developed a simple method for determining the amount of protein precipitated by tannins. In this assay, a BSA covalently bound with Remazol brilliant blue R was used as a substrate. The protein-labeling procedure was adapted from Rinderknecht et al. (1968). The dye reacts irreversibly with

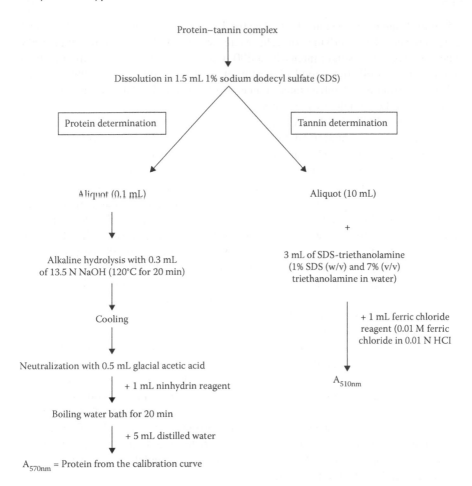

Protein–tannin complex

↓

Dissolution in 1.5 mL 1% sodium dodecyl sulfate (SDS)

| Protein determination | | Tannin determination |

Aliquot (0.1 mL)

Aliquot (10 mL)

+

Alkaline hydrolysis with 0.3 mL
of 13.5 N NaOH (120°C for 20 min)

3 mL of SDS-triethanolamine
(1% SDS (w/v) and 7% (v/v)
triethanolamine in water)

↓

Cooling

+ 1 mL ferric chloride
reagent (0.01 M ferric
chloride in 0.01 N HCl

↓

Neutralization with 0.5 mL glacial acetic acid

+ 1 mL ninhydrin reagent

A_{510nm}

Boiling water bath for 20 min

+ 5 mL distilled water

↓

A_{570nm} = Protein from the calibration curve

FIGURE 11.10   Estimation of both protein and tannin in a tannin-protein complex (Adapted from Makkar, H. P. S., Dawra, R. K., and Singh, B., *J. Agric. Food Chem.,* 36, 523–525, 1988.)

BSA under mild conditions without apparent denaturation. The blue BSA can be stored at 68°C for several months without loss of precipitating activity. In the presence of excess blue BSA, the amount of precipitated protein is proportional to the amount of tannin added. The plant pigments do not interfere with the assay because the maximum absorption of dye-labeled BSA is at longer wavelengths. The assay detects all protein precipitants. Therefore, it should be corroborated by assays for protein-precipitable phenols before the formation of precipitate can be interpreted as indicative of the presence of tannins.

Silber et al. (1998) described a novel dye-labeled protein assay utilizing Coomassie brilliant blue (G-250) BSA complex. Coomassie blue pigment exists in a cation form (red with maximum absorbance at 470 nm), an anion form (blue with maximum absorbance at 595 nm), and a neutral form (green with maximum absorbance at

650 nm) (Matejovicova et al. 1997). The binding of Coomassie blue to protein markedly enhances the absorbance of light (Matejovicova et al. 1997; Silber et al. 1998). The complex absorbs light intensely at 600–620 nm, but the mechanism of its binding is not yet well understood (Atherton et al. 1996; Silber et al. 1998). It takes 3 hours of standing at refrigerated temperatures for the dye-labeled protein–tannin complexes to be completely precipitated.

Hagerman (1987) developed a radial diffusion assay for tannin analysis. In that assay, tannins were quantified by forming a visible ring of precipitate in BSA-containing agarose gel. The tannin-containing solution was placed in a well in a BSA-containing agarose slab. Tannins readily diffuse through the gel to form precipitable BSA–tannin complexes. The carbohydrate medium of the gel does not form precipitable complexes. The acetone extracts of tannins can be analyzed directly, as the acetone used does not affect this assay. The area of the visible ring of precipitate is linearly correlated with the amount of tannin placed in the well. The results obtained with this assay are highly correlated with the results of the protein precipitation method of Hagerman and Butler (1978). The detection limit of the assay is 0.025 mg tannin, and it depends on the concentration of protein in the plates. Nontannin phenolics such as flavonoids, hydroxybenzoic acids, and hydroxycinnamic acids do not interfere with the assay.

Hoff and Singleton (1977) developed a method for determination of tannins in foods by using immobilized BSA on a Sepharose matrix. They found that immobilized proteins bind tannins selectively at pH 4. Monomeric polyphenolic substances such as chlorogenic acid, catechol, and gallic acid did not form complexes with immobilized proteins under these conditions. Thus, this method may be used as a means for separation of tannins from nontannin materials. Tannin–protein complexes were then dissociated in organic solvents such as methanol or dimethylformamide, and the content of tannins was determined spectrophotometrically. However, this method is cumbersome, time-consuming, and does not allow handling of a large number of samples at a time.

11.6.2 COMPETITIVE BINDING ASSAYS

The interactions between tannins and proteins are often measured using assays based on competition of a test protein and labeled protein for binding and coprecipitation with tannins (Asquith and Butler 1985; Bacon and Rhodes 1998; Hagerman and Butler 1981). These assays permit a direct comparison of protein affinities for condensed tannins.

Hagerman and Butler (1981) developed a competition assay based on the assumption that proanthocyanidin–tannin and antigen–antibody interactions were similar. The similarities include comparable sizes of ligand and binding agent that are multivalently bound to form soluble and insoluble complexes. This procedure utilizes an iodine 125-labeled BSA, prepared as described by Hagerman and Butler (1980), as the labeled ligand. In this assay, proanthocyanidins are simultaneously exposed to a mixture of the competitor (unlabeled protein) and labeled protein. The insoluble tannin–protein complex is separated by centrifugation and the amount of unprecipitated labeled protein is determined in supernatant by counting using a Gamma 300 gamma counter.

Asquith and Butler (1985) developed a competition assay based on the dye-labeled protein assay. A dye-labeled protein for this assay was prepared by covalently binding BSA with Remazol brilliant blue R as described by Rinderknecht et al. (1968). The competition procedure used in this assay was similar to that described by Hagerman and Butler (1981). A standard solution of dye-labeled BSA was mixed with a methanolic solution containing enough tannins to precipitate 70%–80% of dye-labeled proteins. Varying amounts of competitor protein were mixed with a standard solution of dye-labeled BSA before tannins are added. The amount of precipitated dye-labeled BSA was measured spectrophotometrically. Results were presented on a semilog plot. Following this, the amount of competitor required to inhibit 50% precipitation of dye-labeled protein due to tannin was established, and it was proposed to serve as a measure of relative affinity of competitor protein for tannins

Bacon and Rhodes (1998) developed a novel competitive tannin–salivary protein binding assay. This assay was developed based on the principle of a competitive ELISA and competition between an enzyme-labeled tannin (HPR-EGC) and a tested tannin to bind to parotid salivary protein immobilized in the well of a microtiter plate. (2)-Epigallocatechin conjugated to horseradish peroxidase via a linker molecule, 1,4-butanediol diglycidyl ether, was used as an enzyme-labeled tannin. To the wells, coated with parotid salivary proteins and loaded with a known quantity of HPR-EGC, different amounts of the tannin under test were added. The plates were incubated for 2 hours at 37°C, and then the peroxidase activity in each well was determined. Following this, the amount of tannin required to displace 50% of HRP-EGC bound to the protein coating on the microtiter plate was calculated and proposed as a measure of tannin affinity for proteins. A control sample (no test tannin added) was used to measure the amount of HRP-EGC conjugated to proteins coating the microtiter plate.

11.7   ENZYMATIC ASSAYS

Enzymatic methods are based on the ability of tannins to form complexes with enzymes, thus inhibiting their activities. Phenolic compounds can affect enzymes by either (1) the formation of insoluble protein–phenolic complexes and lowering the concentration of enzymes in the reaction mixture or (2) the formation of inactive soluble protein–phenolic complexes. Thus, competitive and noncompetitive reaction kinetics as well as mixed kinetics for the inhibition of enzymes by phenolic compounds have to be considered. However, Hagerman and Butler (1978) have suggested that the results of enzymatic assay are difficult to interpret because the relationship between enzymatic activity and the formation of insoluble complexes is not fully understood. Later, Dick and Bearne (1988) studied the inhibition of b-galactosidase activity by polyphenolic compounds. They demonstrated that a *trans* 3-(4-hydroxyphenyl)-propenoyl group, common to 4′-hydroxyflavones and *p*-hydroxyphenolic acid, is important for effective inhibition of enzymatic activity. Mole and Waterman (1987c) found that addition of tannic acid to a standard trypsin solution led to the formation of an insoluble complex. However, trypsin remaining in the solution had an enhanced rate of autolysis. Conformational changes in enzyme brought about by the formation of soluble complexes with tannic acid might have been responsible. However, this phenomenon was not observed when excess BSA was added to the system containing trypsin and tannic acid.

Ittah (1991) developed an enzymatic assay for quantitation of tannins. The assay used the ability of tannins to form complex with more than one protein and was based on determination of the activity of bound alkaline phosphatase. Tannin in this assay was sandwiched between two BSA molecules and alkaline phosphatase. A linear relationship between tannin concentrations and the activity of bound alkaline phosphatase was found. The procedure involved immobilization of BSA on polystyrene microplates. The quantity of immobilized BSA was low in comparison with the amount of tannin applied. Unbound tannins were washed out, and then the microplates were incubated with alkaline phosphate solutions. Following this, the unbound enzyme was washed out, and the activity of the bound alkaline phosphatase was determined using p-nitrophenylphosphate as a substrate.

11.8 CHROMATOGRAPHIC TECHNIQUES

Various chromatographic techniques have been employed for separation, preparative isolation, purification, and identification of phenolic compounds (Karchesy et al. 1989; Jackman et al. 1987; Merken and Beecher 2000). Chromatographic procedures have also been used to study the interaction of phenolics with other food components (Oh et al. 1985).

11.8.1 LIQUID CHROMATOGRAPHY

Many liquid chromatographic methods have been described in the literature for fractionation of tannins (proanthocyanidins) using Sephadex G-25 (McMurrough and McDowell 1978; Michaud and Margail 1977; Somers 1966), Sephadex LH-20 (Asquith et al. 1983; Boukharta et al. 1988; Davis and Hoseney 1979; Lea and Timberlake 1974; Strumeyer and Malin 1975; Thompson et al. 1972); Sepharose CL-4B (Hoff and Singleton 1977), Fractogel (Toyopearl) TSK-HW 40(s) (Derdelinckx and Jerumanis 1984; Mateus et al. 2001; Ricardo da Silva et al. 1991), Fractogel (Toyopearl) TSK 50(f) (Labarbe et al. 1999; Meirelles et al. 1992), inert glass microparticles (Labarbe et al. 1999) as well as C_{18} Sep-Pak cartridges (Guyot et al. 2001; Jaworski and Lee 1987; Salagoity-Auguste and Bertrand 1984; Sun et al. 1998b).

Hoff and Singleton (1977) developed a chromatographic procedure for separation of tannins from nontannin materials using BSA immobilized on Sepharose CL-4B as a column-packing material. The nontannin polyphenolics were first separated by washing the column with an acetate buffer. The protein–tannin complexes were then dissociated by elution of the column with organic solvents such as methanol or dimethylformamide to release and recover tannin substances. The preparative isolation of proanthocyanins is, however, most commonly achieved by employing Sephadex LH-20 column chromatography (Thompson et al. 1972; Strumeyer and Malin 1975; Davis and Hoseney 1979; Asquith et al. 1983). The crude extract was applied to the column, which was then washed with ethanol to elute the nontannin substances. Following this, proanthocyanins were eluted with acetone–water or alcohol–water. For purification of proanthocyanins, the crude extract of phenolic substances was applied to a Sephadex LH-20 column, which was then eluted with water containing increasing proportions of methanol (Nonaka et al. 1983).

Derdelinckx and Jerumanis (1984) employed Fractogel (Toyopearl TSK HW-40(s) gel) for separation of malt and hop proanthocyanidin dimers and trimers after chromatography of polyphenols on Sephadex LH-20 with methanol. The proanthocyanidin fractions were applied to the Fractogel column, which was then eluted with methanol. The four major peaks of hops polyphenolics corresponded to the B3 and B4 procyanidin dimers and two unidentified procyanidin oligomers. On the other hand, malt polyphenolics were separated into a mixture of four proanthocyanidin trimers, two procyanidin trimers, and an unknown procyanidin oligomer. According to Derdelinckx and Jerumanis (1984), Fractogel (Toyopearl TSK HW-40(s) gel) allows one to obtain the proanthocyanidins in an advanced state of purity

Labarbe et al. (1999) applied an inert glass powder (Pyrex microparticles, 200–400 μm) for fractionation of grape (seed or skin) proanthocyanidins according to their degree of polymerization. Proanthocyanidins were extracted from grape with acetone–water (60:40, v/v). Following this, the nontannin phenolics and proanthocyanidins were fractionated on a Fractogel (Toyoperl TSK HW-50(f) gel) column (35 × 8 cm) as described by Souquet et al. (1996). The nontannin phenolics were washed out from the column with two bed volumes of water followed by five bed volumes of ethanol–water–trifluoroacetic acid (55:44.05:0.05, v/v/v). Subsequently, proanthocyanidins were washed from the column with three bed volumes of acetone–water (60:40, v/v). Purified proanthocyanidins were dissolved in methanol and then applied onto the column filled with glass microparticles and equilibrated with methanol–chloroform (25:75, v/v) and massively precipitated on the top of column with chloroform. Proanthocyanidins were sequentially eluted from the column by increasing proportions of methanol in a methanol–chloroform solvent system.

Fulcrand et al. (1999) fractionated wine phenolics into simple (phenolic acids, anthocyanins, flavonols, and flavanols) and polymeric phenolics using a Fractogel (Toyopearl) HW-50(f) column (bed 12 × 120 mm). Simple phenolics were eluted from the column with ethanol–water–trifluoroacetic acid (55:45:0.005, v/v/v), while polymeric phenolics were then recovered with 60% (v/v) acetone. Later, Mateus et al. (2001) used a Fractogel (Toyopearl) HW-40(s) column for fractionation of anthocyanin-derived pigments in red wines. Two liters of wine were directly applied onto the Toyopearl gel column (200 × 16 mm i.d.) at a flow rate of 0.8 mL/min. The anthocyanins were subsequently eluted from the column with water–ethanol (20%, v/v). The elution of wine phenolics from Toyopearl column yielded malvidin 3-glucoside and three derived pigments, namely malvidin 3-glucoside pyruvic adduct, malvidin 3-acetylglucoside pyruvic adduct, and malvidin 3-coumarylglucoside pyruvic adduct. These glucosides accounted for 60% of the total content of monoglucosides.

Oh et al. (1980, 1985) employed a gel chromatographic technique to study the tannin–protein interaction. Tannins were immobilized on Sepharose-4B via epoxy activation. Protein was then applied to the column at pH 4, and elution of protein from the column was achieved using a pH gradient or by anionic and nonionic detergents.

Salagoity-Auguste and Bertrand (1984) as well as Jaworski and Lee (1987) demonstrated that a C_{18} Sep-Pak cartridge can be used to separate grape phenolics into acidic and neutral fractions. Later, Sun et al. (1998b) successfully used a C_{18} Sep-Pak cartridge for fractionation of grape proanthocyanidins according to their polymerization degree. The procedure involved passing the extract of grape phenolics through two preconditioned neutral C_{18} Sep-Pak cartridges connected in series. The phenolic acids were then washed out with water; catechins and oligomeric proanthocyanidins were subsequently eluted with ethyl acetate and anthocyanidins and polymeric proanthocyanidins with methanol. The ethyl acetate fraction was redeposited on the same C_{18} Sep-Pak cartridges, and catechins were first eluted with diethyl ether and then oligomeric proanthocyanidins with methanol.

11.8.2 HIGH-PERFORMANCE LIQUID CHROMATOGRAPHY

HPLC techniques are now most widely used for both separation and quantitation of phenolic compounds. Various supports and mobile phases are available for the analysis of anthocyanins, procyanidins, flavonones and flavonols, flavan-3-ols, procyanidins, flavones, and phenolic acids (Senter et al. 1989). Introduction of reversed-phase columns has considerably enhanced the HPLC separation of different classes of phenolic compounds (Hostettmann and Hostettmann 1982). Several reviews have described the application of HPLC method for the analysis of phenolics (Daigle and Conkerton 1983, 1988; Karchesy et al. 1989; Robards and Antolovitch 1997; Merken and Beecher 2000).

Tables 11.9 through 11.14 summarize some modern HPLC procedures applied for the analysis of various classes of food phenolics. Food phenolics are commonly detected using UV-visible and photodiode array detectors (DAD) (Carando et al. 1999; Edenharder et al. 2001; Siess et al. 1996; Tomas-Barberan et al. 2001; Wang et al. 2000a,b; Zafrilla et al. 2001). Other methods used for the detection of phenolics include electrochemical coulometric array detector (EC) (Mattila et al. 2000; Sano et al. 1999), chemical reaction detection technique (de Pascual-Teresa et al. 1998), and fluorimetric detector (Arts and Hollman 1998; Carando et al. 1999).

A combination of HPLC technique and voltammetry has been successfully applied for detection, identification, and quantification of flavonoid and nonflavonoid phenolics in wine. Positive identification may be obtained by comparing the capacity factor (k') and electrochemical behavior of wine phenols with those of standard solutions containing pure phenolics (Lunte et al. 1988a; Mahler et al. 1988; Woodring et al. 1990).

Mass spectrometry (MS) detectors coupled to high-performance liquid chromatography (HPLC-MS tandem) have been commonly used for structural characterization of phenolics. Electrospray ionization mass spectrometry (ESIMS) has been employed for structural confirmation of phenolics in plums, peaches, nectarines (Tomas-Barberan et al. 2001), grapeseeds (Peng et al. 2001), soyfoods (Barnes et al. 1998), and cocoa (Hammerstone et al. 1999). Satterfield and Brodbelt (2000) demonstrated that complexation of flavonoids with Cu^{12} enhanced the detection of flavonoids by ESIMS. Mass spectra obtained for metal–flavonoid complexes were more intense and simpler for interpretation than that of corresponding flavonoid. Identification of phenolics collected after HPLC analysis was also carried out using fast atom bombardment mass spectrometry (FABMS) (Bakker et al. 1992; Edenharder et al. 2001; Sano et al. 1999) and electron impact mass spectrometry (Edenharder et al. 2001).

TABLE 11.9

Some HPLC Procedures for Determination of Isoflavones in Soy Foods

Food	Sample Preparation	Stationary Phase	Mobile Phase	References
Seed	Extraction with 80% MeOH; centrifugation	Aquapore C_8 (220 × 4.6 mm)	A: 10% acetonitrile in H_2O with 0.1% TFA; B: 90% acetonitrile in H_2O with 0.1% TFA; linear gradient: 100% A, 0% B, 0 min; 70% A, 30% B, 30 min; step gradient to 0% A, 100% B	Simonne et al. (2000)
Soy foods	Extraction with 80% MeOH; centrifugation, evaporation of MeOH, extraction of lipids with hexane	Brownlee Aquapore C_8 reversed-phase (300 × 4.5 mm)	A: 0.1% TFA in H_2O; B: acetonitrile; linear gradient: 100% A, 0% B, 0 min; 53.6% A, 46.4% B, 20.6 min	Coward et al. (1993)
Seeds, soy foods	Extraction with 0.1% HCl acetonitrile (1:5), filtration	YMC-pack ODS-AM-303 (250 × 4.6 mm, 5 µm)	A: 0.1% acetic acid in water, B: 0.1% acetic acid in acetonitrile; 85% A, 15% B, 0 min; 85% A, 15% B, 5 min; 71% A, 29% B, 31 min; 65% A, 35% B, 8 min; 85% A, 15% B, 3 min	Murphy et al. (1997)
Seeds	Extraction with 80% MeOH	YMC-Pack ODS-AQ 303 (250 × 4.6 mm; 5 µm)	A: 0.1% acetic acid in H_2O; B: 0.1% acetic acid in acetonitrile; modified gradient: 85% A, 15% B, 0 min; 69% A, 31% B, 45 min	Wang et al. (2000a)
Soy sauce	Direct injection	Wakosil II $5C_{18}$ HG (250 × 4.6 mm) fitted with a precolumn (30 × 4.6 mm) packed with the same material	A: 0.05% TFA in H_2O; B: 90% acetonitrile with 0.05% TFA; 100% A, 0% B, 0 min; 100% A, 0% B, 20 min; 75% A, 25% B, 270 min, linear gradient	Kinoshita et al. (1997)
Soy foods	Extraction with 80% MeOH, centrifugation	NovaPak C_{18} reversed-phase (150 × 3.9 mm, 4 µm) coupled to Adsorbosphere C_{18} column guard	A: 10% acetic acid in H_2O; B: MeOH-acetonitrile-dichloromethane (10:5:1 v/v/v). 95% A, 5% B, 0 min; 95% A, 5% B, 5 min; linear gradient: 45% A, 55% B, 20 min; 30% A, 70% B, 6 min; 95% A, 5% B, 3 min	Franke et al. (1999)
Soy foods	Extraction with 96% EtOH, centrifugation, filtration	NovaPak C_{18} (150 × 3.9 mm; 4 µm)	Isocratic: acetonitrile-water (33:67, v/v)	Hutabarat et al. (2001)

Source: Merken, H. M., and Beecher, G. R., *J. Agric. Food Chem.*, 48, 577–599, 2000.

TABLE 11.10
Some HPLC Procedures for Determination of Catechins and Proanthocyanidins (PA) in Selected Foods

Food	Sample Preparation	Stationary Phase	Mobile Phase	References
Grape seed	Extraction with EtOH; fractionation of PA using Sephadex LH-20	Exsil 100 ODS C_{18}, reversed-phase (250 × 4.6 mm, 5 μm) coupled to C_{18} column guard	A: 0.2% phosphoric acid (v/v); B: 82% acetonitrile with 0.4% phosphoric acid; gradient: 100% A, 0% B, 0 min; 85% A, 15% B, 15 min; 84% A, 16% B, 25 min; 83% A, 17% B, 5 min; 57% A, 43% B, 3 min; 48% A, 52% B, 1 min; isocratic 52% B, 7 min; 57% A, 43% B, 1 min; 83% A, 17% B, 1 min; 100% A, 0% B, 2 min	Peng et al. (2001)
Beverages	Direct injection	Spherisorb ODS2 (150 × 46 mm, 3 μm)	A: H_2O; B: MeOH; C: 4.5% aqueous formic acid; D: 4.5% aqueous formic acid-MeOH (90:10, v/v); gradient: 0–10 min, 100% A to 100% C; 10–20 min, 0–15% D in C; 20–30 min, 15% D in C, isocratic; 30–40 min, 15%–35% D in C; 40–45 min, 35% D in C, isocratic; 45–60 min, 35%–45% D in C; 60–75 min, 45%–100% D in C; 75–175 min, 0–50% B in D; 175–180 min, 50%–80% B in D	de Pascual-Teresa et al. (1998)
Apples, grapes, beans	Extraction with 90% MeOH (apples, grapes) or 70% MeOH (beans); filtration	Inertsil ODS2 (150 × 4.6 mm, 5 μm) coupled with Opti-Guard PR C_{18} Violet A guard	A: 5% acetonitrile in 0.025 M phosphate buffer pH 2.4; B: 25% acetonitrile in 0.025 M phosphate buffer pH 2.4; isocratic, 0–5 min, 10% B; 5–20 min, linear gradient: 5–20 min, 10%–80% B; 20–22 min, 80%–90% B; isocratic 22–25 min, 90% B; linear gradient, 25–28 min, 10% B; isocratic, 28–37 min, 10% B	Arts and Hollman (1998)

continued

Food	Sample preparation	Column	Mobile phase/gradient	Reference
Wine	Dealcoholized under vacuum; fractionation of polymeric fraction using Tyopearl TSK gel HW-50 (f), thiolysis	Nucleosil 120 (125 × 4 mm, 3 μm)	A: 2% HCOOH in H_2O; B: CH_3CN-H_2O-HCOOH (80:18:2, v/v/v); linear gradient: 15%–75% B, 0–15 min; 75%–100% B, 15–20 min	Fulcrand et al. (1999)
Wine	Dealcoholized under vacuum; fractionation of procyanidins and catechins using two C_{18} Sep Pak cartridges in series	Superspher 100 RP18 (250 × 4 mm; 4 μm)	A: H_2O; B: H_2O-acetic acid (90:10, v/v); catechins: 10%–80% B, 0–5 min; 80%–100% B, 5–29 min; 100% B, 29–45 min; procyanidins: 10%–70% B, 0–40 min; 70%–85% B, 40–55 min; 85%–100% B, 55–74 min	Sun et al. (1998)
Wine	Filtration, direct injection	Nucleosil 100 C_{18} (250 × 4 mm, 5 μm) coupled to C_{18} column guard	A: 2 mM $NH_4H_2PO_4$, adjusted to pH 2.6 with H_3PO_4; B: 20% A with acetonitrile; C: 0.2 M H_3PO_4 adjusted to pH 1.5 with ammonia; gradient: 100% A, 0–5 min; 0–4% B, 5–15 min; 4–8% B, 15–25 min; 8% B, 92%C, 25.1 min; 8–20% B, 25.1–45 min; 20%–30% B, 45–50 min; 30%–40%, 50–55 min; 40%–80% B, 55–60 min	Carando et al. (1999)
Cocoa	Extraction of defatted seeds with 70% acetone (v/v), followed by extraction with 70% MeOH (v/v), fractionation of phenolic from nonphenolics on column packed with Baker octadecyl for flash chromatography	Phenomenex Luna (250 × 4.6 mm, 5 μm)	A: dichloromethane; B: MeOH; C: acetic acid-H_2O (1:1,v/v); gradient: 0 min, 14% B in A; 0–30 min, 14%–28.4% B in A; 28.4–50% B in A, 30–60 min; 50%–86% B in A, 60–65 min; isocratic 65–70 min	Hammerstone et al. (1999)

TABLE 11.10 (continued)
Some HPLC Procedures for Determination of Catechins and Proanthocyanidins (PA) in Selected Foods

Food	Sample Preparation	Stationary Phase	Mobile Phase	References
Oolong tea	Extraction with acetonitrile-water (1:1, v/v)	Devolosil PhA-5 (250 × 46 mm)	0.1 M NaH$_2$ PO$_4$ buffer (pH 2.5)-acetonitrile (85:15, v/v) with 0.1 mM EDTA2 Na, isocratic	Sano et al. (1999)
Green tea beverage	Washed with chloroform; extraction with ethyl acetate, ethyl acetate layers combined, evaporated, residue dissolved in 50% acetonitrile	μm-Bondapak C$_{18}$ (300 × 3.9 mm)	Acetonitrile-ethyl acetate-0.05% phosphoric acid (12: 2:86, v/v/v), isocratic	Wang et al. (2000b)
Green tea	Extraction with hot water, filtration, washing with chloroform, extraction of water layer with ethyl acetate, ethyl acetate layers combined and evaporated, residue dissolved in water	Hypersil ODS (250 × 4.6 mm, 5 μm)	0.05% H$_2$SO$_4$ aqueous-acetonitrile-ethyl acetate (86:12:2, v/v/v), isocratic	Chen et al. (1998)

Source: Merken, H. M. and Beecher, G. R., *J. Agric. Food Chem.*, 48, 577–599, 2000.

TABLE 11.11

Some HPLC Procedures for Determination of Anthocyanins and Anthocyanidins in Selected Foods

Food	Sample Preparation	Stationary Phase	Mobile Phase	References
Red onions	Extraction with MeOH containing 0.1% HCl, filtration.	Prodigy ODS2 (250 × 4.6 mm, 5 μm)	A: 10% formic acid in H_2O (v/v); B: MeOH-H_2O-formic acid (50:40:10, v/v/v); isocratic: 80% A, 20% B, 0–4 min; linear regression: 20%–80% B in A, 4–26 min	Gennaro et al. (2002)
Red wine, fruit juices	Dealcoholization of wine under vacuum; dilution of fruit juice with water; separation of anthocyanins using C_{18} Sep-Pak	SPLC-18-DB (250 × 10 mm, 5 μm) preparative reversed-phase coupled with preinjection C_{18} saturator with silica-based packing (75 × 4.5 mm, 12 μm) and guard with Supelco LC-18 reversed-phase packing (50 × 4.6 mm, 20–40 μm)	A: 5% formic acid in H_2O; B: formic acid-H_2O-MeOH (5:5:90, v/v/v); linear gradient: 5–20% B in A, 0–1 min; 20%–25% B in A, 1–12 min; 25%–32% B in A, 12–32 min; 32%–55% B in A, 32–38 min; 55%–100% B in A, 38–44 min; 100% B, 44–46 min; 100%–5% B in A, 46–47 min	Wang and Sporns (1999)
Red wine	Direct injection	Ultrasphere (C18) ODS (250 × 4.6 mm)	A: H_2O-formic acid (9:1, v/v); B: CH_3CN-H_2O-formic acid (3:6:1, v/v/v); gradient: 20–85% B in A, 0–70 min; 85%–100% B in A, 70–75 min; isocratic: 100% B, 75–85 min	Mateus et al. (2001)
Red wine	Centrifugation, addition of formic acid to 1.5%, filtration	Supelcosik LC-18 (250 × 2.1 mm), reversed-phase	A: 5% formic acid in H_2O (v/v); B: MeOH; gradient: 5% B in A, 0–5 min; 5–65% B in A, 5–55 min; 65%–100% B, 55–58 min; 100%–5% B in A, 58–60 min; 5% B in A, 60–64 min	Mazza et al. (1999)
Red blood, orange juice	Homogenization with (acetone-EtOH-hexane, 25:25:50,v/v/v), centrifugation, concentration of acetone-EtOH layer, separation of anthocyanins using C_{18} Sep-Pak	Prodigy ODS3 (150 × 4.6 mm, 5 μm)	A: 0.1% phosphoric acid in H_2O; B: 0.1% phosphoric acid in acetonitrile; gradient: 10% B in A, 0–2 min; 10%–50% B in A, 2–32 min; 50% B in A, 32–37 min; 50%–70% B in A, 37–57 min;	Lee (2002)

Source: Merken, H. M. and Beecher, G. R., *J. Agric. Food Chem.*, 48, 577–599, 2000.

TABLE 11.12
Some HPLC Procedures for Determination of Flavones and Flavonols in Selected Foods[a]

Food	Sample Preparation	Stationary Phase	Mobile Phase	References
Red onions	Extraction with MeOH stabilized with BHT; dilution with MeOH	Supelcosil LC-18 (250 × 4.6 mm, 5 μm) coupled with a Spherisorb Supelguard LC-18	A: 0.01 M sodium phosphate adjusted to pH 2.5 with H_3PO_4; B: MeOH; linear gradient: 87%–60% A in B, 0–13.5 min; 60%–10% A in B, 13.5–39 min; 10%–0% A in B, 39–42 min; 0%–87% A in B, 42–46 min	Gennaro et al. (2002)
Yellow and green French beans	Extraction with chloroform to remove chlorophyll and carotenoids, drying, extraction with 70% MeOH, evaporation of MeOH, purification of phenolics using polyamide cartridge, filtration	LiChrospher 100 RP-18 (250 × 4, 5 μm) coupled with guard (4 × 4 mm) packed with the same stationary phase	A: acetonitrile; B: 2% acetic acid in H_2O; gradient: 10%–30% A in B, 0–35 min; 30%–45% A in B, 35–37 min; 45% A in B, 37–42 min; 45%–10% A in B, 42–44 min	Hempel and Bohm (1996)
Red beans' seed coats	Extraction with MeOH, separation of tannins using Sephadex LH-20, flavonoid fraction rechromatographed on Sephadex LH-20	Shiseido Capcell Pak C_{18} preparative reversed-phase (250 × 10 mm, 5 μm)	Isocratic: acetonitrile-H_2O (30:70, v/v)	Beninger and Hosfield (1999)
Buckwheat	Extraction with 80% MeOH, filtration, evaporation, dissolving in MeOH-H_2O-oxalic acid (13:36:1, v/v/v) filtration	Capcell Pak C_{18} -SG 120 (100 × 4.6 mm, 3 μm)	A: MeOH-H_2O-acetic acid (13:36:1, v/v/v); B: MeOH-H_2O-acetic acid (73:25:2, v/v/v); gradient: 10%–50% B in A, 0–20 min; 50% B in A, 20–25 min; 50%–10% B in A, 25–30 min	Oomah and Mazza (1996)
Tomatoes, onions, lettuce, celery	Extraction with 1.2 M HCl in 50% MeOH for 2 hours at 90°C; extract adjusted to pH 2.5 with TFA, filtration	C_{18} Symmetry (150 × 3.9 mm, 5 μm) reversed-phase, coupled with C_{18} Symmetry guard	A: acetonitrile; B: H_2O adjusted to pH 2.5 with TFA; gradient: 15%–35% A in B, 0–20 min	Crozier et al. (1997)
Edible tropical plants	Extraction with 1.2 M HCl in 50% MeOH for 2 hours at 90°C, filtration	Nova Pak C_{18} (150 × 3.9; 4 μm)	Isocratic: MeOH-H_2O (1:1, v/v) adjusted to pH 2.5 with TFA	Miean and Mohamed (2001)

Source: Merken, H. M. and Beecher, G. R., *J. Agric. Food Chem.*, 48, 577–599, 2000.

TABLE 11.13

Some HPLC Procedures for Determination of Other Classes of Phenolics in Selected Foods

Food	Phenolics	Sample Preparation	Stationary Phase	Mobile Phase	References
Finger millet	Free phenolic acids	Extraction with 70% EtOH, centrifugation, concentration, adjusting pH to 2–3, extraction with ethyl acetate, evaporation, dilution in MeOH	Shimpak C_{18} (250 × 4.6 mm) reversed-phase	Isocratic: H_2O-acetic acid-MeOH (80:5:15, v/v/v)	Subra Rao and Muralikrishna (2002)
Barley	Phenolic acids	Extractions: hot H_2O; acid hydrolysis; acid and α-amylase hydrolysis; acid and α-amylase and cellulase hydrolysis; centrifugation	Supelcosil LC-18 (150 × 4.6 mm, 5 µm)	A: 0.01 M citrate buffer pH 5.4 adjusted with 50% acetic acid; B: MeOH; gradient: 2–4% B in A, 0–12 min; 4–13% B in A, 12–20 min; 13% B in A, 20–26 min; 13%–2% B in A, 26–30 min	Vasanthan and Temelli (2001)
Citrus fruits	Coumarins	Extraction with acetone, filtration, evaporation dissolving in MeOH-acetone (1:1, v/v), filtration	Hypersil ODS (125 × 4 mm, 5 µm)	Isocratic: MeOH-H_2O (75:25, v/v)	Ogawa et al. (2000)
Rice bran oil	g-Oryzanol	Solubilization of oil in hexane-ethyl acetone (9:1, v/v), removal of lipids using silica column (250 × 25 mm)	Microsorb-MV C_{18} (250 × 4.6 mm)	Isocratic: MeOH-acetonitrile-dichloromethane-acetic acid (50:44:3:3, v/v/v/v)	Xu and Golber (1999)
Flaxseed flour, defatted	Lignans	Extraction with 1,4-dioxane-95% EtOH (1:1, v/v), centrifugation, evaporation, alkaline hydrolysis, acidification to pH 3, removal of salt using C-18 reversed-SPE	Econosil RP C_{18} (250 × 4.6 mm, 5 µm)	A: 5% acetonitrile in 0.01 M phosphate buffer, pH 2.8; B: acetonitrile; gradient: 100%–70% A in B, 0–30 min; 70%–30% A in B, 30–32 min	Johnsson et al. (2000)
Peanuts	Resveratrol	Extraction with 80% EtOH, centrifugation, semipurification Al_2O_3 silica gel 60R$_{18}$ (1:1)	Vydac C_{18} (150 × 4.5 mm) reversed-phase	A: acetonitrile; B: 0.1% TFA; gradient: 0% A in B, 0–1 min; 0–15% A in B, 1–3 min; 15%–27% A in B, 3–23 min; 27%–100% A in B, 23–28 min	Sanders et al. (2000)

Source: Merken, H. M. and Beecher, G. R., *J. Agric. Food Chem.*, 48, 577–599, 2000.

TABLE 11.14
Some HPLC Procedures for Determination of Multiple Classes of Phenolics in Selected Foods

Food	Phenolics	Sample Preparation	Stationary Phase	Mobile Phase	References
Lingonberry, cranberry, onions, broccoli	Catechins, flavonones, flavones, flavonols	Extraction with 1.2 M HCl in 50% MeOH for 2 H at 90°C; filtration	Inertsil ODS (150 × 4 mm; 3 µm) coupled with C-18 guard	A: 50 mM H_3PO_4 pH 2.5; B: acetonitrile; *catechins:* 86% A in B, isocratic; *other flavonoids*-gradient: 95% A in B, 0–5 min; 95%–50% A in B, 5–55 min; 50% A in B, 55–65 min; 50%–95% A in B, 65–67 min	Mattila et al. (2000)
Nectarines, peaches, plums	Phenolic acids, catechins, flavonols, procyanidins	Extraction with 80% MeOH containing 2 mM NaF; centrifugation, filtration	Nucleosil C_{18} (150 × 4.6 mm, 5 µm) reversed-phase coupled with guard containing the same stationary phase	A: 5% MeOH in H_2O; B: 12% MeOH in H_2O; C: 80% MeOH in H_2O; D: MeOH; gradient: 100% A, 0–5 min; 0–100% B in A, 5–10 min; 100% B, 10–13 min; 100%–75% B in C, 13–35 min; 75%–50% B in C, 35–50 min; 50%–0% B in C, 50–52 min; 100% C, 52–57 min; 100% D 57–60 min	Tomas-Barberan et al. (2001)
Red raspberry	Ellagic acids, flavones	Extraction with MeOH, filtration, addition of H_2O, evaporation, semipurification of phenolics using Sep-Pak C_{18}, filtration	Lichrocart 100 RP-18 (250 × 4 mm, 5 µm) reversed-phase	A: 5% formic acid in H_2O; B: MeOH; gradient: 10%–15% B in A, 0–5 min; 15%–30% B in A, 5–20 min; 30%–50% B in A, 20–35 min; 50%–90% B in A, 35–38 min	Zafrilla et al. (2001)
Propolis	Phenolic acids, flavones, flavonones, flavonols	Dilution with EtOH, alkaline hydrolysis, acidification, extraction of phenolic with ethyl acetate, evaporation, dissolving in EtOH	Lichrosorb RP18 (200 × 3, 7 µm) coupled with C-18 guard	A: H_2O adjusted to pH 2.6 with H_3PO_4; B: acetonitrile; gradient: 0%–9% B in A, 0–12 min; 9%–13% B in A, 12–20 min; 13%–40% B in A, 20–40 min; 40%–70% B in A, 40–60 min; 70% B in A, 60–85 min	Siess et al. (1996)
Spinach	Flavonols, flavonones	Extraction with 70% MeOH, removal of carotenoids and chlorophyll using ODS-C_{18} packing material, centrifugation, concentration	YMC ODS-AQ (250 × 4.6, 5 µm)	A: H_2O containing 0.01% TFA; B: acetonitrile containing 0.01% TFA; gradient: 100% A, 0–10 min; 100%–50% A in B, 10–40 min; 50%–0% A in B, 40–50 min	Edenharder et al. (2001)

Source: Merken, H. M. and Beecher, G. R., *J. Agric. Food Chem.*, 48, 577–599, 2000.

Matrix-assisted laser desorption ionization mass spectrometry (MALDI-MS) has also been used for qualitative and quantitative analysis of anthocyanins in foods (Wang and Sporns 1999).

11.8.3 High-Speed Countercurrent Chromatography

High-speed countercurrent chromatography (HSCCC), also known as centrifugal partitioning chromatography (CPC), is an all-liquid chromatographic technique very suitable for preparative isolation of pure compounds (Degenhart et al. 2000a,b). Separation of compounds is based on their partitioning between two immiscible liquids (Conway and Petrovski 1995).

Degenhart et al. (2000c) used HSCCC for preparative isolation of anthocyanins from red wines and grape skins. Anthocyanins were fractionated into four solvent systems based on their polarities. Solvent I, consisting of *tert*-butyl methyl ether–*n*-butanol–acetonitrile–water (2:2:1:5, v/v/v/v) containing 0.1% trifluoroacetic acid (TFA), was used as a medium for fractionation of monoglucosides and acylated diglucosides. Solvent II, consisting of ethyl acetate–*n*-butanol–water (2:3:5, v/v/v) and 0.1%TFA, was used as a medium for separation of visitins and diglucosides. Solvent III, consisting of ethyl acetate–water (1:1, v/v) and 0.1% TFA, was used as a medium for extraction of coumaryl and caffeoyl monoglucosides. Solvent IV, consisting of ethyl acetate–*n*-butanol–water (4:1:5, v/v/v), was employed as a medium for fractionation of acetylated anthocyanins.

Vitrac et al. (2001) also applied HSCCC for fractionation of red wine phenolics. Phenolics were extracted first from red wine into ethyl acetate. Subsequently, the phenolic extract was chromatographed using a 1.5 × 60 cm cation-exchange Dowex (Sigma) column. Nonphenolic constituents were washed out from the column with water, and then phenolics were eluted with aqueous methanol (75%, v/v). Afterwards, the phenolic extract was fractionated using centrifugal partition chromatography in both ascendant and descendant modes. The solvent systems water–ethanol–hexane–ethyl acetate in the ratios of 3:3:4:5 (v/v/v/v) and 7:2:1:8 (v/v/v/v) were used, at a flow rate of 3 mL/min, for elution of phenolics in ascendant and descendant modes, respectively.

Degenhart et al. (2000d) demonstrated that HSCCC can be used for isolation of theaflavins, epitheaflavic acids, and thearubigins from black tea using hexane–ethyl acetate–methanol–water (2:5:2:5 and 1.5:5:1.5:5, v/v/v/v). Theaflavins prior to HSCCC were extracted from tea infusion with ethyl acetate and then cleaned up using a Sephadex LH-20 column to avoid coelution of catechins and theaflavins. On the other hand, isolation of thearubigins required cleaning up of tea infusion on an Amberlite XAD-7 column prior to HSCCC to remove all nonphenolic compounds.

A simple and efficient method for separation of catechin gallates from spray-dried tea extract was developed by Baumann et al. (2001). Tea phenolic extract was first subjected to liquid–liquid partitioning between ethyl acetate and water. The organic layer containing catechins was then submitted to high-speed centrifugal countercurrent chromatography operating in an ascending mode. Favorable partitioning was achieved using *n*-hexanes–ethyl acetate–water (1:5:5, v/v/v) or ethyl acetate–methanol–water (5:1:5 and 5:2:5, v/v/v). Sephadex LH-20 column with methanol as a mobile phase was used for a final purification of catechin gallates.

11.8.4 OTHER CHROMATOGRAPHIC TECHNIQUES

Other chromatographic techniques have also been used for purification and separation of food phenolics. Of these, paper chromatographic (PC) and thin-layer chromatographic (TLC) techniques are still widely used for purification and isolation of anthocyanins, flavonols, condensed tannins, and phenolic acids, using different solvent systems (Chu et al. 1973; Durkee and Harborne 1973; Fenton et al. 1980; Forsyth 1955; Francis et al. 1966; Francis 1967; Harborne 1958, 1967; Jackman et al. 1987; Leung et al. 1979; Mabry et al. 1970).

PC on Whatman No. 3 has been used for separation of anthocyanins using butanol–acetic acid–water, chloroform–acetic acid–water, or butanol–formic acid–water as possible mobile phases (Jackman et al. 1987). On the other hand, two-dimensional PC has been used for the analysis of procyanidin oligomers. Chromatograms were developed using 6% acetic acid as a mobile phase in the first direction and 2-butanol–acetic acid–water as mobile phase in the second direction (Haslam 1966; Thompson et al. 1972).

Azar et al. (1987) have identified phenolics of bilberry juice *Vaccinium myrtillus* using a two-dimensional TLC. Phenolic acids were chromatographed on a 0.1-mm cellulose layer with solvent I: acetic acid–water (2:98, v/v), and solvent II: benzene–acetic acid–water (60:22:1.2, v/v/v). However, TLC analysis of flavonols was carried out on silica gel plates using ethyl acetate–2-butanone–formic acid–water (5:3:1:1, v/v/v/v) or on cellulose plates using solvent I: t-butanol–acetic acid–water (3:1:1, v/v/v) and solvent II: acetic acid–water (15:85, v/v). The phenolic acids were detected by first spraying the chromatograms with deoxidized *p*-nitroaniline and then with a 15% solution of sodium carbonate in water; flavonols were detected by spraying with a 5% aluminum chloride solution in methanol. Two-dimensional cellulose TLC plates have also been employed for separation of procyanidins. *t*-Butanol–acetic acid–water (3:1:1, v/v/v) was used for the development in the first direction while 6% acetic acid was used for the development in the second direction. Detection of polyphenols on TLC was carried out using ferric chloride, potassium ferricyanide, or vanillin–HCl solutions (Karchesy et al. 1989). TLC on silica using ethyl acetate–formic acid–water (90:5:5, v/v/v) or toluene–acetone–formic acid (3:3:1, v/v/v) has been used for monitoring the isolation of procyanidins by column chromatography (Karchesy and Hemingway 1986) and by HPLC (Carando et al. 1999), respectively. On the other hand, phenolic acids have been separated on silica TLC plates using *n*-butanol–acetic acid–water (40:7:32, v/v/v) as the mobile phase (Dabrowski and Sosulski 1984). Various gas chromatographic (GC) methods have been employed for separation and quantitation of phenolic acids (Dabrowski and Sosulski 1984; Krygier et al. 1982), isoflavones (Liggins et al. 1998), capsaicinoids (Thomas et al. 1998), phenolic aldehydes (Friedman et al. 2000), and monomers of condensed tannins (Hemes and Hedges 2000). Novel high-temperature GC columns, electronic pressure controllers, and detectors have significantly improved the resolution and have also led to an increase in the upper range of molecular weights of substances that could be analyzed by GC. Preparation of samples for GC may include the removal of lipids from the extract, liberation of phenolics from ester and glycosidic bonds by alkali (Dabrowski and Sosulski 1984; Krygier et al. 1982), acid (Zadernowski 1987), and enzymatic hydrolysis (Liggins et al. 1998) or acid depolymerization of tannins in the presence of nucleophiles such as phloroglucinol (Hemes and Hedges 2000) or benzyl mercaptan (Guyot et al. 2001; Labarbe et al. 1999). Prior to chromatography, phenolics are usually transformed to more volatile derivatives by methylation (Jurenitsch et al. 1979; Jurenitsch and Leinmuller 1980; Kosuge and Furuta 1970), trifluoroacetylation (Chassagne et al. 1998; Sweeley et al. 1963), conversion to trimethylsilyl derivatives (Dabrowski and Sosulski 1984; Krygier et al. 1982), or derivatization with *N*-(*tert*-butyldimethylsilyl)-*N*-methyltrifluoroacetamide (Hemes and Hedges 2000).

11.9 CAPILLARY ELECTROPHORESIS

Capillary electrophoresis is a novel analytical tool for separation of many classes of compounds (Cikalo et al. 1998; Zeece 1992) based on the electrophoretic migration of charged analytes. Small internal diameter capillary columns minimize the ohmic heating problems that may have an adverse effect on bandwidths. In addition, small sample sizes can be used, and separations required little or no organic solvents (Cikalo et al. 1998).

Hall et al. (1994) used capillary electrophoresis for separation of food antioxidants such as butylated hydroxyanisole (BHA) and butylated hydroxytoluene (BHT). Later, Andrade et al. (1998) used capillary zone electrophoresis to evaluate the effect of grape variety and wine aging on the composition of noncolored phenolics in port wine. Noncolored phenolics were extracted from wine into diethyl ether, then concentrated to dryness, and redissolved in methanol. Subsequently, phenolics were separated on a fused-silica capillary column with a 0.1 M sodium borate (pH 9.5) at 30°C and the voltage of 20 kV producing a current of 90 µA.

Chu et al. (1998) separated pure forms of *cis*- and *trans*-resveratrol isomers from wine using capillary electrophoresis in micellar mode. Direct separation of resveratrols in wine samples was performed with fused silica capillaries in 25 mM sodium borate, 25 mM sodium phosphate, and 75 mM SDS (pH 9.3) at 16 kV and 20°C using a UV detector set at 310 nm. The detection limit of the method was 1.25 mM. On the other hand, Kreft et al. (1999) used capillary electrophoresis with a UV detector for determination of rutin content in different fractions of buckwheat flour and bran. The extraction of rutin from buckwheat fractions was carried out using 60% ethanol containing 5% ammonia in water. Identification of both resveratrol and rutin was confirmed by spiking the samples with standards.

Moane et al. (1998) used capillary electrophoresis for direct detection of phenolic acids in beer. Separation of phenolic acids was performed with fused silica capillary in 25-mM phosphate buffer, pH 7.2 at 25 kV. The sample was injected to capillaries using a reversed-polarity injection technique to remove nonphenolic cationic and neutral compounds. These substances interfere with electrochemical detection of phenolic acids by passivation of the electrode surface. Pan et al. (2001) developed a method for determination of protocatechuic aldehyde and protocatechuic acid by capillary electrophoresis with amperometric detection. Under optimum conditions, these two analytes were completely separated in 8 min with detection limits of 0.10 g/mL for protocatechuic aldehyde and 0.25 µg/mL for protocatechuic acid.

REFERENCES

Adams, J. B. 1972. *Changes in polyphenols of red fruits during processing: the kinetics and mechanism of anthocyanin degradation.* Campden Food Pres. Res. Assoc. Tech. Bull. P22, Chipping Campden, Gloucestershire, UK, pp. 1–185.

Amory, A. M. and Schubert, C. L. 1987. A method to determine tannin concentration by measurement and quantification of protein–tannin interactions. *Oecologia* 73: 420–424.

Andrade, P., Seabra, R., Ferreira, M., Ferreres, F., and Garcia-VigueraC. 1998. Analysis of noncoloured phenolics in port wines by capillary zone electrophoresis. Influence of grape variety and ageing. *Z. Lebensm. Unters Forsch A.* 206: 161–164.

Antolovich, M., Prenzler, P., Robards, K., and Ryan, D. 2000. Sample preparation in the determination of phenolic compounds in fruits. *Analyst* 125: 989–1009.

AOAC Association of Official Analytical Chemists. 1980. *Official Methods of Analysis*, 12th edition, AOAC, Washington, D.C.

Arts, I. C. W. and Hollman, P. C. H. 1998. Optimization of quantitative method for the determination of catechins in fruits and legumes. *J. Agric. Food Chem.* 46: 5156–5162.

Asquith, T. N., Izuno, C. C., and Butler, L. G. 1983. Characterization of the condensed tannin (proanthocyanidin) from group II sorghum. *J. Agric. Food Chem.* 31: 1299–1303.

Asquith, T. N. and Butler, L. G. 1985. Use of dye-labeled protein as spectrophotometric assay for protein precipitants such as tannins. *J. Chem. Ecol.* 11: 1535–1544.

Atherton, B. A., Cunningham, E. L., and Splittgerber, A. G. 1996. A mathematical model for the description of the Coomassie blue protein assay. *Anal. Biochem.* 233: 160–168.

Austin, F. L. and Wolff, I. A. 1968. Sinapine and related esters in seed of Crambe abyssinica. *J. Agric. Food Chem.* 16: 132–135.

Azar, M., Verette, E., and Brun, S. 1987. Identification of some phenolic compounds in bilberry juice, Vaccinium myrtillus. *J. Food Sci.* 52: 1255–1257.

Bacon, J. R. and Rhodes, M. J. C. 1998. Development of competition assay for the evaluation of the binding of human parotid salivary proteins to a dietary complex phenols and tannins using a peroxide-labeled tannin. *J. Agric. Food Chem.* 46: 5083–5088.

Bakker, J., Bridle, P., and Koopman, A. 1992. Strawberry juice color: the effect of some processing variables on the stability of anthocyanins. *J. Sci. Food Agric.* 60: 471–476.

Barnes, S., Coward, L., Kirk, M., and Sfakianos, J. 1998. HPLC-mass spectrometry analysis of isoflavones. *P.S.E.B.M.* 217: 254–262.

Bate-Smith, E. C. 1973a. Tannins in herbaceous leguminosae. *Phytochemistry* 12: 1809–1812.

Bate-Smith, E. C. 1973b. Haemanolysis of tannins: the concept of relative astringency. *Phytochemistry* 12: 907–912.

Bate-Smith, E. C. 1977. Astringent tannins of Acer species. Phytochemistry 16: 1421–1427.

Baumann, D., Adler, S. and Hamburger, M. 2001. A simple isolation method for the major catechins in green tea using high-speed countercurrent chromatography. *J. Nat. Prod.* 64: 353–355.

Bax, A. and Grzesiek, S. 1996. ROESY. In *Encyclopedia of NMR,* edited by Grant, D. M., and Harris, R.K. Wiley, Chichester, UK, pp. 4157–4166.

Bax, A. 1985. A simple description of 2D NMR spectroscopy. *Bull. Magn. Reson.* 7: 167–183.

Beart, J. E., Lilley, T. H., and Haslam, E. 1985. Polyphenol interactions. Covalent binding of procyanidins to proteins during acid catalyzed decomposition: observation on some polymeric proanthocyanidins. *J. Chem. Soc. Perkin Trans.* 2: 1439–1443.

Beebe, K. R. and Kowalski, B. R. 1987. Introduction to multivariate calibration and analysis. *Anal. Chem.* 59: 1007A–1017A.

Belton, P. S., Delgadillo, I., Gil, A. M., Webb, G. A. Eds. 1995. *Magnetic Resonance in Food Science.* The Royal Society of Chemistry, Cambridge, UK.

Beninger, C. W. and Hosfield, G. L. 1999. Flavonol glycosides from Montcalm dark red kidney bean: implications for the genetics of seed coat color in Phaseolus vulgaris L. *J. Agric. Food Chem.* 47: 4079–4082.

Billot, J., Hartmann, C., Macheix, J. J., and Rateau, J. 1978. Les composes phenoliques au cours de la croissance de la poire, Passe-Crassane (Phenolic compounds during Passe-Crassane pear growth). *Physiol. Veg.* 16: 693–714.

Blair, R. and Reichert, R. D. 1984. Carbohydrates and phenolic constituents in a comprehensive range of rapeseed and canola fractions: nutritional significance for animals. *J. Sci. Food Agric.* 35: 29–35.

Boukharta, M., Girardin, M. and Metche, M. 1988. Procyanidines galloylees du sarment de vigne (Vitis vinifera). Separation et identification par chromatographie liquide haute performance and chromatographie en phase gazeuse. *J. Chromatogr.* 455: 406–409.

Brenna, O. V. and Pagliarini, E. 2001. Multivariate analysis of antioxidant power and polyphenolic composition in red wines. *J. Agric. Food Chem.* 49: 4841–4844.

Briandet, R., Kemsley, E. K., and Wilson, R. H. 1996. Discrimination of Arabica and Robusta in instant coffee by Fourier transform infrared spectroscopy and chemometrics. *J. Agric. Food Chem.* 44: 170–174.

Broadhurst, R. B. and Jones, W. T. 1978. Analysis of condensed tannins using acidified vanillin. *J. Sci. Food Agric.* 29: 788–794.

Brune, M., Hallberg, L., and Skanberg, A. B. 1991. Determination of iron-binding phenolic groups in foods. *J. Food Sci.* 56: 128–131 and 167.

Burns, R. E. 1971. Methods for estimation of tannin in grain sorghum. *Agron. J.* 63: 511–512.

Butler, L. G., Price, M.L. and Brotherton, J. E. 1982. Vanillin assay for proanthocyanins (condensed tannins): modification of the solvent for estimation of the degree of polymerization. *J. Agric. Food Chem.* 30: 1087–1089.

Butler, L. G. 1992. Protein–polyphenol interactions: nutritional aspects. In *Proceedings of the 16th International Conference of Groupe Polyphenols.* Volume 16, Part II, pp. 11–18.

Carando, S., Teissedre, P.-L., Pascual-Martinez, L., and Cabanis, J.-C. 1999. Levels of flavan-3-ols in French wines. *J. Agric. Food Chem.* 47: 4161–4166.

Chassagne, D., Crouzet, J., Bayonove, C. L., and Baumes, R. L. 1998. Identification of passion fruit glycosides by gas chromatography/mass spectrometry. *J. Agric. Food Chem.* 46: 4352–4357.

Chavan, U. D., Shahidi, F., and Naczk, M. 2001. Extraction of condensed tannins from beach pea (Lathyrus maritimus L.) as affected by different solvents. *Food Chem.* 75: 509–512.

Chen, Z.-Y., Zhu, Q. Y., Wong, Y. F., Zhang, Z., and Chung, H. Y. 1998. Stabilizing effect of ascorbic acid on green tea catechins. *J. Agric. Food Chem.* 46: 2512–2516.

Chu, Q., O'Dwyer, M., and Zeece, M. G. 1998. Direct analysis of resveratrol in wine by micellar electrokinetic capillary electrophoresis. *J. Agric. Food Chem.* 46: 509–513.

Chu, N. T., Clydesdale, F. M., and Francis, F. J. 1973. Isolation and identification of some fluorescent phenolic compounds in cranberries. *J. Food Sci.* 38: 1038–1042.

Cikalo, M. G., Bartle, K. D., Robson, M. M., Myers, P., and Euerby, M. R. 1998. Capillary electrochromatography. *Analyst* 123: 87R–102R.

Clandinin, D. R. 1961. Effect of sinapine, the bitter substance in rapeseed oil meal, on the growth of chickens. *Poultry Sci.* 40:484–487.

Clifford, M. N. and Wright, J. 1976. The measurement of feruloylquinic acids and caffeoylquinic acids in coffee beans. Development of the technique and its preliminary application to green coffee beans. *J. Sci. Food Agric.* 27: 73–84.

Clifford, M. N. and Wright, J. 1973. Meteperiodate—a new structure-specific locating reagent for phenolic compounds. *J. Chromatogr.* A, 86: 222–224.

Conway, W. D. and Petrowski, R. J. 1995. Modern Countercurrent Chromatography, ACS Symposium Series 593, American Chemical Society, Washington, DC.

Corse, J. W., Sondheimer, E. and Lundin, R. 1962. 3-Feruloylquinic acid. A 3'-methylether of chlorogenic acid. *Tetrahedron* 18: 1207–1210.

Coward, L., Barnes, N. C., Setchell, K. D. R., and Barnes, S. 1993. Genestein, daidzein, and their glucoside conjugates: antitumor isoflavones in soybean foods from American and Asian diets. *J. Agric. Food Chem.* 41: 1961–1967.

Crozier, A., Lean, M. E. J., McDonald, M. S., and Black, C. 1997. Quantitative analysis of the flavonoid content of commercial tomatoes, onions, lettuce and celery. *J. Agric. Food Chem.* 45: 590–595.

Dabrowski, K. J. and Sosulski, F. W. 1984. Quantification of free and hydrolyzable phenolic acids in seeds by capillary gas-liquid chromatography. *J. Agric. Food Chem.* 32: 123–127.

Daigle, D. J. and Conkerton, E. J. 1983. Analysis of flavonoids by HPLC. *J. Liq. Chromatogr.* 6: 105–118.

Daigle, D. J. and Conkerton, E. J. 1988. Analysis of flavonoids by HPLC: an update. *J. Liq. Chromatogr.* 11: 309–325.

Davis, A. B. and Hoseney, R. C. 1979. Grain sorghum condensed tannins. I. Isolation, estimation, and selective absorption by starch. *Cereal Chem.* 56: 310–314.

Davis, K. R. 1982. Effects of processing on composition and Tetrahymena relative nutritive value of green and yellow peas, lentils, and white pea beans. *Cereal Chem.* 58: 454–460.

Dawra, R. K., Makkar, H. P. S., and Singh, B. 1988. Protein binding capacity of microquantities of tannins. *Anal. Biochem.* 170: 50–53.

de Pascual-Teresa, S., Treutter, D., Rivas-Gonzalo, J. C., and Santos-Buelga, C. 1998. Analysis of flavanols in beverages by high-performance liquid chromatography with chemical reaction detection. *J. Agric. Food Chem.* 46: 4209–4213.

Degen, A. A., Becker, K., Makkar, H. P. S., and Borowy, N. 1995. Acacia saligna as a fodder tree for a desert livestock and the interaction of its tannins with fibre fractions. *J. Sci. Food Agric.* 68: 65–71.

Degenhart, A., Knapp, H., and Winterhalter, P. 2000a. Separation and purification of anthocyanins by high speed countercurrent chromatography and screening for antioxidant activity. *J. Agric. Food Chem.* 48: 338–343.

Degenhart, A., Hofmann, S., Knapp, H., and Winterhalter, P. 2000c. Preparative isolation of anthocyanins by high-speed countercurrent chromatography and application of the color activity concept to red wine. *J. Agric. Food Chem.* 48: 5812–5818.

Degenhart, A., Engelhardt, U. E., Wendt, A. S., and Winterhalter, P. 2000d. Isolation of black tea pigments using high-speed countercurrent chromatography and studies on properties of black tea polymers. *J. Agric. Food Chem.* 48: 5200–5205.

Degenhart, A., Engelhardt, U. E., Lakenbrink, C., and Winterhalter, P. 2000b. Preparative separation of polyphenols from tea by high-speed countercurrent chromatography. *J. Agric. Food Chem.* 48: 3425–3430.

Delcour, J. A. and de Verebeke, J. D. 1985. A new colourimetric assay for flavonoids in Pilsner beers. *J. Inst. Brew.* 91: 37–40.

Derdelinckx, G. and Jerumanis, J. 1984. Separation of malt and hop proanthocyandins on Fractogel TSK HW-40 (S). *J. Chromatogr.* 285: 231–234.

Derome, A. E. 1987. *Modern NMR Techniques for Chemistry Research*, Pergamon, Oxford, UK.

Deshpande, S. S. and Cheryan, M. 1987. Determination of phenolic compounds of dry beans using vanillin, redox and precipitation assays. *J. Food Sci.* 52: 332–334.

Deshpande, S. S., Cheryan, M., and Salunkhe, D. K. 1986. Tannin analysis of food products. *CRC Crit. Rev. Food Sci. Nutr.* 24: 401–449.

Deshpande, S. S. and Cheryan, M. 1985. Evaluation of vanillin assay for tannin analysis of dry beans. *J. Food Sci.* 50: 905–910.

Deshpande, S. S. 1985. Investigation of dry beans (Phaseolus vulgaris L.): microstructure, processing and antinutrients. Ph.D. thesis. University of Illinois, Urbana Champaign, cited in Salunkhe, D. K., Chavan, J. K., and Kadam, S. S. 1989. *Dietary Tannins: Consequences and Remedies*, CRC Press Inc., Boca Raton, USA, p. 92.

Dick, A. J. and Bearne, L. 1988. Inhibition of b-galactosidase of apple by flavonoids and other polyphenols. *J. Food Biochem.* 12: 97–108.

Downey, G., Briandet, R., Wilson, R. H., and Kemsley, K. 1997. Near- and Mid-Infrared spectroscopies in food authentication: coffee varietal identification. *J. Agric. Food Chem.* 45: 4357 4361.

Du, C. T. and Francis, F. J. 1975. Anthocyanins of garlic (Allium sativum L.): microstructure, processing and antinutrients. Ph.D. thesis. University of Illinois, Urbana Champaign, cited in Salunkhe, D.K., Chavan, J. K., and Kadam, S. S. 1989. *Dietary Tannins: Consequences and Remedies,* CRC Press Inc., Boca Raton, FL.

Durkee, A. B. and Harborne, 1973. Flavonol glycosides in Brassica and Sinapis. *Phytochemistry* 12: 1085–1089.

Earp, C. F., Akingbala, J. O., Ring, S. H. and Rooney, L. W. 1981. Evaluation of several methods to determine tannins in sorghum with varying kernel characteristics. *Cereal Chem.* 58: 234–238.

Edelmann, A., Diewok, J., Schuster, K. C., and Lendl, B. 2001. Rapid method for the discrimination of red wine cultivars based on mid-infrared spectroscopy of phenolic wine extracts. *J. Agric. Food Chem.* 49: 1139–1145.

Edenharder, R., Keller, G., Platt, K. L., and Unger, K. K. 2001. Isolation and characterization of structurally novel antimutagenic flavonoids from spinach (Spinacia oleracea). *J. Agric. Food Chem.* 49: 2767–2773.

Eskin, N. A. M. 1980. Analysis of rapeseed by TiCl4. In *Proceedings of Analytical Chemistry of Rapeseed and its Product Symposium*, Winnipeg, Manitoba, pp. 171–173.

Fenton, T. W., Leung, J., and Clandinin, D. R. 1980. Phenolic components of rapeseed meal. *J. Food Sci.* 45: 1702–1705.

Ferreira, D. and Li, X-C. 2000. Oligomeric proanthocyanidins: naturally occurring O-heterocycles. *Nat. Prod. Rep.* 17: 193–212.

Ferreira, D. and Brandt, E. V. 1989. New NMR experiments applicable to structure and conformation analysis. In *Chemistry and Significance of Condensed Tannins*, edited by Hemingway, R. W., and Karachesy, J. J. Plenum Press, New York, pp. 153–173.

Fiddler, W., Parker, W. E., Wasserman, A. E. and Doerr, R. C. 1967. Thermal decomposition of ferulic acid. *J. Agric. Food Chem.* 15: 757–761.

Folin, O. and Denis, W. 1912. On phosphotungstic-phosphomolybdic compounds as color reagents. *J. Biol. Chem.* 12: 239–243.

Forsyth, W. G. C. 1955. Cacao polyphenolic substances. *Biochem. J.* 60:108–111.

Francis, F. J., Harborne, J. B., and Barker, W. G. 1966. Anthocyanins in the low bush blueberry, Vaccinium angustifulium. *J. Food Sci.* 31: 583–587.

Francis, F. J. 1967. Criteria for identification of anthocyanins. *HortSci.* 2: 170–171.

Francis, F. J. 1993. Polyphenols as natural colorants. In *Polyphenolic Phenomena,* Scalbert, A. Ed., Institut National de la Rocherche Agronomique. Paris, France, pp. 209–220.

Franke, A. A., Hankin, J.H., Yu, M. C., Maskarinec, G., Low, S.-H., and Custer, L.J. 1999. Isoflavone levels in soy foods consumed by multiethnic populations in Singapore and Hawaii. *J. Agric. Food Chem.* 47: 977–986.

Friedman, M., Kozukue, N., and Harden, L. A. 2000. Cinnamaldehyde content of foods determined by gas chromatography-mass spectrometry. *J. Agric. Food Chem.* 48: 5702–5709.

Fulcrand, H., Remy, S., Souquet, J.-M., Cheynier, V., and Moutounet, M. 1999. Study of wine tannin oligomers by on-line liquid chromatography electrospray ionization mass spectrophotometry. *J. Agric. Food Chem.* 47: 1023–1028.

Fuleki, T. and Francis, F. J. 1968a. Quantitative methods for anthocyanins. 1. Extraction and determination of total anthocyanin in cranberries. *J. Food Sci.* 33: 72–77.

Fuleki, T. and Francis, F. J. 1968b. Quantitative methods for anthocyanins. 2. Determination of total anthocyanin and degradation index for cranberry juice. *J. Food Sci.* 33: 78–83.

Fuleki, T. and Francis, F. J. 1968c. Quantitative methods for anthocyanins. 3. Purification of cranberry anthocyanins. *J. Food Sci.* 33:266–274.

Gennaro, L., Leonardi, C., Esposito, F., Salucci, M., Maiani, G., Quaglia, G., and Fogliano, V. 2002. Flavonoid and carbohydrate contents in Tropea red onions: effects of homelike peeling and storage. *J. Agric. Food Chem.* 50: 1904–1910.

Gerothanassis, I. P., Exarchou, V., Lagouri, V., Troganis, A., Tsimidou, M., and Boskou, D. 1998. Methodology for identification of phenolic acids in complex phenolic mixtures by high-resolution two-dimensional nuclear magnetic resonance. Application to methanolic extracts of two oregano species. *J. Agric. Food Chem.* 46: 4185–4192.

Goldstein, J. L. and Swain, T. 1963b. The quantitative analysis of phenolic compounds. In *Methods of Polyphenol Chemistry*, edited by Pridham, J.B., Macmillan, New York, pp. 131–146.

Goldstein, J. L. and Swain, T. 1963a. Changes in tannins in ripening fruits. *Phytochemistry* 2: 371–383.

Goldstein, J. L. and Swain, T. 1965. The inhibition of enzymes by tannins. *Phytochemistry* 4: 185–192.

Gupta, R. K. and Haslam, E. 1980. Vegetable tannins: structure and biosynthesis. In *Polyphenols in Cereals and Legumes,* edited by Hulse, J.H., International Development Research Center, Ottawa, Canada, pp. 15–24.

Guvindarajan, V. S. and Mathew, A. G. 1965. Anthocyanidins from leucoanthocyanidins. *Phytochemistry* 4: 985–988.

Guyot, S., Marnet, N., Drilleau, J.-F. 2001. Thiolysis-HPLC characterization of apple procyanidins covering large range of polymerization states. *J. Agric. Food Chem.* 49: 14–20.

Haddock, E. A., Gupta, R. K., Al-Shafi, S. M. K., Layden, K., Haslam, E. and Magnaloto, D. 1982. The metabolism of gallic acid and hexahydroxydiphenic acid in plants: biogenetic and molecular taxonomic considerations. *Phytochemistry* 21: 1049–1051.

Hagerman, A. E., Zhao, Y. and Johnson, S. 1997. Methods for determination of condensed and hydrolyzable tannins. In *Antinutrients and Phytochemicals in Food*, edited by Shahidi, F., ACS Symposium Series 662, American Chemical Society, Washington, D.C., pp. 209–222.

Hagerman, A. E. 1987. Radial diffusion method for determining tannins in plant extracts. *J. Chem. Ecol.* 13: 437–449.

Hagerman, A. E. and Butler, L. G. 1980. Determination of protein in tannin–protein precipitates. *J. Agric. Food Chem.* 28: 944–947.

Hagerman, A. E. and Butler, L. G. 1978. Protein precipitation method for the quantitative determination of tannin. *J. Agric. Food Chem.* 26: 809–812.

Hagerman, A. E. and Robbins, C. T. 1987. Implications of soluble tannin–protein complexes for tannin analysis and plant defense mechanisms. *J. Chem. Ecol.* 13: 1243–1259.

Hagerman, A. E. and Butler, L. G. 1981. The specificity of proanthocyanidin-protein interaction. *J. Biol. Chem.* 256: 4494–4497.

Hahn, D. H., Rooney, L., and Earp, C. F. 1984. Tannin and phenols in sorghum. *Cereal Foods World* 29: 776–779.

Hall, C. A., Zhu, A. and Zeece, M. G. 1994. Comparison between capillary electrophoresis and high-performance liquid chromatography separation of food grade antioxidants. *J. Agric. Food Chem.* 42: 919–921.

Hammerstone, J. F., Lazarus, S. A., Mitchell, A. E., Rucker, R., and Schmitz, H. H. 1999. Identification of procyanidins in cocoa (Theobroma cacao) and chocolate using high-performance liquid chromatography/mass spectrometry. *J. Agric. Food Chem.* 47: 490–496.

Harborne, J. B. and Williams, C. A. 2001. Anthocyanins and flavonoids. *Nat. Prod. Rep.* 18: 310–333.

Harborne, J. B. 1967. *Comparative Biochemistry of the Flavonoids.* Academic Press, London.

Harborne, J. B. 1958. The chromatographic identification of anthocyanin pigments. *J. Chromatogr.* 1: 473–488.

Harinder, P., Makkar, S., Gamble, G., and Becker, B. 1999. Limitation of the butanol-hydrochloric acid-iron assay for bound condensed tannins. *Food Chem.* 66: 129–133.

Hartzfeld, P. W., Forkner, R., Hunter, M. D., and Hagerman, A. E. 2002. Determination of hydrolyzable tannins (gallotannins and ellagitannins) after reaction with potassium iodate. *J. Agric. Food Sci.* 50: 1785–1790.

Haslam, E. 1965. Galloyl esters in the Aceraceae. *Phytochemistry* 4: 495–498.

Haslam, E. 1966. *Chemistry of Vegetable Tannins*, Academic Press, New York, pp. 14–30.

Hemes, P. J. and Hedges, J. J. 2000. Determination of condensed tannin monomers in environmental samples by capillary gas chromatography of acid depolymerization extracts. *Anal. Chem.* 72: 5115–5124.

Hemingway, R. W. 1989. Structural variations in proanthocyanidins and their derivatives, In *Chemistry and Significance of Condensed Tannins*, edited by Hemingway, R. W. and Karchesy, J. J., Plenum Press, New York, NY, pp. 83–98.

Hemingway, R. W. and Karchesy, J. J., Eds. 1989. *Chemistry and Significance of Condensed Tannins.* Plenum Press, New York, NY, pp. 153–173.

Hemingway, R. W., Ohara, S., Steynberg, J. P., Brandt, E. V. and Ferrera, D. 1992. C-H HETCOR NMR studies of proanthocyanidins and their derivatives. In *Plant Polyphenols: Synthesis, Properties, Significance,* edited by Hemingway, R. H. and Laks, P. E., Plenum Press, New York, pp. 321–338.

Hempel, J. and Bohm, H. 1996. Quality and quantity of prevailing flavonoid glycosides of yellow and green French beans (Phaseolus vulgaris L.). *J. Agric. Food Chem.* 44: 2114–2116.

Hoff, J. F. and Singleton, K. I. 1977. A method for determination of tannin in foods by means of immobilized enzymes. *J. Food Sci.* 42: 1566–1569.

Hong, V. and Wrolstad, R. E. 1990. Characterization of anthocyanin-containing colorants and fruit juices by HPLC/photodiode array detector. *J. Agric. Food Chem.* 38: 698–708.

Hostettmann, K. and Hostettman, M. 1982. Isolation techniques for flavonoids. In *The Flavonoids: Advances in Research,* edited by Harborne, J. B. and Mabry, T. J., Chapman and Hall, New York, pp. 1–18.

Hutabarat, L. S., Greenfield, H., and Mulholland, M. 2001. Isoflavones and coumestrol in soybeans and soybean products from Australia and Indonesia. *J. Food Compos. Anal.* 14: 43–58.

Inoue, K. H. and Hagerman, A. E. 1988. Determination of gallotannin with rhodanine. *Anal. Biochem.* 169: 363–369.

Ismail, F. and Eskin, N. A. M. 1979. A new quantitative procedure for determination of sinapine. *J. Agric. Food Chem.* 27: 917–918.

Ittah, Y. 1991. Titration of tannin via alkaline phosphatase. *Anal. Biochem.* 192: 277–280.

Jackman, R. L., Yada, R. Y., and Tung, M. A. 1987. A review: separation and chemical properties of anthocyanins used for their qualitative and quantitative analysis. *J. Food Biochem.* 11: 279–308.

Jaworski, A. W. and Lee, C. Y. 1987. Fractionation and HPLC determination of grape phenolics. *J. Agric. Food Chem.* 35: 257–259.

Johnsson, P., Kamal-Eldin, A., Lundgren, L. N., and Aman, P. 2000. HPLC method for analysis of secoisolariciresinol diglucoside in flaxseeds. *J. Agric. Food Chem.* 48: 5216–5219.

Jurd, L. 1957. The detection of aromatic acids in plant extracts by ultra-violet absorption spectra in their ions. *Arch. Biochem. Biophys.* 66: 284–288.

Jurenitsch, J. and Leinmuller, R. 1980. Quantification of nonylic acid vanillylamide and other capsaicinoids in the pungent principle of Capsicum fruits and preparations by gas-liquid chromatography on glass capillary columns. *J. Chromatogr.* 189: 389–397.

Jurenitsch, J., David, M., Heresch, F., and Kubelka, W. 1979. Detection and identification of new pungent compounds in fruits of Capsicum. *Planta Med.* 36: 61–67.

Karchesy, J. J., Bae, Y., Chalker-Scott, L., Helm, R. F., and Foo, L. Y. 1989. Chromatography of proanthocyanidins. In *Chemistry and Significance of Condensed Tannins,* edited by Hemingway, R. W. and Karchesy, J. J., Plenum Press, New York, pp. 139–152.

Karchesy, J. J. and Hemingway, R. W. 1986. Condensed tannins: (4 (7) linked procyanidins in Arachis hypogea L. *J. Agric. Food Chem.* 34: 966–970.

Khanbabaee, K. and Van Ree, T. 2001. Tannins: classification and definition. *Nat. Prod. Rep.* 18: 641–649.

Khanna, S. K., Viswanatham, P. N., Krishnan, P. S., and Sanwai, G. G. 1968. Extraction of total phenolics in the presence of reducing agents. *Phytochemistry* 7: 1513–1517.

Kinoshita, E., Ozawa, Y., and Aishima, T. 1997. Novel tartaric acid isoflavone derivatives that play key roles in differentiating Japanese soy sauces. *J. Agric. Food Chem.* 45: 3753–3759.

Kolodziej, H. 1992. 1H NMR spectral studies of procyanidins derivatives: diagnostic 1H NMR parameters applicable to the structural elucidation of oligomeric procyanidins. In *Plant Polyphenols: Synthesis, Properties, Significance,* edited by Hemingway, R. H. and Laks, P. E., Plenum Press, New York, pp. 295–320.

Kosuge, S. and Furuta, M. 1970. Studies on the pungent compounds in fruits of Capsicum. Part XIV. Chemical constitution of the pungent principle. *Agric. Biol. Chem.* 34: 248–256.

Kozlowska, H., Rotkiewicz, D. A., Zadernowski, R., and Sosulski, F. W. 1983. Phenolic acids in rapeseed and mustard. *J. Am. Oil Chem. Soc.* 60: 1119–1123.

Kramer, R. 1998. *Chemometric Techniques for Quantitative Analysis,* Marcel Dekker, New York.

Kreft, S., Knapp, M., and Kreft, I. 1999. Extraction of rutin from buckwheat (Fagopyrum esculentum Moench) seeds and determination by capillary electrophoresis. *J. Agric. Food Chem.* 47: 4649–4662.

Krygier, K., Sosulski, F., and Hogge, L. 1982. Free, esterified and bound phenolic acids. 1. Extraction and purification. *J. Agric. Food Chem.* 30: 330–333.

Labarbe, B., Cheynier, V., Brossaud, F., Souquet, J.-M., and Moutounet, M. 1999. Quantitative fractionation of grape proanthocyanidins according to their degree of polymerization. *J. Agric. Food Chem.* 47: 2719–2723.

Lea, A. G. H. and Timberlake, C. F. 1974. The phenolics of cider. I. Procyanidins. *J. Sci. Food Agric.* 25: 471–477.

Lee, H. S. 2002. Characterization of major anthocyanins and the color of red-fleshed budd blood orange (Citrus sinensis). *J. Agric. Food Chem.* 50: 1243–1246.

Leung, J., Fenton, T. W., Mueller, M. M., and Clandinin, D. R. 1979. Condensed tannins of rapeseed meal. *J. Food Sci.* 44: 1313–1316.

Li, Y.-G., Tanner, G., and Larkin, P. 1996. The DMACA-HCl protocol and the threshold proanthocyanidin content for bloat safety in forage legumes. *J. Sci. Food Agric.* 70: 89–101.

Liggins, J., Bluck, L. J. C., Coward, A., and Bingham, S. A. 1998. Extraction and quantification of daidzein and genistein in food. *Anal. Biochem.* 264: 1–7.

Limiroli, R., Consonni, R., Ranalli, A., Bianchi, G., and Zetta, L. 1996. 1H NMR study of phenolics in the vegetation water of three cultivars of Olea europea. Similarities and differences. *J. Agric. Food Chem.* 44: 2040–2048.

Little, A. C. 1977. Colorimetry of anthocyanin pigmented products: changes in pigment composition with time. *J. Food Sci.* 42: 1570–1574.

Lunte, C. E., Wheeler, J. F., and Heineman, W. R. 1988a. Determination of selected phenolic acids in beer extracts by liquid chromatography with voltammetric-amperometric detection. *Analyst.* 113: 94–95.

Mabry, T. J., Markham, K. R., and Thomas, M. B. 1970. *The Systematic Identification of Flavonoids,* Springer-Verlag, New York.

Macheix, J.-J., Fleuriet, A., and Billot, J. 1990. *Fruit Phenolics,* CRC Press Inc., Boca Raton, FL.

Maga, J. A. 1978. Simple phenol and phenolic compounds in food flavor. *CRC Crit. Rev. Food Sci. Nutr.* 10: 323–348.

Mahler, S., Edwards, P. A., and Chisholm, M. G. 1988. HPLC identification of phenols in Vidal Blanc wine using electrochemical detection. *J. Agric. Food Chem.* 36: 946–951.

Makkar, H. P. S., Dawra, R. K., and Singh, B. 1987. Protein precipitation assay for quantification of tannins: determination of protein in tannin–protein complexes. *Anal. Biochem.* 166: 435–439.

Makkar, H. P. S., Dawra, R. K., and Singh, B. 1988. Determination of both tannin and protein in a tannin–protein complex. *J. Agric. Food Chem.* 36: 523–525.

Makkar, H. P. S. and Singh, B. 1991. Distribution of condensed tannins (proanthocyanidins) in various fibre fractions in young and mature leaves of some oak species. *Anim. Feed Sci. Technol.* 32: 253–260.

Makkar, H. P. S., Bluemmel, M., and Becker, K. 1997. In vitro rumen apparent and true digestibilities of tannin-rich forages. *Anim. Feed Sci. Technol.* 67: 245–251.

Makkar, H. P. S. 1989. Protein precipitation methods for quantification of tannins: a review. *J. Agric. Food Chem.* 37: 1197–1202.

Mangas, J. J., Rodriguez, R., Suarez, B., Picinelli, A., and Dapena, E. 1999, Study of the phenolic profile of cider apple cultivars at maturity by multivariate techniques. *J. Agric. Food Chem.* 47: 4046–4052.

Marks, D., Glyphis, J., and Leighton, M. 1987. Measurement of protein in tannin–protein precipitation using ninhydrin. *J. Sci. Food Agric.* 38: 55–61.

Martin, J. S. and Martin, M. M. 1983. Tannin essays in ecological studies: precipitation of ribulose-1,5-bisphosphate carboxylase/oxygenase by tannic acid, quebracho, and oak foliage extracts. *J. Chem. Ecol.* 9: 285–294.

Matejovicova, M., Mubagwa, K., and Flameng, W. 1997. Effect of vanadate on protein determination by the Coomassie brillant blue microassay procedure. *Anal. Biochem.* 245: 252–254.

Mateus, N., Silva, A. M. S., Vercauteren, J., and de Freitas, V. 2001. Occurrence of anthocyanin-derived pigments in red wines. *J. Agric. Food Chem.* 49: 4836–4840.

Mattila, P., Astola, J., and Kumpulainen, J. 2000. Determination of flavonoids in plant material by HPLC with diode-array and electro-array detections. *J. Agric. Food Chem.* 48: 5834–5841.

Maxson, E. D. and Rooney, L. W. 1972. Evaluation of methods for tannin analysis in sorghum grain. *Cereal Chem.* 49: 719–729.

Mazza, G. and Miniati, E. 1994. *Anthocyanins in Fruits, Vegetables and Grains.* CRC Press, Boca Raton, FL.

Mazza, G., Fukumoto, L., Delaquis, P., Girard, B., and Ewert, B. 1999. Anthocyanins, phenolics and color of Cabernet Franc, Merlot, and Pinot Noir wines from British Columbia. *J. Agric. Food Chem.* 47: 4009–4017.

McMurrough, I. and McDowell, J. 1978. Chromatographic separation and automated analysis of flavanols. *Anal. Biochem.* 91: 92–100.

Meirelles, C., Sarni, F., Ricardo da Silva, J. M., and Moutounet, M. 1992. Evaluation des pro-cyanidines galloylees dans les vins rouges issue de differents modes de vinification. In *Proceedings of the International Polyphenolic Group Convention,* Lisboa, Vol. 16 (II), pp. 175–178.

Mejbaum-Katzenellenbogen, W. and Kudrewicz-Hubicka, Z. 1966. Application of urea, ferric ammonium sulfate and casein for determination of tanning substances in plants. *Acta Biochim. Polonica* 13(1): 57.

Merken, H. M. and Beecher, G. R. 2000. Measurements of food flavonoids by high-perfor-mance liquid chromatography: a review. *J. Agric. Food Chem.* 48: 577–599.

Merrit, M. C. and Proctor, B. E. 1959. Effect of temperature during the roasting cycle on selected components of different types of whole bean coffee. *Food Res.* 24: 672–680.

Michaud, M. J. and Margail, M. A. 1977. Etude analitique des tanins catechiques. I. Les oligo-meres flavanoliques de l' Actinidia chinesis Planchon. *Bull. Soc. Pharm.* Bordeaux 116: 52–64.

Miean, K. H. and Mohamed, S. 2001. Flavonoid (myrecitin, quercetin, kaempferol, luteoilin, and apigenin) content in edible tropical plants. *J. Agric. Food Chem.* 49: 3106–3112.

Moane, S., Park, S., Lunte, C. E., and Smyth, M. R. 1998. Detection of phenolic acids in beverages by capillary electrophoresis with electrochemical detection. *Analyst* 123: 1931–1936.

Mole, S. and Waterman, P. 1987c. Tannic acid and proteolytic enzymes: enzyme inhibition or substrate deprivation? *Phytochemistry* 26: 99–102.

Monedero, L., Olalla, M., Martin-Lagos, F., Lopez, H., and Lopez. M. C. 1999. Application of chemometric techniques in obtaining macerates with phenolic compound content sim-ilar to that of wines from the Jerez-Shery region subjected to oxidative aging. *J. Agric. Food Chem.* 47: 1836–1844.

Moore, A. B., Francis, F. J., and Jason, M. E. 1982b. Acylated anthocyanins in red onions. *J. Food Protect.* 45: 590–593.

Moore, A. B., Francis, F. J., and Clydesdale, F. M. 1982a. Changes in the chromatographic profile of anthocyanins of red onion during extraction. *J. Food Protect.* 45: 738–743.

Moores, R. G., McDermott, D. L., and Wood, T. R. 1948. Determination of chlorogenic acid in coffee. *Anal. Chem.* 20: 620–624.

Mosel, H. D. and Herrmann, K. 1974b. Die phenolischen Inhalsstoffe der Obstes. IV. Die phenolischen Inhalsstoffe der Brombeeren und Himbeeren und deren Veranderun gen wahrend Wachstum und Reife der Fruchte. *Z. Lebensm. Unters. Forsch.* 154: 324–328.

Mosel, H. D. and Herrmann, K. 1974a. The phenolics of fruits. III. The contents of catechins and hydroxycinnamic acids in pome and stone fruits. *Z. Lebensm. Unters. Forsch.* 154: 6–10.

Moskowitz, A. H. and Hrazdina, G. 1981. Vacuolar contents of fruit subepidermal cells from Vitis species. *Plant Physiol.* 68: 686–692.

Murphy, P. A., Song, T. T., Buseman, G., and Barua, K. 1997. Isoflavones in soy-based infant formula. *J. Agric. Food Chem.* 45: 4635–4638.

Murray, R. D. H., Mendez, J. and Brown, S.A. 1982. *The Natural Coumarins: Occurrence, Chemistry and Biochemistry.* John Wiley-Interscience, Chichester, UK.

Naczk, M., Wanasundara, P. K. J. P. D., and Shahidi, F. 1992b. A facile spectrophotometric quantification method of sinapic acid in hexane-extracted and methanol-ammonia-water treated mustard and rapeseed meals. *J. Agric. Food Chem.* 40: 444–448.

Naczk, M., and Shahidi, F. 1989. The effect of methanol-ammonia-water treatment on the content of phenolic acids of canola. *Food Chem.* 31: 159–164.

Naczk, M., Shahidi, F., and Sullivan, A. 1992a. Recovery of rapeseed tannins by various solvent systems. *Food Chem.* 45: 51–54.

Naczk, M., Amarowicz, R., Pink, D., and Shahidi, F. 2000. Insoluble condensed tannins of canola/rapeseed. *J. Agric. Food Chem.* 48: 1758–1762.

Naczk, M., Amarowicz, R., Zadernowski, R., and Shahidi, F. 2001b. Protein precipitating capacity of condensed tannins of beach pea, canola hulls, evening primrose and faba bean. *Food Chem.* 73: 467–471.

Naczk, M., Amarowicz, R., Zadernowski, R., and Shahidi, F. 2001a. Protein-precipitating capacity of crude condensed tannins of canola and rapeseed hulls. *J. Am. Oil Chem. Soc.* 78: 1173–1178.

Naczk, M. and Shahidi, F. 1991. Critical evaluation of quantification methods of rapeseed tannins. Presented at 8th International Rapeseed Congress, Saskatoon, Canada, July 9–11, 1991.

Newman, R. H. and Porter, L. J. 1992. Solid state 13C-NMR studies on condensed tannins. In *Plant Polyphenols: Synthesis, Properties, Significance,* edited by Hemingway, R. H. and Laks, P. E., Plenum Press, New York, pp. 339–348.

Nonaka, G., Morimoto, S., and Nishioka, I. 1983. Tannins and related compounds. Part 13. Isolation and structures of trimeric, tetrameric, and pentameric proanthocyanidins from cinnamon. *J. Chem. Soc. Perkin Trans.* 1, 2139–2146.

Ogawa, K., Kawasaki, A., Yoshida, T., Nesumi, H., Nakano, M., Ikoma, Y., and Yano, M. 2000. Evaluation of auraptene content in citrus fruits and their products. *J. Agric. Food Chem.* 48: 1763–1769.

Oh, H. I., Hoff, J. E., and Haff, L. A. 1985. Immobilized condensed tannins and their interaction with proteins. *J. Food Sci.* 50: 1652–1654.

Oh, H. I., Hoff, J. E., Armstrong, G. S., and Haff, L. A. 1980. Hydrophobic interaction in tannin–protein complexes. *J. Agric. Food Chem.* 28: 394–398.

Oomah, B. D. and Mazza, G. 1996. Flavonoids and antioxidative activities in buckwheat. *J. Agric. Food Chem.* 44: 1746–1750.

Owades, J. L., Rubin, G., and Brenner, M. W. 1958b. Determination of tannins in beer and brewing materials by ultraviolet spectroscopy. *Proc. Am. Soc. Brew. Chem.* 6: 66–73.

Owades, J. L., Rubin, G., and Brenner, M. W. 1958a. Food tannins measurement, determination of food tannins by ultraviolet spectrophotometry. *J. Agric. Food Chem.* 6: 44–46.

Ozawa, T., Lilley, T. H., and Haslam, E. 1987. Polyphenol interactions: astringency and the loss of astringency in ripening fruit. *Phytochemistry* 26: 2937–2942.

Pan, Y., Zhang, L., and Chen, G. 2001. Separation and determination of protocatechuic aldehyde and protocatechuic acid in Salivia miltorrhrza by capillary electrophoresis with amperometric detection. *Analyst* 126: 1519–1523.

Peng, Z., Hayasaka, Y., Iland, P. G., Sefton, M., Hoj, P. and Waters, E. J. 2001. Quantitative analysis of polymeric procyandins from grape (Vitis vinifera) by reverse phase high-performance liquid chromatography. *J. Agric. Food Chem.* 49: 26–31.

Porter, L. J. 1989. Tannins. In *Methods in Plant Biochemistry,* edited by Harborne, J. B. Academic Press, San Diego, CA, vol. 1, pp. 389–420.

Porter, L. J., Hrtstich, L. N., and Chan, B. G. 1986. The conversion of procyanidins and prodelphinidins to cyanidins and delphinidins. *Phytochemistry* 25: 223–230.

Price, M. L. and Butler, L. G. 1977. Rapid visual estimation and spectrophotometric determination of tannin content of sorghum grain. *J. Agric. Food Chem.* 25: 1268–1273.

Price, M. L., Van Scoyoc, S., and Butler, L. G. 1978. A critical evaluation of the vanillin reaction as an assay for tannin in sorghum grain. *J. Agric. Food Chem.* 26: 1214–1218.

Prior, R. L., Lazarus, S. A., Cao, G., Muccitelli, H., and Hammerstone, J. F. 2001. Identification of procyanidins in blueberries and cranberries (Vaccinium Spp.) using high-performance liquid chromatography/mass spectrometry. *J. Agric. Food Chem.* 49: 1270–1276.

Pyysalo, T., Torkkewli, H., and Honkanen, E. 1977. The thermal decarboxylation of some substituted cinnamic acids. *Lebensm.-Wiss. u.-Technol.* 10: 145–149.

Raskin, J. 1992. Protein–polyphenol interactions: nutritional aspects. In *Proceedings of the 16th International Conference of Groupe Polyphenols.* Volume 16, Part II, pp. 11–18.

Ricardo da Silva, J. M., Rigaud, J., Cheynier, V, Cheminat, A., and Moutounet, M. 1991. Procyanidin dimers and trimers from grape seeds. *Phytochemistry* 4: 1259–1264.

Rinderknecht, H., Geokas, M. C., Silverman, P., and Haverback, B. J. 1968. A new ultrasensitive method for determination of proteolytic activity. *Clin. Chim. Acta* 21: 197–203.

Robards, K. and Antolovitch, M. 1997. Analytical chemistry of fruits bioflavonoids. *Analyst* 122: 11R–34R.

Sacchi, R., Patumi, M., Fontanazza, G., Barone, P., Fiordiponti, P., Mannina, L., Rossi, E., and Segre, A. L. 1996. A high-field 1H nuclear magnetic resonance study of minor components in virgin olive oils. *J. Am. Oil Chem. Soc.* 73: 747–757.

Salagoity-Auguste, M. H. and Bertrand, A. J. 1984. Wine phenolics. Analysis of low molecular weight components by high performance liquid chromatography. *J. Sci. Food Agric.* 35: 1241–1247.

Salunkhe, D. K., Chavan, J. K., and Kadam, S. S. 1989. *Dietary Tannins: Consequences and Remedies.* CRC Press, Boca Raton, FL.

Sanders, T. H., McMichael, R. W., and Hendrix, K. W. 2000. Occurrence of resveratrol in edible peanuts. *J. Agric. Food Chem.* 48: 1243–1246.

Sarkar, S. K. and Howarth, R. E. 1976. Specifity of vanillin test for flavonols. *J. Agric. Food Chem.* 24: 317–320.

Satterfield, M. and Brodbelt, J. S. 2000. Enhanced detection of flavonoids by metal complexation and electrospray ionization mass spectrometry. *Anal. Chem.* 72: 5898–5906.

Scalbert, A., Monties, B., and Janin, G. 1989. Tannins in woods: comparison of different estimation methods. *J. Agric. Food Chem.* 37: 1324–1329.

Scalbert, A. 1992. Quantitative methods for estimation of tannins in plant tissues. In *Plant Polyphenols: Synthesis, Properties, Significance,* edited by Hemingway, R.W. and Laks, P. S., Plenum Press, New York, pp. 259–280.

Schmidtlein, H. and Herrmann, K. 1975. On the phenolic acids of vegetables. IV. Hydroxycinnamic acids and hydroxybenzoic acids of brassica species and leaves of other cruciferae. *Z. Lebensm. Unters. Forsch.* 159: 139–141.

Schultz, J. C., Baldwin, I. T., and Nothnagle, P. J. 1981. Hemoglobin as a binding substrate in the quantitative analysis of plant tannins. *J. Agric. Food Chem.* 29: 823–829.

Schulz, H., Engelhardt, U. H., Wegent, A., Drews, H.-H., and Lapczynski, S. 1999. Application of near-infrared reflectance spectroscopy to the simultaneous prediction of alkaloids and phenolic substances in green tea leaves. *J. Agric. Food Chem.* 47: 5064–5067.

Schuster, B. and Herrmann, K. 1985. Hydroxybenzoic and hydroxycinnamic acid derivatives in soft fruits. *Phytochemistry* 24: 2761–2764.

Senter, S. D., Robertson, J. A., and Meredith, F. I. 1989. Phenolic compounds of the mesocarp of cresthaven peaches during storage and ripening. *J. Food Sci.* 54: 1259–1260 and 1268.

Shahidi, F. 2002. Antioxidants in plants and oleaginous seeds. In *Free Radicals in Food: Chemistry, Nutrition, and Health Effects.* edited by. M. J. Morello, F. Shahidi and C-T. Ho, ACS Symposium Series 807. American Chemical Society. Washington, D.C., pp. 162–175.

Shahidi, F. 2000. Antioxidants in food and food antioxidants. Nahrung 44: 158–163.

Shahidi, F. and Naczk, M. 1989. Effect of processing on the content of condensed tannins in rapeseed meals. A research note. *J. Food Sci.* 54: 1082–1083.

Shahidi, F. and Naczk, M. 1990. Contribution of sinapic acid to the phenolic constituents of solvent extracted cruciferae oilseeds. *Bulletin de Liason du Groupe Polyphenols* 15: 236–239.

Shahidi, F. and Naczk, M. 2004. *Phenolics in Food and Nutraceuticals*, CRC Press, Boca Raton, FL.

Shahidi, F., Chavan, U. D., Naczk, M., and Amarowicz, R. 2001. Nutrient distribution and phenolic antioxidants in air-classified fractions of beach pea (Lathyrus maritimus L.). *J. Agric. Food Chem.* 49: 926–933.

Siess, M.-H., Le Bon, A. M., Canivenc-Lavier, M.-C., Amiot, M.-J., Sabatier, S., Aubert, S. Y., and Suschetet, M. 1996. Flavonoids of honey and propolis: characterization and effects on hepatic drug-metabolizing enzymes and benzo[a]pyrene-DNA binding in rats. *J. Agric. Food Chem.* 44: 2297–301.

Silber, M. L., Davitt, B. B. Khairutdinov, R. F., and Hurst, J. K. 1998. A mathematical model describing tannin–protein association. *Anal. Biochem.* 263: 46–50.

Simonne, A. H., Smith, M., Weaver, D. B., Vail, T., Barnes, S., and Wei, G. I. 2000. Retention and changes of soybean isoflavones and carotenoids in immature soybean seeds (Edamame) during processing. *J. Agric. Food Chem.* 48: 6061–6069.

Sohr, M. and Herrmann, K. 1975a. Die phenolischen Inhaltsstoffe des Obstes. VI. Die phenolischen Inhaltsstoffe der Johannisbeeren, Stachelbeeren und Kulturheidelbeeren Veranderungen der Phenolsauren und Catechine wahrend Wachstum und Reife von Swarzen Johannisbeeren. *Z. Lebensm. Unters. Forsch.* 159: 31–33.

Sohr, M. and Herrmann, K. 1975b. The phenols of fruits. V. The phenols of strawberries and their changes during development and ripeness of fruits. *Z. Lebensm. Unters. Forsch.* 159: 341–343.

Somers, T. C. 1966. Wine tannins-isolation of condensed flavanoid pigments by gel filtration. *Nature* 209: 368–370.

Sondheimer, E. and Kertesz, Z. 1948. Anthocyanin pigments. Colorimetric determination in strawberries and strawberry preserves. *Anal. Chem.* 20: 245–248.

Souquet, J. M., Cheynier, V., Brossaud, F., and Moutounet, M. 1996. Polymeric proanthocyanidins from grape skins. *Phytochemistry* 43: 509–512.

Strumeyer, D. H. and Malin, M. J. 1975. Condensed tannins in grain sorghum: isolation, fractionation, and characterization. *J. Agric. Food Chem.* 23: 909–914.

Subra Rao, M. V. S. S. T. and Muralikrishna, G. 2002. Evaluation of antioxidant properties of free and bound phenolic acids from native and malted finger millet (Ragi, Elusine coracana Indaf-15). *J. Agric. Food Chem.* 50: 889–892.

Sun, B., Ricardo-da-Silva, J. M., Spranger, I. 1998a. Critical factors of vanillin assay for catechins and proanthocyanidins. *J. Agric. Food Chem.* 46: 4267–4274.

Sun, B. S., Leandro, M. C., Ricardo-da-Silva, J. M., and Spranger, M. I. 1998b. Separation of grape and wine proanthocyanidins according to their degree of polymerisation. *J. Agric. Food Chem.* 46: 1390–1396.

Swain, T. and Hillis, W. E. 1959. Phenolic constituents of Prunus domestica. I. Quantitative analysis of phenolic constituents. *J. Sci. Food Agric.* 10: 63–68.

Sweeley, C. C., Bentley, R., Makita, M., and Wells, W. W. 1963. Gas-liquid chromatography of trimethylsilyl derivatives of sugars and related substances. *J. Am. Chem. Soc.* 85: 2497–2507.

Tempel, A. S. 1982. Tannin-measuring techniques. A review. *J. Chem. Ecol.* 8: 1289–1298.

Terrill, T. H., Rowan, A. M., Douglas, G. B., and Barry, T. N. 1992. Determination of extractable and bound condensed tannin concentrations in forage plants, protein concentrate meals and cereal grains. *J. Sci. Food Agric.* 58: 321–329.

Thies, M. and Fischer, R. 1971. New reaction for microchemical detection and the quantitative determination of catechins. *Mikrochim. Acta* 1: 9–13.

Thomas, B. V., Schreiber, A. A., and Weisskopf, C. P. 1998. Simple method for quantitation of capsaicinoids in peppers using capillary gas chromatography. *J. Agric. Food Chem.* 46: 2655–2663.

Thompson, R. S., Jacques, D., Haslam, E., and Tanner, R. J. N. 1972. Plant proanthocyanidins. Part 1. Introduction: the isolation, structure, and distribution in nature of plant procyanidins. *J. Chem. Soc. Perkin Trans.* 1: 1387–1399.

Timberlake, C. F. and Bridle, B. 1967. Flavylium salts, anthocyanidins and anthocyanins. 1. Structural transformations in acid solutions. *J. Sci. Food Agric.* 18: 473–478.

Tomas-Barberan, F. A., Gil, M. I., Cremin, P., Waterhouse, A. L., Hess-Pierce, B., and Kader, A. L. 2001. HPLC-DAD-ESIMS analysis of phenolic compounds in nectarines, peaches, and plums. *J. Agric. Food Chem.* 49: 4748–4760.

Treutter, D. 1989. Chemical reaction detection of catechins and proanthocyanidins with 4dimethylcinnamaldehyde. *J. Chromatogr.* 467: 185–193.

Tzagoloff, A. 1968. Metabolism of sinapine in mustard plants. I. Degradation of sinapine into sinapic acid and choline. *Plant Physiol.* 38: 202–206.

van der Voort, F. R. 1992. Fourier-transform infrared spectroscopy applied in food analysis. *Food Res. Int.* 25: 397–403.

van Sumere, C. F. 1989. Phenols and phenolic acids. In *Methods in Plant Biochemistry.* Volume 1, Plant Phenolics. Harborne, J. B. ed., Academic Press, London, pp. 29–74.

Vasanthan, J. Y. T. and Temelli, F. 2001. Analysis of phenolic acids in barley by high-performance liquid chromatography. *J. Agric. Food Chem.* 49: 4352–4358.

Verzele, M., Delahaye, P., and Damme, F. V. 1986. Determination of the tanning capacity of tannic acids by high performance liquid chromatography. *J. Chromatogr.* 362: 363–374.

Vitrac, X., Castagnino, C., Waffo-Teguo, P., Deleunay, J.-C., Vercauteren, J., Monti, J.-P., Deffieux, G., and Merillon, J.-M. 2001. Polyphenols newly extracted in red wine from southwestern France by centrifugal partition chromatography. *J. Agric. Food Chem.* 49: 5934–5938.

Wang, J. and Sporns, P. 1999. Analysis of anthocyanins in red wine and fruit juice using MALDI-MS. *J. Agric. Food Chem.* 47: 2009–2015.

Wang, C., Sherrard, M., Pagala, S., Wixon, R., and Scott, R. A. 2000a. Isoflavone content among maturity group 0 to II soybeans. *J. Am. Oil Chem. Soc.* 77: 483–487.

Wang, L.-F., Kim, D.-M., and Lee, C. Y. 2000b. Effects of heat processing and storage on flavanols and sensory qualities of green tea beverages. *J. Agric. Food Chem.* 48: 4227–4232.

Waterman, P. G. and Mole, S. 1994. *Analysis of Plant Phenolic Metabolites*, Blackwell, Oxford, U.K.

Willis, R. B. and Allen, P. R. 1998. Improved method for measuring hydrolyzable tannins using potassium iodate. *Analyst* 123: 435–439.

Wilson, T. C. and Hagerman, A. E. 1990. Quantitative determination ellagic acid. *J. Agric. Food Chem.* 38: 1678–1683.

Woodring, P. J., Edwards, P. A., and Chisholm, M. G. 1990. HPLC determination of nonflavonoid phenols in Vidal Blanc wine using electrochemical detection. *J. Agric. Food Chem.* 38: 729–732.

Xu, Z. and Goldber, J. S. 1999. Purification and identification of components of -oryzanol in rice bran oil. *J. Agric. Food Chem.* 47: 2724–8.

Zadernowski, R. 1987. Studies on phenolic compounds of rapeseed flours. *Acta Acad. Agricult. Olst. Technologia Alimentarium*, Supplementum F 21: 1–55.

Zafrilla, P., Ferreres, F., and Tomas-Barberan, F. A. 2001. Effect of processing and storage on the antioxidant ellagic acid derivatives and flavonoids of red raspberry (Rubus idaeus) jams. *J. Agric. Food Chem.* 49: 3651–3655.

Zeece, M. 1992. Capillary electrophoresis: a new analytical tool for food science. *Trends Food Sci. Technol.* 3: 6–10.

12 Sensory Analysis of Foods

Kannapon Lopetcharat and Mina McDaniel

CONTENTS

12.1 SENSORY EVALUATION: IMPORTANCE AND DEFINITION

Now, at the dawn of the twenty-first century, sensory evaluation has become more important than ever: marketplace is consumer driven, and studying foods using machines and chemical reactions is not enough any more. Since the mid-twentieth century, industries and academia have embraced sensory evaluation as an invaluable tool for creating successful products and understanding the sensory properties of materials. Moreover, sensory evaluation lends itself beyond its original domain, human foods, to any materials (e.g., deodorant, fabric) that will be used by any conscious living beings (e.g., humans, cats, dogs), whose responses can be measured appropriately. In this chapter, however, we limit the discussion to human foods only.

Because people are different and influenced by factors such as cultures, experiences, and environments, measuring their responses with precision and accuracy is a difficult task. In addition to the measurement problems, inferring the measurements is another monumental task. Despite these problems, understanding how consumers perceive their products is a requisite for consumer product industries. Therefore, sensory science was established to help industry professionals create products that provide the sensory properties demanded by consumers.

Sensory science is a multidisciplinary science comprising psychophysics, statistics, and other sciences relating to the products of interest; sensory evaluation is a fruit of sensory science. Sensory evaluation comprises a set of methods to stimulate subjects, measure their responses, analyze the data, and interpret the results with minimum or no (if possible) potential biases from other factors besides the products of interest.

In this chapter, we introduce many well-established sensory methods, followed by three important fundamental aspects of sensory science: human physiology, psychophysics, and statistical analysis. At the end, we review some new technologies in gas chromatography–olfactometry and instruments attempting to mimic human chemical senses: electronic nose and electronic tongue.

12.2 CLASSIFICATIONS OF SENSORY METHODOLOGIES

Different problems require different solutions. The same applies to sensory problems. There are two general questions in sensory research: preference and difference questions; and these questions dictate what types of sensory methodologies should be used to solve problems. If the question is about how consumers prefer products, then affective tests should be used. On the other hand, if the question involves the differences between products regardless of consumer preferences, then the analytic sensory method is a right choice.

12.2.1 AFFECTIVE TEST

As the name implies, affective tests are based on how consumers feel about products. There are two classes of affective tests: overall and specific-attribute hedonic (liking) tests. Acceptance test, overall-liking test, and choice test are in the former category and just-about-right (JAR) and specific-attribute-liking tests are in the latter.

In much of the literature, acceptance test and overall-liking test are implied to be the same test.[1,2] However, the authors differentiate between tests because of the meaning of the words *"acceptance"* and *"overall liking."* An acceptance test measures acceptance levels, and researchers use an acceptance scale; an overall-liking test measures how much the consumer likes the product, and researchers use a 9-point hedonic scale.[3] In practice, an acceptance test is useful when a product is new, and there are no data relating overall-liking status with acceptance levels of the product. For example, a brand new product is tested, and there is no relevant product to be compared with in a targeted market. In this scenario, a sensory scientist should run both acceptance and overall-liking tests.

In addition to the acceptance and overall-liking tests that rely on hedonic ratings, a choice test is an alternative that relies on hedonic discrimination and classification. In the choice test, consumers have to choose the most preferred product from a set of products given to them. This type of test serves the objectives of product improvement and parity with competitor products[2]; however, it does not provide the degree of liking of products. Therefore, researchers should know the prior hedonic status of at least one of the products tested. The examples of choice tests are pair preference, rank preference, or multiple paired preferences (all pairs or selected pairs).[2]

In addition to overall hedonic tests, specific-attribute hedonic tests are useful when the questions of interest are about consumer affection toward certain sets of key attributes. There are two general questions asked in the specific-attribute hedonic tests: JAR and specific-attribute-liking tests. The former employs the JAR scale, which comprises five categories: much too low, somewhat too low, just about right, somewhat too much, and much too much. The specific-attribute-liking test employs a 9-point hedonic scale[3] that comprises nine categories: like extremely, like very much, like moderately, like slightly, neither like nor dislike, dislike slightly, dislike moderately, dislike very much, and dislike extremely.

Using the JAR scale is a delicate task because of its categorical responses and its attributes dependency. The distribution of data obtained from the JAR scale needs careful examination before researchers perform any statistical analysis. Moreover, not all attributes have an optimum hedonic level when their strength increases (e.g., sweetness, saltiness). Chocolate flavor in ice cream is a good example. Increasing chocolate flavor in chocolate ice cream may never elicit any answers except "just about right" for two possible reasons. First, increasing chocolate flavor to too much intensity is impossible to achieve in practice. Second, for some people, there is no optimum level for chocolate flavor; therefore, they will always give a "just about right" answer because they like the flavor very much regardless of the chocolate-flavor strength. Another example is bitterness; bitterness is not a pleasant attribute in most food products, and therefore, consumers will give "somewhat too much" or "much too much" responses all the time, regardless of actual bitterness strengths.

Although a 9-point hedonic scale provides categorical response, the relationship from one category to the next category was proved to be equal.[3] Because of this, researchers can assume that the response obtained from the scale is an interval one; and, as a result, standard parametric statistical analyses (e.g., *t*-test, analysis of variance [ANOVA]) can be used. The explanation of the construction of the 9-point hedonic scale is found elsewhere.[4]

12.3   SENSORY EVALUATION AS ANALYTICAL TOOL

Researchers use analytical sensory evaluation when the questions of interest regard only the differences in sensory properties between products. Therefore, information obtained from the method should be treated as if it were a response from instrumental measurements. There are two analytical sensory methods: difference test and descriptive analysis.

12.3.1   DIFFERENCE TEST

Difference test is a set of sensory tools established on the basis of classical psychophysics, a class of psychophysics relating external stimuli to internal sensation.[5] The test sometimes is known as a discrimination test[4;] however, theoretically speaking, the discrimination test is a subset of the difference test.[6] Discrimination test is a task in difference tests that involve two stimuli on a perceptual dimension, mainly overall difference and two responses (yes/no). Moreover, if one of the stimuli is a "null" stimulus, the task is a detection task; and if the objective of the task is to measure subject recognition memory, the task is a recognition one.[6] Besides the discrimination test, there are two more tasks in difference test: identification and classification. The number of stimuli and categories involved in identification task are equal; a classification test involves fewer categories than the number of stimuli.[6]

On the other hand, the number of perceptual dimensions involved in the decision process is another criterion for classifying difference tests. If an overall perceptual dimension is examined, it is called an overall difference test; however, if a specific perceptual dimension is examined, it is called alternative force-choice (AFC) test. The number of stimuli and response categories dictate the name of force-choice tests: the 2-AFC test uses two stimuli and asks for the difference between the stimuli on a specific attribute (a perceptual dimension)—for example, sweetness.

At this point, readers may be confused about choice test and difference test. These two methods are mathematically identical; however, the differences lie in the subjects who participate in the tests and the objectives of the tests. If consumers are the subjects of the tests, a choice test is used. The objective of a choice test is to prove whether the differences between products are perceivable by consumers. Analytic difference tests, on the other hand, involve trained, experienced, or methodology-oriented subjects; therefore, researchers cannot infer the conclusion from this test to consumers. The objectives of the test are, for example, to screen prototypes, provide guidance to other research programs, and screen panelists for descriptive analysis.[4]

12.3.2 Descriptive Analysis

The most sophisticated technique in sensory evaluation is descriptive analysis.[4] The technique provides information about the qualities that differentiate products and how great the differences are. Descriptive analysis consists of eight general steps: (1) recruiting panelists, (2) developing descriptive terms, (3) training the panelists, (4) evaluating their performance, (5) conducting a good test, (6) analyzing data, (7) interpreting results, and (8) communicating the outcome. Different descriptive methods focus on different aspects mentioned, and these differences determine the advantages and disadvantages of the methods. This section will introduce readers to several established, descriptive techniques.

12.3.2.1 Flavor Profile

Flavor Profile® (FT) is a qualitative technique developed originally in the early 1950s. By the combination of intensive training for a specific product category and consensus profiling of products by using well-defined terminologies, the method generates reproducible consensus profiles of products. However, the major disadvantage of FT is its qualitative results. However, the introduction of a numerical scale allows FT to generate quantitative results, and the quantitative version of FT is called Profile Attribute Analysis.[4,7]

12.3.2.2 Quantitative Descriptive Analysis

Realizing the disadvantages of FT, pioneer sensory scientists Dr. Herbert Stone, Dr. Joel Sidel, and their colleagues at the University of California at Davis developed Quantitative Descriptive Analysis® (QDA).[8] Distinct differences between QDA and FT include the role of panel leader, scale usage, language development, and result presentation. The panel leader acts as a facilitator in a QDA panel but as a panelist in a FT panel. Moreover, using a linear graphic scale or visual analog scale with the line extended beyond its verbal endpoints is unique to QDA.[8,9] Based on functional measurement theory,[10–12] extending the line beyond its verbal endpoints and removing the center anchor remove variability in scale usage.[9]

In QDA, panelists individually generate and define terms on the basis of their perception of products; however, the final list of terminologies, their definitions, and evaluation procedures must be consensus results from the panelists. Panelist-derived terminology is an advantage of QDA because it is culturally relevant.[9] QDA avoids using reference products in training and uses them only when it is necessary. The results from QDA are statistically analyzed and presented in a spider plot format. The inventors of QDA noted that researchers should not make any inference about relationships between the intensities of terminologies used in QDA to any factors such as product formulas, because of the multidimensional characteristics of human perception, unless a well-defined experimental design is used.[9] Due to individual differences between panelists and short training period, relative differences between products tested in QDA are more important than absolute values.[4]

12.3.2.3   Texture Profile

In the 1960s, workers at General Mills developed the texture profile (TP) method to relate texture perceptions to physical parameters.[13–15] The philosophy of TP is quite different from that of QDA in the way of standardization of terminologies using several references that are physically defined by and conform to objective measurements. However, panelists in TP generate and agree upon the terminologies.

12.3.2.4   Spectrum Method™

Working closely with TP, Gail Civille developed Sensory Spectrum™, which inherits many characteristics of TP. In Sensory Spectrum, the panel leader functions as an instructor, not as a facilitator as in QDA. Moreover, panelists participating in the Spectrum Method do not generate terminologies; instead, the panel leader provides the terminologies. The list of terminologies or lexicon is standardized, and references used for each term are predefined similar to those used in TP. However, all the references do not need to be quantified by objective measurements as they are in TP. In the case of novel perceptions not listed in standard lexicons, panelists will generate terms, define them, and select references upon consensus; after several applications, the terms that survive will be added to the standard lexicons. Because of its nonproduct–specific terms and references, the Spectrum Method is applicable for any relevant product categories; however, the terms and the references may not be appropriate for cross-cultural studies because of language and cultural barriers.[9] Moreover, the terms are less likely to relate to consumer language than the terms used in QDA.[4,9] Nevertheless, the standard terms used in the Spectrum Method are more closely related to objective measurements and more tangible than those in QDA.[2]

The Spectrum Method does not limit the scale used in the method only to visual-analog (line with labels) scales: it permits the use of any scale that provides at least interval characteristics to the responses such as magnitude estimation and categories scales with 30 or more categories.[2] Because the Spectrum Method uses multiple standardized references to anchor the scale for each term and extensive training in scale usage (much more training time than other methods), its proponents claim that the results from the method are absolute.[4] However, Lawless and Heymann argued that the absolute properties of the response from the Spectrum Method is impossible in practice because of individual differences, unproven equi-interval properties of references used on scales, and universal scale strategy. Nevertheless, based on the response transformation of magnitude estimates,[16] universal scale strategies may be appropriate for certain sets of modalities as it has been proved by many works on cross-modality matching,[17–20] but others are still under scrutiny.

12.3.2.5   Free Choice Profiling

Because of cumbersome procedures in language development and training, conventional descriptive analysis is expensive and time consuming. British sensory scientists developed free choice profiling (FCP) in the 1980s.[21–26] To rate the intensities of attributes, this method requires only highly experienced panelists without any training in scaling technique. Moreover, the attributes are generated freely from each panelist independently.

The results generated from this method require a special statistical treatment, generalized procrustes analysis (GPA).[27] GPA fits individual data sets by forcing them into a consensus space.[27] A technical description of GPA is given in detail elsewhere.[4,27,28] A major disadvantage of FCP is the interpretation of results, and it is intimately involved with the experience of sensory scientists. In addition, attributes used by FCP panelists are not really discriminative and reproducible when the panelists are not really experienced with tested products.[29]

12.3.2.6   Flash Profiling

Flash profiling (FP) is a new descriptive analysis technique. FP inherits many aspects from FCP, such as freely generated terms, no requirement for training, and the use of GPA or similar programs. Therefore, it suffers from the same disadvantage as FCP: complicated results. However, there are significant differences in underlying philosophies between FCP and FP. FP panelists generate their own attributes; however, the attributes must be discriminative ones. Moreover, the discriminative attributes have to be the attributes that the panelists are able to use to rank products in the order of their intensities. The technique allows ties in the ranking task. Therefore, attributes that discriminate products into few groups will not be used in the task. Because of the ranking task, FP panelists will eventually drop redundant attributes and focus only on discriminative attributes. As a result, fewer attributes with more discriminative power are used in FP, which is faster than other methods, even FCP.

Because ranking requires panelists to see all the products, the fatigue effect is prominent in FP, although panelists can spend as much time as they need to finish the task. Some sensations have a long recovery time (e.g., astringency, bitterness, spiciness); thus, the application of FP is limited. Moreover, because of the application of comparative strategies, FP requires all the products to be tested at once; therefore, the conclusion is limited within a set of the products. In other words, the conclusion generated from FP may change if other products are added to the set of the products. On the other hand, other descriptive methods assume that the positions of products are fixed in perceptual space and, by adding more products, it will not change the configuration of the perceptual space because of rating techniques used in the methods.

According to the inventor of FP, this method cannot replace conventional descriptive analysis (CDA) techniques such as QDA or Spectrum, but it increases the efficacy of the language generation step in CDA by shortening time and producing nonredundant and discriminative terms. Panelists with some experience in CDA are more appropriate for FP than new panelists and specialists; moreover, FP is not suitable for consumer study because of the extremely confusing results.

12.4 PHYSIOLOGY OF SENSORY ORGANS

Sense receptors detect stimuli; sensation is produced and transduced through neural networks to the brain to produce perception. After perceiving the sensation, the brain intricately evaluates the sensation and decides to respond to the stimuli or not. Therefore, sense receptors are the gates connecting the external environment to the internal perceptual systems in humans. To stimulate sense receptors, the differences between internal and external potentials of receptor cell membranes must be altered to activate ionic flows that trigger neural transduction processes. Over the course of evolution, sense receptors in humans have evolved to respond to specific stimuli by forming organs that can filter stimuli based on their nature: photons, chemicals, and physical forces. Moreover, the receptors also developed specific detection mechanisms that respond to certain ranges of the filtered stimuli. Consequently, the stimuli react with the receptors to produce specific sensations.

Sensory analysis relies on the ability to detect, recognize, and discriminate stimuli. Therefore, understanding how the sense organs work is very critical for sensory scientists to the same degree as flavor chemists need to understand how gas chromatography works. Many publications organize their contents by sense modalities: vision, olfaction, gustation, hearing, tactile, and irritation. However, this section is organized by the nature of stimuli that stimulate the sense modalities.

12.4.1 PHOTOCHEMICAL SENSES

Visual system is a system devoted to detect photons—the quantum units of radiant electromagnetic energy. The amount of photons radiated from a light source specifies its intensity, and its psychological counterpart is brightness. Photons travel in a continuous cycle of waves, which can be described in units of wavelength, and the wavelength is responsible for the perception of color hues. The photon receptors of humans and most vertebrates have evolved to detect wavelengths of light ranging from about 380 to about 760 nm.

The human visual system contains a specific organ to filter and gather information encrypted in photons: the eyeballs. The sclera is the white layer covering the front of the eye (white of the eye), and there is a translucent layer in the middle of the front of the eye called the cornea. The cornea refracts light that enters the eye. The amount of light is regulated by the iris, a modified part of choroid layer that provides nutrients to the eye. The opening of the iris is called the pupil—a structure analogous to the diaphragm of a camera. The light that passes through the pupil is focused by a transparent and flexible lens. The cilliary muscles are attached to the lens by ligaments, the zonal fibers. The ligaments control the curvature of the lens during focusing.

The focused light falls on the retina in the back of the eyeball. The retina covers about 2008 of the inside of the eyeball. There are two types of photoreceptors in the retina: rods and cones. Both rods and cones contain a light-absorbing pigment called rhodopsin at different concentrations. Because rods contain higher concentrations of rhodopsin than cones, rods are more sensitive to light and are responsible for black-and-white perception; cones are responsible for color perception. At this stage, the information encrypted in photons is transduced or transformed into neural form by the photochemical mechanisms of rhodopsin.

However, the detection of light by rods and cones with different sensitivity does not give a meaning to visual perception, especially color perception. Color perception is a result of neural interpretation in the brain. There are three aspects of light, which are identified physically and psychologically. Wavelength, intensity, and spectral purity are the physical dimensions of light, and their psychological counterparts are hue, brightness, and saturation, respectively.

12.4.2 FREQUENCY SENSES

Odd but true: auditory, touch, and kinesthetic senses share the same basic principle—they detect and discriminate frequencies. The stimuli for auditory senses are sound frequencies, and the stimuli for touch and kinesthetic senses are mechanical frequencies. However, auditory senses are housed in a specific organ, the ear; touch and kinesthetic senses are not restricted to any specific organ.

12.4.2.1 Auditory System: Sound Frequency Detector

Two physical aspects of sound waves directly influence our hearing: frequency and amplitude. Frequency, which is inversely related to wavelength, has pitch as its psychological attribute. Young humans can hear sound frequencies ranging from 20 to 20,000 Hz. Amplitude is the amount of changes in sound pressure, and its psychological counterpart is loudness. Sound amplitude is physically measured as force per unit area, and its unit is decibel (dB). However, daily sound is not a single wave; on the contrary, it is a complex wave created by interaction between many sound waves.

Breaking down a complex wave into its basic components is known as a Fourier analysis, and the human auditory system is performing a crude Fourier analysis as stated in Ohm's acoustic law. The law states that the auditory system detects and discriminates basic frequencies composing a complex wave. Tonal quality or timbre is the experience of the harmonization of sound waves in a complex sound, and fundamental frequency (the lowest frequency of a complex sound) mainly determines the pitch of a complex sound.

Sound travels though air, bones, and muscles, which connect sound sources to the eardrum. The eardrum or tympanic membrane is a thin, translucent membrane stretched across the inner end of the auditory canal. It amplifies incoming sound waves, especially 3000 Hz waves. The eardrum vibrates when sound waves reach its surface. The energy and wave information of sound waves is mechanically conducted to the inner ear by three small bones: malleus (hammer), incus (anvil), and stapes (stirrup). The stapes footplate connects to the membrane of the oval window, the entrance to the inner ear. The vibration of the membrane is conducted through the watery fluid in a small tubular structure called the cochlea. In the cochlea is a basilar membrane that houses a special auditory receptor structure called the organ of Corti. The organ of Corti contains the inner hair cells and the outer hair cells. These hair cells bend when pressure vibration moves through the watery fluid and transduce the mechanical vibration into nerve impulse. They are the first stage of neural conduction in the auditory system. There are several theories that explain how the auditory system encodes sound information. The Place theory, frequency-matching mechanisms, and the volley principle are some of the examples; details regarding these theories whose details can be found elsewhere.[30]

12.4.2.2 Tactile and Kinesthetic Senses: Mechanoreceptors

When we touch and grasp an object, we feel complex blended sensations, such as oiliness, roughness, and smoothness, as well as the object's three-dimensional shape. The feelings are the result of tactile and kinesthetic sensations.

Mechanoreceptors innervating the skin, and sometimes, subcutaneous tissues are the end organ that transduces mechanical energy into neuronal signals. These receptors are responsible for touch feelings or tactile perception. Muscles, tendons, and joints contain mechanoreceptors that detect stress and strain information, and kinesthetic sensation is produced. Incorporating tactile and kinesthetic information allows us to perceive three-dimensional shape of objects. The perception of an object's geometrical properties and manipulation of force during holding or gasping is called haptic perception. The perception is the result of processing both tactile and kinesthetic information by the brain simultaneously; however, our understanding of the physiology and neurology of the haptic system is still limited. Therefore, this section focuses on the physiology of the tactile system more than that of the haptic system.

Skin can be classified as glabrous, hairy, or mucocutaneous. Each type has its complement of mechanoreceptors and internal structures. Glabrous skin is hairless skin such as the skin of the palms and lips; hairy skin is typified by the presence of hair such as that found on the arm, face, and the backs of hands. Mucocutaneous skin borders body orifices.

Epidermis and dermis compose the skin. Epidermis is the outer layer of the skin, and it contains no blood vessels or receptors. Stratum corneum is the outermost layer of epidermis, and it is composed of dead cell tissues. The dermis, on the other hand, contains connective tissues and ground substance, a semifluid, nonfibrillary, and amorphous mixture. Many organelles are embedded in the dermis, including sensory receptors. The regions where the dermis and epidermis converge are called dermal pegs, and the physical arrangement of these regions determines mechanoreceptor activations.

Skin is an incompressible viscoelastic medium, and it has measurable mechanical impedance because of its friction, elastic-restoring capabilities, and mass properties. On the other hand, it can be thought of as a water bed.[31,32] The physical properties of skin vary across skin type and its location in body. Many studies indicated a normal strain compressive deformation, either parallel or perpendicular to the skin surface, is an adequate stimulus for transduction in tactile receptors.[33–36]

When a stimulus contacts skin, it creates waves passing through the skin, and the strain is attenuated as the waves travel. Therefore, the strain is restricted to the close surrounding area of the stimulus.[37] Like any waves, waves generated by stimuli that interact with skin possess wave characteristics: frequency and amplitude. Mechanoreceptors in skin—either with or without accessory structures, the nonneural elements—detect and respond to specific ranges of wave frequency, and the specification is partially dependent on the accessory structures of mechanoreceptors.[38]

Based on accessory structures around the nerve endings of mechanoreceptors, the receptors are classified into three types: encapsulated endings (Pacinian corpuscle, Ruffini capsule, and Meissner corpuscles), attaching endings (Merkel cell–neurite complex, Iggo corpuscle, and circumferential and palisade fibers on the shafts of hairs), and endings without accessory structures (free nerve endings, C-mechanoreceptors).[37] The attaching endings are nerve endings that attach to their accessory structure, unlike encapsulated endings, which are encapsulated but not attached by their accessory structures.

Mechanoreceptors innervating the glabrous skin can also be classified based on their responses toward the frequencies of stimuli. There are four types of receptors: SAI, SAII, FAI or RA, and FAII or PC. SAI and SAII are slowly adapting (SA) receptors, and FAI and FAII are fast adapting (FA) receptors. The adaptation is defined as how the receptors respond to a sustained skin indentation.[37] SA receptors actively respond while skin is moving and during the period of sustained indentation; FA receptors respond only when skin is moving.[37] These receptors have different receptive fields and distribute differently; consequently, differences in sensitivity and acuity of different location on the body are observed.

As does glabrous skin, hairy skin possesses all the aforementioned mechanoreceptors, plus additional receptors: slowly conducting myelinated (A-gamma) fibers and rapidly conducting (A-beta) fibers.[37] The additional receptors are located in hair follicles and are sensitive to hair movement. Tactile perceived magnitude and quality are mediated by the stimulation and interaction between these mechanoreceptors.

12.4.3 CHEMICAL SENSES

Chemical senses detect and discriminate chemical signals. There are two specific organs devoted to chemical senses: the tongue and the olfactory epithelium. Both organs possess specific structures that are normally accessible by chemical stimuli. Receptors embedded mostly in the tongue are responsible for taste or gustation; those in the olfactory epithelium located on the upper-innermost section of the nasal cavity are responsible for smell or olfaction. Taste stimuli must be soluble in aqueous solution, and smell stimuli must be volatile and dissolve in aqueous solution.

12.4.3.1 Gustation

Taste receptors are in taste cells comprising taste buds, specialized onion-like structures located in tiny pits and grooves of the tongue and mouth, throat, pharynx, and soft palate. Taste buds cluster within papillae, small visible elevations on the tongue. There are four types of papillae: fungiform, foliate, circumvalate, and filiform. These papillae differ in their shape and location. The filiform papillae do not contain taste buds.

Each taste bud consists of 50–150 taste cells[30] and projects many microvilli into a taste pore. The microvilli directly contact stimuli such as acids, salts, and sugars. Taste cells live only several days but regenerate regularly. The axons of taste receptors innervate four different pairs of nerves: cranial nerve VII, cranial nerve IX, cranial X, and superficial petrosal nerves.[1] Any area on the tongue responds to any of four taste stimuli: sour, salt, sweet, and bitter stimuli; therefore, the theory of a tongue map is not accurate.[1] Umami taste has been proposed to be the fifth basic taste; however, it is still controversial.[30] Monosodium glutamate and other nucleic acid derivatives elicit the umami taste.

12.4.3.2 Olfaction

There are two olfactory organs exposed to the external environment: the nasal cavity and the olfactory mucosa. The nasal cavity acts as an air passage, thermal stabilizer, and agitator. The olfactory mucosa is visually classified into three layers: the olfactory epithelium, basal lamina, and lamina propria.

Olfactory epithelium is a layer of a bigger structure called the olfactory mucosa, covering the uppermost part of the nasal cavity. Basal lamina and lamina propria are the other layers of the olfactory mucosa. The olfactory mucosa in higher primates, including humans, is relatively small ($\sim$2–6 cm^2) compared with other mammal olfactory epitheliums.[39,40] The thickness of human olfactory epithelium is approximately 30–60 μm; and the entire human olfactory mucosa is approximately 480–500 μm thick, depending on the extent of the nasal venous blood flow.[40]

There are three major types of cells in the olfactory epithelium: supporting cells, neurosensory cells (also called sensory neurons), and basal cells.[39,40] Free nerve endings of the trigeminal system are also found in the olfactory epithelium.[40] The nuclei of these three cellular elements are arranged sequentially from the surface inward. The nuclei of the supporting cells are found on the top layer (surface) followed by the nuclei of the olfactory sensory neurons in the middle layer and the nuclei of the basal cells in the bottom layer.

Supporting cells support the neurosensory cells and secrete mucous, whereas, neurosensory cells, also called olfactory sensory neurons or olfactory receptor cells (ORCs), are the first component of the olfactory system that contacts the odorants. The ORCs are generally accepted as bipolar neurons with a single unbranched dendrite and a single axon with columnar cell body and round cytoplasmic body.[39–41] The number of cilia projected from a single olfactory knob has been reported to be from none to 1000, depending on the type of ORCs and the species.[39,40] In humans, 10 cilia/olfactory knobs have been reported.[42] Most cilia projected from each olfactory knob lie horizontally and are embedded in the epithelial mucous, forming a dense mat; however, some cilia align perpendicularly with the olfactory epithelium.[40] Each cilium membrane contains a high density of globular particles, which are believed to be the olfactory receptor sites.[39] Intramembrane particles were first discovered in bovine cilia.[43] The receptor sites are located in the region of the olfactory knob and proximal portion of the cilia,[44] whereas the 5′-AMP-specific olfactory receptor sites are located along the entire outer dendritic segment of the ORCs.[45]

The axons of the ORCs project directly into the olfactory bulb (OB). The ORCs are stimulated by odorants and convert the information to neuronal signals, which are later transduced at higher levels in the olfactory system and the brain, producing perceptions and responses. Kanda and others[42] reported 30,000 ORCs/mm^2 in humans. Using the estimated sizes of the olfactory epithelium (2.4–6.4 cm^2) and the estimated density of ORCs/mm^2, the total number of human ORCs can be extrapolated from 7.2 to 19.2 million cells. Another report found about 6–7 × 10^6 cells in adult humans.[46]

The last cell type of the olfactory epithelium is the basal cell. Basal cells function as regeneration units and constitute the base of the olfactory epithelium. Basal cells have the same ability as stem cells to divide and form new functional neurons. Therefore, the degraded ORCs can be replaced with freshly generated neurons derived from the basal cells throughout life.[39,40]

The basal lamina layer is a membrane separating the olfactory epithelium and the lamina propria. The basal cells lie on this membrane with the end of the supporting cells and the starting of the axons projected from the ORCs. The final layer of the olfactory mucosa is the lamina propria. The lamina propria contains olfactory axon bundles, Schwann cells, blood vessels, and the olfactory gland or Bowman's gland. The nonmyelinated nerve fibers and the olfactory axons bundle together to form nerve bundles projecting directly to the OB. The Schwann cells direct the olfactory nerve bundles to the designated location in the OB. The Bowman's gland, found in the lamina propria, functions as a mucous generator, similar to that found in the supporting cells in the olfactory epithelium.[41] The mucous from Bowman's gland contains only neutral mucopolysaccharides; whereas, the mucous from the olfactory-supporting cells contains both neutral and acidic mucopolysaccharides.[47] The mucous layer protects the olfactory epithelium from dryness, extreme temperature, and particle and pathogenic contamination.[40] In addition, the mucous layer also acts as the first odorant-screening center of the olfactory system.

The mucous is hydrophilic in nature and not homogeneous. It can be divided into several domains.[48] In order to contact olfactory sensory neurons or ORCs, all odorants (mostly hydrophobic molecules) must be able to dissolve in the olfactory mucous layer (hydrophilic layer). The olfactory mucous also provides ions necessary for the activities of the olfactory neurons and supporting cells.[40] Besides its protection and screening roles, the olfactory mucous also removes odorants from the olfactory epithelium by mucous flow, desorption of odorants into the air, and uptaking odorants into the circulatory system.[47,49]

Besides the olfactory epithelium and the olfactory receptors, the OB is an important module in the olfactory system. Unlike other sensory neurons, olfactory neurons converge to form many glomeruli, which act as relay stations for olfactory information processing. The OB is located above the olfactory epithelium. The cribriform plate separates the OB from the olfactory epithelium; however, axons from the ORCs extrude through perforations in the cribriform plate to enter the OB.

The OB has six major anatomical layers: the olfactory nerve layer, glomerular layer, external plexiform layer, mitral cell layer, internal plexiform layer, and granule cell layer. The glomerular layer is the location of the complex synaptic terminals between the ORCs and mitral cells and the ORCs and tufted cells. A spherical region formed by the complex synaptic terminals is called a glomerulus, with a convergence ratio approximately equal to 1000 ORC axons to a single mitral cell arborization.[39] The approximate number of mitral cell arborizations forming a glomerulus is 1000–2000. This means that there are about 25,000 axons from the ORCs that converge on a single glomerulus. There are approximately 8000 glomeruli in each OB in a young adult human.[50]

Many studies report lateral inhibition in the OB. Periglomerulus cells are inhibitory as tuft and mitral cells.[39] Lateral inhibition sharpens sensory signals, which also happen in many sense systems besides the olfactory system. In addition to the inhibition of incoming neuronal signals, some cells such as granule cells also inhibit the inhibiting cells such as tuft and mitral cells.[39,40]

The complicated organization of the OB with specific cell types in each layer is a very important feature, which allows the olfactory system to manage information from the external environment.

12.4.4 MULTISTIMULUS SENSES

All sense systems mentioned above possess specific structures to filter stimuli and have evolved to respond to the stimuli. However, some sense systems respond to a wide range of stimuli. These nerves are known as free nerve endings and are distributed throughout the skin. Stimulation of free nerve endings results in irritation and pain. Chemical irritation and temperature perception are common perceptions caused by the stimulation of free nerve endings.

12.4.4.1 Chemical Irritation

Ammonia, pepper, mustard, and other substances cause irritation on skin, especially mucosal surfaces. Irritants stimulate common chemical senses or trigeminal system, which exposes its receptors in the form of free nerve endings. A classic example of an irritant is capsaicin, which is found in peppers. Capsaicin elicits a burning and stinging sensation.

12.4.4.2 Temperature Perception

Another sensation caused by the stimulation of free nerve endings is temperature perception. However, these free nerve endings do not randomly respond to any stimuli. Different groups of the nerve endings are responsible for feeling of warmth and coolness.[51]

12.5 PSYCHOPHYSICS AND SENSORY ANALYSIS

To appropriately evoke and quantify sensation and perception through responses toward stimuli, sensory evaluation uses many discoveries and inventions from psychophysics.[4,52] In affective tests, a sensory scientist employs many procedures invented by psychophysicists, such as the 3-AFC, triangle test, duo-trio test, as well as scales for measuring the level of affection, such as the 9-point hedonic scale. The roles of psychophysical scaling are prominent in descriptive analysis; however, many sensory scientists take it for granted and some even abuse it.

Many descriptive analyses are different in their fundamental philosophies underlying the methods; however, they have one thing in common: the inventors of all descriptive methods believe in the human ability to rate or at least order stimuli directly on the basis of the intensity of attributes that the stimuli evoke. And studying human ability to rate is an important aspect of psychophysics.

12.5.1 Sensation and Perception

Sensation is the detected and then encoded signals of the energy of external stimuli-stimulating sensory receptors.[30] Consequently, the energy is organized, interpreted, and given meaning by sense systems, and the systems create the awareness of the stimuli. The awareness is perception. Therefore, perception is the result of psychological processes influenced by psychological factors: judgment, relationship, meaning, and so on play a role.[30]

In general, sensation and perception are inseparable and unified processes, even though their definitions are distinct. In order to disengage sensation and perception, well-controlled test conditions must be achieved.

12.5.2 Psychophysics

Psychophysics is derived from two words—psycho and physics. It is a branch of science dedicated to quantifying relationships between physical stimuli and subjective responses or sensory experiences from subjects.[30,53] Psychophysics is the oldest branch of experimental psychology. Its first theorist was a nineteenth-century German physiologist, E. H. Weber, whose observations in 1834 led to the formulation of Weber's law.[4,30]

Weber's law, also called Weber's fraction or Weber's ratio, states that the ratio between the changes in the intensities of the first stimulus to a next intensity that creates just noticeable differences in perception and the intensity of a preceding stimulus is constant. The law can be expressed in a mathematical equation as follows:

$$\left(\frac{DI}{I}\right) = k \tag{12.1}$$

where I is the magnitude of the physical stimulus at starting level, DI is the increment of the physical stimulus intensity, which when added to the stimulus intensity I, produced a just noticeable difference (JND), and k is a constant that varies with the sensory system being measured.

Twenty-six years later, Gustave Theodor Fechner published the first book of psychophysics, *The Elements of Psychophysics*.[4,30] Fechner's law states that the sensation of the difference between two physical stimuli at the low end of the physical intensity scale that are separated by 1 JND is smaller than the difference between two physical stimuli at the high end of the physical intensity scale that are also separated by 1 JND,[30] and it can be expressed mathematically as follows:

$$S = k \, \log(I) \tag{12.2}$$

where S is the magnitude of sensation, I is the physical intensity of the stimulus, and k is a constant that takes into account the specific Weber's fraction for a given sensory dimension. Fechner believed that change in "a unit" of sensation magnitudes is equivalent to addition or subtraction of "a JND."

About 100 years later, S. S. Stevens proposed a new psychophysical law called Stevens' power law or Stevens' power function.[58] Fechner's law is based on a philosophy that the absolute magnitude of sensations is impossible to measure directly. Stevens' power law, however, is based on the belief that humans can express the magnitude of sensations in numeric forms; hence, the sensations are directly measurable. Stevens' law states that the increase in subjective magnitudes is in proportion to the physical intensity of the stimulus raised to a power. This assumption can be mathematically expressed as follows:

$$S = k\,I^b \quad \text{or} \quad \log(S) = b[\log(I)] + [\log(k)] \tag{12.3}$$

where S is the magnitude of sensation, I is the physical intensity of the stimulus, k is a constant that takes into account the choice of units used in a given sensory dimension, and b is the exponent of the equation. The exponent reflects the relationship between sensory magnitude and stimulus magnitude, which is specific for each sensory dimension, and is categorized into three types: <1 (expansion), =1, and >1 (compression). When "b" is greater than 1, expansion of the response dimension occurs, and it reflects the faster increase rate in subjective intensity than the increase rate in physical intensity. On the other hand, when b is less than 1, compression of the response dimension occurs, and it reflects the slower increase rate in subjective intensity than the increase rate in physical intensity. If b is equal to 1, it reflects that the sensory magnitude grows with physical stimulus magnitude. Moskowitz reported the exponents of many sensory dimensions.[54]

12.6 SENSORY RESPONSES AND THEIR MEASUREMENTS

Quantification of sensory experience by assigning numbers to the experience is called sensory response measurements, and the responses collected are numerical values resulting from the measurements.[2,4] There are four general types of responses: nominal, ordinal, interval, and ratio responses.[55] Differences between the responses are the characteristics of the basic empirical operations that can be performed on the responses; thus, the information contained in the four responses are different.[2]

12.6.1 NOMINAL RESPONSE

Determination of equality is the basic empirical operation of nominal responses; therefore, a nominal response contains pure qualitative information, such as name or gender, which is the least informative among the four response types.[55] The qualitative information contained in nominal responses does not have any numerical value. For example, researchers can assign 0 = blue and 1 = red, and it does not mean that blue and red are numerically different and the difference is 1. Moreover, if the assigned number changes to 0 = blue and 500 = red, the meanings are always the same as 0 and 1, which are blue and red, respectively.

Classification, categorization, detection, recognition, and identification are sensory measurements given to the nominal response.[2,56] For example, asking the subjects to express whether they can or cannot detect the stimuli is equivalent to asking the subjects to classify the stimulus into two categories: yes, something exists, or no, there is nothing. These tasks are very useful in sensory evaluation, especially in different tests such as duo-trio test, triangle test, and threshold determination.

12.6.2 ORDINAL RESPONSE

Determination of greater or less is the basic empirical operation of ordinal response; therefore, information contained in the ordinal response is the order of intensities or levels of stimuli.[55] However, it does not contain any numerical magnitudes of differences between stimuli. For example, subjects are asked to order the weight of a 2-kg brick, 1-kg cotton, and 100-kg iron in the order of the lightest (1) to the heaviest (3). The order will be the cotton = 1 (lightest), the brick = 2 (middle), and the iron = 3 (heaviest); however, the rank does not inform the magnitudes of the differences between the objects.

There are three sensory measurements providing ordinal responses: ordering, ranking, and category scaling with unknown magnitude of differences between categories. Ordering and ranking are tasks that subjects perform based on magnitude of a certain quality or attribute of interest; however, in category scaling with unknown magnitude of differences or uneven-interval category scaling, subjects will place stimuli in assigned categories that are preranked and labeled based on the magnitude of a certain quality such as liking or an attribute of interest. The unknown magnitude of differences between any categories differentiates the ordinal responses from interval responses.[1]

Ordering and ranking are commonly used in preference work, but the tasks are very time consuming and tedious and prone to fatigue and changes in discrimination criteria.[1] Nonetheless, the uneven-interval category scaling is the method of choice in marketing and consumer research because it is less tedious than ordering and ranking; however, its interpretation is a little more complicated because of the prelabeled categories.

12.6.3 INTERVAL RESPONSE

Interval responses contain the equality of intervals or differences; therefore, the responses provide the numerical magnitude of differences along with other information gained from the nominal and ordinal responses.[55] Temperature measurement is a good example for interval responses. In temperature measurement, both °C and °F do not have a real zero point, but have their own arbitrary zero points. However, the mathematical expression of the relationship between these two units exists as $(C/5) = ((F-32)/9)$.

The sensory measurement technique for this type of response is category scaling with a known magnitude of differences between categories, such as the 9-point hedonic scale in which differences between intervals are quantified and equal.[1,57] However, other types of scales are assumed to give interval responses, such as the 16-point intensity scale used by the Spectrum Method and the categorized line scale.[2]

12.6.4 RATIO RESPONSE

Ratio response is used to determine the equality of ratios; therefore, the responses contain the following information: equality or identity, rank order, equality of intervals, and equality of ratios. The ratio responses have a true-zero value and linearly increase in intensity. Physical units such as length, mass, and temperature in kelvin are good examples because these physical values have their true-zero points and the relationship between units in the same continuum is linear. For example, the relationship between centimeter and inch is 2.54 cm = 1 in., and it is constant at any length of any object.

A sensory response measurement that is believed to yield the ratio response is magnitude estimation, but the linear property of the ratio responses gathered from magnitude estimation is still controversial because of contextual effects and a biased number assignment process.[4,58]

12.7 MAGNITUDE ESTIMATION AND THE POWER FUNCTION

Stevens and his colleagues invented magnitude estimation in the early 1950s on the basis of a strong belief that humans are capable of expressing the absolute magnitudes of sensations. Magnitude estimation is the unrestricted application of numbers to represent sensation ratios. Assigned-modulus and modulus-free magnitude estimation are two primary variations of magnitude estimations.

Magnitude estimation requires a standard for subjects to compare their sensation to sensations elicited by products. The use of the standard distinguishes the assigned modulus from the modulus-free variations. If the standard is an external reference product with an arbitrarily preassigned numerical value, it is a modulus; however, if the first product is used as a standard, the task is modulus free.[54,59]

The intensity of modulus used in the assigned modulus magnitude estimation distorts the responses obtained from the assigned modulus method, depending on the location of the modulus among the intensity range of products. Modulus-free magnitude estimation with appropriate presentation order should produce the best data, but analysis of the data is tedious.[54]

Because subjects use different ranges of scales, normalization of data is recommended to bring all responses into the same scale before further analysis. Geometric mean normalization is a common method used for this purpose.[4,54] After normalization, the data are usually submitted to log transformation because the data tend to be log-normally distributed or at least positively skewed.[1] However, if the data contain a zero, any transformation or geometric mean normalization will not be possible because log of zero is undefined. ASTM (1997) suggested replacing the zeros for each subject with a small positive value such as one-tenth of the smallest number given by each subject; however, replacing the zeros will influence the data set. Employing arithmetic mean or median in normalization is an alternative to replacing zeros.[1]

In conjunction with an appropriate experimental design, after normalization, the data from magnitude estimation are usually used to construct the power function of the stimuli. According to Stevens' law, if $\log(s)$, y-axis, is plotted against $\log(I)$, x-axis, it yields a linear line with slope = b and y-intercept = $\log(k)$. The meanings of both b and $\log(k)$ are explained in Section 12.5.2. If both exponent and ordering ability between subjects turn out to be different, there is a high probability of misunderstanding the definition of the descriptor of interest. Therefore, more discussion about attribute definitions is required before any scaling task can proceed.

Usually, magnitude estimation is used for only one descriptor, such as overall intensity; however, it can be applied in multiattribute tasks such as descriptive analysis.[1]

12.8 THRESHOLDS AND THRESHOLD DETERMINATION

Threshold or limen means "the level of stimulus intensities at which a physiological or psychological effect begins to be produced."[60] Fechner's psychophysics or classical psychophysics has been focusing on threshold determination since its conception more than a century ago.[4] There are four types of thresholds: (1) detection threshold, (2) recognition threshold, (3) difference threshold, and (4) terminal threshold, based on the applications.[61]

12.8.1 DETECTION THRESHOLD

Hypothetically, the detection or absolute threshold is the minimum physical energy level of a stimulus necessary for perceptual detection or awareness; and it implies that there is a clear-cut level of physical energy at which the existence of stimuli is perceived.[4,30,61,62] In practice, however, the hypothetical level cannot be found. Therefore, many empirical definitions for detection threshold are established. The stimulus level that has a probability of 0.5 of being detected under the condition of the test is an example.[4,63] Consequently, the empirical definitions of detection thresholds are dependent on methods used.[4] In addition to the detection threshold, other thresholds such as recognition, difference, and terminal threshold endure the same hurdle, the congruency between hypothesis and empirical definitions, as well.

Method of limits is a common threshold determination method, but it has inherent disadvantages: fatigue, sensory adaptation, expectation, and habituation. Presenting products in ascending or descending order causes problems.[64] To minimize these effects, the staircase method was introduced in the 1960s.[30] In the staircase method, the experimenter starts a presentation scheme as in the method of limits; however, the experimenter will reverse the order when a subject responds correctly. The experimenter continues the reverse-when-correct scheme until he or she obtains a constant up–down pattern response.[30] The most common example for threshold determination is the standard method developed by ASTM called forced-choice ascending concentration series method of limits.[63]

The method of constant stimuli,[30] semiascending pair-difference method,[71] estimation from dose–response curves,[65] and 2-AFC with five replications[66] are alternatives to the standard method. For a more detailed list of methods, Brown and others[61] and Lawless and Heymann[4] are recommended.

12.8.2 RECOGNITION THRESHOLD

Recognition threshold is defined as the minimum level of physical stimuli at which a stimulus is *correctly identified*. Usually, the recognition threshold is a little higher than the detection threshold; however, the difference depends on the stimulus used.[4] In addition to *detecting* a correct stimulus, subjects must *identify* the stimulus correctly in recognition threshold determination. Therefore, *identification task* distinguishes recognition threshold from detection threshold; otherwise, the rest of threshold determination procedure is the same. The most commonly used recognition threshold determination method is the ASTM procedure E 679–91,[63] also called 3-AFC ascending concentration series method of limits.

12.8.3 Difference Threshold and JND

Both Weber's fraction and Fechner's law require the determination of difference threshold because the difference threshold is a crucial part of both models. Difference threshold or difference limen is the amount of change in physical stimulus necessary to produce a JND in sensation; nevertheless, empirically, the JND is the amount of change in physical stimuli that subjects can detect 50% of the time.[30,67–70]

In the difference threshold determination, the forced-choice element is always used because of the nature of the question asked of subjects. The subjects are asked to *compare* and *find* the *difference* between a given stimulus and a *reference* stimulus. If the reference stimulus is at zero level or blank, the increase in stimulus level responsible for perceptual changes compared to the reference stimulus is the detection threshold. Therefore, the detection threshold is a special case of the difference thresholds when the reference stimulus is at zero-level or at undetectable level.[4] Because of the requirement of the undetectable level of reference stimulus for the detection threshold, in a background media situation (e.g., beer and wine), the difference threshold and detection threshold are the same entities.[61,71]

12.8.4 Terminal Threshold

The terminal threshold is the minimum level of physical stimuli at which no changes are perceived.[4] In practice, this level is rarely approached; however, for some products, such as hot sauce, perfumery compounds, and very sour or very sweet candy, it would be an exception. There are many problems when the stimulus level approaches the terminal threshold, especially when many other sensations are evoked automatically. For example, irritation and pain are usually perceived along with concentrated odorants and results in mucous production. In turn, mucous obscures perception of descriptors of interest in conjunction with distraction from irritation and pain.[72,73] Moreover, changes in perceived quality at different concentrations are a common phenomenon in olfaction; and this makes studies along the odor perception continuum more difficult.[74]

12.8.5 Individual and Population Thresholds

In addition to the classification based on its applications, threshold can be classified by its inference targets: individual and population. Nevertheless, the definition of individual and population thresholds still depends on which aspect, hypothetical or empirical, is adopted by researchers.

If the hypothetical definition is adopted, the individual threshold can be defined in many ways such as (1) the geometric mean of the ending level and the next (higher or lower) level step that would have been given had the series been extended, (2) the lowest correct stimulus level or the geometric mean of the last incorrect concentration step, and (3) the first correct concentration step.[4,63] After obtaining individual thresholds, the population detection threshold is defined as the geometric mean of the individual threshold.[63]

If the empirical definition is adopted, ASTM (1997) suggests using both graphical and logistic modeling methods obtained from a relationship between the percentage correct above chance and log of stimulus level, which can be found in ASTM procedure E 1432–91.[62] The individual threshold is the level in which the probability of correct responses is 0.5 for detection threshold and 0.75 for recognition threshold. The group threshold for the empirical case is also the geometric mean of the individual thresholds or is calculated from the rank/probability method.[62]

Besides presentation schemes such as ascending, descending, staircase, forced-choice presentation, 2-AFC, 3-AFC, and two out of five, another concern is the number of stimulus levels presented to the subjects. Lawless and Heymann discussed this problem statistically and pointed out the importance of the number of the products within the series, which they called the "stopping rule." The number of the products within the series depends on which forced-choice presentation method is selected, the number of replications, and the confidence level the researcher wants to keep.[1]

12.9 STATISTICAL ASSESSMENTS IN SENSORY SCIENCE

Without exception, validity of sensory studies depends on how the studies and data gathered have been conducted. Sensory studies usually generate a large amount of data and require statistical techniques to understand and represent them. There are three general purposes in sensory studies: describing the data, drawing an inference from experiments, and studying data structures and their relationships.[4] In this section, we suggest some statistical tools and explain underlying restrictions for the tools. Details on how to derive statistics for statistical testing are explained elsewhere.[2,4,52,75,76]

12.9.1 PRESENTING SENSORY DATA USING DESCRIPTIVE STATISTICS

Because there are variations inherent in any measurement, statistics are important in sensory evaluation. Statistics provides two important pieces of information to describe the characteristics of data: the best estimation of measurements and the variation around the estimate. Statistics that are representing the estimates and the variation of the estimates are called descriptive statistics.

The best estimation of measurements is most likely a single value representing the center of the measurement distributions. There are three commonly used measures of central tendency: mean, median, and mode. If the distributions of data are symmetrical, the *mean* of the data is a parameter that can be estimated by a statistic called *average*, the sum of all observations divided by the total number of observations. If the distributions are not symmetrical and contain some outliers or extreme observations, median or 50th percentile is an appropriate estimate. Median is the middle value after all observations are ranked, and outliers do not influence the median. Median is an appropriate estimate of a center value of rank data. If the data are nominal responses, *mode* is an appropriate estimate because it is the most frequent value observed.

In addition to the estimation of center values, how the data disperse is another concern. Standard deviation is a measure of the spread of the data around their center. The standard deviation of samples is denoted by S and its square value is called variance (S^2). S is an unbiased estimator of the standard deviation of a population. Another useful measure is the coefficient of variation (CV). CV is a percent ratio between S and average in a sample. CV allows researchers to compare the variation from different methods, scales, experiments, or studies.

12.9.2 DRAWING INFERENCES BY HYPOTHESIS TESTING

When researchers want to test an assumption, they devise an experiment or experiments under well-controlled conditions and compare the results from different treatments in the experiments in order to reach a conclusion from the studies. Statisticians develop a protocol called hypothesis testing to aid researchers to reach a conclusion regarding their experiments with confidence.

Similar to the scientific method, hypothesis testing in statistics (statistical testing) involves observation and theory, against which the observations will be tested. In classical statistics, actual observations from a product provide a statistic (an estimate of a parameter of the population of interest), and the statistic is tested against an appropriate parameter obtained from a theoretical population, of which distribution has been well characterized.

There are four elements required for statistical testing: null hypothesis (H_0), alternative hypothesis (H_a), test statistic, and rejection region. H_0 is a theoretical hypothesis that the researcher wants to disprove, and it determines the distribution of parameters of interest. H_a is an alternative hypothesis that the researcher will accept if H_0 is rejected. Test statistic is an estimate of the parameter of interest. Selecting an appropriate test statistic is important, and it dictates what type of statistical tools the researcher will use. Rejection region is a predefined value of test statistics for which H_0 is rejected, and researchers define the region based on how confident the researchers want their conclusion to be.

There are two types of errors that researchers should consider when conducting statistical testing: type I and type II errors. If researchers reject H_0 when it is true, they commit type I error; and if the researchers accept H_0 when it is false, they commit type II error. The probabilities associated with type I and II errors are a and b, respectively. Size of a or confidence level will dictate what the rejection region will be and, in conjunction with b, the power of experiment and product size required for an appropriate power.

Many statistical tools were developed to aid researchers when they want to apply statistics to their studies. Selecting appropriate statistical tools to test a hypothesis depends on what sensory study has been conducted. Types of responses and experimental design also influence the decision. In practice, hypothesis testing is critical for many activities such as claim substantiation and quality control. This section will introduce some common statistical assessments in sensory studies.

12.9.2.1 Hypothesis Testing for Interval and Ratio Data

Data obtained from interval and ratio scales, such as 9-point hedonic scale, visual analog scale, magnitude estimation, labeled magnitude scale, and any scales that have been proved to possess interval and ratio scale properties, are qualified to be analyzed with parametric statistical methods, such as t-test and ANOVA.

t-Tests are suitable for comparing two products; they are based on comparing averages of two products using *t*-distribution. There are two main variations of *t*-test widely used in sensory studies: independent and dependent *t*-tests. Dependent *t*-test is appropriate for dependent data. If one subject tests and rates two products, the results for both products are dependent data because one subject produces both ratings. If two different groups of subjects are selected from the same population and each group tests only one product, the ratings for the products are independent from each other. In this case, independent *t*-test is appropriate. If more than 50 subjects are used in testing, *t*-tests are equivalent to testing the hypothesis using normal distribution.

In reality, testing products is an expensive exercise; therefore, when a company has a chance to test products, several products are usually tested in the same study. If two or more products are tested, ANOVA is the method of choice. ANOVA is based on the partitioning of variance associated with variation sources identified and controlled by researchers and comparing the variations with random variation, sometimes called residue or random error, or appropriate variation depending on experimental design used. The ratios between two variances, *F*-statistics, or *F*-ratios, are the statistics of interest; and they have *F*-distributions with numerator and denominator degree of freedoms associated with the ratios. Testing the hypothesis that the ratios are unity implies that the variations due to the sources are not significantly different from a random error or appropriate errors.

Different experimental designs have different models for ANOVA. The simplest form of ANOVA is one-way ANOVA for univariate responses: only one type of response is collected in an experiment. Experimental design accompanying the one-way ANOVA is completely randomized design (CRD). Only the products tested in the study are the source of variations. When only two products are tested in CRD, the ANOVA is equivalent to an independent *t*-test. If each subject in a study tests all products, completely randomized block design (CRBD) is used. CRBD has two sources of variations: subjects and products. Subjects are the random-blocking factors, and the products are fixed factors in the analysis. If two products are tested in CRBD, the ANOVA for CRBD is equivalent to a dependent *t*-test.

There are several linear models developed to suit many situations encountered in sensory studies, such as factorial designs, factorial-structures within CRBD, and incomplete block design. A controversial issue in using ANOVA for sensory studies is the repeated measurement issue. Because subjects rate many products over a time course, the rating is repeated. Therefore, the responses are not totally independent— they may be dependent on previous ratings. A special ANOVA model–repeated measurement analysis is capable of handling this situation. In a simple case such as one-way or CRBD, ANOVA models for split-plot designs are equivalent to repeated measurement analysis. The split-plot design ANOVA contains an interaction term between blocking factor (subjects) and product factor and compares the variation due to both main effects with the variation associated with the interaction term.

Repeated measurement analysis is a form of multivariate analysis of variance (MANOVA) because each subject gives several responses—rating one attribute several times or rating many attributes at once. Because responses are not independent, drawing conclusions from many ANOVAs will result in the inflation of type I error of whole study, called experiment-wise error, and lead to the invalidation of its conclusions. MANOVA is a solution for this case, and it is appropriate to answer the question of *overall differences* between products. Running many ANOVAs after MANOVA is appropriate as long as the results from ANOVAs are used as a suggestion, but not a conclusion.

12.9.2.2 Hypothesis Testing for Nominal and Ordinal Data

Many sensory studies produce data that do not have interval or ratio properties. Thus, *t*-test and ANOVA are not appropriate because both tests require the data to be at least interval data. Nonparametric statistics are developed to handle nominal and ordinal measurements. Examples of such tests are pair comparison, triangle test, and duo-trio test for nominal responses and preference ranking for ordinal responses. Nonparametric statistics assume that the distribution of responses gathered is continuous, but the techniques do not require the shape of the distribution.[75] Hence, it is not essential that the parameters of distributions must be in certain condition as parametric techniques (e.g., *t*-test and ANOVA) demand. Therefore, hypothesis testing is parameter free or nonparametric. Nonparametric statistics are more robust than parametric statistics and less likely to lead to erroneous conclusions and estimations when assumptions regarding underlying distribution of responses are violated. Moreover, the techniques are more sensitive to deviation from random responding than parametric ones.[4]

12.9.2.3 One-Product Case and Replicated Difference Tests

There are several nonparametric techniques commonly used in hypothesis testing, and selecting appropriate techniques depends on how many products are involved in the studies because *t*-test is for two products and ANOVA is for two or more products. Even though "one-product" is part of the name, it does not mean that there will be only one product in the testing procedure. "One-product" implies that only one product is tested for significance testing. For example, researchers are interested in how product A is preferred over product B in a population; although there are two products involved in the study, the researchers are interested in the frequency of product A to be "preferred" or "not preferred" or *vice versa*. Thus, there is only one product of interest, product A or product B; moreover, if product A is preferred over product B, then it is unnecessary to test the significance of product B to be less preferred than product A. Replicated difference tests are one-product cases that are performed more than once on the same product. An example for replicated difference tests is when the same group of consumers performs a triangle test 10 times; the test is called a replicated triangle test.

Beginning with the simplest case, the one-product test with dichotomous responses, the responses have only two categories such as yes/no, right/wrong, prefer/not prefer. Because there are two categories, binomial-based tests are appropriate. Results from several methods—such as triangle test, duo-trio test, pair comparison test—can be analyzed by the binomial tests.[2,4,75] In a replicated case with dichotomous responses, beta-binomial was proposed.[77–80] An alternative to the beta-binomial test is the McNemar test, which uses the chi-square statistic to test the significance of the changes in patterns of a 2 × 2 table.[4,52]

When the responses are ordinal, the sign test and the Wilcoxon-signed ranks test are appropriate. The sign test examines the changes in the direction of the differences between pairs as denoted by "1" and "2" signs; however, if the relative magnitude is considered, the Wilcoxon-signed ranks test incorporates the magnitude into account by adding more weight to a pair with large differences between conditions and *vice versa*. For example, a panelist tastes two products and gives scores to both products using a categorical scale. In the sign test, researchers are concerned only with which product receives a higher or lower score; however, the Wilcoxon-signed ranks test measures the differences between scores and then ranks the differences. The significant testing is based on the rank of the differences and the sign of the differences.[4,75]

In addition to the aforementioned methods, the chi-square statistic is also a useful tool for many situations involving classification of a product or products into categories. For a one-product situation, a result from a JAR scale for a product is a good example. There are five categories in a JAR scale, and the categories are not equally spaced. If 50 of 100 consumers rate the sweetness of product A in the "just-about-right" category, 30 in "somewhat too high," 10 in "much too high," 7 in "somewhat too low," and 3 in "much too low," researchers want to confirm that the frequency of 50 did not happen by chance. The chi-square statistic calculates the differences between actual frequency and expected frequency within each category. The expected frequency is a hypothetical frequency derived from the null hypothesis, which, in this case is 20 in all categories.

12.9.2.4 Two and Several Independent Products

In practice, researchers usually compare more than two products and use categorical scales. If products possess a high level of fatiguing factors (e.g., mustard or extreme sour candy), retasting the products may be impossible if the test is to be bias free. Therefore, several groups of consumers are required to taste each product separately; thus, each product is independent from one another. The chi-square statistic is also applicable in these situations as long as all products are evaluated by independent testers from the same population. Employing frequencies of occurrences in each category for each product, the chi-square statistic measures the interaction between the product and categories. If the occurrences of a product in categories significantly differ from random chance, then it is safe to conclude that the products significantly interact with certain categories.[75] In turn, researchers can imply that the products cause the occurrences in particular categories.

When ordinal responses are gathered, there are several methods developed for the situation, depending on the number of products in the studies. For two independent product cases, the median test, the Wilcoxon–Mann–Whitney test, and the Kolmogorov–Smirnov two-product test are appropriate. The median test examines the differences between the medians of two products, assuming that the consumers are drawn from the same population. However, the other tests examine the differences of the distributions of two products under the homogeneous consumer assumption. If the distributions are significantly different, then researchers can conclude that the products cause the differences, depending on experimental settings. However, the tests do not support the conclusion of the differences between the means or medians of the products.

12.9.2.5 Several Related Products

In most testing situations, a panelist tests several products in one visit with an appropriate resting interval between products to minimize and/or eliminate carry-over effect; hence, the responses from each consumer are related. If responses are nominal or categorical, the Cochran Q test is appropriate.[75] This test examines the differences between the frequencies of products, and it is also appropriate for dichotomous ordinal responses.[75] If the responses are ordinal or interval and each consumer tests all products, the Friedman two-way ANOVA by ranks is the method of choice. The Friedman test examines the differences between product medians under a null hypothesis that all products are drawn from the same population or the product medians are similar.[75] The test ranks the ratings of products within each consumer and compares the rank totals.

12.9.2.6 Several Independent Products

For several independent products, the chi-square test, the extension of the median test, and the Kruskal-Wallis (KW) one-way ANOVA are appropriate. The philosophy and basic assumptions of the former two methods are described in two independent product cases mentioned. Thus, they will not be repeated here. One caution regarding applying chi-square statistics when there are more than two rows or columns in analysis: no more than 20% of cells in a contingency table should be more than 5, and no cell should have an expected frequency of less than 1.[81]

The KW one-way ANOVA tests the differences between medians under a null hypothesis of similar medians for all products. The method ranks the observations regardless of their origins and then calculates the rank totals for each product. Then, KW statistic is calculated and used to test the significance of differences in the medians.

12.9.3 Relationships between Variables and Classification

12.9.3.1 Correlation between Variables

In many cases, the objective of studies is to explore rather than to test an assumption. Hence, the relationships between observed variables are of more interest than proving the causes of phenomena. Therefore, if a study is set to explore, drawing any causal inference from the observations of the study is prohibited; however, many researchers forget this restriction.

The simplest case is studying the association between two variables; however, scales used to measure each variable determine what statistic should be used to quantify the association between the variables. In many disciplines, data are usually interval responses; and the Pearson product moment correlation coefficient is appropriate. In sensory science, the data gathered are not usually interval responses, even though, in some cases, responses are believed to be interval or obtained from scales proven to possess at least interval property. Some nonparametric methods are useful.

If both variables are measured on at least ordinal scale, the Spearman rank-order correlation and the Kendall rank-order correlation coefficient are appropriate. If both variables are categorical or nominal, the Cramer coefficient, the kappa statistic, and the lambda coefficient are applicable.[75] When there are more than two variables involved in studies and the variables are ordinal, the Kendal partial rank-order correlation coefficient is appropriate. Partial correlation is correlations between two variables at a constant level of the other variables. The degree of associations among several rank variables can be measured by the Kendal coefficient of concordance and the Kendal coefficient of agreement measure.[75]

12.9.4 REGRESSION ANALYSIS

Correlation analyses mentioned above are appropriate descriptive statistical methods for studying relationships among variables with variations associated with them. In some cases, the prediction of a variable or variables by another variable or another set of variables is of interest and the regression method is more appropriate.[82]

Simple linear regression is the simplest case of regression analysis: it regresses a variable called "dependent variable" on another variable called "independent variable" or "predictor." The variation of predictors is negligible compared to the dependent variable in a regression model.[82] Coefficient of determinations (r^2) is a measure of the degree of variation of the dependent variables measured by the predictor. In simple linear regression, the coefficient of determinations is the square of the Pearson correlation coefficient as shown by r^2.

An extended method of simple linear regression is multiple linear regression. This method uses multiple independent variables to predict a dependent variable, and the predictors are independent from each other. The independency implies that the correlations among the predictors are zero or at least close to zero. The amount of variations in a dependent variable accounted by predictors is represented by multiple linear correlation (R^2). In practice, some researchers neglect this restriction and use predictors that are not independent. The dependency among predictors is called multicollinearity, and it causes unstable models and estimates.[82] In addition to multicollinearity, the number of observations is another factor that contributes to the stability of prediction models and estimates; the recommended number of observations for a good prediction power is 15 observations per predictor in the models.[76]

12.9.5 MULTIVARIATE TECHNIQUES AND THEIR APPLICATIONS

In previous sections, we introduced the concepts of correlation, regression, and multicollinearity. In this section, we introduce readers to more advanced statistical techniques, generally called multivariate techniques. Sensory analysis commonly generates many responses or variables called multivariate data, and the data are usually correlated with one another. Multivariate techniques, except MANOVA and repeated measurement analysis, are used in sensory science mostly to classify or order objects in space and/or to study degree of associations among variables and objects. For classification or ordination, cluster analysis (CA), multidimensional scaling (MDS), and discriminant analysis (DA) have been developed.[76,83–86] If researchers want to study the association between variables, principal component analysis (PCA) or exploratory factor analysis and correspondence analysis (CDA) are appropriate.[76,84] If relationships between two sets of correlating variables are of interest,

DA, canonical variate analysis (CVA), and partial least square regression (PLS) are appropriate.[87-89] When the associations among three or more data sets are of interest, CVA and generalized Procrustes analysis (GPA) are examples of techniques suitable for this situation.[87,90] Results from these multivariate techniques are usually presented in the forms of multidimensional maps.

Beginning with analyses of one data set with many variables, the most commonly used and the oldest technique is PCA or exploratory factor analysis; therefore, it will be explained in more detail than other methods.[91] PCA constructs many orthogonal linear combinations called principal components (PCs) from a set of variables, and the equations explain the variance in the data set as much as possible. The phase "orthogonal linear combinations" means that linear combination equations or the regression equations of PCs on variables in a data set are mathematically independent or perpendicular from one another. In layperson's terms, each orthogonal linear combination contributes unique information to the analysis. PCA uses variance or covariance matrix of variables as a starting data set; therefore, it uses the Pearson correlation coefficient in calculation. If variables in a data set have different scales, correlation matrix is recommended.[92] Besides the interpretability of each PC, eigenvalue and proportion of variance explained by each PC aid the selection of number of PCs. Suggested cutoff point for eigenvalue is 1.0.[76,93] The contribution of each variable in explaining variation in the data is expressed as the communality of each variable. That communality approaches 1.0 indicates high contribution of that variable; however, communality is never more than 1.0. If a result shows that a variable has a communality value of more than 1.0, that result is wrong.

Because PCA uses a linear regression technique in constructing PCs, the correlation coefficients or loadings of each variable on each PC determine how well the variables correlate to one another and how important each variable is to each PC.[76] Generally, loadings of 0.4 are considered important; however, selecting variables for each PC and selecting number of PCs in the analysis are subjected to the experience of researchers in his or her fields.[76] Because of the use of the Pearson coefficient and linear regression techniques, PCA is restricted to interval or ratio responses and suffers the same hurdle as multiple linear regression, which is the ratio between number of predictors or variables and number of independent observations. Even if we relax the 15 observations per variable to 8 observations per variable, a study can be impossible to conduct just because of this concern, especially in sensory profiling, which usually generates at least 10 attributes to describe products. Therefore, at least 80 independent observations are required for such a study. However, if the correlations between variables are strong and few PCs will be extracted from a data set, smaller product size than the recommendation above may be justified. But this situation rarely happens in reality because if one knows the correlations, why bother to run an experiment? Besides product size considerations, PCA will perform well when the relationships among variables are linear; otherwise, PCA cannot capture a nonlinear relationship. Nonlinear versions of PCA exist,[94,95] but they are beyond the scope of this chapter.

If researchers are not sure that variables in a data set have at least interval properties, factorial correspondence analysis or correspondence analysis (CDA) is recommended. CDA (also known by the name of reciprocal averaging [RA])[96] employs a contingency table as a starting data set. Therefore, it uses c^2-distances as similarity indices between row and column variables in the data set (products and attributes in descriptive analysis).[93] However, if a contingency table contains a product with low frequency, CA will inflate the uniqueness of the product.[96]

Another popular method used to study a set of variables by using a similarity matrix or proximity matrix is MDS.[84–86,97] The objective of MDS is to determine the number of dimensions adequate to explain the representation of objects in space as close to the original similarity matrix as possible.[83] The dimensions derived from MDS are not linear equations or any other models but the results of iterative procedures to derive a reduced k-dimensional space that minimizes the departure of rank-order from an original p-dimensional space ($p \cdot k$). Therefore, it is suitable when the responses are questionable in their scale properties or data with mixed scales and nonlinear relationship among variables are evident.

The questions usually asked when MDS is used are overall similarity or difference questions, such as how much this product is different from the other products or a request to sort these products into groups based on their similarities. If results are obtained from an overall similarity or difference task, they can be interpreted as a holistic representation of perception or an overall perceptual map. In many cases, the similarities among products are mathematically derived from many distance models such as Euclidian distance, squared Euclidian distance, and City Block or Manhattan distance.[83] Correlation and covariance matrices are considered a type of similarity matrix because of the logic that the more correlation between attributes, the more similar the attributes are. Application of these techniques on consumer hedonic ratings is commonly called internal preference mapping.

Another method to classify products in one data set is CA. CA uses a distance matrix as starting data as MDS does; however, CA does not derive any dimension to explain the data. The sole objective of CA is classifying a pool of objects based on their similarity or difference. CA accomplishes the classification by maximizing the variation between clusters and minimizing the variation among cluster members, or sequentially merging or partitioning products into groups.[84] The former clustering technique is found under several names, such as K-mean clustering, and it requires researchers to start with an arbitrary number of clusters; the latter is found under the name of agglomerative techniques, or hierarchical techniques, and researchers subjectively identify number of clusters from a dendrogram, a nondimensional representation of clusters.[83,84,98] Cluster solutions from CA need confirmation in order to prevent misclassification, and DA and MANOVA are popular methods used to confirm the results.

DA and MANOVA require two sets of data: one data set contains categories of products, and the other contains a set of variables called predictor variables in DA. DA provides the probability of a product being a member of a class using orthogonal linear combinations called discriminant functions. On the other hand, DA predicts the chance a product belongs to a class based on the information in predictor variables. Besides, MANOVA is also useful to confirm solutions obtained from CA. MANOVA tests whether the variations caused by predetermined classes (usually treatments of interest) are significantly greater than random variations in a set of variables using general linear models for multivariate data. However, MANOVA requires many assumptions in order to perform an analysis appropriately, and results from CA may not meet these assumptions.[76]

Next useful techniques are CVA and partial least square regression (PLSR). CVA and PLSR are suitable for two data sets, and each set contains several variables. CVA tries to maximize correlation between two data sets, but PLSR tries to maximize predictive ability using one of the sets as a predictor set and the other as a regressor set.[88,89] Both methods create latent variables as in PCA, with linear combinations of variables in each set; however, the methods create the variables for both data sets simultaneously. In CVA, the first pair of latent variables will have maximum correlation between each other; and the method will extract another pair that is maximally correlated and orthogonal from the first pair. The pairs of linear combinations are called canonical variates.[76] Relationships between data sets are of interest in CVA; therefore, the interpretation of CVA is the association between important variables in each data set, which can be determined by their loading scores in canonical variates.

In contrast, PLSR generates latent variables from a predictor data set, and the variables explain the variations in both predictor and regressor sets. Thus, the latent variables created in PLSR possess predictive ability because it maximizes the variance in the regressor set.[88,89] Another method called principal component regression (PCR) is also suitable for two data sets when the regressor set contains only one variable. PCR is a two-step method: first PCA, and then multiple linear regression. Because PCs derived by PCA are orthogonal, multicollinearity is not a problem in PCR. PCR is a useful technique, especially in external preference mapping.[99–101] In external preference mapping, consumer preference is regressed on objective measurements such as descriptive profiles or chemical data; however, PCs derived from the objective data are commonly used.

If there are more than two data sets in a study, such as data obtained from Free-Choice Profiling or FP, Generalized Procrustes Analysis (GPA) is a commonly used method to handle such data.[21,90] However, the extensions of other techniques, such as CVA, PLSR, and MDS, are also available to handle several data sets.[86,88,89] GPA mathematically handles variations caused by assessors in sensory evaluation and the variations include using different parts of scales, using different ranges of scales, and understanding attribute differently.[21,90] GPA, first, centers the data sets about a common origin, which arranges the scores into the same common part of scales; second, it matches the configurations of the data sets, which solve the differences in attribute interpretations; and third, it performs isotropic scaling to eliminate the difference range effect.[21,90] Final results from GPA are an average configuration and Procrustes statistic. GPA is useful for determining how similar the data sets are, identifying bad assessors, and screening attributes.

12.10 GAS CHROMATOGRAPHY–OLFACTOMETRY

Gas chromatography–olfactometry (GCO) is sometimes known as GC-sniffing. GCO is an important analytical tool for flavorists, perfumers, and food scientists for understanding the roles of volatile compounds in complex aromas. GCO was first developed in the 1960s.[102] A GCO system comprises a chromatographic part and a human part. The chromatographic part is concerned with sampling, extraction, separation, and instrumental detection of volatiles; the human part focuses on the subjective detection, quality, intensity, and temporal dynamics of the volatiles.

Despite the chromatographic technicalities, GCO techniques can be classified by their procedure and the data obtained. The techniques are dilution analysis, intensity rating, time-intensity rating, and detection frequency methods.[103,104]

There are two major variants of dilution analysis methods: aroma extraction dilution analysis (AEDA)[105] and Charm Analysis.[106] AEDA and Charm Analysis determine the potency of individual volatiles by serially diluting extracts until the volatiles are not detectable by subjects.[107] AEDA uses dilution factors (FD value), the proportions between the concentrations of volatiles in initial extract, and the final concentration of volatiles detected by subjects to determine the potency of volatiles.[108] Aromagrams obtained from AEDA are the plots between FD values (y-axis) and the Kovats Index (x-axis). Based on the same principle, Charm Analysis determines the potency of the volatiles and expresses it as charm values (c). Conceptually, a charm value is the ratio of the total amount of odor-active compounds eluting at a particular retention index to the threshold amount of the compounds in the same mixture, and it is calculated based on the formula $c = d^{n-1}1$, where n is the number of coincident responses and d is the dilution factor.[103,104,106] Charm value incorporates the length of time an odor is perceived, and it distinguishes Charm Analysis from AEDA. Charm aromagram is the plot between charm values (y-axis) and the retention index (x-axis).[103] The potency indices, FD value and charm value, are based on the assumption that volatile compounds with low-detection thresholds will have high intensity at supra-threshold levels. However, this is usually not true because the potency of volatiles at suprathreshold regions is determined by exponents in Stevens' power function, and the exponents vary between volatiles.[109,110]

Recognizing the drawback of dilution analysis methods, Dr. Mina McDaniel, the director of the sensory laboratory at Oregon State University, and her colleagues developed a time-intensity rating method called OSME.[109,111–113] OSME uses only one representative extract, and several subjects must rate the intensity of eluted volatiles they perceive from first notice until the aromas dissipate and then describe the aromas. The subjects are highly trained to perform the task required by the method and to directly rate the intensity using a 16-point intensity scale.[109] Because the responses gathered from OSME are time intense, many parameters of time-intensity curves, such as peak intensity, duration, and area under the odor peak, can be used to present the perception of volatiles. Another variant in time-intensity method is the finger span method.[110] The difference between the finger span method and OSME is rating procedure. The finger span method uses cross-modality matching called "Finger Span," proposed by S.S. Stevens,[20,114] but OSME uses direct scaling on a 16-point scale. The advantage of time-intensity methods is obtaining a complete chromatogram in one chromatographic run; however, the disadvantage is individual differences in scale usage that requires training and complex data treatment before the results can be interpreted meaningfully. Time-intensity methods have been proved to be reliable, and the data obtained from the methods are more precise than that from AEDA.[109–113,115–119]

Another GCO method is the intensity-rating method. This method can be performed right after sniffing, so it is called posterior intensity-rating technique.[103,104]

The data from the intensity-rating method is similar to the peak intensity obtained from the time-intensity method, but there is no time component involved in the responses. On the other hand, these techniques are short versions of time-intensity methods. This technique enjoys the same advantages as time-intensity methods and also suffers from the same disadvantages. However, the rating task in the posterior technique is easier than that in the time-intensity method. The intensity-rating techniques have not appeared often in reports, but several studies showed promising results.[120,121]

Neither detection frequency method nor olfactometry global analysis uses direct scaling techniques as time-intensity and intensity-rating methods require, nor serial dilution techniques. The method uses one representative extract and asks subjects (at least nine subjects) to smell eluted volatiles. Then the subjects indicate the presence of the volatiles. The number of subjects who detect the volatiles will be tallied as the frequency of detection and represents the intensity of the volatiles.[104,122] This method is quicker than the other methods,[117] but it does not represent the real intensity as do time-intensity and intensity-rating methods nor determine threshold concentration of volatiles. This method is not useful when volatiles are detected by every subject; on the other hand, the concentrations of volatiles are well above threshold levels. There are methods called response interval methods, which employ the detection ability of subjects, but instead of recording detection frequency, they record time duration that a volatile is detected. The response interval methods suffer when the eluted peaks of volatiles are not symmetrical or flattened, and they depend on chromatographic conditions rather than real psychophysical properties of volatiles.[103]

12.11   ELECTRONIC NOSE AND ELECTRONIC TONGUE

With the advance of sensor and computer technologies, scientists are able to develop instruments that mimic the olfactory and gustation system, called electronic nose (e-nose) and electronic tongue (e-tongue). At the most basic level, human sense systems comprise receptors and the brain. In olfactory and gustatory systems, the receptors are believed to be a class of protein called G-protein that is genetically tuned to react to certain classes of chemicals.[123] The neuronal signals from the receptors are transduced as patterns and processed by the brain. Similarly, current e-noses and e-tongues use arrays of sensors to detect chemicals and react with distinctive patterns toward different chemicals just as receptors behave in the olfactory and gustatory systems, but the sensor arrays are much less diverse than the receptor arrays in the olfactory and gustatory systems. In the human olfactory system, at least 1000 genes are estimated to govern the expression of G-protein, which means there are at least 1000 types of receptors and millions of receptors on our olfactory epithelium, which is about the size of a postage stamp.[41,123,124] The brain part of e-noses and e-tongues is a computer with pattern recognition programs. The programs are those explained in the multivariate analysis section of this chapter such as PCA, DA, and PLS. Even though the sensor and the brain of e-noses and e-tongues are not as sophisticated as human sense systems, the instruments operate very well for limited functions.

The performance of e-noses and e-tongues depends on types of detection sensors used in the fabrication of the instruments. In e-noses and e-tongues, sensors are components that signal in the presence of chemical (volatiles or nonvolatiles) reversibly in real time (response time as short as possible). The sensors are not destroyed or changed by the act of sensing.[125] There are three types of sensors generally used: metal-oxide semiconducting sensors, conducting polymer sensors, and quartz microbalance sensors.[126–128] Metal-oxide sensors detect the changes in electrical conductivities caused by the deposition of compounds on the surface of sensors. Polymer-based sensors also detect the changes in electrical conductivities, but the changes are caused by the swelling of polymers coating the surface of the sensor. The swelling increases the resistance within the system of the sensor. Quartz sensors operate differently from the other sensors. The deposition of compounds on the sensor causes the changes in oscillating frequency, and different compounds change the frequency in different degrees.[126,127,129] A reference frequency is usually set before the use of the quartz sensor, and the deviations from the reference frequency are the responses. Besides those three sensors, a fluorescent sensor array has also been developed. The fluorescent sensors emit lights (may be visible or nonvisible) in distinctive patterns when the sensors are contacted with chemicals.[130] The sensitivities of sensors range from part per billion (ppb) to part per trillion (ppt), depending on the types of sensors. The quartz base sensor was reported to detect volatile compounds in the ppt-range within 10 seconds.[127,129]

Every new e-nose and e-tongue must be standardized to detect, recognize, and respond to the pattern of a stimulus (volatiles for e-noses and nonvolatiles for e-tongues); therefore, the pattern is a characteristic "fingerprint" of the stimulus, which is stored in a computer.[131,132] The selectivity of e-noses and e-tongues depends on sensor types and pattern recognition programs. Sensor arrays can be manipulated to be highly specific and sensitive to a certain compound and this is an attractive property of e-noses and e-tongues over the human nose. E-noses require much more powerful selectivity and discrimination ability than e-tongues because of the more diverse attributes in olfaction than gustation (not flavor). The four basic tastes—salty, sour, sweet, and bitter—are constant across most taste perceptions, and taste interaction is not as drastic as that of olfaction. In olfaction, new aromas are perceived by changing concentration of an odorant and/or mixing odorant, and the changes are called odor interaction.[133–136]

The applications of e-noses and e-tongues are numerous and are not limited to food. Actually, e-nose and e-tongue technologies are geared toward applications to which the human nose cannot be applied. In food, e-noses and e-tongues have been used in many applications; for example, to differentiate beer batch variations, model bitterness of beer, detect impurities in raw materials, determine food freshness, distinguish vine-ripen tomatoes from green-picked tomatoes, food from different processing, and solvent residues in food package.[126,130,131,137,138] The applications of e-noses and e-tongues beyond foods and food-related products have received much attention because high risk and safety benefits are involved in the applications; and these applications are possible because of the ability to fine-tune e-noses and e-tongues to detect certain compounds that humans cannot perceive or that are not safe for humans to test; for example, detection of land mines, poisons, nerve agents, biological weapons, and infectious deceases.[132,139–142]

REFERENCES

1. Lawless, H. T. Descriptive analysis of complex odors: Reality, model or illusion? *Food Qual. Prefer.*, *10*, 325–332, 1999.
2. Meilgaard, M. S.; Civille, G. V.; Carr, B. T. *Sensory Evaluation Techniques.* 2nd edn., CRC: New York, 1991.
3. Jones, L. V.; Peryam, D. R.; Thurstone, L. L. Development of a scale for measuring soldiers' food preferences. *Food Res.,* *20*, 512–520, 1955.
4. Lawless, H. T.; Heymann, H. *Sensory Evaluation of Food: Principles and Practices.* 2nd edn., Aspen: Maryland, 1999.
5. Baird, J. C.; Noma, E. *Fundamentals of scaling and psychophysics.* John Wiley & Sons: New York, 1978.
6. Wixted, J., ed., *Methodology in Experimental Psychology.* 3rd edn., vol. 4, John Wiley & Sons: New York, 2002.
7. Moskowitz, H. R. *Applied Sensory Analysis of Foods.* vol. I & II, CRC Press: Boca Raton, FL, 1988.
8. Stone, H.; Sidel, J. L.; Oliver, S.; Woolsey, A.; Singleton, R. C. Sensory evaluation by quantitative descriptive analysis. *Food Technol.,* *28*, 24, 26, 28, 29, 32, 34, 1974.
9. Stone, H.; Sidel, J. L. Quantitative descriptive analysis: Developments, applications, and the future. *Food Technol.,* *52*, 48–52, 1998.
10. Anderson, N. H. Functional measurement and psychological judgment. *Psychol. Rev.,* *77*, 153–170, 1970.
11. Anderson, N. H. *Algebraic models in perception.* vol. 2, Academic Press: New York, 1974.
12. Marks, L. E.; Gescheider, G. A. "Psychophysical scaling," in *Methodology in Experimental Psychology.* J. Wixted, Ed., John Wiley & Sons: New York, 2002.
13. Brandt, M. A.; Skiner, E. Z.; Coleman, J. A. The texture profile method. *J. Food Sci.,* *28*, 404–409, 1963.
14. Szczesniak, A. S. Classification of textural characteristics. *J. Food Sci.,* *28*, 385–389, 1963.
15. Szczesniak, A. S.; Brandt, M. A.; Friedman, H. H. Development of standard rating scales for mechanical parameters of texture and correlation between the objective and the sensory method of texture evaluation. *J. Food Sci.,* *28*, 397–403, 1963.
16. Zwislocki, J. J. Group and individual relations between sensation magnitudes and their numerical estimates. *Percept. Psychophys.,* *33*, 460–468, 1983.
17. Stevens, J. C.; Mack, J. D.; Stevens, S. S. Growth of sensation on seven continua as measured by force of handgrip. *J. Exp. Psy.,* *59*, 60–67, 1960.
18. Stevens, J. C.; Marks, L. E. Cross-modality matching of brightness and loudness. *Proc. Natl. Acad. Sci.,* *54*, 407–411, 1965.
19. Stevens, J. C.; Marks, L. E. Cross-modality matching functions generated by magnitude estimation. *Percept. Psychophys.,* *27*, 379–389, 1980.
20. Stevens, S. S. Cross-modality validation of subjective scales for loudness, vibration and electronic shock. *J. Exp. Psy.,* *57*, 201–209, 1959.
21. Arnold, G.; Williams, A. A. "The use of Generalized Procrustes technique in sensory analysis," in *Statistical Procedures in Food Research.* J. R. Piggott, ed.; Elsevier Applied Science: London, pp. 233–254, 1986.
22. Williams, A. A.; Arnold, G. M. A comparison of the aromas of six coffees characterised by conventional profiling, free-choice profiling and similarity scaling methods. *J. Sci. Food Agric.,* *36*, 206–214, 1985.
23. Williams, A. A.; Arnold, G. M. "A new approach to sensory analysis of foods and beverages," in *Progress in Flavour Research: Proceedings of the 4th Weurman Flavour Research Symposium.* J. Adda, ed., Elsevier: Amsterdam, the Netherlands, pp. 33–50, 1985.

24. Williams, A. A.; Langron, S. P. The use of free-choice profiling for the evaluation of commercial ports. *J. Sci. Food Agric.*, *35*, 558–568, 1984.

25. Langron, S. P. "The application of procrustes statistics to sensory profiling," in *Sensory Quality in Foods and Beverages*. A. A. Williams and R. K. Atkin, eds., Ellis Horwood: Chickester, pp. 89–95, 1983.

26. Thomson, D. M. H.; Macfie, H. J. H. "Is there an alternative to descriptive sensory assessment?" in *Sensory Quality in Foods and Beverages: Its Definition, Measurement and Control.* A. A. Williams and R. K. Atkin, eds.; Horwood: Chichester, UK, pp. 96–107, 1983.

27. Gower, J. C. Generalized procrustes analysis. *Psychometrika, 40*, 33–51, 1975.

28. Schlich, P. SAS European Users Group International (SEUGI) Conference, Cologne, Germany, pp. 529–537.

29. Lachnit, M.; Busch-Stockfisch, M.; Kunert, J.; Krahl, T. Suitability of Free Choice Profiling for assessment of orange-based carbonated soft-drinks. *Food Qual. Pref.*, *14*, 257–263, 2003.

30. Schiffman, H. R. *Sensation and Perception: An Integrated Approach*, 4th edn.; John Wiley & Sons: New York, 1996.

31. Srinivasan, M. A.; Dandekar, K. An investigation of the mechanics of tactile sense using two dimensional models of the primate fingertip. *J. Biomech. Eng.*, *118*, 48–55, 1996.

32. Srinivasan, M. A. Surface deflection of primate fingertip under line load. *J. Biomech. Eng., 22*, 343–349, 1989.

33. Greenspan, J. D. A comparison of force and depth of skin indentation upon psychophysical functions of tactile intensity. *Somatosens. Res., 2*, 33–48, 1984.

34. Philips, J. R.; Johnson, K. O. Tactile spatial resolution. II Neural representation of bars, edges, and gratings in monkey primary afferents. *J. Neurosci., 46*, 1192–1203, 1981.

35. Philips, J. R.; Johnson, K. O. Tactile spatial resolution. III. A continuum mechanics model of skin predicting mechanoreceptor responses to bars, edges, and gratings. *J. Neurophys., 46*, 1204–1224, 1981.

36. van Doren, C. L. A model of spatiotemporal tactile sensitivity linking psychophysics to tissue mechanics. *J. Acoust. Soc. Am., 85*, 2065–2080, 1989.

37. Greenspan, J. D.; Bolanowski, S. J. "The psychophysics of tactile perception and its peripheral physiological basis," in *Pain and Touch*. L. Kruger, ed., Academic Press: San Diego, CA, pp. 25–104, 1996.

38. Ilyinski, O. G.; Akoev, G. N.; Krasnikova, T. L.; Elman, S. I. K and Na ion content in the Pacinian corpuscle fluid and its role in the activity of receptors. *Pflugers Archiv. 361*, 279–285, 1976.

39. Smith, C. U. M. "Olfaction," in *Biology of Sensory Systems*. John Wiley & Sons: West Sussex, pp. 176–193, 2000.

40. Takagi, S. F. "The olfactory reception organ," in *Human Olfaction*. University of Tokyo Press: Tokyo, pp. 101–146, 1989.

41. Buck, L. B. Information coding in the vertebrate olfactory system. *Annu. Rev. Neurosci., 19*, 517–544, 1996.

42. Kanda, T.; Kitamura, T.; Kaneko, T.; Iizumi, O. Scanning electron microscopical observation of the olfactory epithelium of the guinea pig and man. *Nippon Jibiinkoka Gakkai Kaiho., 76*, 739–743, 1973.

43. Menco, B. P. M.; Dodd, G. H.; Davey, M.; Bannister, L. H. Presence of membrane particles in freeze-etched bovine olfactory cilia. *Nature, 263*, 597–599, 1976.

44. Getchell, T. V.; Heck, G. L.; DeSimone, J. A.; Price, S. The location of olfactory receptor sites: Inferences from latency measurements. *Biophys. J., 29*, 397–412, 1980.

45. Blaurstein, D. N.; Simmons, R. B.; Burgess, M. F.; Derby, C. D.; Nishikawa, M.; Olson, K. S. Ultrastructural localization of 5'AMP odorant receptor sites on the dendrites of olfactory receptor neurons of the spiny lobster. *J. Neurosci., 13*, 2821–2828, 1993.

46. Moran, D. T.; Rowley, J. C. I.; Jafek, B. W.; Lovell, M. A. The fine structure of the olfactory mucosa in man. *J. Neurocytol., 11*, 721–746, 1982.

47. Getchell, M. L.; Rafols, J.; Getchell, T. V. Histological and histochemical studies of the secretory components of the salamander olfactory mucosa: Effects of isoproterenol and olfactory nerve section. *Anat. Rec., 208*, 553–565, 1984.

48. Ache, B. W.; Restrepo, D. "Olfactory transduction," in *The Neurobiology of Taste and Smell*. T. E. Finger, W. L. Silver and D. Restrepo, eds., John Wiley & Sons: New York, pp. 159–177, 2000.

49. Hornung, D. E.; Mozell, M. M. Factors influencing the differential sorption of odorant molecules across the olfactory mucosa. *J. Gen. Physio., 69*, 343–361, 1977.

50. Meisami, E.; Mikhail, L.; Baim, D.; Bhatnagar, K. P. Human olfactory bulb: Aging of glomeruli and mitral cells and a search for the accessory olfactory bulb. *Annual New York Academy of Science, 855*, 708–715, 1998.

51. Calson, N. R. *Physiology of Behavior.* Allyn and Bacon: Boston, 1991.

52. Stone, H.; Sidel, J. L. *Sensory Evaluation Practices*. 2nd edn., Academic Press: New York, 1993.

53. Smith, C. U. M. "General features of sensory systems," in *Biology of Sensory Systems,* John Wiley & Sons: West Sussex, 2000.

54. Moskowitz, H. Product Testing and Sensory Evaluation of Foods: Marketing and R&D Approaches. Food & Nutrition Press: Connecticut, 1983.

55. Stevens, S. S. On the theory of scales of measurement. *Science, 103*, 677–680, 1946.

56. Macmillan, N. A. "Signal Detection Theory," in *Methodology in Experimental Psychologoy*. J. Wixted, ed., John Wiley & Sons: New York, pp. 43–90, 2002.

57. Peryam, D.; Girardot, N. F. Advanced taste-test method. *Food Eng., 24*, 58–61, 194, 1952.

58. Stevens, S. S. The direct estimation of sensory magnitudes-loudness. *J. Psychol., 119*, 1–25, 1956.

59. American Society for Testing and Materials. "ASTM Standard test method for unipolar magnitude estimation of sensory attributes. Method E-1697–95," in *Annual Book of ASTM Standards*, American Society for Testing and Materials: Conshoocken, Philadelphia, pp. 114–121, 1999.

60. Anonymous, *Webster's Illustrated Dictionary*. Publication International: Lincoln Wood, IL, 1994.

61. Brown, D. G. W.; Clapperton, J. F.; Meilgaard, M. S.; Moll, M. Flavor thresholds of added substances. *J. Am. Soc. Brew. Chem., 36*, 73–80, 1978.

62. American Society for Testing and Materials. "ASTM Standard practice for defining and calculating individual and group sensory thresholds from forced-choice data sets of intermediate size. E-1432–91," in *Annual Book of ASTM Standards*, American Society for Testing and Materials: Philadelphia, pp. 75–82, 1999.

63. American Society for Testing and Materials. "ASTM Standard practice for determination of odor and taste thresholds by a forced-choice ascending concentration series method of limits. E-679–91," in *Annual Book of ASTM Standards,* American Society for Testing and Materials: Philadelphia, pp. 35–39, 1999.

64. Fechner, G. T. *Elemente der Psychophysik*. Breikopf und Hartel: Leipzig, 1864.

65. Marin, A. B.; Barnard, J.; Darlington, R. B.; Acree, T. E. Sensory thresholds: Estimation from dose-respopnse curves. *J. Sensory Studies*, 205–225, 1991.

66. Stevens, D. A.; O'Connel, R. J. Individual differences in thresholds and quality reports of human subjects to various odors. *Chem. Senses., 16*, 57–67, 1991.

67. Stone, H. Determination of odor difference limens for three compounds. *J. Exp. Psychol., 66*, 466–473, 1963.

68. Stone, H. Behavioral aspects of absolute and differential olfactory sensitivity. *Annual New York Academy of Science, 116*, 527–534, 1964.

69. Stone, H.; Ough, C. S.; Pangborn, R. M. Determination of odor difference thresholds. *J. Food Sci., 27,* 197–202, 1962.

70. Stone, H.; Bosley, J. J. Olfactory discrimination and Weber's law. *Percept. Mot. Skills, 20,* 657–665, 1965.

71. Lundahl, D. S.; Lukes, B. K.; McDaniel, M. R.; Henderson, L. A. A semi-ascending paired difference method for determining sensory thresholds of added substances to background media. *J. Sensory Studies, 1,* 291–306, 1986.

72. Getchell, M. L.; Zielinski, B.; Getchell, T. V. "Odorant and autonomic regulation of secretion in the olfactory mucosa," in *Molecular Neurobiology of the Olfactory System: Molecular, Membranous, and Cytological Studies.* F. L. Margolis and T. V. Getchell, eds., Plenum Press: New York, pp. 71–98, 1988.

73. Cain, W. S.; Murphy, C. L. Interaction between chemoreceptive modalities of odor and irritation. *Nature, 284,* 255–257, 1980.

74. Pause, B. M.; Sojka, B.; Ferstl, R. Central processing of odor concentration is a temporal phenomenon as revealed by chemosensory event-related potentials (CSERP). *Chem. Senses., 22,* 9–26, 1997.

75. Siegel, S.; Castellan, N. J. J. *Nonparametric Statistics for the Behavioral Sciences,* 2nd edn., McGraw-Hill: New York, 1988.

76. Stevens, J. *Applied Multivariate Statistics for the Social Sciences,* 3rd edn.; Lawrence Erlbaum Associates: Mahwah, NJ, 1996.

77. Ennis, D. M.; Bi, J. The Beta-Binomial model: Accounting for inter-trial variation in replicated difference and preference test. *J. Sensory Studies, 13,* 389–412, 1998.

78. Bi, J. The double discrimination methods. *Food Qual. Prefer., 12,* 507–513, 2001.

79. Bi, J.; Ennis, D. M. A Thurstonian variant of the Beta-Binomial model for replicated difference tests. *J. Sensory Studies, 13,* 461–466, 1998.

80. Bi, J. Comparison of correlated proportions in replicated product tests. *J. Sensory Studies, 17,* 105–114, 2002.

81. Chochran, W. G. Some methods for strengthening the common chi-square tests. *Biometrics, 10,* 417–451, 1954.

82. Steel, R. G. D.; Torrie, J. H.; Dickey, D. A. *Principles and Procedures of Statistics: A Biometrical Approach.* 3rd edn.; McGraw-Hill: Boston, MA, 1997.

83. Everitt, B. S. *Cluster Analysis,* 3rd edn., Helsted Press: New York, 1993.

84. Kachigan, S. K. *Statistical Analysis: An Interdisciplinary Introduction to Univariate & Multivariate Methods,* Radius Press: New York, 1986.

85. Romney, A. K.; Shepard, R. A.; Nerlover, S. B., eds., *Multidimensional Scaling: Theory and Applications in the Behavioral Sciences.* vol. II: Applications, Seminar Press: New York, 1972.

86. Schiffman, S. S.; Reynolds, M. L.; Young, F. W. *Introduction to Multidimensional Scaling: Theory, Methods, and Applications*; Academic Press: New York, 1981.

87. Piggott, J. R., ed., *Statistical Procedures in Food Research.* Elsevier Applied Science Publishers: Essex, England, 1986.

88. Martens, H.; Martens, M. *Multivariate Analysis of Quality: An Introduction.* John Wiley & Sons: West Sussex, England, 2001.

89. Martens, M.; Bredie, W. L. P.; Martens, H. Sensory profiling data studied by partial least squares regression. *Food Qual. Prefer., 11,* 147–149, 2000.

90. Dijksterhuis, G. B. *Multivariate Data Analysis in Sensory and Consumer Science.* Food & Nutrition Press: Trumbull, CT, 1997.

91. Pearson, K. On lines and planes of closest fit to systems of points in space. *Phil. Mag., Sixth Series, 2,* 559–572, 1901.

92. Borgognone, M. G.; Bussi, J.; Hough, G. Principal component analysis in sensory analysis: Corvariance or correlation matrix? *Food Qual. Prefer,. 12,* 323–326, 2001.

93. Piggott, J. R.; Sharman, K. "Methods to aid interpretation of multidimensional data," in *Statistical Procedures in Food Research.* J. R. Piggott, ed., Elsevier Applied Science: New York, pp. 181–232, 1986.

94. De Leeuw, J.; Young, F. W.; Takane, Y. Additive structure in qualitative data: An alternating least squares method with optimal scaling features. *Psychometrika, 41,* 471–503, 1976.

95. De Leeuw, J.; van Rijckevorse, J. "HOMAL and PRINCALS-Some generalizations of principal components analysis," in *Data Analysis and Informatics.* E. E. A. Diday, ed., North-Holland: Amsterdam, 1980.

96. Breenacre, M. J. Correspondence analysis of multivariate categorical data by weighted least-squares. *Biometrics, 75,* 457–467, 1988.

97. Schiffman, S. S.; Beeker, T. G. "Multidimensional scaling and its interpretation," in *Statistical Procedures in Food Research.* J. R. Piggott, ed., Elsevier Applied Science: New York, pp. 255–291, 1986.

98. Jacobsen, T.; Gunderson, R. W. "Applied cluster analysis," in *Statistical Procedures for Food Research*; J. R. Piggott, ed., Elsevier Applied Science: New York, pp. 361–408, 1986.

99. MacFie, H. J. H.; Thomson, D. M. H. "Preference mapping and multidimensional Scaling," in *Sensory Analysis of Foods.* J. R. Piggott, ed., Elsevier Science Publishers Ltd.: Essex, pp. 381–409, 1988.

100. Faber, N. M.; Mojet, J.; Poelman, A. A. M. Simple improvement of consumer fit in external preference mapping. *Food Qual. Prefer., 14,* 455–461, 2003.

101. Carroll, J. D. "Individual difference and multidimensional scaling," in *Multidimensional Scaling: Theory and Applications in the Behavioral Sciences.* R. N. Shepard, A. K. Romney, and S. B. Nerlover, eds., Seminar Press: New York, pp. 105–155, 1972.

102. Acree, T. E. GC/Olfactometry. *Anal. Chem. News Featur.,* 170A–175A, 1997.

103. Acree, T. E.; Barnard, J. "Gas chromatography–olfactometry and Charm Analysis™," in *Trends in Flavour Research.* H. Maarse and D. G. van der Heij, eds., Elsevier: Amsterdam, pp. 211–220, 1994.

104. van Ruth, S. M. Methods for gas chromatography–olfactometry: A review. *Biomol. Eng., 17,* 121–128, 2001.

105. Ullrich, F.; Grosch, W. Identification of the most intense odor compounds formed during autoxidation of linoleic acid. *Z. Lebensm Unters Forsch, 184,* 277–282, 1987.

106. Acree, T. E.; Barnard, J.; Cunningham, D. G. A procedure for the sensory analysis of gas chromatographic effluents. *Food Chem., 14,* 273–286, 1984.

107. Blank, I. "Gas chromatography–olfactometry in food aroma analysis," in *Techniques for Analyzing Food Aroma.* R. T. Marsili, ed., Marcel Dekker: New York, pp. 293–329, 1997.

108. Mistry, B. S.; Reineccius, T.; Olson, L. K. "Gas chromatography–olfactometry for the determination of key odorants in foods," in *Techniques for Analyzing Food Aroma.* R. T. Marsili, ed., Marcel Dekker: New York, pp. 265–292, 1997.

109. da Silva, M. A. A. P.; Lundahl, D. S.; McDaniel, M. R. "The capability and psychophysics of OSME: A new GC-olfactometry technique," in *Trends in Flavours Research.* H. Maarse and D. G. Van der Heij, eds., Elsevier: Amsterdam, pp. 191–208, 1994.

110. Guichard, H.; Guichard, E.; Langlois, D.; Issanchou, S.; Abbott, N. GC sniffing analysis: Olfactive intensity measurement by two methods. *Z. Lebensm Unters Forsch, 201,* 344–350, 1995.

111. McDaniel, M. R.; Miranda-Lopez, R.; Watson, B. T.; Micheals, N. J.; Libbey, L. M. "Pino noir aroma: A sensory/gas chromatographic approach," in *Flavors and Off-Flavors.* G. Charalambous, ed., Elsevier Science Publishers: Amsterdam, the Netherlands, pp. 23–25, 1990.

112. Miranda-Lopez, R.; Libbey, L. M.; Watson, B. T.; McDaniel, M. R. Identification of additional odor-active compounds in Pinot Noir wines. *Am. J. Enol. Vitic., 43,* 90–92, 1992.

113. Sanchez, N. B.; Lederee, C. L.; Nickerson, G. B.; Libbey, L. M.; McDaniel, M. R. "Sensory and analytical evaluation of beers brewed with three varieties of hops and an unhopped beer," in *Food Science and Human Nutrition*. G. Charalambous, ed.; Elsevier Science Publisher: Amsterdam, pp. 403–426, 1992.

114. Stevens, S. S. On the psychophysical law. *Psychol. Rev., 64*, 153–181, 1957.

115. Etievant, P. X.; Callement, G.; Langlois, D.; Issanchou, S.; Coquibus, N. Odor intensity evaluation in gas chromatography-olfactometry by Finger Span method. *J. Agric. Food Chem., 47*, 1673–1680, 1999.

116. Callement, G.; Buchet, M.; Langlois, D.; Etievant, P.; Salles, C. "Odor intensity measurements in gas chromatography–olfactometry using cross modality matching: evaluation of training effects," in *Gas Chromatography-Olfactometry: The State of the Art*. J. V. Leland, P. Schieberle, A. Buettner, and T. E. Acree, eds., American Chemical Society: Washington, D.C., 2001.

117. Le Guen, S.; Demaimay, M. Characterization of odorant compounds of mussels (Mytilus edulis) according to their origin using gas chromatography-olfactometry and gas chromatography-mass spectometry. *J. Chrom., 896*, 361–371, 2000.

118. Plotto, A.; Mattheis, J. P.; Lundahl, D. S.; McDaniel, M. R. "Validation of gas chromatography olfactometry results for 'Gala' apples by evaluation of aroma-active compound mixtures," in *Flavor Analysis: Developments in Isolation and Characterization*. C. J. Mussinan and M. J. Morello, eds., American Chemical Society: Washington, D.C., pp. 290–302, 1998.

119. Etievant, P. X. "Odour intensity evaluation in GC-Olfactometry by finger span method," in *Analysis of Taste and Aroma*. J. F. Jackson and H. F. Linskens, eds. Springer-Verlag Berlin Heidelberg: New York, pp. 223–237, 2002.

120. Casimir, D. J.; Whitfield, F. B. Flavour impact values: A new concept for assigning numerical values for potency of individual flavour components and their contribution to overall flavour profile. *Ber. Int. Fruchtsaftunion, 15*, 325–345, 1978.

121. Chamblee, T. S.; Clark, B. C. "Lemon and lime citrus essential oils analysis and organoleptic evaluation," in *Biovolatile Compounds in Plants. ACS Symposium Series 525*. R. Teranishi and H. Sugisawa, eds., American Chemical Society, pp. 88–102, 1992.

122. Pollien, P.; Ott, A.; Montigon, F.; Baumgartner, M.; Munoz-Box, R.; Chaintreau, A. Hyphenated headspace-gas chromatography-sniffing technique: Screening of impact odorants and quantitative aromagram comparisons. *J. Agric. Food Chem. 45*, 2630–2637, 1997.

123. Lopetcharat, K. Sub-threshold effects on the perceived intensity of recognizable odorants: the roles of functional groups and carbon chain lengths. PhD thesis, Oregon State University, 2002.

124. Buck, L. B.; Axel, R. A novel multigene family may encode odorant receptors: A molecular basis for odor recognition. *Cell, 65*, 175–187, 1991.

125. Czarnick, A. W. Desperately seeking sensors. *Chem. Biol., 2*, 423–428, 1995.

126. van Deventer, D.; Mallikarjunan, P. Comparative performance analysis of three electronic nose systems using different sensor technologies in odor anaysis of retained solvents on printed packaging. *J. Food Sci., 67*, 3170–3183, 2002.

127. Allan, R. Saw-based electronic nose accurately characterizes vapors in seconds. *Electron. Des., 48*, 30, 32, 34, 2000.

128. Zambov, L. M.; Copov, C.; Plass, M. F.; Bock, A.; Jelinek, M.; Lancok, J.; Masseli, K.; Kulisch, W. Capacitance humidity sensory with carbon nitride detecting element. *Appl. Phys., 70*, 603–606, 2000.

129. Amato, I. Getting a whiff of PittCon. *Science, 251*, 1431–1432, 1991.

130. Perkins, S. Eau, brother! Electronic noses provide a new sense of the future. *Sci. News, 157*, 125–127, 2000.

131. Fremantle, M. Instrument develops its 'eye' for smell. *C&EN*, 30–31, 1995.

132. Shope, R.; Fisher, D. Enose is enose is enose. *Technol. Teach., 60*, 14, 15, 16, 17, 2000.

133. Berglund, B.; Berglund, U.; Lindvall, T. Psychological processing of odor mixtures. *Psychol. Rev., 83*, 432–441, 1976.

134. Berglund, B.; Berglund, U.; Lindvall, T. On the principle of odor interaction. *Acta Psychologica., 35*, 255–268, 1971.

135. Berglund, B.; Berglund, U.; Lindvall, T.; Svensson, L. T. A quantitative principle of perceived intensity summation in odor mixtures. *J. Exp. Psychol., 100*, 29–38, 1973.

136. Lopetcharat, K.; McDaniel, M. R. Functional group dependent sub-threshold suppression of the perceived intensity of recognizable odorants in binary mixtures. Ph.D Thesis, Oregon State University, 2002.

137. Tan, T.; Schmitt, V.; Isz, S. Electronic tongue: A new dimension in sensory analysis. *Food Technology, 55*, 44, 46, 48, 50, 2001.

138. Sprout, A. L. A new way to sniff products. *Fortune, 130*, 150, 1994.

139. Keshri, G.; Voysey, P.; Magnan, N. Early detection of spoilage moulds in bread using volatile production patterns and quantitative enzyme assays. *J. Appl. Microbiol., 92*, 165–172, 2002.

140. Cassagrande, R. Bioterrorism; biochips, biological warfare; terrorism. *Sci. Am., 287*, 82–87, 2002.

141. Dalton, L. W.; Hishberg, C.; Sinha, G. Gee, your car smells terrific! You feel OK?: A scientist stalks the scent of a new machine. *Popular Sci., 260*, 38, 2001.

142. Sinha, G. That's one smart schnozz: How attractive is your genetic codes? Ask a nose. *Popular Science, 259*, 42, 2001.

13 Determination of Food Allergens and Genetically Modified Components

Kristina M. Williams, Mary W Trucksess,
Richard B. Raybourne, Palmer A.
Orlandi, Dan Levy, Keith A. Lampel,
and Carmen D. Westphal

CONTENTS

13.1 INTRODUCTION TO FOOD ALLERGENS

A food allergen is any component of food that induces an adverse immunologic response when ingested by a sensitized individual. The immune mechanisms involved are classified as IgE-mediated or non-IgE-mediated and can result in a wide range of cutaneous, respiratory, gastrointestinal, or cardiovascular symptoms.[1] The majority of immediate allergic responses to foods are related to food proteins that are absorbed undigested through the intestinal epithelium, where they bind to specific IgE and trigger the release of mediators from intestinal mucosal and submucosal mast cells.[2] A proportion of food allergens may become disseminated by gaining access to mucosal vasculature, resulting in generalized symptoms such as urticaria or anaphylaxis. Food allergies are more common in the first few years of life, probably due to immaturity of the intestinal barrier and the mucosal immune system. Although accurate epidemiological surveys are lacking, it is estimated that food allergies affect approximately 6% of children under 3 years of age and about 2% of adults in the United States.[3,4] However, the prevalence of allergic reactions to specific foods varies geographically and may depend on the frequency of consumption, the age of introduction into the diet, and the genetic makeup of the sensitized population. There has been an apparent increase in the incidence of certain food allergies over that in the past decade, but only a few of these reports are well documented.[5-7] The common misidentification of other food intolerances as allergic reactions may explain some of these observations.

Although food allergens have been identified from diverse plant and animal sources, similarities in the physiochemical characteristics of many food allergens have prompted an intensive search for properties that are predictive of allergenic potential. Most food allergens are proteins having molecular weights ranging from 10,000 to 70,000 Da. Most allergenic proteins are soluble in aqueous solvents, have acidic isoelectric points, and are subjected to posttranslational modifications such as glycosylation, phosphorylation, or N-termination modification.[8-11] Many allergenic proteins are resistant to heat and chemical denaturation[12,13] and enzymatic digestion, as demonstrated by using "simulated" gastric and intestinal fluids.[14] Efforts to determine a structural basis for allergenicity have focused on using comparative alignment of allergen amino acid sequences, sequential IgE-binding sequences, or three-dimensional modeling.[15-17] Although a high degree of sequence homology and similarity of tertiary structure have been demonstrated by some food allergens,[18,19] no conserved structures or sequence motifs that are predictive of allergenic potential have emerged. Unfortunately, the structural and functional heterogeneity of allergenic proteins, as well as the diversity of the human immune response to them, has hindered efforts to identify any universal allergenic characteristic.

Most allergenic foods contain more than one allergenic protein, which can be subdivided into major (recognized by 50% of sensitized patients) or minor (recognized by 50% of sensitized patients) allergens. Although many major allergens represent a majority of the total protein of a particular food, other major allergens constitute only a minor component of the food product. More than 160 allergenic foods, each containing one or more allergenic proteins, have been identified.[20] Each protein may contain multiple IgE-binding epitopes.[21-26] The specificity of IgE binding is variable among allergic individuals, and the level of binding is not necessarily representative of the severity of the clinical response.[27-29] This has complicated the identification, purification, and biochemical characterization of allergenic proteins, which may be useful for diagnostic purposes or detection methods. Over 90% of severe allergic reactions to foods are caused by eight foods or food groups: milk, eggs, peanuts, soybeans, tree nuts, wheat, fish, and crustacea. The Codex Alimentarius Commission, established by the Food and Agriculture Organization (FAO) and the World Health Organization (WHO) to develop food standards, has elaborated a standard for the declaration of these eight food groups on the label of prepackaged foods regardless of the content. Three additional food allergens—sesame seeds, celery, and mustard—have been identified as allergens of emerging international importance and have been included in the labeling proposal from the Commission of the European Union. As a consequence, the development of commercial allergen detection methods has focused on these foods. Because of the geographic variation in the incidence of specific food allergies, many countries have developed labeling standards specific to the needs of their allergic populations (Table 13.1).

Although novel approaches for prophylaxis and immunotherapy of IgE-mediated food allergy are under investigation, protection is not absolute.[30] At present, allergic individuals must rely on strict avoidance of allergens to prevent symptomatic reactions. Despite better awareness of the sensitized consumer and efforts of the healthcare and food industries, accidental consumption of food allergens leading to severe reactions remains a common occurrence. Of major concern to the consumer and the food industry is the presence of potentially allergenic material in foods in which it is not normally present or in situations where labeling ambiguities exist. This can occur by (1) mislabeling of a raw material or the finished product, (2) cross-contact of different foods when processed on shared industrial equipment, (3) increased antigenicity or the introduction of "neoantigens" by process-dependent modification of food proteins,[31–33] (4) complex ingredient terminology (i.e., caseinate instead of milk protein and "may contain" warnings), or (5) collective ingredient labeling (i.e., natural

TABLE 13.1
Mandatory Labeling of Allergenic Ingredients

The United States[a]	European Union	Canada	Japan[b]	Australia/New Zealand
Crustacea	Gluten-containing	Peanuts	Eggs	Gluten-containing cereals
Eggs	cereals and products	Tree nuts	Milk	Crustacea
Fish	Crustacea	Sesame	Wheat	Eggs
Milk	Eggs	Milk	Buckwheat	Fish
Peanuts	Fish	Eggs	Peanut	Milk
Soy	Peanuts	Fish		Nut and sesame seeds
Tree nuts	Soybeans	Crustacea		Peanuts and soybeans
Wheat	Milk and dairy products	Shellfish		
	Tree nuts	Soy		
	Celery	Gluten-containing		
	Mustard	cereals		
	Sesame seed			

[a] Effective January 1, 2006.

[b] Japanese regulations recommend the declaration in the ingredient list of abalone, squid, salmon roe, shrimp/prawn, orange, crab, kiwi fruit, beef, tree nuts, salmon, mackerel, soybeans, chicken, pork, Matsutake mushrooms, peaches, yams, apples, and gelatin.

flavors, spices). An improvement of food labeling regulations could offer some preventive protection for allergic consumers. Unfortunately, the majority of documented recall actions in the United States involve food labeling errors and omissions, and cross-contact during manufacture.[34] Clearly, to protect highly sensitized individuals from accidental consumption, it is necessary to develop and implement assays to detect trace amounts of undeclared allergenic substances in food.

13.1.1 LIMITS OF DETECTION

Analytical methods for food allergens should be able to detect the lowest dose of the allergen that could induce symptoms in a sensitized individual. This detection limit should also allow for a margin of safety for those exquisitely sensitive persons who are more likely to have severe reactions to lower doses. The establishment of a valid "threshold dose" for common allergenic foods depends on clinical data obtained through standardized oral challenge studies. The most reliable data have been obtained for peanuts, eggs, and cow's milk, where double-blind placebo-controlled food challenges have been performed on more than 500 subjects.[35–37] Although there was a lack of uniformity between test protocols (different forms of challenge food and variable criteria for selection of subjects), some useful quantitative data was obtained. The lowest doses provoking symptoms in the respective sensitized test groups ranged from 0.1 to 1000 mg of peanut protein, 0.13 to 200 mg of egg protein, and 1.5 to 180 mg of milk protein. The process of reaching a consensus threshold dose for the most common food allergens is complicated by the fact that most challenge studies exclude the most sensitive subjects, and the no-observed-adverse-effect level (NOAEL) was usually not established. Also, the response of the individual test subjects to specific offending allergenic protein(s) and the bioavailability of these proteins in the challenge material could affect the estimation of the threshold dose. Nonetheless, when examining the available challenge data, it appears that the threshold values for the three common food allergens will fall in the microgram range of food protein. Using peanut as an example, a dose of 0.1 mg of peanut protein equates to 0.4 mg of whole peanut. In an average 50 g portion of food, this level of peanut is equivalent to 8 ppm (mg/g) of peanut material. Based on these limited data, detection levels in the low ppm range should offer protection to all but the most sensitive peanut-allergic consumers. Additional data from standardized clinical challenge studies is prerequisite to establishing legal guidelines for quantitative declaration of allergen content. Such guidelines would benefit the food industry by allowing producers to focus on quality control procedures and reduce the use of precautionary labeling and would assist in the establishment of target detection levels for methods development.

13.2   DETECTION OF ALLERGENIC COMPONENTS OF FOOD

Detection methods can be used to evaluate the presence of allergenic material in raw ingredients, finished products, and equipment wash solutions. Testing for food allergens should become part of the food industry's Good Laboratory Practice (GLP) and Hazard Analysis and Critical Control Points (HACCP) plans. Moreover, testing is a requirement for routine government inspections and other regulatory actions. In addition to the potential adverse health effects on sensitized individuals, the presence of an undeclared allergen can be the origin of an expensive recall action.

Protein- and DNA-based methods have been developed for the qualitative or quantitative assessment of allergens in foods. Immunoassays rely on the use of specific antibodies for the detection of proteins in complex matrices. Immunoassays have been widely used to analyze complex food proteins because of their specificity, sensitivity, and the ability to analyze complex food samples without the need for pre-assay purification steps. DNA-based methods do not analyze the protein directly but detect the gene that encodes for that protein. DNA-based methods for food allergens have been slower to enter the marketplace, probably because of the need to identify and sequence an appropriate target gene. Methods of detection have been described for a variety of allergenic foods, for both in-house use and commercial distribution.

13.2.1   IMMUNOCHEMICAL METHODS FOR THE DETECTION OF ALLERGENIC FOODS

Immunoassays are analytical methods that use antibodies as the main reagent. The most important features of antibodies as analytical reagents are that (1) they can be produced against any biological molecule, including proteins, (2) they have the ability to bind a particular protein in a complex mixture of molecules (specificity), and (3) they can be produced using immunization parameters that will optimize the strength of binding between the antibody and the analyte (sensitivity). There are two different methods of producing antibodies for use in immunoassay. Polyclonal antibodies are those that are produced *in vivo* by animal immunization and serum collection. A polyclonal reagent is characterized by containing a mixture of antibodies with multiple specificities and a range of affinities. Monoclonal antibodies are those that are produced *in vitro* by the fusion of antibody-producing B lymphocytes and immortalized myeloma cells, resulting in "hybridomas." Monoclonal antibodies produced by a single hybridoma cell line will have a singular specificity and affinity. The selection of a polyclonal versus a monoclonal antibody depends on the amount of antibody required, the application, and whether it is desired to have an antibody that detects multiple proteins or epitopes of a single protein (polyclonal) or one that recognizes a single epitope (monoclonal). Most of the detection methods for allergens have used polyclonal antibodies because of their inherent multiple-specificity feature. However, a highly specific and sensitive monoclonal antibody-based immunoassay can also be developed through careful antibody selection.

There are several important issues that must be considered when using immunoassays for the detection of food allergens. First, there are no characterized, standardized food extracts for use in immunoassay development. Each laboratory uses its own food extracts for production of antibodies, for use as control samples, and for development of serial calibrators for quantitative assays. Whole food extracts are difficult to characterize because they include not only allergenic proteins but also nonallergenic proteins and other unknown components. As stated previously, an allergenic food may contain one or more allergenic proteins. The raw material used to develop food extracts, as well as the extraction procedure, can lead to considerable lab-to-lab variability in extract composition. This makes it difficult to draw a comparison between methods developed in different laboratories. To illustrate this point, Hurst et al.[38] evaluated four commercial kits for the detection of peanut in chocolate samples. Because the analysis of the same sample gave different results for each kit, Hurst concluded that these kits were suitable only for qualitative screening rather than for quantitative analysis. Some laboratories have chosen to use purified allergenic proteins as markers for the presence of the allergenic material in food. Examples are the peanut allergen Ara h 1,[39] the egg allergen ovalbumin,[40] the milk allergen β-lactoglobulin,[41,42] Brazil nut 2S protein,[43] and shrimp allergens Pen a 1 or tropomyosin.[44]

The effects of processing on different allergenic proteins must be considered. Allergenic proteins can be degraded, bound by other compounds, or rendered insoluble, thereby reducing their extraction efficiency. Processing can affect the allergenicity and/or immunogenicity of individual proteins through alteration of tertiary structure. This increases the possibility that the allergen may become undetectable using antibodies that were produced against the unprocessed food. A solution to this problem has been to produce antibodies using the processed form of the allergen. Yeung and Collins[45] and Yeung et al.[46] found that antibodies produced against denatured soy, cooked whole egg, and cooked egg white were more suitable for the detection of the corresponding food allergens.

Because food matrices are complex systems, the effects of interfering compounds in the sample extract must be considered. These compounds may include fatty acids, phenolic compounds, surfactants, and endogenous enzymes,[47] all of which may lead to suboptimal assay performance. The complexity of the matrix may also affect sample preparation by reducing the extractability of the analyte. Sample preparation needs to be optimized to maximize recovery from a particular matrix. As an example, dark chocolate is one of the most problematic matrices for the detection of peanuts because of its high content of protein-binding tannins. In this case, the inclusion of additional protein, such as dry-powdered milk or gelatin, in the extraction buffer helps to improve recovery.[48] An acceptable range of recovery is considered to be from 70% to 100%.[49] Matrix components may also be a cause of nonspecific binding to the antibody reagents, resulting in artificially elevated results. This problem can often be reduced by the use of affinity-purified antibodies. Because of the heterogeneity of food matrices, one approach to the validation of the detection method would be to develop in the food matrix standardized reference materials that are similar to that of the test sample.

Immunochemical methods can be qualitative, semiquantitative, or quantitative. For qualitative assays, only one cut-off value is needed, and samples are compared with negative and positive controls. Two-level positive controls are required for semiquantitative assays. In the case of quantitative assays, sample values are compared with that of serial calibrators and a negative control. Currently, there is no limit of tolerance set for any of the food allergens. However, there is a "practical" limit set by the detection limit of a particular assay. There is some controversy concerning the desired limits of detection for food allergens. Allergic individuals and consumer protection groups often desire a detection limit "as low as possible," but this is not a practical solution for the food industry. As a consequence of the lack of theoretical limits of detection, the necessity of quantitative assays is often questioned. Some have suggested that these assays should be used only as indicators of the presence of the allergen rather than as accurate measures of the allergen.[50]

There are numerous immunoassay formats, all of them based on the binding of an antibody to the target analyte. While any of these formats could be adapted for the detection of food allergens, only a few formats have generally been used. To ensure that an immunoassay is suitable for the detection of food allergens (or any analyte), the assay must be evaluated for ruggedness, specificity, recovery, sensitivity, and matrix effects. To ensure performance, full interlaboratory validation studies should be conducted.

13.2.1.1   Enzyme-Linked Immunosorbent Assay

The most common immunoassay for the detection of food allergens is the enzyme-linked immunosorbent assay (ELISA), which uses antibodies bound to a solid support. The solid support is typically a 96-well plate or a polystyrene strip of 8–12 wells, but other materials such as Polyvinylidene fluoride (PVDF) or nitrocellulose membrane can serve this function. In addition to the sensitivity and specificity, other advantages of ELISA are portability, high throughput, and the flexibility offered by the different formats. The ELISA *sandwich* is the format most commonly used for food allergen detection (Figure 13.1). The allergen is bound by two specific antibodies, and the first antibody (capture antibody) is attached to the solid support. In the *direct* ELISA sandwich system, the allergen is directly detected by a second antibody (detector antibody), which is conjugated to an enzyme (peroxidase or phosphatase). In the *indirect* ELISA sandwich assay, a third enzyme-labeled antibody binds to the second antibody, which in this case is unlabeled. A third type of assay is called the *enhanced* ELISA (biotin–streptavidin system), in which the second antibody is attached to a molecule of biotin. A solution of enzyme-coupled streptavidin is then added. Enhancement of sensitivity results because of the ability of the biotin to bind

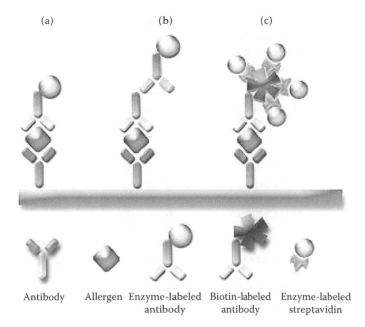

(a)   (b)   (c)

| Antibody | Allergen | Enzyme-labeled antibody | Biotin-labeled antibody | Enzyme-labeled streptavidin |

FIGURE 13.1   ELISA sandwich schemes: (a) direct ELISA, (b) indirect ELISA, and (c) enhanced ELISA.

four molecules of streptavidin. Bound analyte is detected by the formation of a measurable color compound resulting from the catalysis of the substrate by the coupled enzyme. These three assay formats (direct, indirect, and enhanced) can be further divided into *competitive* or *noncompetitive*, depending on whether the capture antibody has to compete for the allergen with antibodies added to the sample extract. In a noncompetitive assay, the concentration of the allergen is directly proportional to the intensity of the developed color. In the case of a competitive assay, the color development is inversely proportional to the concentration of the analyte.

Several laboratories have developed their own ELISAs, which are summarized in Table 13.2. Additionally, there are commercial assays available for several allergens, including peanut, hazelnut, almond, milk, egg, soy, wheat, and sesame seed

TABLE 13.2
ELISA Detection Methods for Food Allergens

Allergen Measured	Format	Limit of Detection/Sensitivity	References
Almond 　Purified amandin	ELISA inhibition Polyclonal Ab	1–100 ng/ml (noncompetitive), 　300 ng/ml (competitive)	[51]
Almond 　Meal from roasted 　almonds	ELISA Polyclonal Ab	1 μg/g	[52]
Cow's milk 　Casein, whey proteins, 　and whole milk	ELISA inhibition Polyclonal Ab	10 ng/ml	[53]
Cow's milk 　β-Lactoglobulin	ELISA indirect competitive Rabbit polyclonal Ab Mouse polyclonal Ab Sheep polyclonal	Sensitivity 0.1–1000 ng/ml 4–50 ng/ml 0.26–50 μg/ml	[54]
Milk Reduced 　β-lactoglobulin Carboxymethylated 　β-lactoglobulin	ELISA Monoclonal Ab	Sensitivity 30 pg/ml 200 ng/ml	[41]
Milk 　Native/denatured 　β-Lactoglobulin	ELISA enhanced Sandwich polyclonal Ab	Sensitivity 0.3 ng/ml	[42]
Hazelnut 　Corylin fraction	ELISA sandwich Polyclonal Ab	0.1–0.2 μg/g Sensitivity 1–600 ng/ml	[55]
Hazelnut Protein extract	ELISA sandwich Polyclonal Ab	1 μg/g Sensitivity 0.5 ng/ml to 1 μg/ml	[56]
Hazelnut Protein extract	ELISA sandwich Polyclonal chicken Ab	Sensitivity 0.03 μg/ml	[57]
Brazil nut	ELISA Polyclonal chicken Ab		[58]

continued

TABLE 13.2 (continued)
ELISA Detection Methods for Food Allergens

Allergen Measured	Format	Limit of Detection/Sensitivity	References
Brazil nut	ELISA competitive	1 µg/g	[43]
Protein extract	Polyclonal Ab		
Hen's egg	ELISA enhanced	0.03% dry powder	[59]
Egg white	Sandwich polyclonal Ab	0.125% sterilized products	
Hen's egg	ELISA competitive	0.2 µg/g	[46]
Cooked whole egg	Polyclonal Ab		
Hen's egg	ELISA sandwich	1 µg/g dried whole egg	[40]
Egg white/ovalbumin	Polyclonal Ab		
Peanut	ELISA	200 µg/g	[60]
Whole extract	Monoclonal/polyclonal	0.5–3.5 µg/ml	
Peanut	ELISA polyclonal Ab	Peanut protein	[61]
Semipurified Ara h 1		1 µg/g	
		Sensitivity 5–1000 ng/ml	
Peanut	ELISA competitive	400 ng/g	[62]
Whole extract	Polyclonal Ab	Sensitivity 1–63 ng/ml	
Peanut	ELISA	2 µg/g (4 µg/g)	[48]
Whole extract	Polyclonal Ab	Sensitivity 24–1000 ng/ml	
Peanut	ELISA	2 µg/g	[48]
Conarachin A	Polyclonal Ab	Sensitivity 10–180 ng/ml	
Peanut ara h 1	ELISA	Sensitivity 30 ng/ml	[39]
	Monoclonal Ab		
Soybean	ELISA sandwich	Sensitivity 10–500 ng per well	[63,64]
Gly m Bd 30k	Monoclonal Ab	2–200 ng for carboxymethylated	
		allergen	
Soybean	ELISA competitive	2 µg/g	[45]
Denatured soy proteins	Polyclonal Ab		
Soybean	ELISA sandwich	Sensitivity 0.2–20 ng per well	[65]
Gly m Bd 28k	Monoclonal Ab		
Soybean	ELISA sandwich	Sensitivity 3 ng/ml	[42]
Kunitz soy trypsin	Polyclonal antibodies		
inhibitor			
Gluten-containing cereals	ELISA monoclonal Ab	0.02% sensitivity	[66]
Gliadin		0.16–5 µg/ml	
Gluten-containing	ELISA sandwich	1.5 ng/ml gliadins	[67]
cereals	Monoclonal Ab	0.05 ng/ml hordeins	
Gliadins, secalins,		0.15 ng/ml secalins	
hordeins, avenins		12 ng/ml avenins	
Shrimp	ELISA enhanced	4–125 ng/ml	[44]
Pen a 1 (tropomyosin)	Sandwich monoclonal Ab		
Peach	ELISA indirect	Sensitivity 0.1 ng/ml	[68]
Pru p 3	Monoclonal/polyclonal Ab		

Source: Besler, M. Determination of allergens in foods, *Trends Anal. Chem., 20,* 662, 2001.

(Table 13.3). Because of the increasing awareness and importance of the food allergen issue, additional assays will be developed for an increasing number of allergens. Along with improvements to current methodology, new technologies continue to be applied to allergen detection.

13.2.1.2 Dipstick and Lateral Flow

The dipstick is any device that can be immersed in a sample or other reagent solution. It uses nitrocellulose or polyester cloth as the solid support. It is a qualitative sandwich-type immunoassay. The assay procedure comprises the same steps as the conventional microwell ELISA. Among the advantages of the dipstick are the following: it is rapid, easy to use, inexpensive, and portable, making it suitable for screening or point-of-sale applications. The sensitivity of dipstick methods can rival that of the conventional ELISA. The literature describes the development of

TABLE 13.3
Commercial Kits for the Detection Food Allergens

	Tepnel BioSystems[a]	R-Biopharm[b]	Neogen Corp[c]	ELISA Systems[d]	Pro-Lab Diagnostics[e]
ELISA	Peanut	Peanut	Peanut	Peanut	Peanut
	Casein	β-Lactoglobulin	Egg	β-Lactoglobulin	Milk
	β-Lactoglobulin	Egg	Milk	Casein	Egg
	Egg	Hazelnut	Almond	Egg	
	Soy	Almond	Gliadin	Hazelnut	
	Wheat	Wheat		Almond	
	Sesame seed			Crustacean	
				Soy	
				Sesame seed	
Rapid methods	Gluten	Gluten	Peanut		
	Peanut	Peanut	Egg		
		Egg	Milk		
		Almond	Almond		
		β-Lactoglobulin	Gliadin		
PCR	Peanut	Soy			
		Hazelnut			
		Almond			
		Peanut			
PCR-ELISA		Soy			
		Hazelnut			
		Almond			
		Peanut			

[a] http://www.tepnel.com
[b] http://www.r_biopharm.com
[c] http://www.neogen.com
[d] http://www.elisas.com
[e] http://www.pro-lab.com

TABLE 13.4
Cloth and Dipstick Methods for Food Allergens

Allergen Measured	Support	Limit of Detection/ Sensitivity	References
Peanut (extract)	Polyester cloth	0.03 µg/ml	[69]
Multiplex	Polyester cloth		[72]
Peanut (extract)		0.01 µg/ml	
Hazelnut (extract)		0.03 µg/ml	
Brazil nut (extract)		0.03 µg/ml	
Peanut (extract)	Immunosticks (Nunc)	1 µg/g	[70]
Hazelnut (extract)			
Egg	Nitrocellulose	20 µg/kg	[71]
Egg white	membrane on a plastic support		

dipsticks for peanut,[69,70] egg,[71] and hazelnut[70] as well as a multiassay system that can simultaneously detect peanut, hazelnut, and Brazil nut.[72] (Table 13.4).

The lateral flow or strip test is a variation of the dipstick ELISA. As described for other immunoassays, the detection of the analyte is based on antibody–antigen recognition. But in this case, the analyte is first recognized by the second or detector antibody, which is coupled to a colored reactant and embedded at one edge of a nitrocellulose strip. The analyte–antibody sandwich migrates along the membrane support to the other end of the strip, which has two capture zones. One zone contains a capture antibody that is specific to the analyte (test line), and the other zone contains antibodies specific to the detector antibody only (control line) (Figure 13.2). To simplify the assay, the detector antibody can be attached to colored particles, such as gold or latex, which make the antibody–antigen reaction visible without the use of a substrate. Although this assay is more difficult to develop, it is easier to run because it requires minimal handling.

13.2.1.3 Biosensors

One of the new technologies in the field of food allergen detection is biosensor technology. Surface plasmon resonance (SPR) immuno(bio)sensors are being developed for this purpose. SPR measures changes of the refractive index value, which is modified when the antibody linked to the surface sensor binds the allergen present in the sample. One of the most important features of the technology is the real-time monitoring of the samples. Quantitative methods have been developed for peanut,[73] soy,[74] ovomucoid (egg), β-lactoglobulin (milk), and corylin (hazelnut).[75]

13.2.2 DNA-BASED METHODS

13.2.2.1 Polymerase Chain Reaction

Polymerase chain reaction (PCR) allows the selective amplification of a specific DNA sequence. The two outstanding features of this technique are its specificity—the

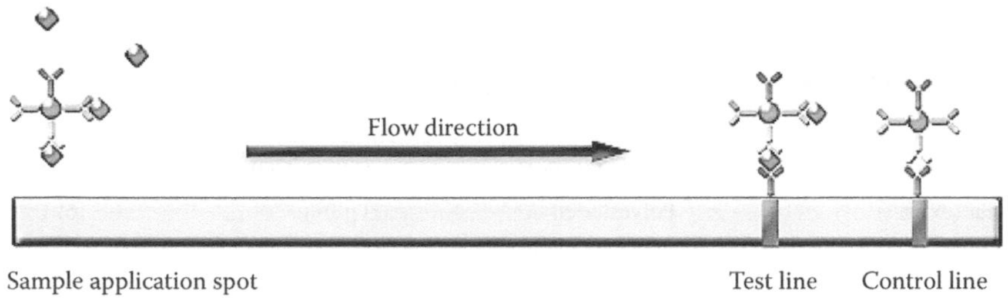

FIGURE 13.2   Lateral flow scheme.

ability to distinguish among minor variations in DNA sequence—and its sensitivity—the ability to detect extremely small quantities of the target sequence in a background of other DNAs, even DNA with similar sequences. PCR has been widely used to characterize, clone, and produce recombinant food allergens using a DNA copy of the allergen-encoding mRNA.[76] In general, DNA amplification of a gene or gene fragment by PCR requires two primer sequences that are chosen to hybridize on opposite DNA strands of a segment of the target DNA. The specificity of the technique relies on reaction conditions that limit hybridization to the exact target sequence. Repeated cycles of reaction amplify the product of the reaction in an exponential fashion (2^{30} copies for 30 two-minute cycles), which explains the sensitivity of the technique.

The success of PCR methodologies is dependent on several practical but critical considerations associated with their use. These include the elements of primer design, the theoretical size of the resulting PCR amplicon, the food matrices to be examined, the sampling size, and the quality of the DNA template. Because the detection of food allergens requires a high level of specificity and sensitivity, proper selection of DNA primer pairs is vital. The pairs must also be chosen so as to produce an amplicon even when a highly fragmented DNA template is used. DNA templates are prepared from an eclectic variety of starting materials, ranging from the original plant or animal to processed food. The suitability of this DNA as a marker for the presence of a particular allergenic protein will dictate how well the assay performs.

Two considerations in the isolation of DNA templates are template purity and DNA integrity. DNA must be free of potential PCR inhibitors that can dramatically influence the amplification of target DNA. These inhibitors range from chemicals in plant extracts and raw materials to substances added during food processing. Removal of these inhibitors during DNA template preparation and differences in the efficiency of sample cleanup can affect both the sensitivity and specificity of detection of the same DNA target in different foods.

Processing steps associated with food production may also be a factor in the integrity of DNA templates. Milling processes (wet- or dry-milled), steeping techniques, and manufacturing conditions may physically shear or chemically modify the target DNA, reducing its efficiency as a template in the PCR. Alternatively, DNA can be degraded to undetectable levels, leading to a false negative assay result.[77] It is more difficult to obtain significant amounts of amplifiable DNA from processed food because heat treatment, pressure, enzymatic activities, or acidic pH can lead to fragmentation and other DNA modifications.[78,79] If the DNA is partially degraded, and the amplification site is affected to some degree, the PCR can underestimate the real presence of an allergen in that particular food. This point was demonstrated by Holzhauser et al.[80] using soured and fermented foods. On the other hand, processing of foods could degrade the allergenic protein itself, leaving the DNA relatively intact. In this case, a positive DNA analysis could lead to unnecessary labeling or a recall action. The best approach is to use PCR methods only in situations where the detection of DNA correlates with the presence of the allergenic protein.[80] This requires the selection of primers to amplify the genes of stable allergenic proteins. One advantage to the use of PCR methods as opposed to the use of antibody-based methods is that PCR reagents are stable and homogeneous, thus minimizing lot-to-lot reagent variation.

PCR assays are designed to be either qualitative (what DNA sequences are present, if any) or quantitative (what percent of the target DNA is present). For qualitative analyses, conventional or traditional PCR uses a thermal cycler to amplify target DNA, with the resulting amplicons visualized by agarose gel electrophoresis and ethidium bromide staining. Any attempt to quantify results in such a manner requires intensive standardization of the amplification parameters (i.e., cycle number, standard curve construction), gel densitometric measurements, and regression analysis of the target with standard DNA. Besides the extra manual manipulations, the time required for completion is in the order of hours. The sensitivity and accuracy of such data processing may be of limited value for the quantification of minute levels of allergens.

In recent years, several manufacturers have developed instrumentation that permits detection of product formation during the PCR, often called *real-time PCR*. The two advantages of this technique are the possibility of quantifying the initial template concentration and the elimination of detection as a separate step. In real-time PCR, product formation is detected during each cycle. During PCR, production of the product is exponential and is usually limited by the quantity of reagents in the reaction mixture. Thus, changing the amount of the target analyte frequently has no effect on the amount of product present after 30 cycles. In real-time PCR, monitoring the number of cycles required before the product can be detected allows for an accurate determination of the original level of the target sequence. Detection of product formation during real-time PCR considerably speeds up the analysis by eliminating the need to transfer samples to an agarose gel for electrophoresis. Furthermore, most real-time platforms rely on one of a variety of hybridization probe techniques to confirm the sequence of the product. Although the precise details vary from technique to technique,[81] all of them rely on the specificity of DNA–DNA hybridizations at carefully controlled temperatures coupled with fluorescent dyes that are activated only when hybridization occurs.

Currently, there are several methods developed for the detection of food allergens by PCR, most of them being qualitative. The UK Central Science Laboratory, commissioned by the UK Food Standards Agency, has developed a PCR test for the detection of peanuts, which can detect 1 part in 10 million. PCR assays have been developed for soy[82] and hazelnut allergens.[80] Among the commercial vendors, Tepnel BioSystems has developed a kit for the detection of peanut, and R-Biopharm has PCR kits available for the detection of soy, almond, and hazelnut (Table 13.3).

13.2.2.2 PCR-ELISA

PCR-ELISA is similar to traditional PCR methods in that specific DNA sequences are amplified using pairs of primers and DNA polymerase. However, amplicons are detected by ELISA using a complementary DNA probe that is subsequently detected using an enzyme-labeled antibody. The ELISA portion of the test confirms that the amplified DNA contains the desired sequence by use of a probe that is complementary to that particular sequence. In this way, false positive results are minimized. Among other advantages of the assay, PCR-ELISA is very sensitive and specific, does not take longer to run than a conventional ELISA, makes the conventional PCR a quantitative technique, and allows several samples to be analyzed simultaneously. With regard to food allergen detection, R-Biopharm has commercialized a SureFood® ELISA-PCR for almond, peanut, soy, and hazelnut (Table 13.3). The test has been shown to avoid the cross-reactivity seen with hazelnut ELISA methods.[80]

13.2.3 New Technologies

Because of the complexity of food matrices, new technologies are often used in the pharmaceutical field before being applied to food analysis. Nonetheless, new technologies are being evaluated for detection of food contaminants or food components, including potential allergens. An important step would be the ability to automate the analytical process, thus permitting the acquisition of results more quickly and efficiently than with conventional procedures. Accuracy of results would improve due to less sample handling and pipeting, which are usually significant sources of error. The use of new technologies associated with the fields of genomics and proteomics, such as microarrays, would allow the analysis of multiple allergens present in the same sample or the analysis of thousands of samples simultaneously, with minimal consumption of reagents. At present, there is no method for microarray analysis of food allergens, but the technique has already been applied to the testing of serum from allergic patients to determine IgE reactivity profiles.[83]

13.3 GENETICALLY MODIFIED FOODS

Genetic modification of foods involves the *in vitro* introduction of one or more foreign genes into the recipient genome of a food plant or animal. The integrated gene(s) can be derived from related or unrelated genotypes to include viral, bacterial, fungal, animal, or plant coding sequences. DNA may be introduced into plant cells by bacterial transfer using disarmed strains of *Agrobacterium tumefaciens*, by DNA particle gun technology, or by electroporation of cultured plant cells whose cell walls had been enzymatically digested.[84] The cotransfer of the marker gene along with the gene of interest allows for selective growth of the transformed plant cells. The advantage of these methods of genetic modification over traditional plant breeding is the ability to control the discrete gene(s) being transferred, to target the time or level of transgene expression, or to eliminate undesirable traits through gene suppression technology. Genetically modified (GM) plant foods are selected for improved agronomic and human health traits, including herbicide tolerance, disease and insect resistance, and enhanced nutritional content. The products of agricultural biotechnology are increasingly common in the worldwide marketplace. Biotechnology-derived crops account for 34% of soybean production, 16% of cotton production, and 7% of corn production. In 2001, the majority of GM crops were produced in the United States (68%), followed by Argentina (11.8%), Canada (6%), and China (3%).[85,86] Reliable detection methods are critical for food producers who must take into consideration whether a GM product is subject to the import restrictions of a particular country.

Guidelines for the safety assessment of foods derived from recombinant-DNA plants are being considered for adoption by the Codex Alimentarius Commission.[87] Included in these guidelines are criteria for the assessment of potential allergenicity of proteins expressed through the introduction of recombinant DNA. The recommended criteria for evaluation were (1) allergenic potential of the source of the transferred gene, (2) amino acid sequence homology to known allergenic proteins, (3) stability to digestion with pepsin or other digestive enzymes, (4) specific IgE screening using sera from patients with known allergies to the source material (where sera are available and methods can be standardized), and (5) additional validated methods as science and technology evolve. Although these criteria cannot provide a definitive assessment of protein allergenicity, the implementation of this strategy should provide sufficient information to evaluate the potential for human health risk.

13.3.1 DETECTION OF COMPONENTS OF GENETICALLY MODIFIED FOODS

Analytical methods for the identification and quantitation of raw GM agricultural products and highly processed or refined GM ingredients are necessary for the purposes of product tracing and compliance with appropriate labeling regulations. There are two general scientific approaches for the detection of genetically engineered materials. Protein-based methods detect the presence of the protein expression product of the inserted gene and are typically immunochemical methods that follow an ELISA format. DNA-based methods detect the specific DNA sequence of the inserted gene using PCR techniques. These methods are similar in format to those described for food allergen detection, and issues described previously regarding sample preparation, specificity, matrix effects, standardization, and validation also apply to detection methods for GM foods. Numerous qualitative and quantitative immunochemical and DNA-based methods for GM foods are widely available in the commercial market, and the reader is referred to published reviews of available methods[88-90] or to specific vendors.

13.3.2 THE STARLINK™ CORN EXPERIENCE

Starlink™ (Aventis CropSciences) is a variety of GM corn that contains a gene from the soil organism *Bacillus thuringiensis* (Bt), which encodes a pesticidal toxin known as Cry9C. Many Bt toxins occur naturally and, though insecticidal to various species, are considered to be harmless to humans. The Cry9C protein is more resistant to heat and digestion than other Bt toxins. These characteristics suggested that the Cry9C protein might be an allergen. In 1998, the Environmental Protection Agency (EPA) approved Starlink corn as a registered pesticide[91,92] for use in animal feed and for other industrial uses. Approval of Starlink corn for human use was not granted by the EPA on the grounds that the issue of potential allergenicity of the Cry9C protein had not been resolved. In September 2000, the discovery of *cry9C* DNA in taco shells from US markets led to the recall of over 300 corn-based products and invoked interest in the development of scientific methods to support compliance activities.

13.3.2.1 Immunochemical Methods for Determination of Cry9C

In 2001, four manufacturers—Agdia (Elkhart, IN), Strategic Diagnostics, Inc. (Newark, DE), Neogen Corporation (Lansing, MI), and EnviroLogix, Inc. (Portland, ME)—introduced ELISA test kits for the detection of Cry9C protein in corn, corn meal, and grits. The first three kits express their results in terms of percentage of Starlink corn contamination present in the test matrix. The claimed limit of determination of these kits ranges from 0.01% to 0.125% of Starlink corn. The EnviroLogix kit expresses results in terms of percentage of Starlink and nanogram Cry9C protein per gram of corn product, with a claimed limit of determination of approximately 1 ng/g using a polyclonal antibody-based technique.

An interlaboratory study (six laboratories) was conducted to determine the accuracy, repeatability, and reproducibility parameters of the EnviroLogix ELISA kit method for the determination of Cry9C protein in processed foods.[93] In this study, blind duplicates of control samples (blank material prepared from non-Starlink corn), spiked samples (blank material with the addition of Cry9C protein), and incurred samples (products prepared with 100% Starlink corn) were included. The matrices included flour, starch, and oil; minimally processed foods such as muffins and bread; medium processed foods such as corn puffs and corn flake breakfast cereal; and highly processed foods such as soft tortillas and tortilla chips. Cry9C proteins from two different sources were used to spike the food products. Cry9C protein produced and purified from a bacterial host was used to prepare spiked test samples at 2.72 and 6.8 ng/g. Cry9C protein from Starlink corn flour was used to prepare spiked samples at 1.97 ng/g. Average recoveries for samples spiked with corn flour Cry9C protein at 1.97 ng/g ranged from 73% to 122%, within-laboratory relative standard deviations (RSD_r) ranged from 6% to 22%, and between-laboratory standard deviations (RSD_R) ranged from 16% to 56%. Average recoveries for samples spiked with bacterial Cry9c protein at 2.72 and 6.8 ng/g were 27%–96% and 32%–113%, respectively; the RSD_r ranged from 10% to 35% and 7% to 38%, respectively; and the RSD_R ranged from 28% to 84% and 15% to 75%, respectively. The incurred test samples were found to contain Cry9C protein at levels ranging from 0.8 to 3,187 ng/g, depending on the product. The RSD_r ranged from 5% to 16% and RSD_R from 11% to 71% for the incurred samples. Results of statistical analysis indicate that this method is applicable to determine Cry9C protein in the eight types of corn-based products containing Cry9C protein (from Starlink) at levels 1.97 ng/g.

The ELISA test kits of EnviroLogix, Inc. and Strategic Diagnostics, Inc. were evaluated for Cry9C protein in corn flour and corn meal.[94,95] Blind duplicates at 0%, 0.01%, 0.025%, 0.05%, and 0.075% Starlink flour and corn meal were sent to 28 collaborators. For corn meal, the RSD_R ranged from 17% to 50% and 25% to 70% and for corn flour, the RSD_R ranged from 12% to 38% and 15% to 20%, using the SDI kit and EnviroLogix kit, respectively. The higher range of the RSD_R for meal was found to be due to sampling variability. The sampling variability of meal is about twice that of flour because of the larger particle size.[96]

Statistical modeling on the theoretical effect of sampling variability on analytical test results has demonstrated the importance of an appropriate sampling procedure in the testing of GM products. Sampling variability can be reduced by increasing sample size; variability in sample preparation can be reduced by grinding the seed or meal into smaller particle sizes and/or increasing the subsample size; and the analytical variability can be reduced by increasing the number of aliquots quantitated.[97] In the United States, the Department of Agriculture, Grain Inspection, Packers and Stockyards Administration (GIPSA)[98] and the US Food and Drug Administration (FDA)[99] have developed guidelines for grain sampling to test for the presence of biotechnology-derived products. GIPSA recommended the use of immunoassays because of their ability for batch analysis and short analytical time to handle the large scale of agriculture production in the United States. Subsequently, GIPSA has been actively involved in verification of the performance of some of the commercial ELISA test kits for Roundup Ready protein CP4 EPSPS in soybean or corn and Cry9C protein in Starlink from four manufacturers: Agdia, EnviroLogix, Inc., Neogen Corporation, and Strategic Diagnostics, Inc.[100]

13.3.2.2 PCR Methods for Detection of Cry9C

In 1999, flour, meal, and processed foods made from yellow corn or its derivatives were analyzed for adulteration by Starlink corn.[101] Samples were crushed and extracted to isolate DNA that was then purified and tested for the presence of modified *cry9C* DNA using PCR. For most samples, the extraction protocol provided DNA suitable for PCR. However, the amount and quality of DNA obtained varied greatly with the type of product being analyzed. Factors known to influence DNA extraction include the type of the milling process (wet- or dry-milled), steeping techniques used, and the processing conditions used to manufacture each individual food type.[77] Processing conditions such as high temperature or extreme pH can degrade DNA as a template for PCR, as indicated by low-molecular-weight products seen in an agarose gel electrophoretic analysis of the sample. As seen in Figure 13.3, purified DNA obtained from food extracts using agarose gel electrophoresis indicated varying degrees of DNA degradation. In most instances, however, the extraction and isolation procedure delivered sufficient DNA, that is, 400 bp suitable for use as template in PCR analysis, as judged by the successful PCR amplification of the corn aldolase gene. Targeting this gene provided an indication of the quality and quantity of DNA extracted from samples.

Although 0.1 g of most commodities was sufficient to yield DNA for PCR analysis, extracts from corn flakes and cornstarch did not produce detectable levels of DNA. Moreover, increasing the amount of product extracted to 5 g and using an alternative DNA purification system employing magnetic beads did not substantially improve the DNA yield. Thus, the presence of Starlink could not be determined in these products.

PCR analysis relied on the specificity of the primers to detect the *cry9C* transgene associated with Starlink. One primer used was in the region 5′ to the coding sequence for the gene, annealing to regulatory DNA sequences derived from sources

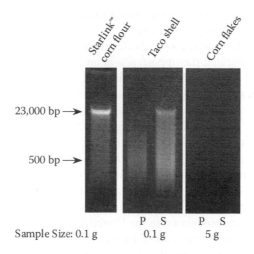

FIGURE 13.3 Agarose gel electrophoresis analysis of extraction of genomic DNA. Starlink corn flour produces relatively high molecular weight DNA (M, DNA size markers) compared with DNA extracted from 0.1 g taco shells. No DNA was visible from 5 g of corn flakes. *P* is product and *S* is product spiked with 10 mg Starlink corn flour. Because of dilution, the spiked lane of the corn flake sample cannot be seen in this reproduction.

other than bacteria or maize. The other primer targeted the coding region for the modified *cry9C* gene. Thus, successful amplification relies on the proximity of these sequences, from disparate sources, to one another in the extract being analyzed. However, because of the presence of inhibitors and the potential for low target concentration, the PCR comprises several cycles using relatively "relaxed" PCR conditions, increasing the possibility of all amplification events, including amplification of an inappropriate target. To exclude nonspecific amplification events, all potentially positive samples were subjected to a confirmatory test.

Of the 63 products analyzed, only four (Taco Bell taco shells) contained the target transgene as seen by the resulting 174-bp amplicon using *cry9C*-transgenic specific primers. No amplification of the aldolase gene occurred during analysis of corn flake and cornstarch samples, correlating with the absence of genomic DNA analysis previously described. One baby food product failed to produce a PCR product when spiked with Starlink flour, suggesting the presence of inhibitors that would prevent detection of the modified *cry9C* transgene.

Sequencing of putative PCR-positive amplicons provided the most definitive confirmation of specific amplification. However, this approach is time consuming and costly for large-scale surveys. Restriction endonuclease digestion had been used as the confirmatory test. *NcoI* digestion of PCR products of the *cry9C* transgene yielded two fragments, 128 and 46 bp. However, some PCR products were not suitable for Restriction fragment length polymorphism (RFLP) analysis. For example, these samples contained faint bands that might have been amplicons but that were too faint to be expected to generate a visible product on agarose gels after dilution and restriction analysis.

To provide a more robust confirmatory test, nested PCR primers based on the sequence of the PCR product were developed. These primers, like the original ones, hybridized to both coding and regulatory DNA sequences and were specific to the chimeric nature of the *cry9C* transgene. Agarose gel electrophoresis analysis of positive control samples subjected to nested PCR revealed multiple bands. The amplified products were generated due to the presence of not only the nested PCR primers but also the original primers carried over from the initial PCR amplification. The extraneous bands were eliminated by dilution of the template before PCR, thus reducing the concentration of the primers used in the initial PCR reaction. Dilution of a positive control sample 10,000-fold still produced a distinct 111-bp band using nested PCR.

Use of the nested primers allowed definitive rejection of the hypothesis of potential Starlink adulteration in several samples with faint bands seen by agarose gel electrophoresis analysis of the initial PCR products. Although use of nested PCR may increase the overall sensitivity of the assay, no samples analyzed were found to be adulterated, based solely on the results of the nested PCR reactions.

13.3.2.3 Post-Market Cry9C Exposure Assessment

Details regarding the circumstances surrounding the presence of Starlink in the human food supply have been previously reported.[102] In September 2000, *cry9C* DNA was detected first in taco shells and then in other corn products. This resulted in several widely publicized food recalls. These reports lent credibility to the possibility of extensive comingling of Starlink with other varieties of corn. Subsequently, the FDA began receiving reports of adverse health effects from consumers who had eaten food products containing corn. All adverse health events reported to the FDA between July 1 and November 30, 2000, involving the consumption of a corn product were considered and were screened based on a working case definition that included signs and symptoms consistent with systemic anaphylaxis. Adverse event reports were included for further analysis provided they occurred within 12 hours of product consumption, involved only one individual among meal companions, and could not be explained by a preexisting medical condition. Twenty-nine of 51 event reports met these criteria. Ultimately, 18 of those 29 provided serum samples under informed consent, which were used for further testing. Based on the symptoms reported, development of a laboratory method to detect serum antibodies was undertaken to further assess the relation between allergic manifestations and the Cry9C protein. The goal was to determine if we could detect a difference in immune responsiveness to the Cry9C protein among differentially exposed groups using an ELISA method. Results of the serum test would indicate if individuals had specific IgE antibodies to Cry9C, suggestive of hypersensitivity.

Several obstacles were encountered in this process, including obtaining adequate control sera (both positive and negative) and the availability of an adequately characterized source of antigen. The most significant handicap was the lack of a positive control human serum known to contain IgE against the Cry9C protein. A hyperimmune serum against recombinant Cry9C (r-Cry9C) produced in a goat, as well as r-Cry9C protein, was made available by the producer of Starlink. This enabled the development of optimal conditions to detect bound Cry9C antibody in the ELISA. SDS-Page and Western blot analysis of the r-Cry9C revealed two bands that reacted with the goat antiserum (Figure 13.4). In addition, an extract of corn flour made from

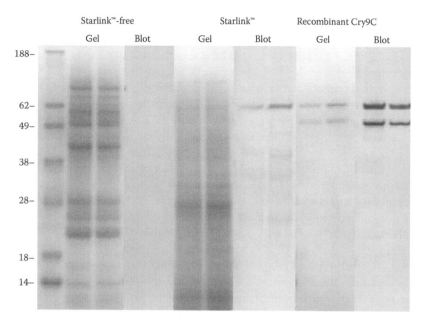

FIGURE 13.4 SDS-Page and Western blot analysis of recombinant Cry9C, corn flour prepared from Starlink, and corn flour prepared from non-Starlink corn.

Starlink contained a protein band of similar molecular weight (MW) that reacted with the goat antibody. An extract of Starlink-free flour did not react. The identity of the two bands from the r-Cry9C gel was confirmed by electrospray mass spectrometry and partial sequencing. Data from this analysis indicated that tryptic peptide fragments from both bands were consistent with sequence data from Cry9C (accession no. CAC36454), suggesting that the 56 kDa band is an N-terminal truncated form of the larger molecule.

Negative control sera were obtained from serum samples banked prior to the licensing of Starlink. An additional control, intended to address the potential problem of nonspecific binding of IgE, involved testing of sera from patients with unrelated allergies, which contained high levels of total IgE. Testing of adverse event and control sera did not show any enhanced reactivity in any of the adverse event sera versus control sera.[103]

Although we did not have a positive human serum control that reacted with Cry9C protein, it is probable that in any scenario in which possible allergic reactions to a novel food protein are reported, no such serum will be readily available. The approach used to circumvent this problem was to incorporate controls for as many aspects of the ELISA as possible. Accordingly, the coating of ELISA plates with r-Cry9C was confirmed by binding of goat anti-Cry9C antiserum. To verify the ability of the anti-IgE detection reagent to detect bound human IgE, ELISA procedures were developed to test for IgE antibodies to known environmental and food allergens. These ELISA methods were capable of detecting IgE antibodies to other allergens (cat, grass, and peanut) in every control sample tested. This suggests that the Cry9C ELISA would likely have been able to detect IgE antibodies had they been present in serum at levels comparable to those seen in other allergic individuals. It is possible that glycosylation of Cry9C as it is expressed in Starlink corn might affect reactivity if any IgE epitopes were to involve carbohydrate moieties because such epitopes would not be present in bacterially expressed r-Cry9C. SDS-PAGE/Western blot analysis suggests that large differences in MW do not exist between corn and bacterially expressed Cry9C; however, some glycosylation cannot be ruled out, and it remains possible that IgE directed against a carbohydrate-containing epitope could have gone undetected in the ELISA.

The accidental appearance of Starlink corn in the human food supply presented a number of problems related to developing methods intended to produce evidence of exposure, or reactivity in individuals with possible exposure. These are problems that could be encountered in a similar event in the future. Thus, this episode and the response to it may serve as a guide for a future response. The focus on IgE-specific antibodies is valid because of their demonstrated role in clinical allergy; however, the presence of serum IgE alone may not correlate with clinical allergy. In addition, detection of IgG antibodies might provide evidence of exposure but would not correlate with allergic reactions. The presence of specific serum IgE in this, and in any future study, would provide participants with information that they could then use to seek further guidance from their healthcare provider to further define their personal allergy potential. With respect to protein standards, it would be useful if the producers of bioengineered products maintained sufficient stocks of the purified protein for the development of ELISA tests or other detection methods. Ideally, a quantity of the protein should also be available in its plant-produced form. As was the case here, availability of a hyperimmune serum would also facilitate assay development. It is also assumed that producers can serve as a source of information on the molecular characteristics of proteins such as sequence data and especially on glycosylation patterns in the plant host.

REFERENCES

1. Sicherer, S. Food allergy, *Lancet, 360*, 701, 2002.
2. Helm, R. M. and Burks, A. W. Mechanisms of food allergy, *Curr. Opin. Immunol., 12*, 6647, 2000.
3. Bock, S. A. Prospective appraisal of complaints of adverse reactions to foods in children during the first 3 years of life, *Pediatrics, 79*, 683, 1987.
4. Sampson, H. A. Food allergy. Part 1: Immunopathogenesis and clinical disorders, *J. Allergy Clin. Immunol., 103*, 717, 1999.
5. Grundy, J. et al. Rising prevalence of allergy to peanut in children: Data from 2 sequential cohorts, *J. Allergy Clin. Immunol., 110*, 784, 2002.
6. Sampson, H. A. Epidemiology of food allergy, *Pediatr. Allergy Immunol., 7* (suppl 9), 42, 1996.
7. Van Odijk, J. et al. Specific IgE antibodies to peanut in western Sweden—Has the occurrence of peanut allergy increased without an increase in consumption? *Allergy, 56*, 573, 2001.
8. Burks, W. et al. Food allergens, *Curr. Opin. Allergy Clin. Immunol., 1*, 243, 2001.
9. Taylor, S. L. and Lehrer, S. B. Principles and characteristics of food allergens, *Crit. Rev. Food Sci. Nutr., 36* (Suppl), S91, 1996.
10. Bredehorst, R. and David, K. What establishes a protein as an allergen, *J. Chromatogr. B, 756, 33*, 2001.

11. Bayard, C. and Lottspeich, F. Bioanalytical characterization of proteins, *J. Chromatogr. B.*, *756*, 113, 2001.
12. Hefle, S. L. Impact of processing on food allergens, *Adv. Exp. Med. Biol.*, *459*, 107, 1999.
13. Besler, M., Steinhart, H., and Paschke, A. Stability of food allergens and allergenicity of processed foods, *J. Chromatogr. B. Biomed. Sci. Appl.*, *756*, 207, 2001.
14. Fu, T. J., Abbott, U. R., and Hatzos, C. Digestibility of food allergens and nonallergenic proteins in simulated gastric fluid and simulated intestinal fluid—a comparative study, *J. Agric. Food Chem.*, *50*, 7154, 2002.
15. Betzel, C. X-ray structure analysis of food allergens, *J. Chromatogr. B*, *756*, 179, 2001.
16. Rösch, P. Nuclear magnetic resonance studies of allergens, *J. Chromatogr. B*, *756*, 165, 2001.
17. Reese, G. et al. Characterization and identification of allergen epitopes: Recombinant peptide libraries and synthetic, overlapping peptides, *J. Chromatogr. B*, *756*, 157, 2001.
18. Tichá, M., Pacáková, V., and Štulík, K. Proteomics of allergens, *J. Chromatogr. B*, *771*, 343, 2002.
19. Hileman, R. E. et al. Bioinformatic methods for allergenicity assessment using a comprehensive allergen database, *Int. Arch. Allergy Immunol.*, *128*, 280, 2002.
20. Hefle, S. L., Nordlee, J. A., and Taylor, S. L. Allergenic foods, *Crit. Rev. Food Sci. Nutr.*, *36* (Suppl.), S91, 1996.
21. Burks, A. W. et al. Epitope specificity of the major peanut allergen, Ara h II, *J. Allergy Clin. Immunol.*, *95*, 607, 1995.
22. Burks, A. W. et al. Recombinant peanut allergen Ara h I expression and IgE-binding in patients with peanut hypersensitivity, *J. Clin. Invest.*, *96*, 1715, 1995.
23. Burks, A. W. et al. Mapping and mutational analysis of the IgE-binding epitopes on Ara H 1, a legume vicilin protein and a major allergen in peanut hypersensitivity, *Eur. J. Biochem.*, *245*, 334, 1997.
24. Rabjohn, P. et al. Molecular cloning and epitope analysis of the peanut allergen Ara h 3, *J. Clin. Invest.*, *103*, 535, 1999.
25. Robotham, J. M. et al. Linear IgE epitope mapping of the English walnut (*Juglans regia*) major food allergen, Jug r 1, *J. Allergy Clin. Immunol.*, *109*, 143, 2002.
26. Stanley, J. S. et al. Identification and mutational analysis of the immunodominant IgE binding epitopes of the major peanut allergen *Ara h 2*, *Arch. Biochem. Biophys.*, *342*, 244, 1997.
27. Hourihane, J. O. et al. Clinical characteristics of peanut allergy, *Clin. Exp. Allergy.*, *27*, 634, 1997.
28. Hourihane, J. O., Smith, P. K., and Strobel, S. Food allergy in children, *Indian J. Pediatr.*, *69*, 61, 2002.
29. Sampson, H. A. and Ho, D. G. Relationship between food-specific IgE concentrations and the risk of positive food challenges in children and adolescents, *J. Allergy Clin. Immunol.*, *100*, 444, 1997.
30. Leung, D. Y. M. et al. Effect of anti-IgE therapy in patients with peanut allergy, *N. Engl. J. Med.*, *348*, 986, 2003.
31. Moneret-Vautrin, D. A. Modifications of allergenicity linked to food technologies, *Allerg. Immunol.*, *30*, 9, 1998.
32. Davis, P. J., Smales, C. M., and James, D. C. How can thermal processing modify the antigenicity of proteins? *Allergy*, *56* (Suppl 67), 56, 2001.
33. Maleki, S. J. et al. The effects of roasting on the allergenic properties of peanut proteins, *J. Allergy Clin. Immunol.*, *106*, 763, 2000.
34. Vierk, K. et al. Recalls of foods containing undeclared allergens reported to the U.S. Food and Drug Administration, fiscal year 1999, *J. Allergy Clin. Immunol.*, *109*, 1022, 2002.

35. Taylor, S. L. et al. Factors affecting the determination of threshold doses for allergenic foods: How much is too much?, *J. Allergy Clin. Immunol.*, *109*, 24, 2002.
36. Bindslev-Jensen, C., Briggs, D., and Osterballe, M. Can we determine a threshold level for allergenic foods by statistical analysis of published data in the literature? *Allergy*, *57*, 741, 2002.
37. Wensing, M. et al. The distribution of individual threshold doses eliciting allergic reactions in a population with peanut allergy, *J. Allergy Clin. Immunol.*, *110*, 915, 2002.
38. Hurst, W. J. et al. A comparison of commercially available peanut ELISA test kits on the analysis of samples of dark and milk chocolate, *J. Immunoassay Immunochem.*, *23*, 451, 2002.
39. Pomés, A. et al. Monitoring peanut allergen in food products by measuring Ara h 1, *J. Allergy Clin. Immunol.*, *111*, 640, 2003.
40. Hefle, S. L., Jeanniton, E., and Taylor, S. L. Development of sandwich enzyme-linked immunosorbent assay for the detection of egg residues in processed foods, *J. Food Prot.*, *64*, 1812, 2001.
41. Negroni, L. et al. Two-site enzyme immunometric assays for determination of native and denatured β-lactoglobulin, *J. Immunol. Methods*, *220*, 25, 1998.
42. Rosendal, A. and Barkholt, V. Detection of potentially allergenic material in 12 hydrolyzed milk formulas, *J. Dairy Sci.*, *83*, 2200, 2000.
43. Clemente, A. et al. Use of the indirect competitive ELISA for the detection of Brazil nut in food products, *Food Control*, *15*, 65, 2004.
44. Jeoung, B. J. et al. Quantification of the major brown shrimp allergen Pen a 1 (tropomyosin) by a monoclonal antibody-based sandwich ELISA, *J. Allergy Clin. Immunol.*, *100*, 229, 1997.
45. Yeung, J. M. and Collins, P. G. Determination of soy proteins in food products by enzyme immunoassay, *Food Tech. Biotech.*, *35*, 209, 1997.
46. Yeung, J. Newsome, W. H., and Abbott, M. A. Determination of egg proteins in food products by enzyme immunoassay, *J. AOAC Int.*, *83*, 139, 2000.
47. Hefle, S.L. Immunoassay fundamentals, *Food Tech.*, *February*, 102, 1995.
48. Keck-Gassenmeier, B. et al. Determination of peanut traces in food by a commercially-available ELISA test, *Food Agric. Immunol.*, *11*, 243, 1999.
49. Besler, M. Determination of allergens in foods, *Trends Anal. Chem.*, *20*, 662, 2001.
50. Crevel, R. W. R., Kerkhoff, M. A. T., and Koning M. M. G. Allergenicity of refined vegetable oils, *Food Chem. Toxicol.*, *38*, 385, 2000.
51. Acosta, M. R. et al. Production and characterization of rabbit polyclonal antibodies to almond (*Prunus amygdalus* L.) major storage protein, *J. Agric. Food Chem.*, *47*, 4053, 1999.
52. Hlywka, J. J., Hefle, S. L., and Taylor, S. L. A sandwich enzyme-linked immunosorbent assay for the detection of almonds in foods, *J. Food Prot.*, *63*, 252, 2000.
53. Gern, J. E. et al. Allergic reactions to milk-contaminated nondairy products, *N. Engl. J. Med.*, *324*, 976, 1991.
54. Mariager, B. et al. Bovine BETA-lactoglobulin in hypoallergenic and ordinary infant formulas measured by an indirect competitive ELISA using monoclonal and polyclonal antibodies. *Food Agric. Immunol.*, *6*, 73, 1994.
55. Holzhauser, T. and Vieths, S. Quantitative sandwich ELISA for determination of traces of hazelnut (*Corylus avellana*) protein in complex food matrixes, *J. Agric. Food Chem.*, *47*, 4209, 1999.
56. Koppelman, S. J. et al. Comparison of different immunochemical methods for the detection and quantification of hazelnut proteins in food products, *J. Immunol. Methods*, *229*, 107, 1999.
57. Blais, B. W. and Phillippe, L. M. Detection of hazelnut proteins in foods by enzyme immunoassay using egg yolk antibodies, *J. Food Prot.*, *64*, 895, 2001.

58. Blais, B. W., Omar, M., and Phillippe, L. Detection of Brazil nut proteins in foods by enzyme immunoassay, *Food Agric. Immunol.*, *14*, 163, 2002.
59. Leduc, V. et al. Immunochemical detection of egg-white antigens and allergens in meat products, *Allergy*, *54*, 464, 1999.
60. Hefle, S. L. et al. A sandwich enzyme-linked immunosorbent assay (ELISA) for the quantitation of selected peanut proteins in foods, *J. Food Prot.*, *57*, 419, 1994.
61. Koppelman, S. J. et al. Detecting peanut allergens. The development of an immuno-chemical assay for peanut proteins. *World Ingredients*, *12*, 35, 1995.
62. Yeung, J. and Collins, P. G. Enzyme immunoassay for determination of peanut proteins in food products, *J. AOAC Int.*, *79*, 1411, 1996.
63. Tsuji, H. et al. Preparation and application of monoclonal antibodies for a sandwich enzyme-linked immunosorbent assay of the major soybean allergen, Gly m Bd 30K. *J. Nutr. Sci. Vitaminol.*, *39*, 389, 1993.
64. Tsuji, H. et al. Measurement of Gly m Bd 30K, a major soybean allergen, in soybean products by a sandwich enzyme-linked immunosorbent assay. *Biosci. Biotechnol. Biochem.*, *59*, 150, 1995.
65. Bando, N. et al. Quantitative analysis of Gly m Bd 28K in soybean products by a sand-wich enzyme-linked immunosorbent assay, *J. Nutr. Sci. Vitaminol.*, *44*, 655, 1998.
66. Skerrit, J. H. and Hill, A. S. Enzyme Immunoassay for determination of gluten in foods: Collaborative study, *J. AOAC Int.*, *74*, 257, 1991.
67. Sorell, L. et al. An innovative sandwich ELISA system based on an antibody cocktail for gluten analysis, *FEBS Lett.*, *439*, 46, 1998.
68. Duffort, O. A. et al. Immunoassay to quantify the major peach allergen Pru p 3 in foodstuffs. Differential allergen release and stability under physiological conditions, *J. Agric. Food Chem.*, *50*, 7738, 2002.
69. Blais, B. W. and Phillippe, L. M. A cloth-based enzyme immunoassay for detection of peanut proteins in foods, *Food Agric. Immunol.*, *12*, 243, 2000.
70. Stephan, O. et al. Development and validation of two dipstick type immunoassays for determination of trace amounts of peanut and hazelnut in processed foods, *Eur. Food Res. Technol.*, *215*, 431, 2002.
71. Baumgartner, S. et al. Towards the development of a dipstick immunoassay for the detection of trace amounts of egg proteins in food, *Eur. Rood Res. Technol.*, *214*, 168, 2002.
72. Blais, B. W., Gaudreault, M., and Phillippe, L. M. Multiplex enzyme immunoassay for the simultaneous detection of multiple allergens in foods, *Food Control*, *14*, 43, 2003.
73. Mohammed, I. et al. Is biosensor a viable method for food allergen detection?, *Ann. Chim. Acta*, *444*, 97, 2001.
74. Haasnoot, W. et al. Direct biosensor immunoassays for the detection of nonmilk proteins in milk powder, *J. Agric. Food Chem.*, *49*, 5201, 2001.
75. Jonsson, H. and Hellenäs, K. E. Optimizing assay conditions in the detection of food allergens with Biacore's SPR technology, *Biacore J.*, *2*, 16, 2001.
76. Lorenz, A. R. et al. Recombinant food allergens, *J. Chromatogr. B. Biomed. Sci. Appl.*, *756*, 255, 2001.
77. Gawienowski, M. C. et al. Fate of maize DNA during steeping, wet-milling, and pro-cessing, *Cereal Chem.*, *76*, 371, 1999.
78. Wurz, A. et al. Quantitative analysis of genetically modified organisms (GMO) in pro-cessed food by PCR-based methods, *Food Control*, *10*, 385, 1999.
79. Chiter, A., Forbes, J. M., and Blair, E. DNA stability in plant tissues: Implications for the possible transfer of genes from genetically modified food, *FEBS Lett.*, *481*, 164, 2000.
80. Holzhauser, T., Wangorsh, A., and Vieths, S. Polymerase chain reaction (PCR) for detec-tion of potentially allergenic hazelnut residues in complex food matrixes, *Eur. Food Res. Technol.*, *211*, 360, 2000.

81. Lampel, K. A. and Levy, D. D. Analysis of milk and dairy products: DNA-based assays, in *Encyclopedia of Dairy Sciences,* Rojinski, H. et al., Eds. Academic Press, New York, 2002.

82. Meyer, R. and Candrian, U. PCR-based DNA analysis for the identification and characterization of food components. *Lebensm.-Wiss. u.-Technol.*, *29*, 1, 1996.

83. Hiller, R. et al. Microarrayed allergen molecules: Diagnostic gatekeepers for allergy treatment, *FASEB J.*, *16*, 414, 2002.

84. Day, P. R. Genetic modification of proteins in food, *Crit. Rev. Food Sci. Nutr.*, *36* (Suppl), S49, 1996.

85. Schilter, B. and Constable, A. Regulatory control of genetically modified (GM) foods: Likely developments, *Toxicol. Lett.*, *127*, 341, 2002.

86. Pan, T. M. Current status and detection of genetically modified organisms, *J. Food Drug Anal.*, *10*, 229, 2002.

87. ftp.//ftp.fao.org/codex/alinorm03/Al03_34e.pdf

88. Anklam, E. et al. Analytical methods for detection and determination of genetically modified organisms in agricultural crops and plant-derived food products, *Eur. Food Res. Technol.*, *214*, 3, 2002.

89. Ahmed, F. E. Detection of genetically modified organisms in foods, *Trends Biotechnol.*, *20*, 215, 2002.

90. Stave, J. W. Protein immunoassay methods for detection of biotech crops: Applications, limitations, and practical considerations, *J. AOAC Int.*, *85*, 780, 2002.

91. http://www.epa.gov/pesticides/biopesticides/index.htm

92. Office of Federal Register, Code of Federal Regulations, Title 21, sec.342(a)(2)(B), U.S. Government Printing Office, Washington, D.C., 1998.

93. Trucksess, M. W. Determination of Cry9C protein in corn-based foods by enzyme-linked immunosorbent assay: Interlaboratory study, *J. AOAC Int.*, *84*, 1891, 2001.

94. Wehling, P. EnviroLogix Cry9C collaborative study data summary (Technical Report of Association of American Cereal Chemists), 2002.

95. Wehling, P. SDI Cry9C collaborative study data summary (Technical Report of Association of American Cereal Chemists), 2002.

96. Whitaker, T. B. et al. Evaluation of sampling plans to detect Cry9C protein in corn flour and meal, *J. AOAC Int.*, *87*, 950, 2004.

97. Whitaker, T. B. et al. Sampling grain shipments to detect genetically modified seed, *J. AOAC Int.*, *84*, 1941, 2001.

98. http://www.usda.gov/gipsa/biotech/sampling_grains_for_biotechnolog.htm

99. http://www.cfsan.fda.gov/~dms/starguid.html

100. http://www.usda.gov/gipsa/tech-servsup/metheqp/testkit.htm

101. Orlandi, P. A. et al. Analysis of Flour and Food Samples for *cry9C* from Bioengineered Corn, *J. Food Prot.* *65*, 426, 2002.

102. Bucchini, L. and Goldman, L. R. Starlink corn: A risk analysis, *Environ. Health Perspect.*, *110*, 5, 2002.

103. Raybourne, R. B. et al. Development and use of an ELISA test to detect IgE antibody to Cry9c following possible exposure to bioengineered corn, *Int. Arch. Allergy Immunol.*, *132*, 322, 2003.

14 Determination of Pesticide Residues

Steven J. Lehotay and Katerina Mastovska

CONTENTS

14.1 INTRODUCTION

Why bother to analyze foods for pesticide residues? If good agricultural and manufacturing processes are followed, then violative levels of pesticide residues are not going to occur. In theory, all farmers and food processors would follow proper procedures, and analyses of the food would be superfluous. In the real world, though, appropriate practices are not always followed and pesticide residue violations do occur. Also, how can one verify that good practices were followed or no mistakes were made unless routine monitoring is conducted? In this instance, ignorance *is not* bliss; it is a liability for lawsuits and/or enforcement actions.

Ideally, any analysis of foods should provide value-added information to justify the cost of the analysis. Usually, the samples do not contain violative residues, in which case the analysis only serves to remove samples from the shipment, delay distribution, and potentially reduce shelf life of the food, cost a lot of money to transport and analyze the samples in the laboratory (which usually increases the price of the food to the consumer), and create perception problems among some consumers when pesticides are detected (even at safe levels). Thus, pesticide residue analysis adds to the cost of the food, but it does not necessarily add value or provide a higher quality food product.

Then again, a shipload of fruit may be worthless in some countries until it has been analyzed for pesticide residues, or conversely, the shipment may be rendered worthless if it is found to have violative levels of pesticides. Producers and importers/exporters have a strong financial interest to ensure that pesticide residue violations do not occur in their shipments. Even then, the violation rate is approximately 1% in imported foods from around the world.[1]

Whether pesticide residues are truly harmful to human or environmental health is immaterial to the current situation. There is sufficient evidence cited by proponents and opponents alike about whether modern agrochemicals (or plant protection products as some call them) are significant risks to human or ecological health. The uncertainty of these risks has led to a concept known as the "precautionary principle." Essentially, this tenet follows the maxim that it is better to be safe than sorry and err on the side of caution. The current situation dictates that pesticide residues be determined in foods because consumers perceive pesticides to be a health risk, and laws have been established worldwide to regulate residue levels. In this situation, it is a risk for importers/exporters *not* to measure their raw food commodities for residues.

14.1.1 HISTORICAL PERSPECTIVE

During the 1940s and 1950s, scientists were highly regarded and trusted by the public, and the revolutionary advances made in the agrochemical industry to combat pests and disease were welcomed by governments, farmers, and consumers. Indeed, the development of relatively inexpensive, mass-produced synthetic pesticides has enabled the ability to sustain a rapidly growing current world population of six billion people with agricultural capacity to spare. However, notable transgressions were made, particularly with respect to persistent organochlorinated pesticides such as DDT, chlordane, and toxaphene. They are not necessarily very acutely toxic to humans or nonarthropodal wildlife, but they cause serious detrimental effects to the ecosystem, as famously described by Rachel Carson in *Silent Spring* in 1962.[2] These types of pesticides (and others) were also implicated as endocrine disrupters, which was described in *Our Stolen Future* published in 1996.[3] The use of many persistent polychlorinated pesticides was banned during the 1970s (except DDT may be used in some tropical countries to combat malaria), but they are still commonly found at ultratrace levels in humans, wildlife, the environment, and the food supply due to their long-term persistence.

Due to such transgressions, the credibility of scientists and the agricultural industry and perceptions about pesticides have been immeasurably changed since the 1950s. Now, the media and public do not blindly trust scientists. Most people are concerned about pesticide residues in their food, and a growing percentage of consumers are willing to pay higher prices for organically grown food. Despite the accumulated knowledge we have now and the greater amount of testing and stricter controls placed on the registration of pesticides for use in agriculture, there is still no guarantee that mistakes are not being made. After all, scientists are only human, and agribusinesses are motivated by money; thus, the only guarantee is that mistakes *are* being made, and the only question is the severity of the mistakes.

Indeed, a string of food safety scandals or crises has been occurring and aggressively reported by the media. Debates among scientists, consumer groups, industry, and government decision makers raise concerns among the public about a diverse range of topics, such as genetically modified organisms, irradiation, subtherapeutic use of antibiotics in livestock, organic farming practices, and slaughterhouse operations. Although pesticides have not been implicated in a major crisis in the past decade, the food industry and government entities must remain informed and diligent.

14.1.2 A QUESTION OF RISKS

Risk of one action or activity can only be judged with respect to risk of other actions. When it comes to pesticide residues in food, the adverse effects of having pesticide residues in the diet should be judged versus the situation of not having them in the diet. Taken in context of the unknown toxicology of natural components present in foods, and the tremendously wide variability of risks of one individual from another, the additional relative risk (if any) of pesticides at trace levels in the food supply is not known. Short-term acute toxicity can be determined in toxicological experiments, and epidemiology can sometimes find correlations between some causes and effects, but when it comes to the long-term chronic effects involving cancer, reduced intelligence, endocrine system effects, and a variety of health syndromes, then current science cannot fully answer this question of risk. It may be that we will never be able to determine the true risk of pesticide residues in food due to the many variables involved and low levels that occur. However, there is evidence that infants and children during critical developmental changes are highly susceptible to pesticide residues in their diet,[4] and the European Union has recommended that baby foods should not contain more than 10 ng/g of any pesticide residue. On the whole, however, we know that people are living healthier, longer, and more prosperous lives than any other time in human history and that pesticide use in agriculture has contributed to this beneficial outcome.

In the case of current-use pesticides, direct cost-benefit analysis shows that the appropriate use of pesticides on the market is worthwhile, or else the pesticide product would not be on the market. However, long-term costs/benefits cannot be determined until the pesticides have been used over a period. Monitoring results provide data that can be used to better assess the risks associated with pesticides in our food. Undoubtedly, pesticide residues in food are not a high human health priority, and some people would argue that analyzing for them is a waste of time and money. However, food is central to life itself, and with this in mind, one could argue that nothing is *more* important than having a safe food supply (there are *equally* important issues that sustain life, but not *more* important ones).

Since we have the means to measure, monitor, and control these potential hazardous residues in our food, then we should do so, within reasonable cost. Just as in the cost-benefit scenario for using pesticides requires that the costs should not exceed the benefits, the cost of the analysis should not exceed its worth. Even though the benefits of residue monitoring are considered to be high, it still adds to the cost of food, thus the costs of analysis should be as low as possible.

14.1.3 Monitoring of Pesticide Residues

So, why bother to analyze for pesticide residues in food? Simply put, because they are there. The findings reported by monitoring programs show high percentages of some commodities have pesticide residues. For example, the U.S. Department of Agriculture's Agricultural Marketing Service sponsors the Pesticide Data Program in the United States. Table 14.1 lists pesticides that have been found in more than 20% of certain commodities over the course of the program since 1992. The table also gives the U.S. tolerance or action level for pesticides in the given commodity and the number of violative samples. Many pesticides have been found in many raw commodities, which provides fodder to environmentalists and consumer groups that pesticides should not be used. However, very few violations occur, which provides support for the food and agrochemical industry that pesticide usage is under control. Independent of enforcement actions, the data generated by these and other monitoring programs around the world are used by pesticide regulators to perform risk assessments and help make reregistration decisions involving currently used pesticides.

Nearly all governments of the world have monitoring programs that check for pesticide residues in fruits, vegetables, grains, meat and poultry, seafood, beverages, eggs, nuts, tobacco, and nearly every other type of agricultural commodity. Food manufacturers also regularly conduct pesticide residue analysis of ingredients in their food products, often contracting out the work to private laboratories. Consumer groups and research organizations also contract laboratories to conduct analyses for a variety of purposes to look for pesticide residues in food. It is safe to estimate that hundreds of thousands of analyses are conducted worldwide per year to monitor for pesticide residues in food.

TABLE 14.1
Pesticide Residues with High Incidences in Food Commodities as Determined by the Pesticide Data Program in the United States, 1992–2001

Commodity	Pesticide	Pesticide Group	Tolerance (mg/kg)	No. of Violations	Samples with Detected Residues (Range in %)
Apples	Azinphos methyl	I	1.5	0	23–66
	Carbaryl	I	10	0	4–22
	Chlorpyrifos	I	1.5	0	8–26
	Diphenylamine	F	10	0	35–86
	Propargite	I	NT (3)[a]	1 (0)[a]	0–0.1 (25–35)[a]
	Thiabendazole	F	10	0	34–70
Apple juice	Carbaryl	I	10	0	25–33
	Dimethoate	I	2	0	9–25
	Thiabendazole	F	10	0	26–38
Bananas	Imazalil	F	0.2	3	8–15
	Thiabendazole	F	3	3	26–55
Broccoli	Dacthal	H	5	0	16–22
Cantaloupes	Endosulfan sulfate	I[b]	2	0	10–34
Carrots	p,p'-DDE	I[b]	3[c]	0	22–43
	Iprodione	F	5	0	25–37
	Linuron	H	1	0	2–69
	Trifluralin	H	1	0	22–62
Celery	Acephate	I	10	0	22–51
	Chlorothalonil	F	15	0	36–68
	Dicloran	F	15	0	21–52
	Methamidophos	I[b]	1	0	11–28
	Oxamyl	I	0.1	0	17–35
	Permethrins	I	5	0	23–42
Cherries	Azinphos methyl	I	2	0	59–62
	Carbaryl	I	10	0	44–61
	Iprodione	F	20	0	15–28
	Myclobutanil	F	5	0	51–60
	Propiconazole	F	1	0	28–30
	Tebuconazole	F	4	0	19–25
Cucumbers	Endosulfans	I	2	0	34–42
	Endosulfan sulfate	I[b]	2	0	53–55
Grapes	Captan	F	50	0	23–48
	Iprodione	F	60	0	17–45
	Myclobutanil	F	1	0	19–28
	Vinclozolin	F	NT (6)[a]	3 (0)[a]	0.1–0.3 (4–44)[a]
Grape juice	Carbaryl	I	10	0	38–42
Grapefruit	Imazalil	F	10	0	11–29
	Thiabendazole	F	10	0	39–63

continued

TABLE 14.1 (continued)

Pesticide Residues with High Incidences in Food Commodities as Determined by the Pesticide Data Program in the United States, 1992–2001

Commodity	Pesticide	Pesticide Group	Tolerance (mg/kg)	No. of Violations	Samples with Detected Residues (Range in %)
Green beans	Acephate	I	3	2	19–48
	Endosulfans	I	2	0	10–40
	Endosulfan sulfate	I[b]	2	0	23–25
Lettuce	Acephate	I	10	0	7–15
	Endosulfans	I	2	0	1–21
	Permethrins	I	20	0	3–15
Milk	p,p′-DDE	I[b]	1.25[d]	0	14–17
Nectarines	Dicloran	F	20	0	19–32
	Fenbuconazole	F	2	0	15–21
	Fludioxonil	F	5	0	22–49
	Iprodione	F	20	0	57–69
	Phosmet	I	5	0	48–75
	Propargite	I	4	0	32–48
Oranges	Imazalil	F	10	0	27–64
	Thiabendazole	F	10	0	31–62
Peaches	Azinphos methyl	I	2	0	14–33
	Benomyl[e]	F	15	0	2–27
	Dicloran	F	20	0	19–43
	Fludioxonil	F	5	0	35
	Iprodione	F	20	0	42–79
	Parathion methyl	I	1	0	1–30
	Phosmet	I	10	0	9–81
	Propargite	I	NT (7)[a]	27 (0)[a]	1.9–3.2 (18–33)[a]
Pears	Azinphos methyl	I	2	0	52–68
	Diphenylamine	F	NT[f]	288	15–24
	o-Phenylphenol	F	25	0	21–28
	Phosmet	I	10	0	14–26
	Thiabendazole	F	10	0	71–80
Potatoes	Chlorpropham	H	50	0	61–75
Rice	Malathion	I	8	0	11–17
	Piperonyl butoxide	I	20	0	14–18
Soybean	Chlorpyrifos	I	0.3	0	10–80
	Malathion	I	8	0	33–41
Spinach	p,p′-DDE	I[b]	0.5[c]	0	41–58
	Permethrins	I	20	0	53–84
Strawberries	Benomyl[e]	F	5	0	11–26
	Captan	F	25	0	54–62
	Carbaryl	I	10	0	18–36
	Iprodione	F	15	0	34–49

continued

TABLE 14.1 (continued)
Pesticide Residues with High Incidences in Food Commodities as
Determined by the Pesticide Data Program in the United States, 1992–2001

Commodity	Pesticide	Pesticide Group	Tolerance (mg/kg)	No. of Violations	Samples with Detected Residues (Range in %)
	Malathion	I	8	0	16–24
	Myclobutanil	F	0.5	2	13–22
	Vinclozolin	F	10	0	3–34
Sweet bell	Endosulfans	I I^b I^b I^b	2 2 1 1	0 0 0 0	14–23 13–22 30–36
peppers	Endosulfan sulfate				30–36
	Methamidophos				
	Methamidophos				
Tomatoes	Endosulfans	I	2	0	13–36
	Endosulfan sulfate	I^b	2	0	14–35
	Methamidophos	I^b	1	0	22–37
Wheat	Chlorpyrifos methyl	I	6	0	54–73
grains					
	Malathion	I	8	0	68–71
Winter squash	Endosulfan sulfate	I^b I I^b	2 0.1^c 0.01^c	0 1 2	21–25 60–77 2–29
(fresh)	Dieldrin Heptachlor				
(frozen)	epoxide				

Note: I 5 insecticide; H 5 herbicide; F 5 fungicide; NT 5 no tolerance level set for that pesticide/commodity pair.

^{a} Tolerance for propargite withdrawn from apples and peaches and for vinclozolin from grapes; data in parentheses obtained before the withdrawals.

^{b} Metabolite and/or degradation product (methamidophos can also be a parent compound).

^{c} Action level established by FDA.

^{d} Tolerance on fat basis.

^{e} Analyzed as carbendazim.

^{f} Tolerance application for diphenylamine as a postharvest fungicide for pears was submitted to EPA in 1998 (pears monitored in years 1997–99).

14.1.4 NEEDS FOR PESTICIDE RESIDUE ANALYSIS

A good analytical chemist provides results to meet the needs of the analysis. A great analytical chemist provides results to meet the needs of the analysis quickly and efficiently. Because pesticide residue analysis generally adds to the cost and delays distribution of the goods being analyzed, the analysis should be as fast and inexpensive as possible, provided that acceptable quality of results is obtained. The criteria that define "acceptable quality of results" depend on the need for the data, and this relationship is at the core of the "fit for purpose" concept in analytical methods.[5]

Some of the reasons that pesticide residue analysis of foods is conducted include the following:

- Compliance and enforcement monitoring
- Protection of the environment

- International trade issues
- Data for risk assessment
- Registration and reregistration of pesticides
- Scientific study (e.g., endocrine disrupting effects, or efficacy of pesticides)
- Marketing of residue-free products
- Verification of organic food labeling
- Deterrence of deliberate acts of contamination
- Instilling confidence in the safety of food

From our perspective, the greatest need of pesticide residue analytical methods for all purposes has been much the same since pesticides were found to be a concern: *to achieve accurate and precise results for a wide concentration range of a maximum number of analytes in the appropriate matrices in a rapid, inexpensive, safe, and simple manner.* We shall focus on this central need, and the aim of this chapter is to discuss current approaches in the multiresidue analysis of pesticides to meet this overall need, and in the process, to present the main issues related to determining pesticide residues in food. Several books and review chapters have already been written on this topic,[6–10] and we do not wish to duplicate what is already published. Instead, we hope to provide a more critical assessment of the current situation in order to help improve it.

14.1.5 CHEMICAL PROPERTIES OF PESTICIDES

Multiclass, multiresidue analysis of pesticides in food is a major challenge to the analytical chemist due to the sheer number and diversity of pesticide analytes and sample types. Tables 14.1 and 14.2 name some of the pesticide residues commonly found in raw commodities and their types of usages. A highly trained and experienced analyst knows which pesticides are problematic and why, which matrices give difficulties for certain pesticides, the effect of pH and water content, and a myriad of other factors that can occur in analyses. Devising standard operating procedures in a monitoring laboratory is necessary for quality assurance purposes, but no set of rules or guidelines can cover all contingencies, and there is no substitute for knowledgeable and intelligent chemists.

There are numerous sources of information on the chemistry, applications, and regulation of pesticides in the scientific literature, and in the past decade, the Internet has become a valuable additional way to find information for analysis. For example, the U.S. Department of Agriculture (USDA) National Agricultural Statistics Service provides annual reports, which are accessible through the Internet, concerning pesticide usage on field crops in the United States (http://www.usda.gov/nass/pubs/pubs.htm), and the Agricultural Marketing Service provides monitoring results for risk assessment and other purposes as part of the Pesticide Data Program (http://www.ams.usda.gov/science/pdp/). The Food and Drug Administration (FDA) is responsible for regulatory enforcement monitoring of pesticides in many foods in the United States, and these results (along with monitoring results from the Total Diet Study, which measures pesticide levels in prepared foods in typical meals) are posted on the Internet (http://www.cfsan.fda.gov/~lrd/pestadd.html). Another interesting database

TABLE 14.2

Major Chemical Classes of Fungicides, Herbicides, and Insecticides/Acaricides

Pesticide Group	Chemical Class	Representatives
Fungicides	Chlorine-substituted aromatics	Chlorothalonil, dicloran, quintozene (PCNB), tecnazene (TCNB), pentachloroaniline
	Dithiocarbamate	Ferbam, mancozeb, maneb, metiram, propineb, thiram, zineb, ziram
	Dicarboximide	Chlozolinate, iprodione, procymidone, vinclozolin
	Imidazole	Benomyl, carbendazim, imazalil, prochloraz, thiabendazole, thiophanate-methyl
	N-Trihalomethylthio	Captafol, captan, folpet, dichlofluanid, tolylfluanid
	Triazole	Bitertanol, fenbuconazole, hexaconazole, metconazole, myclobutanil, penconazole
Herbicides	Chlorophenoxy acid	2,4-D, 2,4-DB, dichlorprop, MCPA, MCPB, mecoprop, 2,4,5-T
	Aryloxyphenoxypropionate	Cyhalofop-butyl, diclofop-methyl, fenoxaprop-ethyl, fluazifop-butyl, haloxyfop-methyl
	Chloroacetamide	Acetochlor, alachlor, metazachlor, metolachlor
	Dinitroaniline	Benfluralin, dinitramine, pendimethalin, trifluralin
	Imidazolinone	Imazamox, imazapyr, imazaquin, imazethapyr
	Quaternary ammonium	Diquat, paraquat
	Phenylurea	Chlorotoluron, diuron, fenuron, isoproturon, linuron, metoxuron, monolinuron, neburon
	Sulfonylurea	Chlorimuron-ethyl, chlorsulfuron, metsulfuron-methyl, sulfometuron-methyl, triasulfuron
	Thiocarbamate	Butylate, cycloate, EPTC, molinate, pebulate, thiobencarb, tri-allate, vernolate
	Triazine	Ametryn, atrazine, cyanazine, prometryn, propazine, simazine, terbutryn
	Uracil	Bromacil, lenacil, terbacil
Insecticides	Carbamate	Aminocarb, carbaryl, carbofuran, fenoxycarb, methiocarb, methomyl, oxamyl, propoxur
	Organochlorine	DDE, DDT, dicofol, dieldrin, endosulfan, endrin, heptachlor epoxide, lindane, mirex
	Organophosphorus	Acephate, azinphos-methyl, chlorpyrifos, diazinon, malathion, parathion, phosmet
	Neonicotinoid	Acetamiprid, imidacloprid, thiamethoxam, clothianidin, dinotefuran, nitenpyram
	Pyrethroid	Acrinathrin, bioallethrin, cyfluthrin, cyhalothrin, deltamethrin, fenvalerate, permethrin

can be found at http://www.pesticides-online.com/. In print, *The Pesticide Manual*[11] and *Farm Chemicals Handbook* (now called the *Crop Protection Handbook*)[12] are considered essential reference books by many pesticide chemists.

The physiochemical properties of the analyte(s) determine the type of possible approaches that lead to a successful extraction, cleanup, and detection procedure. An important aspect is the polarity of the analyte, which can be estimated through its solubility in water and/or its octanol/water partitioning coefficient ($K_{o/w}$). In the analysis of pesticides that are weak acids and bases, pH and ionic strength also become critical aspects in the approach. Other factors of note include volatility and stability, which indicate what precautions must be made to avoid analyte losses. Several chemical properties of interest for many pesticides are listed in the USDA Pesticide Properties Database, which is also accessible on the Internet (http://www.arsusda. gov/ppdb2.html). The polarity range of the pesticides recovered by an analytical method often sets the scope of analysis and number of pesticides that can be determined at the same time.

14.1.6 Food Matrices

Foods are often complex matrices with widely varying compositions, but many foods consist of four major components: carbohydrates, lipids, proteins, and moisture. The capability of analytical methods can be heavily dependent on matrix, but the great variety of foods limits practicality to validate new analytical methods for all food matrices. For this reason, scientists have proposed that certain types of foods could serve as a reference for other types of foods of a similar nature. For example, the Dutch monitoring program chooses orange, apple, grape, lettuce, Chinese cabbage, and wheat as representative commodities for validation studies. Another possibility is to mix a variety of food types together in one composite matrix.

In a very useful service, the U.S. Department of Agriculture provides a wide-ranging food composition database that is accessible in searchable format on the Internet (http://www.nal.usda.gov/fnic/foodcomp/). Knowing the composition of the different foods is very important so that trends in pesticide recoveries and interferences can possibly be correlated with respect to water, sugars, lipids, or other factors in the sample types.

14.2 THE ANALYTICAL PROCESS

The analytical process consists of the following main steps:
1. Collection of a meaningful sample
2. Transportation and storage of the sample
3. Sample preparation
   - Homogenization and subsampling
   - Extraction
   - Cleanup
   - Concentration
4. Analysis (quantitation and confirmation)
5. Data processing and quality review
6. Reporting of the results

Just as a chain is only as strong as its weakest link, an analytical scheme is only as good as its worst component. Thus, each step in the process must be carefully optimized. Of course, cost constraints must also be observed, but if any step is unable to be conducted properly in order to save costs, such as not obtaining a representative sample, then the entire analysis and all other costs associated with it are wasted because the analytical result will not suit the need for analysis.

14.2.1 DESIRABLE TRAITS OF AN ANALYTICAL METHOD

The main desirable traits in an analytical method include (1) highly accurate (true and precise result), (2) sensitive (low detection limits), (3) widely applicable, (4) sufficiently selective or confirmatory, (5) rapid, (6) easy, (7) inexpensive, (8) reliable, (9) safe, (10) automated, (11) portable or small, and (12) waste-free. Unfortunately, some of these traits are opposites, and for a method to possess all these features, such as automated, cheap or compact, is almost impossible. However, sometimes seem ingly opposite traits can be merged, such as the widely applicable yet very selective detection method using MS. Usually, though, sacrifices and compromises are nearly always necessary, and the selection of an approach should be made to provide the most advantages within constraints of analytical needs and resources.

The first consideration in choosing an analytical method is to define the needs for the analysis. Once a list of the analytes and matrices has been prepared, then one should set the desired analytical figures of merit (limits of detection, recoveries, reproducibility). Typically, the quality of the analytical result for common applications should generally meet acceptability criteria of 70%–120% recoveries, 10%–25% relative standard deviations, and limits of detection,10%–100 ng/g (depending on the pesticide, matrix, and a variety of other factors). These desired goals must be realistic and meet constraints in resources, time, and technology. For enforcement monitoring, the scope of pesticides to be detected should be as wide as possible, and the qualitative nature of the findings should ensure that no false positives should occur.[5] In the case of risk assessment, false negatives should be kept to a minimum, and detection limits should be as low as possible.

Oftentimes in multiresidue methods, sacrifices in detectability of a particular analyte must be made to gain sensitivity or extend analytical scope to other analytes. Priorities must be made depending on the importance of the problem and resources available.

14.2.2 FIELD-BASED ANALYSIS

One of the largest costs in monitoring programs is the handling and transport of the samples to the laboratory. In a typical program, food and environmental samples are collected by field personnel throughout a region and are sent to a central analytical laboratory for analysis. For regulatory applications involving perishable foods, the results should be returned as rapidly as possible to avoid spoilage (if the commodity is held until results are received). However, a high percentage of the total time to obtain results is taken up in transport (and storage) of the sample.

In many cases, it would be more cost-effective and efficient if the field personnel could conduct the analysis at the location of the sample rather than shipping it to a lab. Ideally, the same quality of result would be achieved, but the procedure would have to be rapid, inexpensive, rugged, and simple. At present, the techniques for field-based analysis sacrifice analytical quality and range of analytes for speed, ease, and expense. In general, single-analyte screening methods have been developed rather than the more desirable multiresidue analytical methods. Thin-layer chromatography is capable of high-throughput screening of multiple pesticides, but analysis is usually conducted in the laboratory and sensitivity is rather low.

14.2.3 IMMUNOCHEMICAL ASSAYS

Multiresidue approaches that do exist take advantage of a mode of action common to a particular group of pesticides, such as cholinesterase inhibition. However, these approaches often lack sensitivity, give unequal responses depending on the analyte, and cannot distinguish between pesticides.

Immunological screening approaches have been touted since the 1980s as being potentially useful for field-based analysis of pesticides and environmental contaminants.[13] Enzyme-linked immunosorbent assay (ELISA) has been the most common format for immunochemical techniques.[14] With a few exceptions, ELISA has failed to meet initial promising claims in pesticide residue applications. Despite the high degree of selectivity of antibodies and the great sensitivity of analysis, matrix effects and cross reactivities in complex samples can lead to excessive rates of false positives and/or negatives in real applications. Even in the successful water-based applications, the samples typically need to be collected in the field and taken to the laboratory for replicate analysis using relatively long incubation times and instrumental-based detection.

Field-based readers are available, which typically have lower sample throughput, but in the analysis of complex matrices such as foods, sample preparation (extraction/cleanup) is usually needed to get acceptable results, and these steps are often more easily conducted by an analyst in a lab. Another common drawback with antibody methods is the detrimental effect of organic solvents on the analysis. Organic solvents are needed in nearly all separation methods for pesticides, and evaporation steps are too time-consuming and inconvenient in field-based methods. Some antibodies can withstand 5%–25% of organic solvent with moderate loss of capabilities, but others are significantly affected by only 1%–2% organic content. Ionic strength, pH, time, and temperature are other factors that can lead to degradation of antibodies and variable results in some assays.

The low detection limits, lower costs, reduced sample volumes, and high sample throughput with multiwell immunoassay kits for *individual* analytes in aqueous matrices are still excellent advantages. However, the time, effort, and costs are multiplied as each analyte is added to the analysis (assuming a good antibody kit is available). Also, the inversely proportional, semilogarithmic relationship between signal and concentration and narrow linear dynamic range of many immunoassays make quantitation more problematic than with most analytical methods. The ELISA results are not always trustworthy; thus, confirmation of analyte identity and its concentration is often required. Then, the simultaneous quantitation and confirmation and other advantages of gas chromatography/mass spectrometry (GC/MS) or liquid chromatography/tandem mass spectrometry (LC/MS-MS) overtake those of the immunoassay. Unlike ELISA, the time, effort, and costs remain much the same for multiple analytes in LC/MS-MS and GC/MS, and usually, wider analytical scope is important; thus, these techniques can be used instead of ELISA in the first place rather than just as confirmatory approaches.

14.2.4 MOBILE MINILAB APPROACH

To conduct analysis in the field, the ease of use, speed of analysis, and size of the device take precedence over more complicated methods that may be possible (but not necessarily preferable) in the lab. Ideally, the method used in the field would also be multiresidue, quantitative, confirmatory, inexpensive, and rugged, but no current approach has reached that stage of development. Although portable GC instruments are available, gas and power needs still make them too bulky for convenient applications in the field. Micro-technological innovations may someday permit lab-based chemical methods to be useful in the field, but practical considerations of sample handling and representative sample sizes may limit their utility.

In the meantime, typical lab instruments placed in a van (mobile minilab) may be the best option for multiresidue field-based screening analysis. Rapid and simple sample preparation steps may be used in combination with fast chromatographic analysis to obtain quick multiresidue results, but costs of such an approach could be prohibitive.

14.3 EXTRACTION

Extraction of pesticides from the sample matrix is a fundamental aspect in the analytical process. At this time, no method can adequately detect pesticide residues *in situ* in a food item. It is possible to thermally desorb analytes directly from the sample into a GC,[15] but quantitation is not straightforward, samples are too small to be representative, and the lack of cleanup steps places a difficult burden on detection.

The following list of parameters constitutes the main factors that are involved in the extraction process:

Sample matrix
Comminution
pH
Extraction solvent(s)
Water content
Temperature
Sample:solvent ratio
Extraction method
Time of extraction
Type(s) of salt added
Amount of salt(s)
Pressure

Each of these factors can have an effect on pesticide recoveries, stability, and selectivity in the extraction, and these effects on the method being used should be known by the analyst.

14.3.1 Sample Comminution and Subsampling

The simplest way to improve efficiency of an analytical method is to reduce sample size to the minimum amount that still provides reliable results to suit the needs for the analysis. Minimizing the sample size minimizes solvent usage and waste, effort, and expense. A little extra time and effort put into proper sample comminution techniques can save a lot of time and effort later in the analytical process. A number of informative articles have been published on this critical, but often ignored, topic.[16–19]

Appropriate devices and techniques must be used during the chopping procedure, which entails using the correct sample amount to volume of the chopper container, possibly frozen conditions (via refrigeration, dry ice, or liquid N_2), and sufficient time. For example, a large sample collected in the field (e.g., 10 kg) that is homogenized in a large chopper can be subsampled quickly (e.g., 250 g portion), which is further homogenized in a smaller blender. Then, a quite small sample (e.g., 10 g) can nicely represent a quite large sampling of the food product. The sampling constant (K_s) of the homogenizing procedure and equipment used in a lab should be determined experimentally to find the smallest subsample size that meets the sample homogeneity requirements for the purpose of analysis.[17] The subsample size extracted should also be large enough to achieve the desired limit of detection (LOD) for the final extract (assuming that matrix is not the limiting interference). When proper procedures are followed and good equipment is used, a 10-g sample is often acceptable for analysis of pesticides in fruits and vegetables.[16–17]

14.3.2 Choice of Extraction Solvent

The most common multiclass, multiresidue methods in use today for fruit and vegetable analysis use acetone,[20–22] ethyl acetate (EtAc),[23,24] or acetonitrile (MeCN)[25–28] for extraction. Each solvent has been shown to achieve high recoveries of a wide variety of GC-amenable pesticides in fruits and vegetables, but in practical terms, each solvent has advantages and disadvantages with respect to each other. In several respects, MeCN has advantages over the other solvents because it does not coextract as many lipids, proteins, sugars, and salts as the other common solvents when it is partitioned from water in a salting-out procedure. EtAc is also very selective in non-fatty foods in combination with a drying agent (e.g., Na_2SO_4), but it coextracts lipids to a large extent. MeCN is more selective in the case of commodities such as avocados, eggs, milk, shellfish, and meats, which have moderate lipid content, but for very fatty foods, such as lard, a nonpolar solvent that dissolves and penetrates the fat is needed to extract the highly lipophilic pesticides that can concentrate in those matrices. In this situation, gel-permeation chromatography (GPC) is needed for cleanup to separate the pesticides from coextracted fats. Acetone is the least selective of the three common solvents, and its combination with nonpolar cosolvent(s) is needed to partition acetone extracts from water. Yet, acetone is the most commonly used extraction solvent due to its extensive validation by the FDA.[29–31]

As a result of the growing number of thermally labile and polar pesticides applied in modern agriculture, it is likely that MeCN will become the most common extraction solvent because of its better compatibility with LC-based analysis and other advantages over acetone and EtAc. A recently developed approach called the quick, easy, cheap, effective, rugged, and safe (QuEChERS) sample preparation method uses MeCN for extraction for a number of reasons as described in the publication.[28] Table 14.3 compares the features of the QuEChERS method with some other multiclass, multiresidue methods for pesticide analysis of nonfatty foods.

14.3.3 MEANS OF EXTRACTION

Shaking should be the first choice in extraction as long as it provides acceptable results for incurred samples and/or standard reference materials (if they are available). The advantages of shaking over blending include the following: (1) the sample is not exposed to the active metal surfaces of the blender; (2) shaking can be done by hand if needed (in the field or lab); (3) no cleaning of the blender jar/probe is

TABLE 14.3

Comparison of Practical Features of Common Multiclass, Multiresidue Methods for Analysis of Pesticides in Foods

Parameter	FDA[20]	Swedish[23]	German[21]	Canadian[25]	Dutch[22]	QuEChERS[28]
Sample size	100 g	75 g	50 g	50 g	15 g	10 g
Solvents						
Acetone	367 mL		100 mL	40 mL	30 mL	
MeCN				115 mL		10 mL
EtAc		240 mL	115 mL			
CH_2Cl_2	300 mL				34 mL	
Petroleum ether	23 mL				30 mL	
Isooctane					4.5 mL	
Toluene				5 mL	0.5 mL	
Cyclohexane		40 mL	65 mL			
n-hexane						
Total	690 mL	280 mL	280 mL	160 mL	99 mL	10 mL
Estimated time	60 min	70 min	80 min	50 min	40 min	30 min
Nondisposable labware	10 items	5 items	10 items	10 items	5 items	1 item
Equipment	Blender SPE setup Steam bath filters	Blender GPC system Rotovap N₂ evap. filters	Blender GPC system Rotovap N₂ evap. filters	Blender SPE setup Centrifuge Rotovap N₂ evap.	Blender SPE set-up Centrifuge steam bath	Centrifuge
Chemicals	5 g Na₂SO₄, SPE cartridges (C-18, NH₂ and QMA)	60 g Na₂SO₄	19 g NaCl Na₂SO₄ 1.5 g	10 g NaCl Na₂SO₄ 10 g SPE cartridges (C-18, GCB and NH₂)	SPE cartridge (NH₂)	4.15 g MgSO₄ 1 g NaCl PSA 25 mg
Material cost	<$13	<$5	<$5	<$9	<$2	<$1

needed between samples; (4) extraction is conducted in a closed vessel, thus it is safer; (5) chance of carryover from one sample to another, and thus the inconvenient cleaning of the blender between usages, is eliminated; (6) no extra solvent from rinsings is added to the sample and no extract or sample is removed by the mixing approach, unlike the case with blenders; (7) only a single container is used for each extraction; (8) a batch of samples can be extracted more easily in parallel rather than sequentially as with blenders; (9) cost of mixers/shakers is less than blenders and they need less maintenance; (10) shaking makes less noise than blenders; and (11) no frictional heat is generated during extraction (especially when solids are added).

Shaking is acceptable for the extraction of many pesticide residues from many crops, but there are possible concerns for systemic pesticides in certain pasty foods in which shaking does not allow easy penetration into the matrix. This is why sample comminution procedures are more important in a method that uses shaking for extraction. Studies using radioactive labeling and other forms of measurement for pesticides applied in the field have shown that, in general, pesticides in high-moisture foods, such as fruits and vegetables, do not strongly interact with the matrix. Thus, shaking with an organic solvent is a reasonable approach for extraction of most pesticides from fruits and vegetables.

In the case of dry foods, stronger measures are often needed for complete extraction of bound residues, particularly in the case of systemic pesticides. Water is nearly always helpful in the extraction of a dry food to swell the matrix and open pores for better access of the solvent. Blending procedures, high temperatures, and/or use of strongly acidic or basic conditions are often required to achieve complete extraction of bound pesticide residues in dry food matrices. The use of an alternate extraction procedure as described in the next section may be very suitable in these cases.

14.3.4 ALTERNATIVE TECHNIQUES

Alternative extraction techniques include microwave-assisted solvent extraction (MASE),[32,33] supercritical fluid extraction (SFE),[34,35] and pressurized liquid extraction (PLE),[36,37] which is also known as accelerated solvent extraction or pressurized solvent extraction , depending on the manufacturer. There was a great deal of activity during the 1990s to investigate and evaluate alternatives to the liquid-based extraction procedures for pesticide analysis. SFE, for example, has many advantages including (1) higher degree of selectivity, (2) ability to automate, (3) reduced or eliminated solvent usage, (4) elimination or reduction of solvent evaporation steps, and (5) commonly convenient hyphenation with cleanup and/or detection methods. However, SFE is too selective to extract both polar and nonpolar pesticides simultaneously, takes longer than blending methods, may give recoveries dependent on matrix, requires bulky, expensive instruments and gas cylinders, automates extraction of samples sequentially rather than in parallel, and often involves complicated method development. SFE can be more convenient than a few of the liquid-based extraction methods commonly used, but it cannot match the capability of the most convenient liquid-based methods.

PLE and MASE use heated and pressurized liquids to potentially increase the speed of extraction, but this also acts to reduce selectivity, and the application of heat increases the chance of analyte degradation. In many pesticide residue applications, recoveries with solvents at room temperature and pressure are 100% with a 1-minute shaking procedure. PLE and MASE have advantages over Soxhlet and reflux procedures if they are necessary, but even then, conditions can often be found to achieve acceptable recoveries at room temperature and pressure if the instrumental techniques are not available in a laboratory.

Matrix solid-phase dispersion (MSPD) is another alternative extraction approach that has been evaluated for pesticide residue analysis.[38,39] MSPD essentially incorporates a small portion of sample into a sorbent, and cleanup is performed at the same time as extraction. It has some advantages of convenience over the conventional approach to separately extract the sample, evaporate solvent, and then conduct cleanup. The very small sample size (0.1–2 g) can be an advantage of MSPD if limited sample is an issue, but in most residue applications, it is a crucial disadvantage with the approach due to the difficulty of getting a sufficiently representative, homogeneous subsample.

Another type of alternative extraction technique is to use a sorptive extraction device. At this time, two forms of sorptive extraction have been commercialized: solid-phase microextraction (SPME) and stir-bar sorptive extraction (SBSE), which are the subject of several reviews.[40–43] In another possibility, the coating is contained in a tube, as in a short piece of a capillary GC column.[44] All three techniques are actually forms of the same approach in which a material, such as polydimethylsiloxane, is coated over a fiber or stir-bar to semiselectively extract chemicals from an aqueous or gaseous sample. The type and amount of chemicals that partition into the coating depend on the partitioning coefficient, coating volume, sample volume, time, temperature, matrix effects, pH, ionic strength, solvent composition, and mechanical factors.

The general advantages of sorptive extraction include the following: (1) compact device, (2) multiresidue capability, (3) simple operation, (4) possibility of short analysis times, (5) potential for selective extraction/cleanup, (6) rugged method of injection in GC (i.e., contaminants do not touch the liner or enter the column), (7) automated, (8) quantitative with appropriate precautions, (9) low LODs for some analytes, (10) elimination of organic solvents, (11) inexpensive, and (12) use of small sample size.

However, a simple listing of the general advantages of sorptive extraction is deceiving. The most appropriate applications of SPME relate to the qualitative analysis of volatile components in gaseous samples, and SBSE is most useful for the analysis of semivolatiles in liquids, mainly water. Sorptive extraction devices are compact, but they are not stand-alone analytical techniques, and thus are no more portable than the detection system to which they are coupled. It is possible to collect the sample in the field and transport the small device to the lab for analysis, but storage of the devices is not straightforward, and accuracy of results is poor unless the exact same conditions are used for calibration standards as in the sampling procedure. Usually, the commercialized approaches are used with an autosampler in the lab, and injection into a GC is made immediately after the extraction.

Other concerns are that coatings can be prone to memory effects and can become contaminated with nonvolatile matrix components. Thus, in residue methods, sorptive extraction methods best meet their potential advantages in the analysis of clean water and air matrices. The potential for selective sampling of analytes and cleanup exists by using different solid-phase materials, but in practice this is not easily accomplished and only a few coating materials are available. Unlike the capillary GC approach, SPME and SBSE are not cheap enough for disposal after a single use. Analyses are sequential; thus, sample throughput is limited by the equilibration and analysis times. The sensitivity of analysis can be very high (very low detection limits for those relatively nonpolar pesticides that readily partition into the coating phase), but this often translates into longer equilibration times (some equilibrations take hours to reach a maximum).

Preferably, extraction of analytes from the sample to the extracting medium should be complete, instantaneous, and selectively exclude matrix. The fundamental difficulty with sorbent extraction is that different analytes have different partitioning ratios and rates between the sample matrix and the fiber coating. Matrix components often affect the equilibration process and lead to variable results. Also, water is essentially the only liquid medium with which the coatings can be used because analytes do not partition into the fiber from organic solvents. Similarly, polar compounds do not partition into the coatings from water. Different temperatures, phases, volumes, time, sample treatments (e.g., addition of salt) can increase recoveries or speed the equilibration process, but in reality, the fundamental nature of the sorptive extraction process limits its usefulness.

Despite all these alternative extraction options, the most common extraction method by far is to simply mix an organic solvent with a solid sample. This approach is rapid, simple, reproducible, cheap, commonly gives high recoveries, and uses compact and rugged devices. Simply blending or shaking for a minute followed by a short centrifugation step is hard to beat in practice.

14.4 CLEANUP

Ideally, the extraction method achieves 100% recoveries of the pesticides of interest and contains no interfering coextractives from the matrix in the analysis. Amazingly, this ideal comes true when a very selective detector (such as MS-MS) is used for analysis of relatively uncomplicated food matrices, such as melon or cucumber. However, the ruggedness of the analytical system must also be considered, and even if matrix coextractives do not directly interfere in the detection, they often indirectly cause signal suppression or enhancement effects, and/or lead to a greater need for instrument maintenance. Thus, it often pays to conduct a cleanup step of the raw extracts prior to analysis. Approaches for extract cleanup involve some sort of separation process based on molecular size (GPC or membrane filtration), volatility (distillation), electrochemistry (dialysis or ion exchange), solubility (precipitation), or partitioning (liquid–liquid or solid–liquid).

In pesticide residue methods, GPC is not an uncommon precautionary procedure to remove large molecules from extracts, which avoids buildup of nonvolatiles in the analytical instruments and improves ruggedness. However, GPC is a time-consuming, solvent-intensive, and relatively expensive instrument-based procedure that should be avoided if possible. Unfortunately, extraction of fatty samples often requires GPC to remove coextracted lipids. Fundamentally in GPC, the extracts need to be concentrated prior to injection (often exchanged into an appropriate solvent such as EtAc/cyclohexane), and then reconcentrated after the pesticide fraction has been collected from the GPC column. Some pesticides, such as pyrethroids, elute near the lipids in GPC, thus it is sometimes difficult to get complete recovery of those pesticides and still perform adequate cleanup. Although GPC works well to improve the ruggedness of the analysis, it does not necessarily help to remove interfering components, which tend to have the same molecular weight as the pesticides. Thus, a partitioning type of cleanup procedure is also frequently conducted in combination with GPC.

Liquid–liquid partitioning is commonly used for cleanup in pesticide residue analysis of high-moisture foods when a water-miscible solvent is used for extraction. In the case of acetone, traditional methods call for the initial aqueous extract to be mixed with dichloromethane and petroleum ether or EtAc/cyclohexane. This forms a nonpolar solvent combination that separates from the water phase, leaving the most polar coextractives behind (along with some polar pesticides). The addition of salt to the system helps to force more of the polar pesticides into the organic solvent phase. In the case of MeCN-based extraction, salt alone is enough to induce the phase separation between water and MeCN, thus the addition of a nonpolar solvent (and the concomitant dilution of the extract) is not necessary. In extensive experiments to develop the QuEChERS method, anhydrous MgSO$_4$ was found to be a salt with excellent features to induce liquid–liquid partitioning between MeCN and water and still achieve high recoveries of relatively polar pesticides.[28] Its combination with NaCl served to modify the partitioning so that sugars tended to remain in the aqueous layer. Another advantage of MeCN is that it is not miscible with alkane solvents; thus, liquid–liquid partitioning can be used with solvents such as hexane or iso-octane to help remove coextracted lipids (but nonpolar pesticides will also partition into the nonpolar solvent).[45]

14.4.1 Solid-Phase Extraction

As in many applications, solid-phase extraction (SPE) has become the most common and useful cleanup approach in pesticide residue analysis. Conventionally, SPE uses plastic cartridges containing 100–1000 mg of a sorbent material, vacuum manifolds, column preconditioning, solvent waste fractions, collection fractions, solvent evaporation steps, manual operation, and multiple solvents.[46–48] Other formats utilize sorbents contained in disks, pipette tips, or 96-well plates. SPE is one of the most common techniques used for cleanup in pesticide residue analysis for many good reasons. SPE cartridges and disks are compact, disposable, easy to use, relatively selective, and widely applicable yet versatile. They can concentrate analytes from solution, have low solvent consumption, can be used in parallel to increase sample throughput, and can be automated.

A wide variety of sorbents are commercially available in SPE, and the most common in pesticide residue analysis include C$_{18}$, silica, Florisil, Alumina, graphitized carbon black (GCB), aminopropyl (-NH$_2$), primary secondary amine (PSA), and divinylbenzene/polystyrene.

In the past, analytical pesticide methods often used Florisil columns with fractionation of the pesticides with different elution solvents,[31] but recently, the use of a weak anion-exchange sorbent, such as $-NH_2$ or PSA in combination with GCB has been shown to provide effective removal of fatty acids, chlorophyll, and sterols from foods.[25] Unfortunately, GCB strongly retains planar pesticides, such as hexachlorobenzene, thus its usefulness is reduced in multiclass, multiresidue methods. C_{18} can be helpful in removing a small amount of lipids from extracts, but otherwise PSA alone often provides enough cleanup of extracts of nonfatty foods.

The traditional column-based SPE has nice advantages over alternate cleanup approaches, but it is still not ideal. Instead, the QuEChERS method employs a very simple cleanup approach called dispersive-SPE.[28] In this approach, an aliquot of the sample extract (1 mL) is added to a vial containing a small amount of the SPE sorbent (25-mg PSA), and the mixture is vortexed briefly to mix the SPE material with the extract. Then, the sorbent is separated by centrifugation, and the final extract is taken for analysis. This approach is most convenient when the SPE sorbent only removes matrix components and not the analytes, and the extract itself serves as the elution solvent.

Dispersive-SPE saves time, labor, money, and solvent compared with the traditional approach, and it has other practical and effective advantages, such as (1) no SPE apparatus needed, (2) no SPE cartridges, (3) no vacuum, (4) no pretreatment of sorbent, (5) no channeling, (6) no drying out, (7) no collection tube, (8) no flow control, (9) no elution solvent, (10) no dilution of extract, (11) no solvent evaporation, (12) no knowledge of SPE theory or skill needed by the analyst, (13) uses less sorbent, (14) takes less time, (15) costs less, and (16) provides better interaction with the extract for cleanup.

14.5   ANALYSIS

Just as the combination of the extraction and cleanup steps is considered the "sample preparation" procedure, the analytical separation and detection steps are referred to as the "determinative step," "analytical procedure," or "analysis." In the analysis of organic substances such as pesticides, an analytical separation step is almost always needed, even if only one analyte is being detected because a detector response at a given retention time yields more information about the identity of the substance than just a detector signal. Simply infusing a sample extract into an MS-MS instrument may provide some information about the identity of an analyte, but matrix effects during ionization and increased chance of false positives and negatives limit the effectiveness of this approach in practice. An ion cyclotron resonance–Fourier transform mass spectrometer may be able to do the job, but the cost of such an approach is prohibitive for pesticide monitoring purposes.

GC and LC have long been established as exceptional methods to separate chemicals in complex mixtures, and at present, there are no better overall alternatives for analytical separations than GC or LC coupled with the appropriate detection system. Capillary electrophoresis (CE) has shown some promise for the analysis of ionic pesticides,[49] but ultimately, the better ruggedness and larger sample injection volumes in LC give it strong advantages over CE. In pesticide residue analysis, analytical procedures are often divided into pesticide groups that are most effectively analyzed by GC or LC (usually reversed-phase). In multiclass, multiresidue methods, GC with capillary columns is generally preferred because it gives better separations, has typically lower detection limits, and has more diverse detectors. Thus, LC is generally reserved for ionic, thermally labile, and less volatile pesticides. Due to the fewer number of theoretical plates of separation in LC, and the mode of separation based on polarity, LC methods are typically designed for single classes of pesticides rather than the more diverse range of analytes possible in a single method by GC, in which separation is largely a function of volatility. However, due to the recent advancements in LC/MS-MS instruments, LC will likely become the primary approach for the majority of pesticides, and GC will be used primarily for the thermally stable, nonpolar, and semivolatile pesticides.

LC and GC are the best available analytical separation approaches at this time. However, if one were to evaluate these chromatographic methods with the ideal in mind, then analysis times are too long; injection of extracts sequentially reduces sample throughput; not all analytes can be monitored with a single method; cleanup is often needed to improve ruggedness and to avoid matrix interferences; the instruments are not sufficiently small or portable; sample capacity is low; and the costs of instruments, columns, and maintenance are high. Sections 14.5.1 through 14.5.6 discuss possibilities in improving the current state of the art in chromatographic analysis.

14.5.1 Matrix Effects

The real-world analysis of pesticide residues in food samples is associated with adverse phenomena called matrix effects, which are caused by the unavoidable presence of coextracted matrix components in the final extract.[50] In GC, matrix effects may impact all steps in the analysis (injection, separation, and detection) leading to inaccurate quantitation, decreased analyte detectability, reduced method ruggedness, and/or reporting of false positive/negative results.

Serious matrix effects occur during sample introduction in GC where degradation and/or adsorption of certain analytes take place, which was first described by Erney et al.[51] as "matrix-induced response enhancement." When a food extract is injected, the matrix components tend to fill (block) active sites in the inlet and column (mainly free silanol groups), thus reducing losses of susceptible analytes due to irreversible adsorption and/or degradation. This phenomenon results in a higher transfer of these analytes to the GC column and consequently in their higher signals in matrix compared with matrix-free solutions. If analyte standard solutions prepared only in solvent are used for calibration, the calculated concentrations of the affected analytes in food extracts become overestimated. The extent of the matrix-induced enhancement effect relates to both the chemical structure and concentration of the analyte and type and content of the matrix components.[52–54] Thermally labile pesticides and those capable of hydrogen bonding, such as pesticides with hydroxy (–OH) and amino (R–NH–) groups, imidazoles (–N5), carbamates (–O–CO–NH–), urea derivatives (–NH–CO–NH–), and certain organophosphates (–P5O), are the analytes most susceptible to this effect.[50–54]

In theory, the root causes of the matrix-induced response enhancement (i.e., active sites or matrix components) would be eliminated to solve this problem. In practice, however, it is virtually impossible to perform complete GC system deactivation and/or thorough cleanup. The use of alternative injection techniques that decrease analyte thermal degradation and/or residence time in the injection port, such as programmed temperature vaporizing (PTV) or pulsed splitless injection, may lead to a significant reduction in the matrix-induced enhancement, but rarely to its elimination.[55–58] Currently the most widely used method to compensate for matrix effects is to prepare calibration standards in blank matrix extracts (matrix-matched standardization) rather than in pure solvent.[59] Nevertheless, this approach has several drawbacks, including a rather time-consuming and laborious preparation of matrix-matched standards, the unavailability of appropriate blanks, the limited stability of certain pesticides in matrix solutions,[60] and the increased amount of injected matrix in an overall sequence of samples, which can lead to the increased contamination of the inlet and front part of the analytical column. This results in formation of new active sites and gradual decrease in analyte responses, thus in a "matrix-induced response diminishment effect."[61]

The concept of "analyte protectants" (compound additives) can offer a less laborious, more convenient and effective solution.[62,63] Analyte protectants are compounds that strongly interact with active sites in the GC system (inlet and column); thus, they do not allow access to the analytes most susceptible to the effects. Compounds capable of hydrogen bonding, such as those with multiple hydroxy groups (e.g., sugar derivatives), represent the most effective analyte protectants.[63] When added to sample extracts and matrix-free standards alike, the analyte protectants can induce the same response enhancement in both instances, resulting in effective equalization of the matrix-induced response enhancement effect. Therefore, this approach takes advantage of the response enhancement and optimizes it rather than trying to eliminate it.

14.5.2 Large Volume Injection and Direct Sample Introduction

In order to decrease LODs, particularly when no concentration step is conducted in sample preparation, it is desirable to use a large-volume injection (LVI) technique for sample introduction into GC. In recent years, a number of commercial techniques and inlets have been introduced to permit LVI through the control of pressure and temperature during vaporization.[64] However, the wide volatility and polarity range of pesticides makes LVI a difficult proposition. The volatile pesticides, such as dichlorvos, may be partially or completely lost during the solvent evaporation step, and the analysis of some low-volatile pyrethroids may cause the introduction of some undesirable nonvolatiles into the column. Certain pesticides interact with active sites and/or degrade on surfaces, and LVI may inherently increase this problem when extended resident times occur in the GC inlet.

Contamination is bound to occur in nearly all LVI approaches, and routine maintenance items are often more expensive, as is the typical cost of the injectors. Solid-phase materials may reduce the contamination of the GC system by nonvolatile matrix compounds; however, despite manufacturer claims, the chances that a sorbent material can selectively retain nonvolatile components without affecting analyte results are slim. Chemical pretreatment of some sorbents can help reduce interactions of either acidic or basic compounds, but not both. CarboFrit material added to GC liners in typical PTV injectors permits injection of <10 µL of relatively dirty extracts.[65,66] If the problematic pesticides in GC are moved to LC/MS-MS for analysis, then LVI with CarboFrit (with full-scan MS detection) may become a very widely used analytical approach.

DSI is a novel LVI approach, which also has a major advantage in preventing nonvolatile matrix components from accumulating in the GC system. Thus, direct sample introduction (DSI) potentially eliminates the matrix-induced response diminishment effect, reducing the maintenance, extending column life, and generally improving method ruggedness.[67–70] In DSI, typically 10–15 µL of the sample extract is placed in a small, glass microvial, which is then placed in a probe or injector liner. The probe is then inserted into the GC liner, which is kept at reduced temperature long enough for the solvent to evaporate. The inlet temperature is then rapidly increased to volatilize the analytes, which are focused at the head of the analytical column, and then the GC analysis proceeds normally. Afterwards, the spent microvial is removed along with the nonvolatile matrix components that normally would contaminate the GC system. A very selective detection system is required for analysis of dirty extracts, but MS-MS and certain element-selective detectors have been successfully employed in the DSI/GC approach.[15,68,69] Recently, the manual DSI device (ChromatoProbe) has been commercially automated in a technique marketed as difficult matrix injection (DMI).[70]

14.5.3 Fast Gas Chromatography

In GC, the retention times of analytes (analysis times) can be decreased in a number of ways,[71–74] such as by (1) reducing column length, (2) increasing carrier gas flow rate, (3) reducing film thickness, (4) increasing column diameter, and/or (5) increasing column temperature more quickly. Unfortunately, the separation efficiency is also reduced in each case, and selectivity losses can be too severe. In practice, the GC conditions should be designed to give the shortest analysis time while still providing the necessary selectivity (both analyte-analyte and matrix-analyte) to conduct the analysis. The use of element-selective detectors may improve matrix-analyte selectivity, but analyte-analyte selectivity must be addressed solely by the separation. MS detection often improves both types of selectivity, and by reducing the reliance on the GC separation, faster analysis times can often be achieved for a given list of analytes and matrices.

Currently, practical approaches to fast GC/MS[72] applicable in pesticide residue analysis employ (1) short, microbore capillary GC columns (i.d., 0.2 mm) coupled with time-of-flight (TOF)-MS or other high-duty cycle detectors for analysis; (2) fast temperature programming; (3) lowering the pressure in the analytical column to improve separation and increase flow rate low-pressure GC-MS; (4) supersonic molecular beam for MS at high carrier gas flow rates; and (5) pressure-tunable (also called stop-flow) GC-GC.

14.5.4 DETECTION IN GAS CHROMATOGRAPHY AND LIQUID CHROMATOGRAPHY

A variety of detectors are available and have been extensively evaluated in chromatographic analysis.[8] In GC of organochlorinated pesticides, detection is often accomplished using the electron-capture detector (ECD), electrolytic conductivity detector (ELCD or Hall detector), or halogen-specific detector (XSD). The XSD is more selective for Cl and Br; ELCD and ECD not only detect some organonitrogen pesticides at reasonably low levels but also have the potential for more interferences. For the analysis of organophosphorus (OP) pesticides (and sulfur-containing pesticides), the flame photometric or pulsed flame photometric detectors are most common. The pulsed flame photometric detectors have higher detectivity and selectivity (able to distinguish P from S in detection). The nitrogen phosphorus detector (NPD or thermionic selective detector) can also be used for analysis of both organonitrogen and OP pesticides, but despite its high sensitivity, it is somewhat less selective due to the possible presence of nitrogen interferences in typical matrices. Other than MS, few other GC detectors are commonly employed in pesticide residue analysis. The atomic emission detector has the ability to distinguish many elements simultaneously, but its cost is higher than MS, it cannot provide confirmation, and it does not achieve low LOD for many pesticides.

In LC, UV-visible absorbance detection is generally not selective or sensitive enough for general pesticide residue applications. The use of diode array detection often improves selectivity and has better applicability. However, the most common approach is fluorescence detection after postcolumn derivatization in pesticide methods using LC (e.g., N-methyl carbamates, phenylureas, glyphosate). For analytes that fluoresce naturally, such as thiabendazole, LC/fluorescence is an excellent method due to the high degree of selectivity, low LOD, and wide linear dynamic range. The combination of the nondestructive detection techniques of UV-vis absorbance and fluorescence in a single method increases selectivity and expands the range of analytes in LC. These detectors can also be used in series prior to MS detection to provide more information, but MS can also be quantitative and confirmatory at adequate LOD by itself.

14.5.5 GAS CHROMATOGRAPHY/MASS SPECTROMETRY

GC/MS is an indispensable approach in the modern pesticide residues monitoring laboratory. MS detection is the only adequately sensitive and affordable technique that is able to simultaneously provide quantitative and qualitative results. The types of MS instruments used for pesticide analysis include quadrupole, ion trap, and TOF systems, and each type has advantages and disadvantages with respect to each other. Different ionization techniques are also possible in GC/MS to provide higher degrees of information and lower LODs, but in multiresidue analysis by GC/MS, electron ionization (EI) is most commonly used. Chemical ionization (positive or negative modes) is a softer ionization approach that tends to give lower LODs depending on the pesticide, but it is not as widely applicable in multiclass pesticide methods and does not provide as much structural information about the analyte as EI.

LODs in GC/MS with modern instruments, especially in selected ion monitoring (SIM) and MS-MS operation, have reached the levels of selective GC detectors in many cases. In both SIM and MS-MS, though, time segments are needed, which limits the number of targeted analytes that can be detected in a given time. The selectivity of detection is also increased, especially in MS-MS, but full-scan operation generally provides sufficient information for confirmation (depending on the pesticide) and can monitor hundreds of analytes,[22,75] whereas SIM and MS-MS are typically limited to about 100 pesticides in a single chromatogram.[25,65,66] This becomes especially important in fast-GC analysis (provided that the data acquisition rate is fast enough).[72] Other trade-offs with SIM relate to the difficulty of identifying analytes due to fewer ions monitored and higher chance of matrix interferences than in MS-MS.

Despite the "universal" selectivity in MS, for certain pesticides in complex matrices, there are instances when coeluting interferences swamp the MS in any mode of operation (indirect effects can affect MS-MS quantitation using ion trap instruments).[69] Selective detectors are often useful to overcome these effects and help to provide critical retention time and elemental composition information to aid in both confirmation and quantitation when needed.

14.5.6 Liquid Chromatography/Mass Spectrometry

The commercial introduction of atmospheric pressure ionization (API) techniques, which commonly include electrospray ionization (ESI) and atmospheric pressure chemical ionization (APCI), is leading to a revolution in LC methods.[76,77] Previous approaches to couple LC with MS (thermospray and particle beam) have been noted for poor sensitivity, high costs, and poor instrument reliability. The cost of liquid chromatography/mass spectrometry (LC/MS) (and MS-MS) is decreasing slowly but is still high, and many laboratories have begun to use the approach routinely for pesticide analysis because there are no other approaches that can match the combination of selectivity, sensitivity, and broad analytical scope. The majority of new pesticides being developed by agrochemical companies are best analyzed using LC/MS techniques.

Due to the added selectivity of MS-MS in particular, the speed of LC separations can be increased because coeluting analytes are not as much of a problem. Maximum response in API/MS depends on many factors (ESI/APCI, ± mode, pH, ionic strength, flow rates, solvent composition, temperatures, gas flows, capillary voltage, etc.), but indications are that an acidified methanol/water gradient for ESI in the positive mode works well for the large majority of LC-type pesticides. In GC, the peaks tend to be too narrow (1–2 seconds), to give enough MS-MS points across a peak, but LC gives much wider peaks (10–30 seconds) which permits numerous MS-MS data acquisition events to still achieve quantitative results. Thus, hundreds of pesticides can be more easily analyzed by LC/MS-MS than by GC/MS-MS (or SIM).

The soft API techniques in LC/MS are not ideal for analyte identification. Oftentimes, only a single ion occurs in the spectrum and fragment ions arise from losses of water or carbon dioxide, which are less useful for confirmation than more distinctive structural fragmentation patterns. Thus, in most applications, the use of MS-MS becomes essential for achieving both low LODs (increased signal/noise ratios due to improved selectivity) and almost unambiguous identification.

Matrix effects impact not only GC but also LC/API-MS analysis.[50] Direct spectral interferences are relatively rare, especially in MS-MS, but coeluting matrix components may interfere with the analyte ionization process leading to signal suppression or (less often) enhancement. As in GC, matrix-matched standardization is mostly used to compensate for the matrix effects when it is not possible to overcome them by improving LC separation, increasing sample preparation selectivity (both extraction and cleanup), and using a different ionization technique (APCI is usually less prone to matrix effects than ESI). In the case of a limited number of analytes, the use of internal standards may represent less laborious compensation methods than matrix-matched standardization. The internal standards must elute with the same or very similar retention times as analytes to be influenced by matrix effects to the same extent. This requires the use of deuterated standards or application of an echo-peak technique[78]—injection of a standard solution shortly after the sample, resulting in closely eluting peaks of the same analyte.

14.6 CONCLUSIONS

Because pesticide residue analysis increases food costs to the consumer, an important goal in monitoring programs is to achieve maximal efficiency while still meeting the data quality needs. Let us set the standard that an approach should be able to perform a quantitative and confirmatory analysis of hundreds of diverse pesticides in complex food samples at levels 10 ng/g with rugged instruments and the entire sample preparation and analysis procedure should be accomplished within 30 minutes by a single technician in the field using less than US$1 of materials per sample. In theory, this goal could be met using the QuEChERS method for sample preparation followed by automated DMI/LVI/fast-GC/MS and LC/ESI/MS-MS for analysis. To achieve field portability, the extraction devices and instruments would be have to be operated in a mobile minilab (a van) with an independent power generator. Even though the cost of materials would be low, the capital cost of this "ultimate" approach would be at least $500,000 plus high maintenance costs. It is very unlikely that such an approach would be used in the field, but several top monitoring labs (government and contract) already have the experience and means to use these state-of-the-art techniques in the laboratory for routinely monitoring pesticide residues in foods. Someday, nearly all labs may use such a sophisticated approach, because indeed, the savings in laboratory efficiency and benefits of expanded analytical scope should help pay for the capital expense of the instruments.

REFERENCES

1. http://www.cfsan.fda.gov/~dms/pesrpts.html
2. Carson, R. *Silent Spring*, Houghton Mifflin, Boston, 1962.
3. Colburn, T., Dumanoski, D., and Myers, J. P. *Our Stolen Future*, Dutton, New York, 1996.

4. National Research Council, *Pesticides in the Diets of Infants and Children*, National Academy Press, Washington, D.C., 1993.
5. Bethem, R. Boison, J., Gale, J., Heller, D., Lehotay, S., Loo, J., Musser, S., Price, P., and Stein, S. Establishing the fitness for purpose of mass spectrometric methods, *J. Am. Soc. Mass Spectrom.* 14, 528–541, 2003.
6. Fong, W. G., Moye, H. A., Seiber, J. N., and Toth, J. P. *Pesticide Residues in Foods: Methods, Techniques, and Regulations*, John Wiley & Sons, New York, 1999.
7. Lee, P. W., Ed. *Handbook of Residue Analytical Methods for Agrochemicals*, John Wiley & Sons, Chichester, 2002.
8. Lehotay, S. J. Multiclass, multiresidue analysis of pesticides, strategies for, in *Encyclopedia of Analytical Chemistry*, Meyers R. A., Ed., John Wiley & Sons, Chichester, 2002, pp. 6344–6384.
9. Mastovská, K. Food & nutritional analysis; (q) pesticide residues, in *Encyclopedia of Analytical Science, 2nd Edition*, Worsfold, P., Townshend, A., and Poole, C., Eds., Elsevier, 2004.
10. Sherma, J. Pesticide residue analysis (1999–2000): A review, *J. AOAC Int.*, 84 1303–1312, 2001.
11. Tomlin, C. D. S., Ed. *The Pesticide Manual, 12th Edition*, British Crop Protection Council, Surrey, 2000.
12. *Crop Protection Handbook 2003*, Meister Publishing, Willoughby, 2003.
13. Vallejo, R. P., Bogus, E. R., and Mumma, R. O. Effects of hapten structure and bridging groups on antisera specificity in parathion immunoassay development, *J. Agric. Food Chem.*, 30, 572–580, 1982.
14. Nunes, G. S., Toscano, I. A., and Barceló, D. Analysis of pesticides in food and environmental samples by enzyme-linked immunosorbent assays, *Trends Anal. Chem.*, 17, 79–87, 1998.
15. Jing, H. and Amirav, A. Pesticide analysis with the pulsed-flame photometer detector and a direct sample introduction device, *Anal. Chem.*, 69, 1426–1435, 1997.
16. Young, S. J., Parfitt, Jr., C. H., Newell, R. F., and Spittler, T. D. Homogeneity of fruits and vegetables comminuted in a vertical cutter mixer, *J. AOAC Int.*, 79, 976–80, 1996.
17. Maestroni, B., Ghods, A., El-Bidaoui, M., Rathor, N., Jarju, O. P., Ton, T., and Ambrus, Á. Testing the efficiency and uncertainty of sample processing using ^{14}C-labelled chlorpyifos, in *Principles and Practices of Method Validation*, Fajgelj, A. and Ambrus, Á., Eds., Royal Society of Chemistry, Cambridge, 2000, pp. 49–74.
18. Hill, A. R .C., Harris, C. A., and Warburton, A. G. Effects of sample processing on pesticide residues in fruit and vegetables, in *Principles and Practices of Method Validation*, Fajgelj, A. and Ambrus, Á., Eds., Royal Society of Chemistry, Cambridge, UK, 2000, pp. 41–48.
19. Fussell, R. J., Jackson, A. K., Reynolds, S. L., and Wilson M. F. Assessment of the stability of pesticides during cryogenic sample processing. 1. Apples, *J. Agric. Food Chem.*, 50, 441–448, 2002.
20. Cairns, T., Luke, M. A., Chiu, K. S., Navarro, D., and Siegmund, E. G., Multi-residue pesticide analysis by ion-trap technology: a clean-up approach for mass spectral analysis, *Rapid Commun. Mass Spectrom.*, 7, 1070–1076, 1993.
21. Specht, W., Pelz, S., and Gilsbach, W. Gas-chromatographic determination of pesticide residues after clean-up by gel-permeation chromatography and mini-silica gel-column chromatography, *Fresenius J. Anal. Chem.*, 353, 183–190, 1995.
22. General Inspectorate for Health Protection, *Analytical Methods for Pesticide Residues in Foodstuffs, 6th Edition*, Ministry of Public Health, Welfare and Sport, The Netherlands, 1996.
23. Andersson, A. and Palsheden, H. Comparison of the efficiency of different GLC multi-residue methods on crops containing pesticide residues, *Fresenius J. Anal. Chem.*, 339, 365–367, 1991.

24. Fernández-Alba, A. R., Valverde, A., Agüera, A., and Contreras, M. Gas chromatographic determination of organochlorine and pyrethroid pesticides of horticultural concern, *J. Chromatogr.*, 686, 263–271, 1994.

25. Fillion, J., Sauvé, F., and Selwyn, J. Multiresidue method for the determination of residues of 251 pesticides in fruits and vegetables by gas chromatography/mass spectrometry and liquid chromatography with fluorescence detection, *J. AOAC Int.*, 83, 698–713, 2000.

26. Cook, J., Beckett, M. P., Reliford, B., Hammock, W., and Engel, M. Multiresidue analysis of pesticides in fresh fruits and vegetables using procedures developed by the Florida Department of Agriculture and Consumer Services, *J. AOAC Int.*, 82, 1419–1435, 1999.

27. Lee, S. M., Papathakis, M. L., Hsiao-Ming, C. F., and Carr, J. E. Multipesticide residue method for fruits and vegetables: California Department of Food and Agriculture, *Fresenius J. Anal. Chem.*, 339, 376–383, 1991.

28. Anastassiades, M., Lehotay, S. J., Štajnbaher, D., and Schenck, F. J. Fast and easy multiresidue method employing acetonitrile extraction/partitioning and "dispersive solid-phase extraction" for the determination of pesticide residues in produce, *J. AOAC Int.*, 86, 412–431, 2003.

29. Luke, M., Froberg, J. E., and Masumoto, H. T. Extraction and cleanup of organochlorine, organophosphate, organonitrogen, and hydrocarbon pesticides in produce for determination by gas-liquid chromatography, *J. Assoc. Off. Anal. Chem.*, 58, 1020–1026, 1975.

30. Sawyer, L. D. The Luke et al. method for determining multipesticide residues in fruits and vegetables: collaborative study, *J. Off. Anal. Chem.*, 68, 64–71, 1985.

31. Food and Drug Administration, *Pesticide Analytical Manual, 3rd Edition*, U.S. Department of Health and Human Services, Washington, D.C., 1994.

32. Lopez-Avila, V., Young, R., Benedicto, J., Ho, P., Kim, R., and Beckert, R. F. Extraction of organic pollutants from solid samples using microwave energy, *Anal. Chem.*, 67, 2096–2102, 1995.

33. Eskilsson, C. S. and Bjorklund, E. Analytical-scale microwave-assisted extraction, *J. Chromatogr A.*, 902, 227–250, 2000.

34. Lehotay, S. J. Supercritical fluid extraction of pesticides in foods, *J. Chromatogr. A*, 785, 289–312, 1997.

35. Lehotay, S. J. Determination of pesticide residues in nonfatty foods by supercritical fluid extraction and gas chromatography/mass spectrometry: collaborative study, *J. AOAC Int.*, 85, 1148–1166, 2002.

36. Richter, B. E., Jones, B. A., Ezzell, J. L., Porter, N. L., Avdalovic, N., and Pohl, C. Accelerated solvent extraction: a technique for sample preparation, *Anal. Chem.*, 68, 1033–1039, 1996.

37. Adou, K., Bontoyan, W. R., and Sweeney, P. J. Multiresidue method for the analysis of pesticide residues in fruits and vegetables by accelerated solvent extraction and capillary gas chromatography, *J. Agric. Food Chem.*, 49, 4153–4160, 2001.

38. Barker, S. A. Matrix solid-phase dispersion, *J. Chromatogr. A*, 885, 115–127, 2000.

39. Barker, S. A. Applications of matrix solid-phase dispersion in food analysis, *J. Chromatogr. A*, 880, 63–68, 2000.

40. Pawliszyn, J. *Solid Phase Microextraction Theory and Practice*, Wiley-VCH, New York, 1997.

41. Beltran, J., López, F. J., and Hernández, F. Solid-phase microextraction in pesticide residue analysis, *J. Chromatogr. A*, 885, 389–404, 2000.

42. Baltussen, E., Cramers, C. A., and Sandra, P. J. Sorptive sample preparation—a review, *Anal. Bioanal. Chem.*, 373, 3–22, 2002.

43. Kataoka, H. Automated sample preparation using in-tube solid-phase microextraction and its application—a review, *Anal. Bioanal. Chem.*, 373, 31–45, 2002.

44. Gordin, A. and Amirav, A. SnifProbe—a new method and device for vapor and gas sampling, *J. Chromatogr A*, 903, 155–172, 2000.
45. Argauer, R. J., Lehotay, S. J., and Brown, R. T. Determining lipophilic pyrethroids and chlorinated hydrocarbons in fortified ground beef using ion-trap mass spectrometry, *J. Agric. Food Chem.* 45, 3936–3939, 1997.
46. Tekel, J. and Hatrik, S. Pesticide residue analyses in plant material by chromatographic methods: clean-up procedures and selective detectors, *J. Chromatogr. A*, 754, 397–410, 1996.
47. Wells, M. J. and Yu, L. Z. Solid-phase extraction of acidic herbicides, *J. Chromatogr. A*, 885, 237–250, 2000.
48. Liška, I. Fifty years of solid-phase extraction in water analysis—historical development and overview, *J. Chromatogr A*, 885, 3–16, 2000.
49. Menzinger, F., Schmitt-Kopplin, P., Freitag, D., and Kettrup, A. Analysis of agrochemicals by capillary electrophoresis, *J. Chromatogr. A*, 891, 45–67, 2000.
50. Hajšlová, J. and Zrostlíková, J. Matrix effects in (ultra)trace analysis of pesticide residues in food and biotic matrices, *J. Chromatogr. A*, 1000, 181–197, 2003.
51. Erney, D. R., Gillespie, A. M., and Gilvydis, D. M. Explanation of the matrix-induced chromatographic response enhancement of organophosphorus pesticides during open tubular column gas chromatography with splitless or hot on-column injection and flame photometric detection, *J. Chromatogr.*, 638, 57–63, 1993.
52. Hajšlová, J., Holadová, K., Kocourek, V., Poustka, J., Godula, M., Cuhra, P., and Kempn´y M. Matrix-induced effects: a critical point in gas chromatographic analysis of pesticide residues, *J. Chromatogr. A*, 800, 283–295, 1998.
53. Anastassiades, M. and Scherbaum, E. Multiresidue method for determination of pesticide residues in citrus fruits by GC-MSD. Part 1. Theoretical principles and development of methods, *Deutsche Lebensmittel Rundschau*, 93, 316–327, 1997.
54. Schenck, F. J. and Lehotay, S. J. Does further clean-up reduce the matrix enhancement effect in gas chromatographic analysis of pesticide residues in food? *J. Chromatogr. A*, 868, 51–61, 2000.
55. Wylie, P. L., Klein, K. J., Thomson, M. Q., and Hermann, B. W. Using electronic pressure programming to reduce the decomposition of labile compounds during splitless injection, *J. High Resolut. Chromatogr.*, 15, 763–768, 1992.
56. Godula, M., Hajšlová, J., and Alterová, K. Pulsed splitless injection and the extent to matrix effects in the analysis of pesticides, *J. High Resolut. Chromatogr.*, 22, 395–402, 1999.
57. Godula, M., Hajšlová, J., Mastovská, K., and K_ivánková, J. Optimization and application of the PTV injector for the analysis of pesticide residues, *J. Sep. Science*, 24, 355–366, 2001.
58. Zrostlíková, J., Hajšlová, J., Godula, M., and Mastovská, K. Performance of programmed temperature vaporizer, pulsed splitless and on-column injection techniques in analysis of pesticide residues in plant matrices, *J. Chromatogr. A*, 937, 73–86, 2001.
59. Erney, D. R., Pawlowski, T. M., and Poole, C. F. Matrix-induced peak enhancement of pesticides in gas chromatography: is there a solution? *J. High Resolut. Chromatogr.*, 20, 375–378, 1997.
60. Kocourek, V., Hajšlová, J., Holadová, K., and Poustka, J. Stability of pesticides in plant extracts used as calibrants in the gas chromatographic analysis of residues, *J. Chromatogr. A*, 800, 297–304, 1998.
61. Soboleva, E., Rathor, N., Mageto, A., and Ambrus, Á. Estimation of significance of "matrix-induced" chromatographic effects, in *Principles and Practices of Method Validation,* Fajgelj, A. and Ambrus, Á., Eds., Royal Society of Chemistry, Cambridge, 2000, pp. 138–156.

62. Erney, D. R. and Poole, C. F. A study of single compound additives to minimize the matrix induced chromatographic response enhancement observed in the gas chromatography of pesticide residues, *J. High Resolut. Chromatogr.*, 16, 501–503, 1993.

63. Anastassiades, M., Mastovská, K., and Lehotay S. J. Evaluation of analyte protectants to improve gas chromatographic analysis of pesticides, *J. Chromatogr. A*, 1015, 163–184, 2003.

64. Teske, J. and Engewald, W. Methods for, and applications of, large-volume injection in capillary gas chromatography, *Trends Anal. Chem.*, 21, 584–593, 2002.

65. Sheridan, R. S. and Meola, J. R. Analysis of pesticide residues in fruits, vegetables, and milk by gas chromatography/tandem mass spectrometry, *J. AOAC Int.*, 82, 982–990, 1999.

66. Gamón, M., Lleo, C., Ten, A., and Mocholi, F. Multiresidue determination of pesticides in fruit and vegetables by gas chromatography/tandem mass spectrometry, *J. AOAC Int.*, 84, 1209–1216, 2001.

67. Amirav, A. and Dagan, S. A direct sample introduction device for mass spectrometry studies and gas chromatography mass spectrometry analyses, *Eur. Mass Spectrom.*, 3, 105–111, 1997.

68. Lehotay, S. J. Analysis of pesticide residues in mixed fruit and vegetable extracts by direct sample introduction/gas chromatography/tandem mass spectrometry, *J. AOAC Int.*, 83, 680–697, 2000.

69. Lehotay, S. J., Lightfield, A. R., Harman-Fetcho, J. A., and Donoghue, D. J. Analysis of pesticide residues in eggs by direct sample introduction/gas chromatography/ tandem mass spectrometry, *J. Agric. Food Chem.*, 49, 4589–4596, 2001.

70. Fussell, R. and Nicholas, D. Multi-residue analysis of pesticides in samples of lettuce and peas using large volume-difficult matrix introduction-gas chromatography-mass spectrometry (LV-DMI-GC-MS), ATAS Chromatography Note No. 33, ATAS UK, 2002.

71. Matisová, E. and Dömötörová, M. Fast gas chromatography and its use in trace analysis. *J. Chromatogr. A*, 1000, 199–221, 2003.

72. Mastovská, K. and Lehotay, S. J. Practical approaches to fast gas chromatography-mass spectrometry, *J. Chromatogr. A*, 1000, 153–180, 2003.

73. Korytár, P., Janssen, H.-G., Matisová, E., and Brinkman, U. A. T. Practical fast gas chromatography: methods, instrumentation and applications, *Trends Anal. Chem.*, 21, 558–572, 2002.

74. Cramers, C. A., Janssen, H.-G., van Deursen, M. M., and Leclercq, P. A. High-speed gas chromatography: an overview of various concepts, *J. Chromatogr. A*, 856, 315–329, 1999.

75. Sandra, P., Tienpont, B., and David, F. Multi-residue screening of pesticides in vegetables, fruits and baby food by stir bar sorptive extraction-thermal desorption-capillary gas chromatography-mass spectrometry, *J. Chromatogr. A*, 1000, 299–309, 2003.

76. Niessen, W. M. Progress in liquid chromatography-mass spectrometry instrumentation and its impact on high-throughput screening, *J. Chromatogr. A*, 1000, 413–436, 2003.

77. Hogendoorn, E. and van Zoonen, P. Recent and future developments of liquid chromatography in pesticide trace analysis, *J. Chromatogr. A*, 892, 435–453, 2000.

78. Zrostlíková, J., Hajšlová, J., Poustka, J., and Begany, P. Alternative calibration approaches to compensate the effect of co-extracted matrix components in liquid chromatography—electrospray ionisation tandem mass spectrometry analysis of pesticide residues in plant materials, *J. Chromatogr. A*, 973, 13–26, 2002.

15 Determination of Pollutants in Foods

Douglas G. Hayward

CONTENTS

15.1 INTRODUCTION

Polyaromatic compounds (PACs) originate from both man-made and natural sources. The high toxicity of some compounds, combined with their persistence, bioaccumulation, environmental mobility, and diverse sources, has provoked intensive research efforts for decades, which show little sign of slowing down. The polyhalogenated contaminants of greatest interest in foods are grouped into several classes based on their carbon skeleton. A common feature of all the classes is the large number of potential isomers producing a large number of isomers with similar physical and chemical properties, while conversely possessing widely varying potencies toward biological systems.[1] Although these chemicals exhibit very similar chemical characteristics, the individual isomers must be identified to determine their potential importance to a biological system. Table 15.1 lists some classes of polyhalogenated or polyaromatic contaminants, which have been studied as a result of food contamination incidences or occupational exposures. Table 15.1 omits several polyaromatic pollutant classes including metabolites, sulfur analogs for oxygenated compounds, alkylated polyaromatics, mixed halogenated compounds, and heterocyclic compounds.

TABLE 15.1

Some Classes of Polyhalogenated Aromatic Compounds That Are
Synthetically and Environmentally Derived Contaminants

Chemical Class	Range (X)	Possible Congeners
Polyaromatic hydrocarbons (PAHs)	0	—
Polychlorinated dibenzo-*p*-dioxins (PCDDs)	1–8 (Cl)	75
Polychlorinated dibenzofurans (PCDFs)	1–8 (Cl)	135
Polychlorinated naphthalenes (PCNs)	1–8 (Cl)	75
Polychlorinated biphenyls (PCBs)	1–10 (Cl)	209
Polybrominated biphenyls (PBBs)	1–10 (Br)	209
Polybrominated diphenyl ethers (PBDEs)	1–10 (Br)	209
Polychlorinated diphenyl ethers (PCDEs	1–10 (Cl)	209
Polybrominated dibenzo-*p*-dioxins (PBDDs)	1–8 (Br)	75
Polybrominated dibenzofurans (PBDFs)	1–8 (Br)	135

Estimated potencies often vary by many orders of magnitude for chemicals[1] with fairly similar boiling points, melting points, vapor pressures, and octanol/water partitioning coefficients[2–4] (Table 15.2). The large differences in potencies have driven analytical approaches to be isomer specific. High toxicity and low environmental levels of some congeners require that the analytical methods must use the most sensitive and selective mass spectrometry (MS) techniques and must employ large test portions and extensive sample preparation and concentration. Another complicating factor is a general lack of agreement over what constitutes a safe exposure to these chemical classes. Because of the large number of sources of the chemicals and pathways and scenarios of food contamination, analytical chemists must confront a wide variety of foods, which require routine monitoring and frequently require limits of determination in the nanogram per kilogram range or below.

15.2   MATRIX SEPARATION AND ANALYTE PURIFICATION

15.2.1   MATRIX SEPARATION

Foods requiring analysis are usually solids with varying water and lipid content or liquids that require a matrix-specific extraction. Several solvent systems including cyclohexane/dichloromethane, toluene, hexane/acetone, hexane or petroleum ether/ diethyl ether/ethanol, hexane, ethanol/saturated ammonium sulfate, toluene/ethanol, dichloromethane, chloroform/methanol, toluene/isopropanol, dichloromethane/ isopropanol, and pentane or hexane/diethyl ether/methanol are used commonly to remove analytes. Solvent systems using an alcohol are normally used with liquids, while solids typically are dried first with sodium sulfate or other drying agent before a 16- to 24-hour Soxhlet extraction, blending, or column extraction. Some methods use HCl digestion prior to extraction with hexane, whereas others use KOH digestion. Nonpolar extraction solvents employ large amounts of drying agents, large volumes, and tedious sample manipulation followed by solvent exchange and concentration.

TABLE 15.2
Some Physical Properties and TEFs for Polyhalogenated Pollutants

Class or Member	mp	bp	Vp[a]	log(Kow)	TEF
Polychlorinated dibenzo-*p*-dioxins (PCDDs)					
2,3,7,8-TetraCDD	305	412.2	1.4E-09	6.2–7.3	1
1,2,7,8-TetraCDD	—	—	—	—	0
1,2,3,7,8-PentaCDD	240	—	—	6.8	1
1,2,3,4,7,8-HexaCDD	—	—	—	—	0.1
1,2,3,6,7,8-HexaCDD	~286	—	—	7.6	0.1
1,2,3,7,8,9-HexaCDD	~244	—	—	7.6	0.1
1,2,3,4,6,7,8-HeptaCDD	—	—	—	—	0.01
1,2,3,4,6,8,9-HeptaCDD	140	—	—	—	0
1,2,3,4,6,7,8,9-OctaCDD	150	—	—	—	0.0001
Arochlor 1254 or 1260	—	275–420	7.7E-05[b]	6.8	—
3,3′,4,4′-TetraCB	—	—	1.1E-05[b]	—	0.0001
3,3′,4,4′,5-PentaCB	—	—	2.1E-06[b]	—	0.1
3,3′,4,4′5,5′-HexaCB	—	—	4.0E-07[b]	—	0.01
2,3,3′,4,4′-PentaCB	—	—	5.3E-06[b]	—	0.0001
2,3,′4,4,′5-PentaCB	—	—	7.2E-06[b]	—	0.0001
2,3,4,4′,5-PentaCB	—	—	3.3E-05[b]	—	0.0005
2′,3,4,4′,5-PentaCB	—	—	6.8E-06[b]	—	0.0001
2,3,3′,4,4′,5,-HexaCB	—	—	5.5E-06[b]	—	0.0005
2,3,3′,4,4′,5′-HexaCB	—	—	1.0E-06[b]	—	0.0005
2,3′,4,4′5,5′-HexaCB	—	—	1.4E-06[b]	—	0.00001
2,3,3′,4,4′,5,5′-HeptaCB	—	—	1.08E-06[b]	—	0.0001
2,2′,3,3′,4,4′5-HeptaCB	—	—	2.8E-06[b]	—	0.0
2,2′,3,4,4′,5,5′-HeptaCB	—	—	3.8E-06[b]	—	0.0
Polychlorinated naphthalenes (PCNs)					
1,2,3,5,6,7/1,2,3,4,6,7-HexaCN	—	234/205	—	—	0.002(EROD)[c]
1,2,3,4,5,7/1,2,3,5,6,8-HexaCN	—	—	—	—	0.0002(EROD)[c]
1,2,3,5,7,8-HexaCN	—	148	—	—	0.002(EROD)[c]
1,2,4,5,6,8-HexaCN	—	153	—	—	0.000007(EROD)[c]
1,2,3,4,5,6,7-HeptaCN	—	—	—	—	0.003(EROD)[c]

[a] Vp in mm Hg and mp/bp in °C.
[b] *Source:* Holmes, et al., *J., Environ. Sci. and Technol.*, 27, 725, 1993.
[c] *Source:* Hanberg, A. et al., *Chemosphere,* 20, 1161, 1990.

Several alternative approaches have been proposed and some are gaining wider use. Accelerated solvent extraction (ASE) or pressurized liquid extraction (PLE) uses elevated temperatures and pressures to reduce extractions times to 30 minutes or less for most matrices, including foods.[5] Besides increase in speed, an added advantage of this procedure is that much less solvent is used and fewer toxic solvents are needed to extract the analytes.[5] Often foods are freeze-dried prior to extraction, which reduces the sample size and toxicity of the solvents used.[6] One potential difficulty is that fortification of the food with certain compounds will produce large errors if it is done prior to freeze-drying.[7]

The solvent choice can be important for automation. For example, if hexane or a similar solvent is used, then the extract can be passed directly into an automated cleanup system without solvent exchange. Microwave-assisted extraction (MAE) accomplishes similar goals except that it uses a relatively inexpensive microwave oven with extraction vessels designed to withstand the required temperatures and pressures.[5] An automated online extraction and cleanup is probably not possible with MAE. Supercritical fluid extraction (SFE) has been investigated for foods, for both off-line and online cleanup. SFE procedures for foods often use carbon dioxide without modifiers as is sometimes needed with other environmental matrices. SFE conditions can be adjusted to extract analytes while removing much less matrix than in a conventional hot solvent extraction process, by selecting conditions or using retainers.[8,9]

The major drawbacks of both ASE (PLE) and SFE are the cost and complexity of the equipment. Some other approaches have proved quite useful but only for some types of foods or biological materials. Solid-phase extraction (SPE) has been tried with both liquids and solids, using nonpolar C_{18} modified silica or, more recently, modified styrene divinyl benzene polymers.[10] Liquid samples such as blood serum have been most successful when precipitates have been minimized by using formic acid as a denaturant, but tissues have also been attempted.[11] Solids usually require solvent extraction and solvent exchange before application to the SPE columns. However, if the extraction and concentration steps are combined with other reagents or integrated with further cleanup using SPE columns, then SPE can be an effective and economical approach.[12]

15.2.2 ANALYTE PURIFICATION

Purification strategies can be broadly categorized into destructive and nondestructive (or less destructive) methods. Matrix removal often begins at extraction as is the case for some polychlorinated dibenzo-p-dioxin and polychlorinated dibenzofuran (PCDD)/F isolation procedures that are still part of the official methods (HCl digestion, etc.) of the Environmental Protection Agency (EPA).[13] Methods designed to isolate halogenated aromatics normally include steps using large amounts of strongly acidic and basic chemicals to remove coextracted biogenic compounds such as lipids. There are less destructive approaches that use high-efficiency carbon (AX21 or carbosphere)[14–16] to remove the coextracted matrix. AX21 carbon can remove PCDD/Fs from 30 g of lipid with very little carryover.[15] Liem et al.[16] reported an elegant method for milk that used only 2 g of carbosphere and a 5 g basic Woelm super I alumina column to purify solvent extracts from milk before gas chromatography/mass spectrometry (GC/MS) analysis. Florisil or alumina will absorb fats even in the presence of strongly polar solvent systems such as hexane/dichloromethane. Another nondestructive approach for human milk samples involves the use of the lipophilic gel Lipidex® 5000.[17] Norén and Sjövall[17] could remove lipids with Lipidex® 5000 from small samples of milk (10 mL) and purify and fractionate persistent pesticides, polychlorinated biphenyls (PCBs), and PCDD/Fs using deactivated alumina, activated alumina, and silica gel with small solvent elution volumes (1–30 mL). 2,3,7,8-tetrachlorodibenzo-p-dioxin (TCDD) was separated from all other PCDD/Fs on highly activated Woelm alumina.

Gel permeation chromatography (GPC) has long been used to remove lipids and high molecular weight oils from environmental samples.[18] GPC has the advantage of being easily automated, but it typically handles less than 1 g of coextracted material.[6] Semipermeable membrane devices (SPMDs) can handle large amounts of lipids or other biogenic material.[19,20] A polyethylene tube is filled with the concentrated extract, sealed, and placed in contact with 0.5–1 L of pentane or cyclohexane/dichloromethane. The tube equilibrates for 24–72 hours, with the analytes of interest migrating into pentane, resulting in as much as 99% lipid removal in high-fat foods.[19] Besides requiring larger amounts of solvent and time, SPMDs may not effectively recover compounds such as some PBDEs that have molecular weight (MW) more than 550. SPMDs are also not easily automated.

15.3 CLASS SEPARATIONS AND AUTOMATION

15.3.1 CLASS SEPARATIONS

Once biogenic material is removed, further fractionation of the polyaromatic and/or halogenated PACs is often desired, particularly when bioassay screening is used and isomer-specific GC/MS determinations are required. Florisil and alumina have been used to separate pesticides from PCBs or PCDD/Fs from PCBs and chlorinated diphenylethers.[21] Open low-pressure gravity-driven liquid chromatographic columns require calibration of individual batches of chromatographic-grade alumina or Florisil. Water content is usually specified for methods using alumina or Florisil as water content will alter the performance of the material. For example, fully activated Florisil (0% water) will separate dioxins from most PCBs with only hexane elution, but dioxin-like PCB-77, 126, and 169 are recovered with the PCDD/Fs. PCDD/Fs can be separated from all PCB congeners with Florisil that is deactivated with 6% water.[6]

Carbon columns are extensively used to fractionate PACs.[22,23] The conditions used are highly dependent on the type and amount of carbon used.[6,14,22–24] Aromatic compounds are placed on these columns in nonpolar solvents such as hexane as with alumina or Florisil and then step-eluted with increasingly polar and/or aromatic solvents.[23,24] Porous graphic carbon has been used to pack high-performance liquid chromatography (HPLC) columns so that the separation of PCDD/Fs from PCBs can be controlled and automated and fractionation of PCDD/Fs and PCBs can be achieved using a single solvent.[22] PCBs are separated based on the number of ortho-chlorine substitution steps.[23,24] Congeners with 2 or 3 ortho-chlorines elute first with nonpolar solvent; mono-ortho substituted congeners elute next as polarity increases, and nonortho-substituted PCBs elute last with toluene. An automated HPLC system using two columns packed with AX21 carbon dispersed on C_{18} silica gel was used to clean up and separate PCDD/Fs, nonortho PCBs, mono-ortho PCBs, and di-ortho PCBs. The system provided high capacity for analytes on the two columns, with a total run time of 3 hours.[24] All HPLC-based procedures require expensive equipment that is normally not replicated in a single laboratory such that very few samples can be run simultaneously, but an automated system could be run unattended to overnight. Carbon columns can show batch-to-batch variation and tendency for tailing of the eluting congeners.[25] The use of aromatic solvents rules out UV detection of eluting fractions. An electron donor–acceptor HPLC column using 2-(1-pyrenyl) ethyldimethylsilylated silica (PYE column) stationary phase has been proposed for difficult PCB separations in fat samples with high amounts of PCBs.[25]

Other HPLC column phases have been tested for separating PAHs, PCNs, PCBs, and PCDD/Fs. Amino-derivatized silica packing was used for separating PAHs into mono-, di-, and polyaromatic fractions that can be further fractionated,[26] and the PYE column has been successfully used for separation of PCBs and for separation of PCNs, which is difficult.[27] HPLC reversed-phase or normal phase columns, with methanol or hexane as the mobile phase, respectively, can be used to separate homolog groups. This can be particularly useful when a sample contains very high concentrations of one homolog group, while determination is needed on a different homolog group found at a much lower concentration, as with 2,3,7,8-TCDD in the presence of excess amounts of OCDD.[29]

15.3.2 Automation

The preparation of foods for PCDD/F or PCB/BDE determination requires a number of labor-intensive steps that greatly increase the cost of analysis. Modern instrumentation used for detection has become cheaper, automated, and computer controlled, while sample preparation has remained laborious and often done by less-skilled technicians. Prepackaged columns and commercially prepared reagents have helped speed up manual sample preparations.[11] Nevertheless, sample preparation and concentration steps contribute to major costs of analysis.

Lapeza et al.[30] achieved semiautomation of a method reported by Smith et al.[14] for use with adipose and blood, thereby saving 50% of the time required but still requiring approximately 3 hours/sample. A refinement of this system [manufactured by Fluid Management Systems (FMS)] that fully automated blood preparation from the extraction step to the GC/MS produced a purified human serum extract for PCDD/Fs and PCBs, requiring approximately 1.5 hours/sample.[31] A manually operated optimized procedure, that is, a modified method of that followed by Smith et al.,[14] using four carbon columns running simultaneously, could also prepare samples in 1.5 hours/sample.[15] A recent version of the FMS automated system can run five samples in parallel, requires less than 0.5 hour/sample, and could exceed the capacity of a single high-resolution mass spectrometry (HRMS).[31]

More recent developments of the FMS automation system include combining the extraction of solids using PLE with the increased time savings of these automated systems.[32] Sample preparation in automated systems or scaled-down manual systems with prepacked columns can be so efficient that researchers have begun to look for faster ways to separate and detect the analytes of interest or to use screening methods for further reducing overall costs per sample.

GPC can also be automated, saving time when handling large amounts of coextracted material. Preparation of PCBs and persistent pesticides using an automated system comprising SPE columns, automated GPC, and automated silica column to remove matrix and fractionate PCBs from pesticides in blood serum has been recently reported.[9] Another method uses selective SFE extraction combined with alumina in the extraction cell for absorbing fat.[8] This method can be used with an online carbon column to fractionate the SFE-extracted polyhalogenated aromatics in a single step.[28]

15.4 ISOMER-SPECIFIC SEPARATIONS

15.4.1 HIGH-RESOLUTION GAS CHROMATOGRAPHY

Ryan et al.[33] published a report on the separation of all 136 tetra- to octa-CDD/Fs on nine different stationary phases and also reported retention times on all the nine phases applied on narrow bore (0.22–0.32 mm I.D.) fused silica capillary columns that were either 30, 50, or 60 m in length. The columns' phases ranged from nonpolar methyl silicone and the widely used 95% methyl, 5% phenyl polysiloxane (DB-5, RTX-5, equivalent, etc.) to very polar phenylcyanopropyl biscyanopropyl polysiloxane (SP2331) and 100% biscyanopropyl (CP-Sil-88) and a more recently introduced SB-smectic liquid crystalline methyl (80%) diphenyl carboxylic ester (20%) phase. Polar phases generally separated more of the 2,3,7,8-substituted isomers with fewer numbers of congeners coeluting, especially with 2,3,4,7,8-PeCDF and 2,3,7,8-TCDF, while leaving some hexachlorodibenzofuran (HxCDFs) or hexachlorodibenzo-p-dioxin (HxCDDs) unresolved. Columns containing DB 210 (methyl trifluropropyl polysiloxane) or DB 225 (50% methylcyanopropyl/50% methyl-phenyl polysiloxane) equivalent phases are used for resolving the multiple coelutions with 2,3,7,8-TCDF found on nonpolar columns, but also produce coelutions with 2,3,7,8-TCDD, 1,2,3,7,8-PeCDD and certain HxCDFs such that they must be used with a second, usually nonpolar, column.

Two medium-polarity columns that were tested produced separations comparable to or better than that produced by some more-polar phases. A DB-17 column (50% phenyl/50% methyl polysiloxane) separated 2,3,7,8-TCDF completely and separated most other 2,3,7,8-substituted congeners except one isomer coeluting with 2,3,7,8-TCDD, 1,2,3,7,8-PeCDF, and achieved an incomplete separation of 1,2,3,6,7,8-HxCDF. An OV-17 column contains an isomer coeluting with 2,3,7,8-TCDF and 1,2,3,7,8-PeCDD, while it achieves more complete separation of HxCDFs. While some column phases perform much better than others, a combination of at least two columns is needed to identify all 2,3,7,8-chlorine-substituted dibenzo-*p*-dioxin and -furan isomers in complex mixtures. The highly polar phases (SP-2331, CP-Sil-88) achieve separation of all but a few of the twelve 2,3,7,8-substituted tetrachloro-hexachloro-DD/Fs, but require hour-long GC run times or more, leaving significant overlaps between chlorination levels. The performance of these phases will deteriorate faster than that of the nonpolar ones, which can significantly alter the quality of separation and increase maintenance and costs. Some phases have been marketed specifically to produce a complete separation of the critical twelve 2,3,7,8-substituted tetrachloro-through hexachloro-DD/Fs. The "DB-Dioxin" (similar to SP-2331, Cp-Sil-88, and RTX2332) separates all 2,3,7,8-substituted congeners except for 1,2,3,4,7,8-HxCDF, although some congeners have not been completely resolved to baseline from other isomers.

Fraisse et al.[34] reported that an improved separation could be achieved with a nonpolar phase using correct temperature programming. The so-called "DB-5ms" phase had the same composition (95% methyl/5% phenyl polysiloxane) as a DB-5 or RTX-5 equivalent phase but with a phenyl group interspersed in the siloxane polymer chain. This phase separated 2,3,7,8-TCDD and 2,3,7,8-TCDF from 21 or 37 other isomers, respectively, compared with an ordinary DB-5 that could not resolve 2,3,7,8-TCDF from five other isomers and produced only an incomplete separation of 2,3,7,8-TCDD from several closely eluting isomers. All the 2,3,7,8-substituted HxCDD/Fs isomers were also separated, except for a poor separation of 1,2,3,7,8,9-HxCDF from a single other isomer; this contrasts favorably with a standard DB-5 phase, which separates only four of the seven HxCDD/Fs congeners. The 1,2,3,7,8-PeCDD was still resolved well enough but not to baseline as with DB-5. The DB-5 ms phase permits the use of a single, durable high-temperature column for a nearly complete isomer-specific separation of 2,3,7,8-substituted PCDD/Fs in foods, which is suitable enough for most regulatory purposes.

PCB congeners with no orthochlorine substitutions are often isolated for analysis with the dioxins. Other PCB congeners are analyzed separately or sometimes fractionated based on the number of ortho-chlorine substitutions (e.g., mono-ortho-chloro- and di-ortho-chloro-). Isomer-specific PCB methods have been developed to utilize a unique GC separation process for the determination of PCB congeners with assigned toxic equivalence factors (TEFs) (see Table 15.2), which also separated several PCB congeners pairs not separated on a nonpolar phase such as SE-54, by using a 50% methyl/50% octyldecyl polysiloxane column.[35] EPA 1668 Rev. A proposes to separate PCBs 77, 81, 105, 114, 118, 123, 126, 156, 157, 167, 169, and 189 from all other congeners in an environmental sample using SPB-octyl column (Supelco 2–4218). The method requires retention time of decachlorobiphenyl to be greater than 55 minutes, while PCB 156 and 157 are often found as a summed peak.[36] The combination of the SPB-octyl column and a DB-1 is reported in a unique method to separate 180 PCB congeners. Garrett et al.[36] reported that the SPB octyl column produced high bleed and unstable retention times within a few days. Some new columns could not meet the specifications of EPA 1668 Rev. A. An RTX-5 sil MS column was suggested as an alternative. Larson et al.[37] investigated the ability of several column phases to separate various PCB congeners. SPB-octyl and SIL-88 were not practical except for certain separations, due to high bleed and low temperature maximums, but SIL-5 (dimethysilicone) and SIL-19 (14% cyanopropylphenyl/1% vinyl dimethyl silicone) were the best overall and produced good separation of some dioxin-like PCB congeners (PCB-77, 126, 169, 118, etc.). Covaci et al.[62] used a high-temperature column (dicarbaclosodecaborane and 8% phenyl methyl silicone) with human adipose, which can separate 191 PCB congeners at least partially.

The separation and retention behavior of all 72 polychlorinated naphthalene isomers was investigated on six different high-resolution gas chromatography (HRGC) column phases by Järnberg et al.[38] No column could separate all congeners at any chlorination level from dichloro- to hexachlorohaphthalene. A pair of hexachloronaphthalene isomers (1,2,3,5,6,7 and 1,2,3,4,6,7) known to strongly bioaccumulate was not separated on any phase. More recently, complete separation of all 14 pentachloro- and 10 hexachloronaphthalene isomers was demonstrated using a proprietary column phase made by Restek that used per-methylated β-cyclodextrin.[39]

15.4.2   FAST GC

The efficient separation of PCDD/F mixtures has required relatively long columns (30 M or 60 M). In addition, better separations are accomplished on polar stationary phases that require long run times to elute higher chlorinated congeners Hayward et al.[15] demonstrated that a shorter and narrower "minibore" 40 M × 0.18 mm I.D. DB-5 ms column could produce separation of 2,3,7,8-chlorine-substituted PCDD/Fs identical to that by a 0.25 mm I.D. 60 M DB-5 ms column[34] in under 40 minutes using a column head pressure of only 50 PSIG. This column was found to be durable enough for use with a large number of food samples.[15] MacPherson et al.[40] optimized the conditions for PCDD/Fs, coplanar PCBs, and chlorinated pesticides on 40 M × 0.18 mm and 20 M × 0.1 mm DB-5 columns by adjusting temperature ramps and using higher inlet pressures (61 and 100 PSIG),[40] enabling comparable separations and quantifications using either 60 M, 40 M, or 20 M DB-5 columns, using less time with each progressively shorter column. On 20 M columns, all PCDD/F congeners were eluted in as little as 14 minutes.[38] The same approach, using 20 M columns and fast temperature ramps, was applied to a mixture of 56 PCBs and to PAH mixtures.[41]

Microbore columns have also been tested in dual-column applications for determining target PCB congeners and PCDD/F, in the same GC run. MacPherson et al.[42] recently reported the sequential acquisition by HRMS of mono-ortho PCBs (PCBs 105, 118, 156, etc.) eluting from a 20 M × 0.1 mm I.D. column, while PCDD/Fs and nonortho-PCBs 77, 81, 126, and 169 were separated on a 40 M × 0.18 mm I.D. column in the same GC, which was injected immediately after the injection of the PCB fraction into the 20 M column. This way, separate PCB fractions could be determined with the same run, and the same data file could be used for processing. Worrall et al.[43] used a similar method but delayed the injection of the PCDD/F-containing fraction into the 40 M column for 2.4 minutes to help avoid codetermination of chlorinated diphenlyethers that would be present in the mono-ortho PCB fraction eluting in the 20 M column. Dimandja et al.[44] demonstrated a very fast separation of 38 PCBs found in human serum in 5 minutes using a 15 M column and fast spectral acquisition rates with a time-of-flight (TOF) mass spectrometer for detection. The high spectral acquisition rates with TOF have been used to speed up separation on standard-sized columns, allowing simultaneous PCB and pesticide determinations in as little as 9.5 minutes.[45]

15.4.3   MULTIDIMENSIONAL GC

Multidimensional GC or Comprehensive GC×GC refers to the separation of chemicals using two or more independent migrations for a chemical during the same fixed distance (GC) experiment. Multidimensional separations increase the capacity and resolution of a given GC system. Liu and Phillips[46] report achieving two-dimensional GC through use of an on-column thermal modulator system. Two columns are connected through a modulator (short section of column rapidly heated and cooled) that pulses components separated on the first column (30 M, 0.25 mm I.D.) to a second short (0.5 M, 0.1 mm I.D.) column producing a second orthogonal separation. Two-dimensional gas chromatograms could be generated on complex mixes of hydrocarbons using this system. The same approach was reported by Liu et al.[47] for pesticides in human serum. Fast separations for PCBs or PCDD/Fs are often produced with short columns requiring fast mass spectral acquisition rates of a TOF instrument for MS detection. Recent work has focused on robust modulator designs needed to focus compounds from the first column onto the second. Vreuls et al.[48] compared the performance of four modulator systems and reported the separation of 91 PCB congeners and all 17 2,3,7,8-PCDD/Fs with the GC × GC or GC^2/micro-ECD with an ultimate goal of using TOF for the determination of PCBs and PCDD/Fs.

15.5 DETECTION SYSTEMS

A number of GC detectors including electron capture, atomic emission,[49] low-resolution quadrupole mass spectrometry (LRMS), quadrupole ion tandem mass spectrometry (QISTMS), high-resolution magnetic sector mass spectrometry, triple quadrupole mass spectrometry, and matrix isolation Fourier transform infrared spectrometry have been investigated for measuring polyhalogenated aromatic compounds.[50] MS techniques combine high sensitivity, selectivity, and a molar response that is identical for a carbon skeleton regardless of the halogen substitution pattern.[49]

Advances in computer technology have made mass spectral data acquisition fast, easy to use, and affordable by any laboratory. Very fast computing has caused rebirth of an older mass spectrometric technique, time-of-flight, which now shows great promise for the future. The advantages of TOF include compact tabletop design, very high spectral acquisition rates, high mass resolution (with greater cost), and simplicity of hardware and tuning. TOF instruments are now being used to measure many PACs such as PAHs, PCBs, and BDEs in foods. Only PCDD/Fs are still beyond the reach of TOF for most food applications.

PCDD/Fs are usually measured with magnetic sector instruments operating at 10,000 resolution (HRMS). Instruments that are designed to enhance the sensitivity of PCDD/Fs by the choice of ion source (ion boxes, etc.) are available. Since the mid-1980s, research on HRMS has focused on improving the sensitivity of 2,3,7,8-TCDD, a problematic congener if detected using low-resolution MS techniques. The modern sector instruments are more sensitive, reliable, and rugged, with faster computer data processing and control, than the instruments that were used in the early 1980s. Today, HRMS instruments are sold with a guaranteed sensitivity of 3 fg 2,3,7,8-TCDD at 3:1 signal-to-noise ratio at 10,000 resolution. This sensitivity is usually enough for most food analyses. However, this sensitivity is not enough to ensure that an average person is not receiving more than the virtually safe dose according to the current dioxin risk assessment, unless all the PCDD/F and PCB exposures were mainly originating from TCDD alone. This circumstance is almost never the case except in some unusual food exposure situations.[51] HRMS has allowed routine PCDD/F measurement of food background levels that are unachievable by any other technique. HRMS has two disadvantages: The first is the high cost of purchasing and maintaining the instrument (2–5 times than other GC/MS systems) and the second is the skill of the laboratory staff needed to achieve and maintain the very low limits of determination needed. These issues aside, HRMS is the only way to detect PCDD/F background levels in many food matrices (e.g., vegetable matter), particularly in human blood serum.

The costs and complexity involved in HRMS have prompted some researchers to investigate alternatives to HRMS or the less sensitive and specific LRMS quadrupoles for PCDD/Fs. One method investigated before HRMS was in common use is tandem mass spectrometry (TMS). TMS measures a product ion of a characteristic loss from a molecular ion (usually the loss of COCl with PCDD/Fs) to provide greater sensitivity and specificity than LRMS, but it is usually operated at unit mass resolution (low resolution). A great number of specific interferences are removed by MS/MS techniques.[52,53] However, Fraisse et al.[54] found that greater selectivity than that provided by HRMS was not always the case with fly ash; Clement et al.[55] found that HRMS did not remove all the interferences present with LRMS from M12 and M14 ions of TCDD in a fish extract. Although a large number of interferences can be avoided with this technique, the early applications suffered from the fact that the instruments needed for TMS (a triple quadrupole or hybrid HRMS) were expensive and more difficult to operate than sector instruments designed for TCDD measurements. The more important problem was that the sensitivity was always 5–10 times lower than the sensitivity of HRMS.[52,53]

Recently introduced quadrupole ion storage tandem mass spectrometry (QISTMS) greatly reduces cost and is simple to operate. The sensitivity is still at least 5- to 10-fold lower than the sensitivity of high-resolution sector instruments.[56] Molecular ions of PCDD/Fs or PCBs stored in the ion trap are excited through application of a waveform of sufficient amplitude, matching the secular frequency of the ion-producing product ions through collision-induced dissociation with the helium buffer gas of the trap. Routine food monitoring for PCDD/Fs using QISTMS and HRMS in a complementary manner has been reported.[57] QISTMS is finding increasing application in the monitoring of PCBs and BDEs in foods, wherein lower sensitivity is not so important when the target compound levels are higher.[58,59]

Compound classes found at the high ppt to low ppb levels in foods (e.g., PCBs, PAHs, BDEs) can be measured by several GC/MS techniques. For example, BDEs have been determined using HRMS, ECNI MS, EI-LRMS, and TMS in a quadrupole ion trap. HRMS provides somewhat more sensitivity and more mass spectral information, while ECNI methods monitor only Br preventing the use of coeluting labeled standards, but both methods produce similar chromatograms from biological fluids.[60] ECNI shows an advantage in monitoring brominated and mixed halogenated compounds in marine mammals.[61] TMS has been investigated recently, but it demonstrates higher sensitivity than EI-LRMS, while providing more BDE spectral information than ECNI-MS and less potential interference than HRMS.[58] EI-LRMS has been reported in conjunction with high-volume injection to maintain sensitivity with human adipose.[62]

15.6 BIOANALYTICAL METHODS

These methods use a cell-based or protein-based assay system to detect PACs. The methods can be divided into *in vitro* ligand-binding assays and cell-based bioassays.[63] All bioanalytical methods are indirect measures of the presence of PACs and therefore depend on the selectivity of the interaction. They are often used in conjunction with chemical methods such as GC/MS or GC/ECD. Ligand-binding assays use an antibody to a compound or a small group of compounds. A radioimmunoassay for 2,3,7,8-TCDD was developed as early as 1979.[64] Later monoclonal antibody-based enzyme-linked immunoassays to TCDD were developed.[65] Recently a more sensitive polyclonal antibody to TCDD that could detect as little as 1 pg TCDD was reported.[66] Antibodies selective for dioxin-like PCB congeners have also been reported.[67] The assays demonstrated promising results in screening soils[64] and were later used with food matrices.[68] The ligand-binding assays require near-complete extraction and some cleanup of the ligand from a solid sample to prevent false positives/false negatives and solvent exchange to an assay-compatible solvent (DMF, DMSO, methanol, etc.).[63,67] Sample preparation should be efficient; otherwise, the benefits obtained due to the speed of the immunoassay would be muted. Antibody assays are fairly specific for the compound used to generate the antibody, but some cross-reactivity often occurs.[67]

Some bioassays rely on a biochemical response from an appropriately sensitive cell line with an easily detectable biochemical change. Ever since the reporting of a stereospecific high-affinity binding receptor in hepatic cells for 2,3,7,8-TCDD,[69] work proceeded rapidly on the development of quick bioassays for TCDD and TCDD-like chemicals through binding of the receptor and assaying of the induced cytochrome P4501A1 activity.[70] This bioassay uses an immortalized rat hepatoma cell line (H4IIe) with a suitable substrate such as benzo(a)pyrene[70] or 7-ethoxy-resorufin[71] and has been used with food samples.[71] Extraction and extensive cleanup are required to remove matrix and concentrate the sample so that the assay responds to a specific chemical(s) rather than all chemicals that could induce activity. The sensitivity of the bioassay is limited by the bioassay substrate, cell line responsiveness, and substrate inhibition. Despite these limitations, highly sensitive and specific bioassays systems employing chicken embryo liver cells can be used with food analysis yielding a low ED_{50} as low as any other bioassay available.[72]

The discovery and DNA sequencing of a dioxin responsive enhancer (DRE) that binds the TCDD/receptor complex increasing gene transcription[73] provided an opportunity to develop an alternative approach to the construction of a bioassay for dioxin-like compounds.[74] A suitable assay vector could be engineered by splicing a series of DREs to a strong promoter sequence linked to a desirable reporter gene,[74–76] resulting in a plasmid vector that could be transfected into any immortalized cell line from a large number of species.[75] A gene selected was firefly luciferase, which produces light with a proper substrate when expressed in the cells.[75] The stably transfected construct is induced in a time- and dose-dependent manner through the cell's own AH-receptor activation system.[75] The result is a bioassay that is as sensitive as any chemical method[77] that can assay a wide variety of PACs.[75] Samples require pretreatment with acids and carbon chromatography to ensure that the response is due to dioxin-like compounds of a specific class, or a total response of dioxin-like chemicals can be measured with no or more limited cleanup. The Chemical Activated Luciferase Expression (CALUX™) bioassay uses a mouse hepatoma cell line, but a rat hepatoma cell line reporting more sensitivity (DR-CALUX®) has been recently developed.[76] Another sensitive and easily assayed reporter gene, the green fluorescent protein, has also been investigated.[78] This assay (CALFUX) would have difficulty in distinguishing between persistent and less-persistent PACs without fractionation or removal of the less-persistent PACs. The CALUX bioassay system is now being validated by several European researchers for routine screening of food samples.[77]

REFERENCES

1. Ahlborg, U. G. et al. Toxic equivalency factors for dioxin-like PCBs, *Chemosphere*, 28, 1049, 1994.
2. United States Environmental Protection Agency, National Study of Chemical Residues in Fish, Volume 2, Office of Science and Technology (WH-551) Washington, D.C. 20460, EPA 823-r-92–008b, September 1992.
3. Hanberg, A. et al. Swedish dioxin survey: determination of 2,3,7,8-TCDD toxic equivalent factors for some polychlorinated biphenyls and naphthalenes using biological tests, *Chemosphere*, 20, 1161, 1990.
4. Holmes, D. A., Harrison, B. K., and Dolfing, J., Estimation of Gibbs free energies of formation for polychlorinated biphenyls, *Environ. Sci. and Technol.*, 27, 725, 1993.
5. Björklund, E., von Holst, C., and Anklam, E., Fast extraction, clean-up and detection methods for the rapid analysis and screening of seven indicator PCBs in food matrices, *Trends Anal. Chem.*, 21(1), 2002.
6. Malisch, R., et al. Results of a quality control study of different analytical methods of determination of PCDD/PCDF in egg samples, *Chemosphere*, 32, 31, 1996.
7. de Voogt, P., van der Wielen, F.W.M., and Gover, H.A.J., Freeze-drying brings about errors in polychlorinated biphenyl recovery calculations, *Trends Anal. Chem.*, 19(5), 2000.
8. van Bavel, B. et al. Supercritical fluid extraction of PCBs from human adipose tissue for HRGC/LRMS analysis, *Chemosphere*, 30, 1229, 1995.
9. Nam, K. S. et al. Supercritical fluid extraction and procedures for determination of xenobiotics in biological samples, *Chemosphere*, 20, 873, 1990.
10. Sandou, C. D. et al. Comprehensive solid-phase extraction method for persistent organic pollutants. Validation and application to the analysis of persistent chlorinated pesticides, *Anal. Chem.*, 75, 71, 2003.
11. Chen, C.-Y., Hass, J. R., and Albro, P. W., A screening method for PCDD/Fs and non-ortho PCBs in biological matrices, *Organohalogen Compounds*, 40, 5, 1999.
12. Focant, J.-F. and De Pauw, E., Easy access multi-cartridge extraction and cleanup assembly for specific isolation of dioxins in biological fluids, *Organohalogen Compounds*, 55, 77, 2002.
13. US EPA method 1613 revision B. Tetra- through octa-chlorinated dioxins and furans by isotope dilution HRGC/HRMS. 1994.
14. Smith, L. M., Stalling, D. L., and Johnson, J. L., Determination of part-per-trillion levels of polychlorinated dibenzofurans and dioxins in environmental samples, *Anal. Chem.*, 56, 1830, 1984.
15. Hayward, D. G., Hooper, K., and Andrzejewski, D., Tandem-in-time mass spectrometry method for the sub-parts-per-trillion determination of 2,3,7,8-chlorine-substituted dibenzo-p-dioxins and -furans in high-fat foods, *Anal. Chem.*, 71, 212, 1999.

16. Liem, A. K. D. et al. A rapid clean-up procedure for the analysis of polychlorinated dibenzo-p-dioxins and dibenzofurans in milk samples, *Chemosphere*, 20, 843, 1990.

17. Norén, K. and Sjövall, J., Analysis of organochlorine pesticides, polychlorinated biphenyls, dibenzo-p-dioxins and dibenzofurans in human milk by extraction with lipophilic gel Lipidex® 5000, *J. Chromatogr.*, 422, 103, 1987.

18. Norstom, R. J., Simon, M., and Mulvihill, M. J., A gel permeation/column chromatography method for the determination of CDDs in animal tissue. *Intern. J. Environ. Anal. Chem.*, 23, 267, 1986.

19. Srandberg, B., Bergqvist, P.-A., and Rappe, C., Dialysis with semipermeable membrane as an efficient lipid removal method in the analysis of bioaccumulative chemicals, *Anal. Chem.*, 70, 526, 1998.

20. Meadows, J. et al. Large-scale-dialysis of sample lipids using a semipermeable membrane device, *Chemosphere*, 26, 1993, 1993.

21. Ryan, J. J., Lizotte, R., and Newsome, W. H., Study of chlorinated diphenyl ethers and chlorinated 2-phenoxyphenols as interferences in the determination of chlorinated dibenzo-p-dioxins and chlorinated dibenzofurans in biological samples, *J. Chromatogr.*, 303, 351, 1984.

22. Creaser, C. S. and Al-haddad, A., Fractionation of Polychlorinated biphenyls, Polychlorinated dibenzo-p-dioxins, polychlorinated dibenzofurans on porous graphic carbon, *Anal. Chem.*, 61, 1300, 1989.

23. de Boer, J., et al. Method for the analysis of non-ortho substituted chlorobiphenyls in fish and marine animals, *Chemosphere*, 25, 1277, 1992.

24. Feltz, K. P. et al. Automated HPLC fractionation of PCDDs and PCDFs and planar and nonplanar PCBs on C_{18}-dispersed PX-21 carbon, *Environ. Sci. Technol.*, 29, 709, 1995.

25. Haglund, P. et al. Isolation of mono- and non-ortho polychlorinated biphenyls form biological samples by electron-donor acceptor high performance liquid chromatography using a 2-(1-pyrenyl)ethyldimethylsilylated silica column, *Chemosphere*, 20, 887, 1990.

26. Zebür, Y. et al. An automated HPLC separation method with two coupled columns for the analysis of PCDD/Fs, PCBs and PACs, *Chemosphere*, 27, 211, 1993.

27. Asplund, L. et al. Analysis of non-ortho polychlorinated biphenyls and polychlorinated naphthalenes in Swedish dioxin survey samples, *Chemosphere*, 20, 1481, 1990.

28. van Bavel, B. et al. Development of a solid phase carbon trap for simultaneous determination of PCDDs, PCDFs, PCBs and pesticides in environmental samples using SFE-LC, *Anal. Chem.*, 68, 1279, 1996.

29. Hagenmaier, H. and Brunner, H., Isomer specific analysis of pentachlorophenol and sodium pentachlorophenate for 2,3,7,8-substituted PCDD and PCDf at sub-ppb levels, *Chemosphere*, 16, 1759, 1987.

30. Lapeza, C. R., Patterson, D. G. Jr. and Liddle, J. A., Automated apparatus for the extraction and enrichment of 2,3,7,8-tetrachlorodibenzo-p-dioxin in human adipose tissue, *Anal. Chem.*, 58, 713, 1986.

31. Turner, W. E. et al. An improved SPE extraction and automated sample method for serum PCDDs, PCDFs and coplanar PCBs, *Organohalogen Compounds*, 19, 31, 1994.

32. Focant, J.-F., Shirkhan, H., and De Pauw, E., On-line automated PLE and multi-column clean-up system for PCDD/F and PCB analysis in food samples, *Organohalogen Compounds*, 55, 33, 2002.

33. Ryan, J. J. et al. Gas chromatographic separations of all 136 tetra to octa-polychlorinated dibenzo-p-dioxins and polychlorinated dibenzofurans on nine different stationary phases, *J. Chromatogr*, 541, 131, 1991.

34. Fraisse, D. et al. Improvements in GC/MS strategies and methodologies for PCDD and PCDF analysis, *Fresenius J. Anal. Chem.*, 348, 154, 1994.

35. Ballschmiter, K., Mennel, A., and Buyten, J., Long chain alkyl-polysiloxane as non-polar stationary phases in capillary gas chromatography, *Fresenius J. Anal. Chem.*, 346, 396, 1993.
36. Garrett, J. H. GC columns for the determination of specific polychlorinated biphenyl (PCB) congeners in environmental samples using U.S. EPA method 1668, *Organohalogen Compounds*, 45, 21, 2000.
37. Larsen, B. Choice of GC column phase for analysis of toxic PCBs, *Chemosphere*, 25, 1343, 1992.
38. Jarnberg, U., Asplund, L., and Jakobsson, E., Gas chromatographic retention behavior of polychlorinated naphthalenes on non-polar, polarizable, polar and smectic capillary columns, *J. Chromatogr. A*, 683, 385, 1994.
39. Helm, P. A. et al. Complete separation of isomeric penta- and hexachloronaphthalenes by capillary gas chromatography, *J. High Resol. Chrom.*, 22, 639, 1999.
40. MacPherson, K. A. et al. Optimization of gas chromatographic parameters for reduced analysis times of chlorinated organic compounds, *Organohalogen Compounds*, 40, 19, 1999.
41. Reiner, E. J. et al. Analysis of persistent organic pollutants (POPS) using microbore columns, *Organohalogen Compounds*, 45, 17, 2000.
42. MacPherson, K. A., Reiner, E. J., and Kolic, T. M., Dual microbore column GC/HRMS analysis of polychlorinated dibenzo-p-dioxins (PCDDs), polychlorinated dibenzofurans (PCDFs) and dioxin-like polychlorinated biphenyls (DLPCBs), *Organohalogen Compounds*, 50, 40, 2001.
43. Worrall, K. et al. Method for the simultaneous analysis of PCDD/Fs and DLPCBs using dual microbore column GC/HRMS in sinter ash samples, allowing the determination of "totals" group concentrations, *Organohalogen Compounds*, 55, 171, 2002.
44. Dimandja, J.-M. D., Grainger, J., and Patterson Jr., D. G., New fast single and multidimensional gas chromatography separations coupled with high resolution mass spectrometry and time-of-flight mass spectrometry for assessing human exposure to environmental toxicants, *Organohalogen Compounds*, 40, 23, 1999.
45. Focant, J.-F. et al. Time-compressed analysis of PCBs and persistent pesticides in biological samples by isotopic dilution gas chromatography/time-of-flight mass spectrometry, *Organohalogen Compounds*, 50, 25, 2001.
46. Liu, Z. and Phillips, J. B., Comprehensive two-dimensional gas chromatography using an on-column thermal modulator interface, *J. Chrom. Sci*, 29, 227, 1991.
47. Liu, Z. et al. Comprehensive two-dimensional gas chromatography for the fast separation and determination of pesticides extracted from human serum, *Anal. Chem.*, 66, 3086, 1994.
48. Vreuls, R. et al. Modulator selection for comprehensive multi-dimensional GC of dioxins and PCBs, *Organohalogen Compounds*, 55, 127, 2001.
49. Schimmel, H. et al. Molar response of polychlorinated dibenzo-p-dioxins and dibenzofurans by the mass spectrometric detector, *Anal. Chem.*, 65, 640, 1993.
50. Mossoba, M. M., Niemann, R. A., and Chen, J.-Y. T., Picogram level quantitation of 2,3,7,8-tetrachlorodibenzo-p-dioxin in fish extracts by capillary gas chromatography/matrix isolation/Fourier transform infrared spectrometry, *Anal. Chem.*, 61, 1678, 1987.
51. Hooper, K. et al. Analysis of breast milk to assess exposure to chlorinated contaminants in Kazakhstan: Sources of 2,3,7,8-tetrachlorodibenzo-p-dioxin (TCDD) exposures in an agricultural region of southern Kazakhstan, *Environ. Health Prospect.*, 107, 447, 1999.
52. Tondeur, Y., Niederhut, W. N., and Campana, J. E., A hybrid HRGC/MS/MS method for the characterization of tetrachlorinated-p-dioxins in environmental samples, *Biomedical Environ. Mass Spec.*, 14, 449, 1987.
53. Charles, M. J. and Tondeur, Y., Choosing between high-resolution mass spectrometry and mass spectrometry/mass spectrometry: Environmental applications, *Environ. Sci. Technol.*, 24, 1856, 1990.

54. Fraisse, D., Gonnord, M. F., and Becchi, M., High resolution chromatography/tandem mass spectrometry. Analysis of polychlorinated dibenzo-p-dioxins and furans, *Rapid Comm. Mass Spec.*, 3(3), 79, 1989.

55. Clement, R. E., Bobbie, B., and Taguchi, V., Comparison of instrumental methods for chlorinated dibenzo-p-dioxins (CDD) determination — Interim results of a round-robin study involving GC-MS, MS-MS, and high resolution MS, *Chemosphere*, 15, 1147, 1986.

56. Plomley, J. B., Koester, C. J., and March, R. E., Rapid screening technique for tetrachlorodibenzo-p-dioxin in complex environmental matrices by gas chromatography/tandem mass spectrometry with an ion-trap detector, *Organic Mass Spectrometry*, 29, 372, 1994.

57. Hayward, D. G. et al. Quadrupole ion storage tandem mass spectrometry and high-resolution mass spectrometry: complementary application in the measurement of 2,3,7,8-chlorine substituted dibenzo-p-dioxins and dibenzofurans in U.S. foods, *Chemosphere*, 43, 407, 2001.

58. Pirard, C., De Pauw, E., and Focant, J-F, New strategy for comprehensive analysis of polybrominated diphenyl ethers, polychlorinated dibenzo-p-dioxins, polychlorinated dibenzofurans and polychlorinated biphenyls by gas chromatography coupled with mass spectrometry, *J. Chromatogr. A*, 998(1–2), 169, 2003.

59. Malavia, J., Santos, F. J., and Galceran, M. T., Ion trap MS/MS as alternative to HRMS for the analysis of non-ortho PCBs in biota samples, *Organohalogen Compounds*, 55, 103, 2002.

60. Thomsen, C. et al. Comparing electron ionization high-resolution and electron capture low-resolution mass spectrometric determination of polybrominated diphenyl ethers in plasma, serum and milk, *Chemosphere*, 46, 641, 2002.

61. Vetter, W., A GC/ECNI-MS method for the identification of lipophilic anthropogenic and natural brominated compounds in marine samples, *Anal. Chem.*, 73, 4951, 2001.

62. Covaci, A. et al. Determination of polybrominated diphenyl ethers and polychlorinated biphenyls in human adipose tissue by large-volume injection-narrow-bore capillary gas chromatography/electron impact low-resolution mass spectrometry, *Anal. Chem.*, 74, 750, 2002.

63. Sherry, J., Environmental immunoassays and other bioanalytical methods: Overview and update, *Chemosphere*, 34, 1011, 1997.

64. Albro, P. W. et al. A radioimmunoassay for chlorinated dibenzo-p-dioxins, *Toxicol. Appl. Pharmacol.*, 50, 137, 1979.

65. Watkins, B. E., Stanker, L. H., and Vanderlaan M., An immunoassay for chlorinated dioxins in soils, *Chemosphere*, 19, 267, 1989.

66. Sugawara, Y. et al. Development of a highly sensitive enzyme-linked immunosorbent assay based on polyclonal antibodies for the detection of polychlorinated dibenzo-p-dioxins, *Anal. Chem.*, 70, 1092, 1998.

67. Chiu, Y.-W. et al. A monoclonal immunoassay for the coplanar polychlorinated biphenyls, *Anal. Chem.*, 67, 3829, 1995.

68. Harrison, R. O. and Carlson, R. E., Simplified sample preparation methods for rapid immunoassay analysis of PCDD/Fs in foods, *Organohalogen Compounds*, 45, 192, 2000.

69. Poland, A. and Glover, E., Stereospecific, high affinity binding of 2,3,7,8-tetrachlorodibenzo-p-dioxin by hepatic cytosol, *J. Biol. Chem.*, 251, 4936, 1976.

70. Bradlaw J. A. and Casterline Jr., J. L., Induction of enzyme activity in cell culture: a rapid screen for detection of planar polychlorinated organic compounds, *J. A. O. A. C.*, 62, 904, 1979.

71. Hanberg, A. et al. Swedish dioxin survey: Evaluation of the H-4-II E Bioassay for screening environmental samples for dioxin-like enzyme induction, *Pharmacol. Toxicol.*, 69, 442, 1991.

72. Engwall, M. et al. Toxic potencies of lipophilic extracts from sediments and settling particulate matter (SPM) collected in a PCB-contaminated river system, *Environ. Toxicol. Chem.*, 15(2), 213, 1996.

73. Dension, M. S., Fisher, J. M., and Whitlock Jr., J. P. The DNA recognition site for the dioxin-AH receptor complex: Nucleotide sequence and functional analysis, *J. Biol. Chem.*, 263, 17221, 1988.

74. El-Fouly, M. H. et al. Production of a novel recombinant cell line for use as a bioassay system for detection of 2,3,7,8-tetrachlorodibenzo-p-dioxin-like chemicals, *Environ. Toxicol. Chem.*, 13, 1581, 1994.

75. Garrison, P. M. et al. Species-specific recombinant cell lines as bioassay systems for the detection of 2,3,7,8-tetrachlorodibenzo-p-dioxin-like chemicals, *Fund. App. Toxicol.*, 30, 194, 1996.

76. Sonneveld, E. et al. Development of improved DR-CALUX® bioassay for sensitive measurement of aryl hydrocarbon receptor activating compounds, *Organohalogen Compounds*, 58, 369, 2002.

77. Bovee, T. F. H. et al. Validation and use of the CALUX-bioassay for the determination of dioxins and PCBs in bovine milk, *Food Add. Contam.*, 15, 863, 1998.

78. Nagy, S. R., et al. A green fluorescent protein based recombinant cell bioassay for the detection of activators of the aryl hydrocarbon receptor: application for screening of a 1,5 dialkylamino-2,4-dinitrobenzene combinatorial chemical library, *Organohalogen Compounds*, 45, 232, 2000.

.

16 Analysis of Chemical Preservatives in Foods

Adriaan Ruiter and Peter Scherpenisse

CONTENTS

16.1 INTRODUCTION

Since prehistoric ages, man has used several methods to improve the shelf-life of food or food products. Drying procedures are among the oldest methods, and so are cooling and freezing in areas with a moderate or cool climate. The addition of salt too was a shelf-life improvement method used in ancient times. Smoke-curing, although intended as a drying process, can also be considered as an addition of components that have a preservative function (i.e., reactive aldehydes and phenolic substances) to food. In later times, the use of saltpeter to preserve meat products and the generation of sulfur dioxide by burning sulfur in wine casks to preserve wine were practiced. Thus, preservation by means of what we now call "additives" or, more specifically, "preservatives" has a very long history.

But old-time "preservatives" sometimes caused unwanted side effects such as being unfavorable to health. Preservative components that might even lead to acute intoxications, for example, salts of heavy metals or formaldehyde appeared on the scene. This, in turn, made it necessary to restrict the use of food preservatives to components that are generally recognized as safe in the concentrations needed to exert their food preservation function. But even now, some preservatives are allowed only in special food products, which limits their total intake (e.g., boric acid in caviar or hexamethylene tetramine in provolone cheese).

The European Parliament and Council Directive No. 95/2 European Committee (EC) on Food Additives[1] provides a list of preservatives and antioxidants permitted in the European Union (Annex III: conditionally permitted preservatives and antioxidants). This is given in Table 16.1.

The United States has laid down legislation with respect to additives in the Federal Regulation, Part 172.[2] The preservatives and antioxidants permitted in the United States are presented in Table 16.2.

Normally, concentrations of allowed preservatives are below the levels of tolerance, and monitoring programs that are carried out from time to time are often sufficient for ensuring an adequate control of the concentration of allowed preservatives.[3] However, some preservatives (e.g., sulfite) are sometimes abusively applied to foods. In fact, Association of Analytical Communities International has published findings on collaboratively tested methods for a large number of preservatives in foods.[4] Routine methods for the analysis of these components are discussed, in detail, in the following sections. There is a need, however, to check for the presence of nonpermitted preservatives and antioxidants in foods.

16.2   SORBIC ACID, BENZOIC ACID, P-HYDROXYBENZOIC ESTERS (PARABENS), AND DEHYDROACETIC ACID

Sorbic acid is allowed in many drinks such as wine-based flavored drinks, some wines, spirits with less than 15% alcohol, nonalcoholic flavored drinks and juices, liquid soups and broths, and in polenta, to varying levels up to 500 mg/L or mg/kg. Sorbic acid as much as 1000 mg/kg or per liter is allowed in fruit and vegetable preparations, olives and olive-based preparations, in a variety of cheese, milk, and egg products, and in topping syrups, aspic, and fat emulsions. In the latter case, 2000 mg/kg is allowed when the fat content is lower than 60%. This level also holds good for several bakery products. In some of these products, benzoic acid is also allowed, in concentrations equal to or somewhat less than that of sorbic acid. For other products, combinations of sorbic and benzoic acid are legalized. Higher levels of mixtures of sorbic acid and benzoic acid are allowed in cooked shrimps and liquid egg (i.e., 6 and 5 g/kg, respectively). In a few cases, a mixture of sorbic acid and parabens is allowed.[1]

These preservatives show moderate reactivity and can be easily isolated from a food or beverage matrix, for example, by steam distillation after acidification, followed by extraction of the distillate with an appropriate (suitable) solvent. It should be noted, however, that the carbon chain in sorbic acid is diunsaturated and, therefore, it undergoes oxidation in aqueous solutions.[5]

Direct extraction of sorbic acid is possible from some matrices with a not-too-complex composition, such as dried prunes.[6] Modern high-performance liquid chromatography (HPLC) techniques, however, have resulted in a spectacular improvement in speed, resolution, and sensitivity. Since 1970s, this technique has been frequently applied, in various modes, to the analysis of sorbic acid, benzoic acid, and parabens

TABLE 16.1
Conditionally Permitted Preservatives and Antioxidants in the European Union

Part A	Sorbic acid, benzoic acid, and *p*-hydroxybenzoic esters and their Na, K, or Ca salts
Part B	Sulfur dioxide and sulfites
Part C	Other preservatives
	Biphenyl or diphenyl
	o-Phenyl phenol and its Na salt
	Nisin
	Natamycin (pimaricin)
	Hexamethylene tetramine
	Dimethyl dicarbonate
	Boric acid and sodium tetraborate (borax)
	Sodium and potassium nitrite
	Sodium and potassium nitrate
	Propionic acid and its Na and K salts
	Lysozyme
Part D	Antioxidants
	Propyl, octyl, and dodecyl gallate
	Butylated hydroxyanisole
	Butylated hydroxytoluene
	Erythorbic acid and its Na salt

TABLE 16.2
Food Preservatives Permitted in the United States for Direct Addition to Food for Human Consumption

Anoxomer
BHA
BHT
Calcium disodium EDTA
Dehydroacetic acid
Dimethyl dicarbonate
Disodium EDTA
Ethoxyquin
Heptylparaben
4-Hydroxymethyl-2,6-di-*tert*-butyl phenol
Natamycin (pimaricin)
Potassium nitrate
Quaternary ammonium chloride combination
Sodium nitrate
Sodium nitrite
Sodium nitrite used in processing smoked chub
Stannous chloride
TBHQ
THBP

in foodstuffs.[7-22] Solid-phase extraction (SPE) is found to be successful for the cleanup of extracts prior to analysis of these components,[9,10] and it results in higher recoveries than acetone extraction or steam distillation. Microdialysis sampling has also used in this respect.[12] Some methods have taken advantage of extraction by ion-pair formation,[14,15,23,24] mostly in simultaneous determination of preservatives, sweeteners, and antioxidants.

Sorbic acid, benzoic acid, and parabens show strong Ultraviolet (UV) absorption, so UV detection is the method usually chosen for analysis of these components. However, there are considerable differences in maximum absorbance. Sorbic acid shows the highest absorbance, and therefore it can be more sensitively detected by this way. It should also be noted that the UV spectrum of the undissociated acid slightly differs from the spectrum of the ion. Measurements at a wavelength close to the maximum wavelength of benzoic acid (230 nm) allow a comparable sensitivity for both sorbic acid and benzoic acid but, obviously, diode array detection and, of course, Liquid Chromatography-Mass Spectrometry (LC/MS) show a clear advantage as compared to detection at fixed wavelength. Data with respect to parabens are scarce.

Dehydroacetic acid, $C_8H_8O_4$, is allowed in the United States as a preservative for cut or peeled squash. The amount that remains in the squash should not exceed 65 mg/L.[2]

Dehydroacetic acid can be determined by HPLC, either alone[25,26] or in combination with other preservatives.[15,27] The UV absorbance shows a maximum about 310 nm.[25,28] In one method, differential pulse polarography was applied for determination of dehydroacetic acid (Figure 16.1).[29]

Gas chromatographic methods for the analysis of carboxylic acids in food have been used for decades. These procedures were laborious, as these called for esterification of the free acid. New analysis techniques used in this century, however, are combined with gas chromatography. Methylation is possible by an on-line pyrolytic technique.[30] Other methods take advantage of capillary chromatography and mass spectrometric detection.[31,32] Flame ionization detection (FID) is still applied as well.[33] Mass spectrometry (MS) allows sorbic acid, benzoic acid, and parabens to be determined in one run, but it requires derivatization of the analytes to be performed. This, however, is not always a disadvantage, since several convenient derivatization methods are available now. FID is used for the analysis of sorbic and propionic acids in rye bread.

Continuous SPE has been found to be a valuable pretreatment technique in the gas chromatographic determination of benzoic acid, sorbic acid, and phenolic antioxidants.[32] In this method, the components are detected either by flame ionization or by MS.

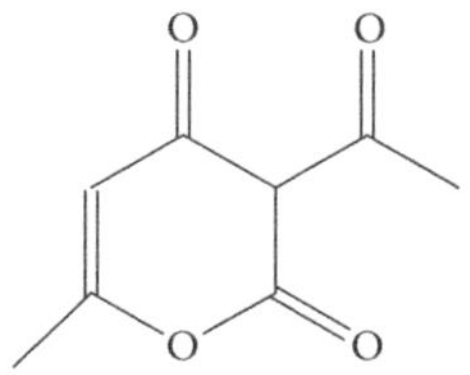

FIGURE 16.1 Dehydroacetic acid.

In 1981, a system was described for performing zone electrophoresis in open-tubular glass capillaries.[33] This system resulted in high separation efficiency and, additionally, a relatively short time of analysis. However, the sensitivity was moderate, which necessitated the use of an on-column fluorescence detector.

Since 1994, capillary electrophoresis (CE) has been applied to preservatives in food, either as capillary zone electrophoresis (CZE),[34] also known as free zone electrophoresis, or as micellar electrokinetic chromatography (MEKC). In this latter technique, micelles act as a pseudostationary phase. CZE methods have been described for sorbic acid[35,36] and MEKC methods for sorbic acid and benzoic acid[37] and later on for a number of compounds including sorbic acid, benzoic acid, and parabens.[38,39] These latter methods clearly demonstrate the separating capacity of CE. It should be noted that an extreme sensitivity is unnecessary for the analysis of preservatives, as these preservatives are added in mg/kg (or mg/L) amounts.

Cyclodextrin-modified CE, which was introduced in 1990,[40] was applied to the analysis of benzoic acid, sorbic acid, and parabens in food products.[38] Recently, several articles that demonstrate the utility of capillary analysis for preservatives have been compiled.[41]

Remarkable is a method for benzoic acid analysis in which a biosensor is based upon a mushroom homogenate.[42] Valuable reviews of this technique, in which the analysis of preservatives also finds its place, have been given by Dong,[43] Frazier et al.,[44] and Boyce.[45]

Other methods for the analysis of preservatives in food include lanthanide-sensitized luminescence[46] determination of benzoic acid (and saccharin) and enzymatic determination of benzoic acid[47] and of sorbic acid.[48] A recent article has described the determination of sorbic acid, benzoic acid, and parabens by a convenient extraction technique, that is, stir-bar sorptive extraction (developed in 1999) and thermal desorption GC-MS.[49]

16.3   SULFUR DIOXIDE AND SULFITES

Sulfites are used both as preservatives and as agents that stop browning reactions. They are further described as "effective bleaching agents, antimicrobials, oxygen scavengers, reducing agents and enzyme inhibitors," and they are also used "to maintain the outward appearance of freshness in salad bar vegetables, assure a desirable color in instant potatoes, and prevent black spot formation on shrimp."[50] In foods and drinks, the HSO_3^- species predominates, but in dehydrated food, it is considered that S(IV) mainly exists as metabisulfite ($HS_2O_5^-$). Wędzicha uses the term S(IV) to describe a mixture of SO_2, HSO_3^-, SO_3^{-2}, and $S_2O_5^{2-}$ when the detailed composition is either unknown or irrelevant.[51] Apart from this, equilibria exist between these species that are pH-dependent.

"Free" S(IV), in the quantitative analysis of sulfites in food or drink samples, is rapidly converted to SO_2 on acidifying the sample. The term "bound sulfite" is used for the proportion that is present as hydroxysulfonate adducts formed by the reaction of carbonyl compounds with HSO_3^-. Bound sulfite, in turn, can be distinguished into two forms: reversibly and irreversibly bound sulfites. The latter category comprises a number of compounds, for example, from pigments and from ascorbic acid. Proteins may take up S(IV) by cleavage of S–S bonds.

In the EC, the content of sulfiting agents allowed in foods and drinks (expressed as SO_2) varies between 10 mg/kg (in some sugars) and 2 g/kg or 2 g/L (in some dried fruits and in concentrated grape juice, respectively).[1] Fazio and Warner refer to the U.S. Federal Register that states that any food containing more than 10 mg/kg of a sulfiting agent must have a label declaration. For this reason, analytical methods were developed to monitor these compounds at this regulatory limit.[50] Apart from this, it has to be realized that high levels of sulfur dioxide should not be present in food products and in beverages because of its odor. As a consequence, application of sulfur dioxide is limited to foods processed in such a way that most of the sulfite is bound.

The classical Monier–Williams method (1927) can be regarded as the basis for accurate methods for the analysis of S(IV), with sufficient sensitivity, in food and drinks,[50] and it has been the official method in many countries for many years. In this method, the sample is refluxed with ca. 4 M hydrochloric acid for 1.75 hours. A stream of nitrogen sweeps the entrained sulfur dioxide through a 3% hydrogen per-oxide trap. The sulfuric acid formed is either titrated or precipitated with a barium salt solution. Fazio and Warner report that many attempts to reduce the time and effort needed for this procedure have led to other problems, as these attempts lack the rigidity and reliability of the Monier–Williams distillation method.

Warner et al., upon evaluation of the Monier–Williams method, found that the distillation procedure yields 90% recovery of sulfite added to foods such as table grapes, honey, dried mangoes, and lemon juice. Less than 85% recovery was obtained with broccoli, soda crackers, cheese-peanut butter crackers, mushrooms, and potato chips.[52]

For legislative purposes, the SO_2 content should be the sum of free sulfite and as much as possible of the bound sulfite. Fortunately enough, very few foods lead to false positives in the existing procedures at this level. *Allium* and *Brassica* vegetables and isolated soy protein, however, are exceptions to this rule.[50] For other purposes such as monitoring for viticulture, separation of free and reversibly bound sulfite is described in several procedures, e.g., those of Lawrence and Chadha[53,54] and Warner et al.[55]

A number of procedures, however, do not utilize distillation but merely require conversion of S(IV) to sulfur dioxide, which then participates in a quantitation reac-tion that is usually preceded by stabilization. Treatment with mercuric chloride results in a stable sulfito-mercuric complex.[56]

It was also recognized that formaldehyde can be used to convert S(IV) into stable hydroxymethyl sulfonate, as indicated by Fazio and Warner.[50] This can be quanti-tated by ion chromatography[57] or by other methods but, for some purposes, these methods lack the required sensitivity.

Oxidation of S(IV) in the samples or during analysis may be another problem, particularly when heavy metal ions catalyze oxidation. Conversion to SO_3^{-2} may over-come this, as this anion is more resistant to oxidation than HSO_3^- and stops or largely reduces the catalytic action of the metal ions. Apart from this, carbonyl-bisulfite addition products dissociate rapidly under alkaline conditions. On the other hand, sulfite ions may cleave disulfide bonds in proteins, which results in losses of S(IV), thus making this procedure less suitable for products rich in proteins. Severe losses, that is, 47%–60%, were observed in an intercomparison study on tropical shrimp spiked with sodium metabisulfite and HMS.[58]

Several other methods of analysis are discussed or mentioned by Fazio and Warner, but most of these are not suitable to determine S(IV) at the 10 mg/kg level.

Since 1980s, ion chromatography has been recognized as a promising technique for the determination of sulfite in foods.[55,59–64] Enzymatic detection, making use of sulfite oxidase, has been frequently applied as well.[65–72] The enzyme is taken up in an enzyme electrode or reactor, and the reaction is, in most cases, usually monitored electrochemically. HPLC techniques have also been applied by others for this purpose.[73–77] It was found that, in white wine, the results obtained by a HPLC method were in close agreement with those from Monier–Williams distillates.[78] HPLC has also been applied to the determination of sulfite in fruits and vegetables[79] and in a variety of other foods.[80] Flow injection analysis has been a frequently used alternative.[65,70,77,78,81–84]

Some methods are based on the absorption of sulfur dioxide, which is evaluated upon acidification of the sample, and can be regarded as headspace techniques.[54,55,85]

Biosensors have been used in the field of sulfite analysis as well. These were based on sulfite oxidase,[86,87] a microbial base,[88–90] or other biological materials.[91] Coupling of a biosensor to HPLC has also been reported.[71]

Other methods are based on fluorimetry with *N*-(9-acridinyl)maleimide as the reagent,[92,93] coulometric titration,[94] CE,[95] and isotachophoresis.[96] Detection with a piezoelectric crystal detector has been applied to sulfite as well.[97] Recently, an amperometric determination has been described.[98]

Two simple determination methods that are based on reactions with *para*-rosaniline and with 5,5′-dithiobis(2-nitrobenzoic)acid, respectively, have been described.[99] And finally, a diffuse reflectance Fourier transform infrared method affords determination in a single drop.[100]

16.4   NITRITE AND NITRATE

Nitrate is present naturally in plant materials and thus in the human diet. Its reduction product (i.e., nitrite) is a potent inhibitor of microorganisms, in particular, clostridia. For that reason, nitrite or nitrate is added to cured meat products, pickled herring and sprat, and some types of cheese as well, in order to protect the products from *Clostridium botulinum* and other *Clostridium* species.

In all probability, nitrite in its undissociated form is not responsible for this inhibition of microorganisms, but nitric oxide (NO) is the active species.[101,102] Nitrate is applied in cases where a slow release of nitrite is wanted. This allows the nitrate to diffuse in the product (e.g., cheese) before it is reduced to the much more reactive nitrite form.

Nitrite reacts with many components in the matrix and, for that reason, determination of the nitrate and/or the nitrite content does not reflect the amount of preservatives added. This particularly holds good when ascorbic or erythorbic acid is added to enhance release of nitric oxide from nitrite during heating. For this reason, the maximum levels of nitrate (50–250 mg/kg, as $NaNO_3$) and of nitrite (50–175 mg/kg, as $NaNO_2$) relate to the residual amount in the product. For cured meat products, indicative ingoing amounts are given for both nitrate (300 mg/kg) and nitrite (150 mg/kg).

The classic method for the determination of nitrite in food is based upon its ability to convert aromatic amines into diazonium ions, which, in turn, are coupled to another aromatic compound to produce an azo dye (the Griess–Romijn reaction). This is the basis for many spectrometric methods[103–108] and also for a recently developed polarographic technique.[109] Nitrate, after reduction by spongy cadmium to nitrite, can be determined by the same type of reaction.[105,106,108,110]

Other spectrometric methods for nitrite are based on the catalytic effect of acridine orange,[111] gallocyanine,[112] and pyrogallol red[113] on oxidation or on an enzymatic procedure.[114] These are applied to flow-injection analysis, and so are two of the methods based on the Griess–Romijn reaction.[107,108]

A novel spectrometric detection is based on reduction of nitrite to nitric oxide and subsequent reaction with bivalent iron and thiocyanate to form $Fe(NO)SCN^+$ whose absorbance is measured at 460 nm.[110]

The use of cadmium in nitrate reduction led to a study to evaluate alternative methods for nitrate based upon ion chromatography.[115] It was concluded that the use of a strong anion exchanger is necessary for assuring reliability. The method proposed by the European Committee for Standardization (CEN)[116] was found to be suitable for the determination of nitrite in various kinds of matrices.

HPLC methods for the detection of nitrite and nitrate have been used since 1980.[117–123] Myung et al. reported that their chromatographic method was superior to the spectrophotometric procedure used by them.[120] Detection was in the low UV region (around 210 nm) or amperometric.[118]

Other methods for the determination of nitrite and nitrate in foodstuffs include kinetic techniques,[124] CE,[125,126,127] anodic stripping,[128] potentiometry,[129,130] amperometry,[131] and polarography.[132]

16.5   BIPHENYL (DIPHENYL) AND ORTHOPHENYLPHENOL AND ITS SODIUM SALT

For the surface treatment of citrus fruits, biphenyl (BP) and o-phenylphenol (2-phenylphenol; OPP) may be added in concentrations not exceeding 70 mg/kg (for BP) or 12 mg/kg (for OPP and/or its sodium salt).

Several methods, either single or combined, exist for the determination of BP and OPP, along with other fungicides. Liquid chromatography is a popular technique for this purpose,[133–135] and so is gas chromatography,[136–138] which may be linked to MS.[139–141] An early spectrophotometric method is worth mentioning,[142] and the same holds for Thin Layer Chromatography (TLC) densitometry that was of interest in the nineties.[143] One of the methods for biphenyl is based upon derivative infrared spectrometry.[144]

16.6   ANTIBIOTICS

Nisin is a polypeptide antibiotic produced by *Streptococcus lactis*. It is composed of 34 amino acid residues, eight of which are rarely found in nature, that is, lanthionine (two alanine molecules bound to sulfur at the β carbon atoms) and β-methyllanthionine. It is widely used as a food preservative, particularly in some cheese products. A concentration of 12.5 mg/kg is allowed in ripened cheese and processed cheese, and 10 mg/kg in clotted cream and in mascarpone. In tapioca puddings and similar products, the maximum content allowed is 3 mg/kg. However, the EC Directive states that nisin may be present naturally in certain cheeses as a result of fermentation processes.

Nisin in processed cheese can be determined by a classic agar diffusion method using *Micrococcus flavus* as a test organism[145] or by immunoassay methods.[146] CZE is applied to the determination of nisin in milk.[147] A monoclonal antibody-based, sequential flow-injection immunoassay system allows determination of nisin in milk in amounts of 6–90 μg/kg, with a recovery of 86%.[148] HPLC, after SPE, is applied to the analysis of nisin in wine.[149]

Natamycin (pimaricin) has a polyene structure and is produced from some *Streptomyces* spp. It can be used for surface treatment of hard, semihard, and semi-soft cheese and of dried cured sausages. It should not be present at a depth of 5 mm or more, and the surface layer of the products mentioned should not contain more than 1 mg/dm².

Studies on the analysis of natamycin are scarce, which illustrates that its application as an antifungal agent does not lead to many problems from the viewpoint of public health. This also holds good for lysozyme that is found in many body fluids and tissues and that can be described as a mucolytic enzyme (MW = 14 kDa) with antibiotic properties. Like nisin and natamycin, it may be applied to cheese for protection against molds. Maximum concentrations are not given by the EC directive, which recommends sufficient amounts only ("quantum satis").

16.7 OTHER PRESERVATIVES

Hexamethylene tetramine (Figure 16.2) (hexa) was originally developed for medical applications. Since the beginning of the twentieth century, it has been applied to the preservation of certain food products such as fish and marinated mussels. It acts by slow hydrolysis in an acid medium, thus yielding formaldehyde, which is the active principle. Because of its relatively high toxicity, its application is restricted now to provolone cheese in which 25 mg/kg, expressed as formaldehyde, of hexa is allowed. Hexa can be easily determined by acidification of an aqueous extract, followed by a characteristic reaction, for example, with chromotropic acid (1,8-dihydroxy naphthalene-3,6-disulfonic acid) and subsequent spectrophotometry (at 570 nm).

FIGURE 16.2   Hexamethylene tetramine.

Dimethyl dicarbonate (Figure 16.3) is allowed for the preservation of nonalcoholic flavored drinks, alcohol-free wine, and liquid-tea concentrate in ingoing amounts up to 250 mg/L. Because of its reactivity to nucleophilic groups present in proteins, it easily binds to these, which, in the case of enzymes, leads to inactivation of these and hence to inactivation of microorganisms.[150] In an aqueous medium, it undergoes complete hydrolysis to methanol and carbon dioxide within a few hours. The EC Directive provides a remark that dimethyl dicarbonate residues are not detectable in drinks.

For many years, boric acid (or sodium tetraborate) has been used as a preservative for crustacea, in particular, shrimps. Because of its relatively high toxicity it is not allowed in any product, except in sturgeon's roe (ikra or caviar) where it may be applied in concentrations up to 4 g/kg. Mostly, the determination of boric acid is based upon colorimetric or fluorimetric methods after chelation and extracxtion.[151,152,153]

Germinated spores of *Bacillus subtilis* are the cause of ropiness in bread. Propionic acid and salts of this short-chain fatty acid are able to inhibit the growth of *B. subtilis* in bread and in some bakery products. The growth of molds is also retarded but the activity of yeasts is unaffected. The maximum levels allowed vary from 1000 to 3000 mg/kg in bread. Determination of propionic acid is usually performed by a gas chromatographic technique.[154,155]

Quaternary ammonium salts (Figure 16.4) are allowed, in the United States, as antimicrobial agents in raw cane sugar juice. These are added before clarification when further processing of this juice must be delayed. Maximum concentrations are given below (in mg/kg, based upon the weight of the raw cane):

n-Dodecyl dimethyl benzyl ammonium chloride	0.25–1.0
n-Dodecyl dimethyl ethylbenzyl ammonium chloride	3.4–13.5
n-Hexadecyl dimethyl benzyl ammonium chloride	1.5–6.0
n-Octadecyl dimethyl benzyl ammonium chloride	0.25–1.0
n-Tetradecyl dimethyl benzyl ammonium chloride	3.0–12.0
n-Tetradecyl dimethyl ethylbenzyl ammonium chloride	1.6–6.5

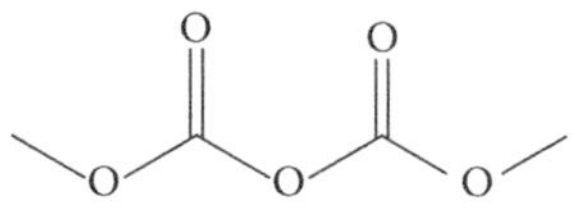

FIGURE 16.3 Dimethyl dicarbonate.

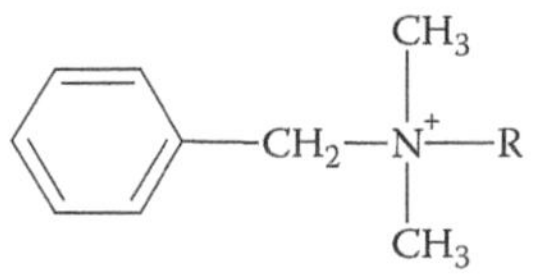

FIGURE 16.4 Benzyl dimethyl alkylammonium compounds.

Ion-pair chromatography is an obvious method for components such as these. However, recoveries were reported to be moderate.[156] As for CE, recoveries are excellent but detection limits are hardly sufficient, that is, between 2 and 10 mg/kg.[157] Liquid chromatography with electrospray ionization mass spectrometry and tandem mass spectrometry (LC-ESI-MS) is a suitable alternative in situations where low detection limits are necessary.[158]

The disodium and calcium disodium salts of ethylenediaminotetraacetate (EDTA) bind traces of heavy metals such as Cu^{++} and Fe^{+++} and, for that reason, can be used as agents that retain color, flavor, and texture of a variety of products such as pickled vegetables, salads, canned crustacea, egg products, and so on. The concentrations allowed vary from 25 mg/kg for distilled alcoholic beverages to 340 mg/kg for canned clams.

EDTA can be determined in canned crustacea by addition of copper chloride and ferric chloride to the extract and determination of the copper complex by HPLC.[159] The use of the two complexes serves to confirm the presence of EDTA. For the determination of EDTA in mayonnaise, a method in which online capillary isotachophoresis is coupled to CZE is described.[160]

Stannous chloride is allowed in the United States for color retention in asparagus packed in glass, in concentrations not exceeding 20 mg/kg. It can be simply determined by atomic absorption spectrometry using a nitrous oxide–acetylene flame.

16.8   ANTIOXIDANTS

Antioxidants protect fats, oils, and lipids in foodstuffs. Most of these act through reaction with free radicals. Formation of these free radicals from lipids is the first step in their oxidative deterioration. Other antioxidants such as ascorbic acid lower the oxidation potential of the foodstuffs to which they are added.

Preservatives in food or in edible oils that bind to free radicals are mostly of a phenolic nature. Both EU and US regulations allow addition of a number of these components. The chemical structures of these components are shown in Figure 16.5.

"Anoxomer" is a polymer of p-hydroquinone from an alkylated p-hydroxyquinone with diethenyl benzene and several phenolic components (CAS registration number 60837–57-2). It is allowed in foods at levels not exceeding 5000 mg/kg on the basis of the fat and oil content of the food. The amount of monomers, dimers, and trimers with molecular weights under 500 should be not more than 1% of the total weight of the food.

Tolerances for the other oxidants are in the range of 10–50 mg/kg of the total food base or 100–200 mg/kg based on oil or fat content. In ingredients or natural dyestuffs, 100–200 mg/kg is allowed. Mixtures of antioxidants are allowed as well.

Many methods have been developed for the quantitative analysis of these compounds,[161] but HPLC has become the dominant analytical procedure. This technique allows simultaneous determination of antioxidants, preservatives, and sweeteners[15]. Sample treatment is simple, the time of analysis is short, reproducibility is high, and the detection limit is sufficient. UV detection at 280 nm is frequently applied,[161,162,163] but electrochemical[164] and fluorimetric detection[165] are used as well. Apart from this,

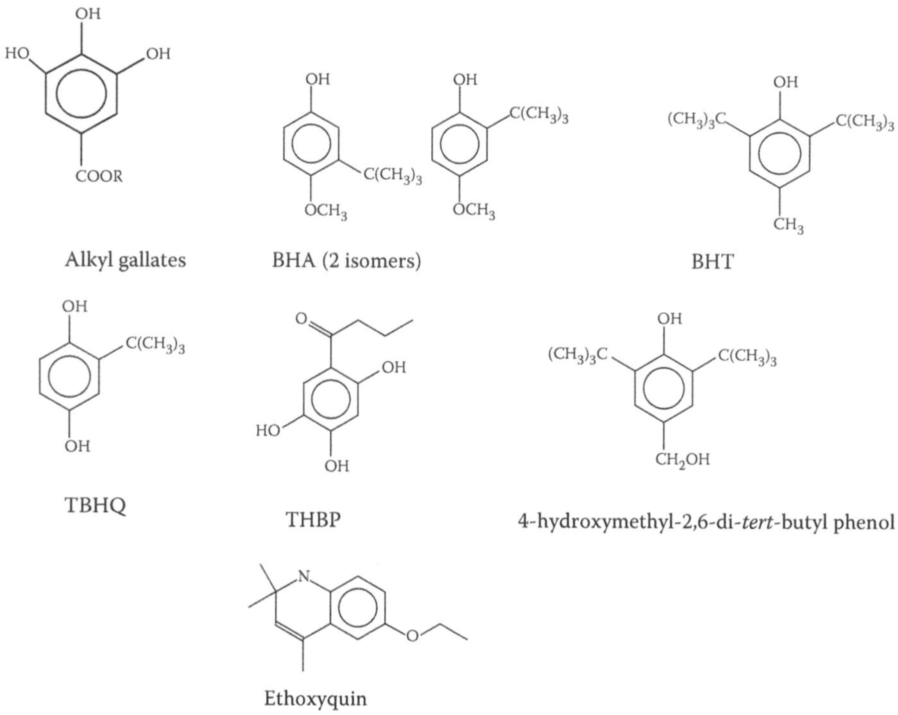

FIGURE 16.5 Antioxidants.

clean-up can be easily performed online. As an example, this proved to be necessary in the analysis of liver pâté, probably because of a complex and interfering matrix.[166]

CE was found to be a good alternative for the analysis of phenolic antioxidants, as it was superior to reversed-phase liquid chromatography with respect to efficiency and separation time. CE also requires smaller samples.[167] This, however, requires a not-too-low concentration of the analytes, and apart from this, the matrix should not be too complicated.

Boyce[39] used micellar electrokinetic capillary electrophoresis for the simultaneous determination of antioxidants, preservatives, and sweeteners.

Earlier methods were based upon gas chromatographic techniques and were applied to the determination of antioxidants in vegetable oils.

16.9 ERYTHORBIC ACID

Erythorbic or isoascorbic acid is a diastereo isomer of ascorbic acid and shows the same antioxidative properties as ascorbic acid. In the human body, it reduces the turnover, the body pool, and the metabolism of ascorbic acid. Its antiscorbutic activity is less than 5% of that of ascorbic acid.[168] It is, nevertheless, allowed in the EU in quantities up to 500 mg/kg in semipreserved and preserved meat products and up to 1500 mg/kg in corresponding fish products. It has to be kept in mind that both compounds are able to (partly) convert nitrite to nitric oxide. Their use in frozen fish with red skin, however, is merely because of antioxidative properties. In products such as these, the concentrations allowed are the same as that in the preserved and semipreserved fish products mentioned above.

Ascorbic acid and erythorbic acid can be separated well by HPLC[168–172] and by CE.[173,174] UV detection is suitable but electrochemical[169] or fluorometric[171] detection is possible as well.

16.10 CONCLUDING REMARKS

Preservatives in food can be easily, quickly, and reliably determined using modern chromatographic techniques coupled to a wide choice of detection methods. Capillary electrophoresis techniques and several specific methods also contribute to the possibility of detecting these substances in foodstuffs.

Chromatography coupled to MS allows detection of illegally used preservatives whose identity is unknown to the inspector. However, the combination of skill and knowledge of analytical science of both inspectors and analysts remains necessary to guarantee safe chemical preservation of food and food products.

REFERENCES

1. European Parliament and Council Directive No. 95/2/EC of 20 February 1995 on food additives other than colours and sweeteners. Last (7th) Amendment by Directive 2006/52/EC (5 July 2006).
2. U.S. Federal Regulations, 21st Code, Vol. 172, April 24th, 2003.
3. Parke, D.V., and Lewis, D.V.F., Safety aspects of food preservatives. *Food Addit. Contam.* 9, 561, 1992.
4. W. Horwitz, Ed., Official methods of analysis of AOAC International, 17th edn., 1st revision, AOAC International, Gaithersburg, MD, 2002. Available at www.aoac.org.
5. Arya, S.S., Stability of sorbic acid in aqueous solutions. *J. Agric. Food Chem.* 28, 1247, 1980.
6. Stafford, A.E., Rapid analysis of potassium sorbate in dried prunes by ultraviolet or colorimetric procedures. *J. Agric. Food Chem.* 24, 894, 1976.
7. Karovičová, J., and Šimko, P., Preservatives and antioxidants. *Food Sci. Technol.* (NewYork) 100, 575, 2000.
8. Park, G.L., and Nelson, D.B., HPLC analysis of sorbic acid in citrus fruit. *J. Food Sci.* 46, 1629, 1981.
9. Sher Ali, M., Rapid quantitative method for simultaneous determination of benzoic acid, sorbic acid, and four parabens in meat and nonmeat products by liquid chromatography. *J. Assoc. Off. Anal. Chem.* 68, 488, 1985.
10. Terada, H., and Sakabe, Y., Studies on the analysis of food additives by high-performance liquid chromatography. V. Simultaneous determination of preservatives and saccharin in foods by ion-pair chromatography. *J. Chromatogr. A* 346, 333, 1985.
11. Bui, L.V., and Cooper, C., Reverse-phase liquid chromatography of benzoic and sorbic acids in foods. *J. Assoc. Off. Anal. Chem.* 70, 892, 1987.
12. Chen, B.H., and Fu, S.C., Comparison of extraction methods and column types for the determination of additives by liquid chromatography. *J. Liq. Chrom. & Rel. Technol.* 19, 625, 1996.

13. Mannino, S., and Cosio, M.P., Determination of benzoic and sorbic acids in food by micro-dialysis sampling coupled with HPLC and UV detection. *Ital. J. Food Sci.* 4, 311, 1996.
14. Puttemans, M.L., Branders, C., Dryon, L., and Massart, D.L., Extraction or organic acids by ion-pair formation with tri-n-octylamine. Part 6. Determination of sorbic acid, benzoic acid, and saccharin in yogurt. *J. Assoc. Off. Anal. Chem.* 68, 80, 1985.
15. Chen, B.H., and Fu, S.C., Simultaneous determination of preservatives, sweeteners and antioxidants by paired-ion liquid chromatography. *Chromatographia* 41, 43, 1995.
16. Saito, I., Oshima, H., Kawamura, N., Uno, K., and Yamada, M., Determination of sorbic, dehydroacetic, and propionic acids in cheese by liquid chromatography and gas chromatography. *J. Assoc. Off. Anal. Chem.* 70, 506, 1987.
17. Bruzzoniti, M.C., Mentasti, E., and Sarzanni, C., Divalent pairing ion for ion-interaction chromatography of sulphonates and carboxylates. *J. Chromatogr.* A 770, 51, 1997.
18. Mandrou, B., Nolleau, V., Gastaldi, E., and Fabre, H., Solid-phase extraction as a clean-up procedure for the liquid chromatographic determination of benzoic and sorbic acids in fruit-derived products. *J. Liq. Chrom. & Rel. Technol.* 21, 829, 1998.
19. Mihyar, G.F., Yousif, A.K., and Yamani, M.I., Determination of benzoic and sorbic acids in labaneh by high-performance liquid chromatography. *J. Food Compos. Anal.* 12, 53, 1999.
20. Ferreira, I.M.P.L.V.O., Mendes, E., Brito, P., and Ferreira, M.A., Simultaneous determination of benzoic and sorbic acids in quince jam by HPLC. *Food Res. Int.* 33, 113, 2000.
21. Pylypiw, H.M. Jr., and Grether, M.T., Rapid high-performance liquid chromatography method for the analysis of sodium benzoate and potassium sorbate in foods. *J. Chromatogr.* A 883, 299, 2000.
22. Saad, B., Bari, M.F., Saleh, M.I., Ahmad, K., khairuddin, M., and Talib, M., Simultaneous determination of preservatives (benzoic acid, sorbic acid, methylparaben and propyl-paraben) in foodstuffs using high-performance liquid chromatography. *J. Chromatogr.* A 1073, 393–397, 2005.
23. Xie, Y.-T., Chen, P., and Wei, W.-Z., Rapid analysis of preservatives in beverages by ion chromatography with series piezoelectric quartz crystal as detector. *Microchem. J.* 61, 58, 1999.
24. Chen, Q.-C., and Wang, J., Simultaneous determination of artificial sweeteners, preservatives, caffeine, theobromine and theophylline in food and pharmaceutical preparations by ion chromatography. *J. Chromatogr.* A 937, 57, 2001.
25. Welling, P.L.M., Duyvenbode, M.C. van, and Kaandorp, B.H., Liquid chromatographic analysis of dehydroacetic acid and its application in wines. *J. Assoc. Off. Anal. Chem.* 68, 650, 1985.
26. Marini, D., and Flori, A., Detection of illegal chemical compounds added to wines. I. HPLC determination of dehydroacetic acid. *Prod. Chem. Aerosol. Sel.* 31, 3, 1990.
27. Ikai, Y., Oka, H., Kawamura, N., and Yamada, M., Simultaneous determination of nine food additives using high-performance liquid chromatography. *J. Chromatogr.* A 457, 333, 1988.
28. Lück, E., *Chemische Lebensmittel-Konservierung (Chemical preservation of foods)*, 2nd Edn. Springer-Verlag, Berlin, 1986, pp. 159–162.
29. Sontag, G., Aziz, I., and Smola, U., Bestimmung von Dehydracetsäure in Lebensmitteln mittels Differential-Pulspolarographie (Determination of dehydroacetic acid in food by differential pulse polarography). *Ernährung* 15, 663, 1991.
30. Pan, Z., Wang, L., Mo, W., Wang, Ch.,Hu, W., and Zang, J., Determination of benzoic acid in soft drinks by gas chromatography with on-line pyrolytic methylation technique. *Anal. Chim. Acta* 545, 218, 2005.
31. Luca, C. De, Passi, S., and Quattrucci, E., Simultaneous determination of sorbic acid, benzoic acid and parabens in foods: a new gas chromatography-mass spectrometry technique adopted in a survey on Italian foods and beverages. *Food Add. Contam.* 12, 1, 1995.

32. González, M., Gallego, M., and Valcárcel, M., Gas chromatographic flow method for the preconcentration and simultaneous determination of antioxidant and preservative additives in fatty foods. *J. Chromatogr. A* 848, 529, 1999.
33. Jorgenson, J.W., and Lukacs, K.D., Zone electrophoresis in open-tubular glass capillaries. *Anal. Chem.* 53, 1298, 1981.
34. Han, F., He, Y.Z., Li, L., Fu, G.N., Xie, H.Y., and Gan, W.E., Determination of benzoic acid and sorbic acid in food products using electrokinetic flow analysis—ion pair solid extraction—capillary zone electrophoresis. *Anal. Chim. Acta* 618, 79, 2008.
35. Kaniansky, D., Masár, M., Madajová, V., and Marák, J., Determination of sorbic acid in food products by capillary zone electrophoresis in a hydrodynamically closed separation compartment. *J. Chromatogr. A* 677, 179, 1994.
36. Dobiášová, Z., Pazourek, J., and Havel, J., Simultaneous determination of *trans*-veratrol and sorbic acid in wine by capillary zone electrophoresis. *Electrophoresis* 23, 263, 2002.
37. Pant, I., and Trenerry, V.C., The determination of sorbic acid and benzoic acid in a variety of beverages and foods by micellar electrokinetic capillary chromatography. *Food Chem.* 53, 219, 1995.
38. Kuo, K., and Hsieh, Y., Determination of preservatives in food products by cyclodextrin-modified capillary electrophoresis with multiwave detection. *J. Chromatogr. A* 768, 334, 1997.
39. Boyce, M.C., Simultaneous determination of antioxidants, preservatives and sweeteners permitted as additives in food by mixed micellar electrokinetic chromatography. *J. Chromatogr. A* 847, 369, 1999.
40. Terabe, S., Miyashita, Y., Shibata, O., Barnhart, E.R., Alexander, L.R., and Patterson, D. G., Separation of highly hydrophobic compounds by cyclodextrin-modified micellar electrokinetic chromatography. *J. Chromatogr. A* 516, 23, 1990.
41. Frazier, R.A., Inns, E.L., Dossi, N., Ames, J.M., and Nursten, H.E., Development of a capillary electrophoresis method for the simultaneous analysis of artificial sweeteners, preservatives and colours in soft drinks. *J. Chromatogr. A* 876, 213, 2000.
42. Sezgintürk, M.S., Göktuğ, T., and Dinçkaya, E., detection of benzoic acid by an amperometric inhibitor biosensor on mushroom tissue homogenate. *Food Technol. Biotechnol.* 43, 329–334, 2005.
43. Dong, Y., Capillary electrophoresis in food analysis. *Trends Food Sci. Tech.* 10, 87, 1999.
44. Frazier, R.A., Ames, J.M., and Nursten, H.E., The development and application of capillary electrophoresis methods for food analysis. *Electrophoresis* 20, 3156, 1999.
45. Boyce, M.C., Determination of additives in food by capillary electrophoresis. *Electrophoresis* 22, 1447, 2001.
46. Aguilar-Caballos, M.P., Gómez-Hens, A., and Pérez-Bendito, Simultaneous determination of benzoic acid and saccharin in soft drinks by using lanthanide-sensitized luminescence. *Analyst* 124, 1079, 1999.
47. Hamano, T., Mitsuhashi, Y., Aoki, N., Semma, M., and Ito, Y., Enzymic method for the spectrophotometric determination of benzoic acid in soy sauce and in pickles. *Analyst* 112, 259, 1997.
48. Hofer, K. and Jenewein, D., Enzymatic determination of sorbic acid. *Eur. Food Res. Technol.* 211, 72, 2000.
49. Ochiai, N., Sasamoto, K., Takino, M., Yamashita, S., Daishima, S., Heiden, A.C., and Hoffmann, A., Simultaneous determination of preservatives in beverages, vinegar, aqueous sauces, and quasi-drug drinks by stir-bar sorptive extraction (SBSE) and thermal desorption GC-MS. *Anal. Bioanal. Chem.* 373, 56, 2002.
50. Fazio, T., and Warner, C.R., A review of sulphites in foods: analytical methodology and reported findings. *Food Addit. Contam.* 7, 433, 1990.
51. Wedzicha, B.L., Chemistry of sulphiting agents in food. *Food Addit. Contam.* 9, 449.

52. Warner, C.R., Daniels, D.H., Joe Jr., F.L., and Fazio, T., Reevaluation of the Monier-Williams method for determining sulfite in food. *J. Assoc. Off. Anal. Chem.* 69, 3, 1986.

53. Lawrence, J.F., and Chadha, R.K., Headspace liquid chromatographic technique for the determination of sulfite in food. *J. Chromatogr.* 398, 355, 1987.

54. Lawrence, J.F., and Chadha, R.K., Determination of sulfite in foods by headspace liquid chromatography. *J. Assoc. Off. Anal. Chem.* 71, 930, 1988.

55. Warner, C.R., Daniels, D.H., Fitzgerald, M.C., Joe Jr., F.L., and Diachenko, G.W., Determination of free and bound sulphite in foods by reverse-phase, ion-pairing high-performance liquid chromatography. *Food Addit. Contam.* 7, 575, 1990.

56. Sullivan, J.J., Hollingworth, T.A., Wekell, M.M., Newton, R.T., and Larose, J.E., Determination of sulfite in food by flow injection analysis. *J. Assoc. Off. Anal. Chem.* 69, 542, 1986.

57. Cooper, P.L., Marshall, M.R., Gregory, J.F. III, and Otwell, W.S., Ion chromatography for determining residual sulfite on shrimp. *Food Sci.* 51, 924, 1986.

58. Luten, J., Bouquet, W., Oehlenschläger, J., Meetschen, U., Etienne, M., Stroud, G., Bykowski, P., Batista, I., Vyncke, W., and Stefansson, G., An intercomparison study of the determination of sulfite in tropical shrimps by the West European Fish Technologists' Association (WEFTA). *Z. Lebensm. Unters. Forsch. A* 204, 237, 1997.

59. Sullivan, D.M., and Smith, R.L., Determination of sulfite in foods by ion chromatography. *Food Technol.* 39, 45, 1985.

60. Anderson, C., Warner, C.R., Daniels, D.H., and Padgett, K.L., Ion chromatographic determination of sulfites in foods. *J. Assoc. Off. Anal. Chem.* 69, 14, 1986.

61. Kim, H.-J., Determination of foods and beverages by ion-exclusion chromatography with electrochemical detection: collaborative study. *J. Assoc. Off. Anal. Chem.* 73, 216, 1990.

62. Borchert, C., Jorge-Nothaft, K., and Krüger, E., Ionenchromatographische Methode zur Bestimmung des Nitratgehaltes in Brauwasser, Malz, Hopfen, Würze und Bier (Ion chromatographic determination of nitrate in brewwater, malt, hops, worts and beer). *Monatsschr. Brauwissensch.* 41, 464, 1988.

63. Ruiz, E., Santillana, M.I., De Alba, M., Nieto, M.T., and Garcia-Castellano. S., High performance ion chromatographic determination of total sulfites in foodstuffs. *J. Liq. Chromatogr.* 17, 447, 1994.

64. Wagner, H.P., Stabilization of sulfite for automated analysis using ion exclusion chromatography combined with pulsed amperometric detection. *J. Am. Soc. of Brew. Chem.* 53, 82, 1995.

65. Matsumoto, K., Matsubara, H., Ukeda, H., and Osajima, Y., Determination of sulfite in white wine by amperometric flow injection analysis with an immobilized sulfite oxidase reactor. *Agric. Biol. Chem.*53, 2347, 1989.

66. Cabre, F., Cascante, M., and Canela, E., Sensitive enzymic method for sulphite determination. *Anal. Lett.* 23, 23, 1990.

67. Fassnidge, D., and Van Engel, E., Evaluation of a sulphite oxidase electrode for determination of sulphite in beer. *J. Am. Soc. Brew. Chem.* 48, 122, 1990.

68. Mulchandani, A., Groom, C.A., and Luong, J.H.T., Determination of sulphite in food products by an enzyme electrode. *J. Biotechnol.* 18, 93, 1991.

69. Nabirahni, M.A., and Vaid, R.R., Development and application of an immobilized enzyme electrode for the determination of sulphite in foods and feeds. *Anal. Lett.* 24, 551, 1991.

70. Yao, T., Satomura, M., and Nakahara, T., Simultaneous determination of sulfite and phosphate in wine by means of immobilized enzyme reactions and amperometric detection in a flow-injection system. *Talanta* 41, 2113, 1994.

71. Patz, C.-D., Galensa, R., and Dietrich, H., Beitrag zur Bestimmung von Sulfit in Fruchtsäften mittels HPLC-Biosensor-Koppling (Sulphite determination in juices using HPLC-biosensor coupling). *Deutsche Lebensmittel-Rundschau* 93, 347, 1997.

72. Littmann-Nienstedt, S., Enzymatische Bestimmung von Sulfit in tiefgefrorenen Krustentieren (Enzymatic determination of sulphite in frozen shrimps). *Deutsche Lebensmittel-Rundschau* 96, 17, 2000.
73. Paino-Campa, G., Pena-Egido, M.J., and Garcia-Morena, C., Liquid chromatographic determination of free and total sulfites in fresh sausages. *J. Sci. Food Agric.* 56, 85, 1991.
74. Williams, D.J., Scudamore-Smith, P.D., Nottingham, S.M., and Petroff, M., A comparison of three methods for determining total sulfur dioxide in white wine. *Am. J. Enol. Vitic.* 43, 227, 1992.
75. Armentia-Alvarez, A., Fernandez-Casero, A., Garcia-Morena, C., and Pena-Egido, M.J., Residual levels of free and total sulfite in fresh and cooked burgers. *Food Addit. Contam.* 10, 157, 1993.
76. Leubolt, R., and Klein, H., Determination of sulphite and ascorbic acid by high-performance liquid chromatography with electrochemical detection. *J. Chromatogr.* 640, 217, 1993.
77. Pizzoferrato, L., Di Lullo, G., and Quattrucci, E., Determination of free, bound and total sulphites in foods by indirect photometry-HPLC. *Food Chem.* 63, 275, 1998.
78. Sullivan, J.J., Hollingworth, T.A., Wekell, M.M., Meo, V.A., Saba, H.H., Etemad-Moghadam, A., and Eklund, C., Determination of total sulfite in shrimp, potatoes, dried pineapple, and white wine by flow injection analysis: collaborative study. *J. Assoc. Off. Anal. Chem.* 73, 35, 1990.
79. McFeeters, R. and Barish, A., Sulfite analysis of fruits and vegetables by high-performance liquid chromatography (HPLC) with ultraviolet spectrophotometric detection. *J. Agric. Food Chem.* 51, 1513, 2003.
80. Wai Cheung Chung, S., Chan, B.T.P., and Chan, A.C.M., Determination of free and reversibly-bound sulfite in selected foods by high-performance liquid chromatography with fluorometric detection. *J. AOAC Int.* 91, 98, 2008.
81. Bartroli, J., Escalada, M., Jimenez Jorquera, C., and Alonso, J., Determination of total and free sulfur dioxide in wine by flow injection analysis and gas diffusion using p-aminobenzenc as the colorimetric reagent. *Anal. Chem.* 63, 2532, 1991.
82. Shi, R., Stein, K., and Schwedt, G., Fliessinjektions-Analysen (FIA) von Carbonat, Sulfat und Essigsäure in Lebensmitteln (Flow-injection analysis of carbonate, sulfite and acetate in food). *Deutsche Lebensmittel-Rundschau* 92, 323, 1996.
83. Fernandes, S.M.V., Rangel, A.O.S.S., and Lima, J.L.F.C., Determination of sulphur dioxide in beer by flow injection spectrophotometry using gas diffusion and the merging zones technique. *J. Inst. Brew.* 104, 203, 1998.
84. Cadwell, T.J., and Christophersen, M.J., Determination of sulfur dioxide and ascorbic acid in beverages using a dual channel flow injection electrochemical detection system. *Anal. Chim. Acta* 416, 105, 2000.
85. Arowolo, T.A., and Cresser, M.S., Automated determination of sulphite by gas-phase molecular absorption spectrometry. *Analyst* 116, 1135, 1991.
86. Adeloju, S.B., Shaw, S.J., and Wallace, G.G., Biosensing of sulfite in wine and beer. *Chem. Aust.* 62, 26–27, 1995.
87. Manihar, S., Hibbert, D.B., Gooding, J.J., and Barnett, D., A sulfite biosensor fabricated using electrodeposited polytyramine: application to wine analysis. *Analyst* 124, 1775, 1999.
88. Matsumoto, T., Fukaya, M., Akita, S., Kawamura, Y., and Ito, Y., Improvement of alkaline hydrolysis treatment for the determination of residual sulfite in liquid foods by a microbial sensor method. *J. Jpn. Soc. Food Sci. Technol.* 42, 926, 1995.
89. Matsumoto, T., Fukaya, M., Kanegae, Y., Akita, S., Kawamura, Y., and Ito, Y., Comparison of the microbial biosensor method with the modified Rankine's method for determination of sulfite in fresh and dried vegetables including sulfur compounds. *J. Jpn. Soc. Food Sci. Technol.* 43, 716, 1996.

90. Matsumoto, T., Fukaya, M., Akita, S., Kawamura, Y., and Ito, Y., Determination of residual sulfite in various foods by the microbial sensor method. *J. Jpn. Soc. Food Sci. Technol.* 43, 731, 1996.

91. Sezgintürk, M.K., and Dinckaya, E., Direct determination of sulfite in food samples by a biosensor based on plant tissue homogenate. *Talanta* 65, 998, 2005.

92. Akasaka, K., Suzuki, T., Ohrui, H., and Meguro, H., Fluorimetric determination of sulphite with N-(9-acridinyl)maleinimide for high-performance liquid chromatography. *Agr. Biol. Chem.* 50, 1139, 1986.

93. Akasaka, K., Matsuda, H., Ohrui, H., Meguro, H., and Suzuki, T., Fluorimetric determination of sulphite in wine by N-(9-acridinyl)maleinimide. *Agr. Biol. Chem.* 54, 501, 1990.

94. Lowinsohn, D., and Bertotti, M., Determination of sulphite in wine by coulometric titration. *Food Addit. Contam.* 18, 773, 2001.

95. Trenerry, V.C., The determination of the sulphite content of some foods and beverages by capillary electrophoresis. *Food Chem.* 55, 299, 1996.

96. Kvasnička, F., and Miková, K., The isotachophoretic determination of sulphite in mustard. *J. Food Compos. Anal.* 15, 585, 2002.

97. Nie, L., He, F., and Yao, S., Microdetermination of sulphite with a piezoelectric crystal detector. *Mikrochim. Acta* 1, 293, 1990.

98. Chakorn, C., Kulwadee, P., Narong, P., Toshihiko. I., and Orawon, C., Amperometric determination of sulfite by gas diffusion-sequential injection with boron-doped diamond electrode. *Sensors* 8, 1846, 2008.

99. Yongjie, L., and Meiping, Z., Simple methods for rapid determination of sulfite in food products. *Food contr.* 17, 975, 2006.

100. Verma, S.K., and Deb, M.K., Single-drop and nanogram level determination of sulfite (SO_3^{2-}) in alcoholic and nonalcoholic beverage samples based on diffuse reflectance Fourier transform infrared spectroscopic (DRS-FTIR) analysis on KBr matrix. *J. Agric. Food Chem.* 55, 8319, 2007.

101. Grever, A.B.G., and Ruiter, A., Prevention of *Clostridium* outgrowth in heated and hermetically sealed meat products - a review. *Eur. Food Res. Technol.* 213, 165, 2001.

102. Ruiter, A., and Moerman, P.C., A mechanism explaining the preservative action of nitrite in cured meat products. *Proc. Euro Food Chem XII*, Bruges, September, 304–307, 2003.

103. Sjöberg, A.-M.K., and Alanko, T.A., Spectrophotometric determination of nitrate in baby foods: collaborative study. *J. AOAC Int.* 77, 425, 1994.

104. Hunt, J., A method for measuring nitrite in fresh vagetables. *Food Addit. Contam. A* 11, 317, 1994.

105. Pringuez, E., Saude, I., and Hulen, C., Improvement of standard method IDF 95A-1984 for determination of the nitrate and nitrite contents of dried milk. *J. Food Comp. Anal.* 8, 344, 1995.

106. Binstok, G., Campos, C.A., and Gerschenson, L.N., Determination of nitrites: an improved procedure. *Meat Sci.* 42, 401, 1996.

107. Staden, J.F. van, and Makhafola, M.A., Spectrophotometric determination of nitrite in foodstuffs by flow injection analysis. *Fresenius J. Anal. Chem.* 356, 70, 1996.

108. Ahmed, M.J. et al., Simultaneous spectrophotometric determination of nitrite and nitrate by flow-injection analysis. *Talanta* 43, 1009, 1996.

109. Li, Q. et al., Determination of nitrite by single-sweep polarography. *J. AOAC Int.* 85, 456, 2002.

110. Andrade, R. et al., A flow-injection spectrophotometric method for nitrate and nitrite determination through nitric oxide generation. *Food Chem.* 80, 597, 2003.

111. Chen, H. et al., Flow-injection catalytic spectrophotometric determination of trace amounts of nitrite. *Anal. Lett.* 32, 2887, 1999.

112. Ensafi, A.A., and Bagherian-Dehaghei, G., Ultra-trace analysis of nitrite in food samples by flow injection with spectrophotometric detection. *Fresenius J. Anal. Chem.* 363, 131, 1999.

113. Kazemzadeh, A., and Ensafi, A.A., Simultaneous determination of nitrate and nitrite in various samples using flow-injection spectrophotometric detection. *Microchem. J.* 69, 159, 2001.

114. Hamano, T., Mitsuhashi, Y., Aoki, N. Semma, M., Ito, Y., and Oji, Y., Enzymic method for the determinaton of nitrite in meat and fish products. *Analyst* 123, 1127, 1998.

115. Merino, L. et al., Liquid chromatographic determination of residual nitrite/nitrate in foods: NMKL collaborative study. *J. AOAC Int.* 83, 365, 2000.

116. European Commission for Standardization (CEN), *HPLC/IC method for the determination of nitrate and nitrite content of meat products.* European Standard pr ENV 12014, Part 2, 1998. Brussels.

117. Schreiner, G. et al., Ionenchromatographische Methode zur gleichseitigen Bestimmung von Nitrit und Nitrat in Fleischerzeugnissen (Ion chromatographic method for simult aneous determination of nitrite and nitrate in meat products). *Arch. Lebensmittelhyg.* 39, 49, 1988.

118. Dennis, M.J. et al., The determination of nitrate and nitrite in cured meat by HPLC//UV. *Food Addit. Contam.* 7, 455, 1990.

119. Kim, H.J., and Conca, K.R., Determination of nitrite in cured meats by ion-exclusion chromatography with electrochemical detection. *J. Assoc. Offic. Anal. Chem.* 73, 561, 1990.

120. Myung, C.O., Chang, K.O., and Soo, H.K., Rapid analytical method of nitrite and nitrate in fish by ion chromatography. *J. Food Sci. Nutr.* 1, 1, 1997.

121. Siu, D.C., and Henshall, A., Ion chromatographic determination of nitrate and nitrite in meat products. *J. Chromatogr. A* 804, 157, 1998.

122. Reece, P., and Hird, H., Modification of the ion-exchange HPLC procedure for the detection of nitrate and nitrite in dairy products. *Food Addit. Contam.* 17, 219, 2000.

123. Butt, S.B., Riaz, M., and Iqbal, M.Z., Simultaneous determination of nitrite and nitrate by normal phase ion-pair liquid chromatography. *Talanta* 55, 789, 2001.

124. Liang, B., Iwatsuki, M., and Fukusawa, T., Catalytic spectrophotometric determination of nitrite using the chlorpromazine - hydrogen peroxide redox reaction in acetic acid medium. *Analyst* 119, 2113, 1994.

125. Marshall, P.A., and Trenerry, V.C., The determination of nitrite and nitrate in foods by capillary ion electrophoresis. *Food Chem.* 57, 339, 1997.

126. Ötzekin, N., Nutku, M.S., and Erim, F.B., Simultaneous determination of nitrite and nitrate in meat products and vegetables by capillary electrophoresis. *Food Chem.* 76, 103, 2002.

127. Shiddiky, M.J.A., Lee, K-S., Son, J., Park, D-S.,and Shim, Y-B., Development of extraction and analytical methods of nitrate ion from food samples: microchip electrophoresis with a modified electrode. *J. Agric.Food Chem.* 57, 4051, 2009.

128. Lu, G., Jin, H., and Song, D., Determination of trace nitrite by anodic stripping voltammetry. *Food Chem.* 59, 583, 1997.

129. Li. J.Z. et al., Cobalt phthalocyanine derivatives as neutral carriers for nitrite-sensitive polyvinyl chloride membrane electrodes. *Analyst* 119, 1363, 1997.

130. Schaller, U. et al., Nitrite-selective microelectrodes. *Talanta* 41, 1001, 1994.

131. Bertotti, M., and Pletcher, D., Amperometric determination of nitrite via reaction with iodide using microelectrodes. *Anal. Chim. Acta* 337, 49, 1997.

132. Ximenes, M.I.N., Rath, S., and Reyes, F.G.R., Polarographic determination of nitrate in vegetables. *Talanta* 51, 49, 2000.

133. Player, R.B., and Wood, R., Methods of analysis: collaborative studies. III. Determination of biphenyl and 2- hydroxybiphenyl (2-phenylphenol) in citrus fruit. *J. Assoc. Public Anal.* 18, 109, 1980.

134. Luckas, B., Methodik zur gemeinsamen Erfassung fungicider Pflanzenbehand-lungsmitteln auf Citrusfrüchten und Obst mit Hilfe der HPLC und selektiver Detektion (Methodology to the joint determination of fungicides and other fruit by HPLC and selective detection). *Z. Lebensm. Unters. Forsch.* 184, 195, 1987.

135. Agneessens, R., Nangniot, P., and Roland, J., Determination of biphenyl residues in orange pulps by high-performance liquid chromatography. *Analusis* 73, 632, 1987.

136. Pyysalo, H., Kiviranta, A., and Lahtinen, S., Extraction and determination of *o*-phenyl-phenol and biphenyl in citrus fruits and apples. *J. Chromatogr. A* 168, 512, 1979.

137. Anklam, E., and Müller, A., A simple method of sample preparation for analysis of biphenyl residues in citrus fruit peels by gas chromatography. *Z. Lebensm. Unters. Forsch.* 198, 329, 1994.

138. Yu, L. et al., Determination of *o*-phenylphenol, diphenylamine, and propargite pesticide residues in selected fruits and vegetables by gas chromatography/mass spectrometry. *J. AOAC Int.* 80, 651, 1997.

139. Yamazaki, Y., and Ninomaya, T., Determination of benomyl, diphenyl, o-phenylphenol, thiabendazole, chlorpyrifos, methidatihon, and methyl parathion in oranges by solid-phase extraction, liquid chromatographt, and gas chromatography. *J. AOAC International* 82, 1474, 1999.

140. Yoshioka, N., Akiyama, Y., and Teranishi, K., Rapid simultaneous determination of *o*-phenylphenol, diphenyl, thiabendazole, imazalil, and its major metabolite in citrus fruits by liquid chromatography-mass spectrometry using atmospheric pressure photo-ionization. *J. Chromatogr. A* 1022, 145, 2004.

141. Kolbe, N., and Andersson, T., Simple and sensitive determination of *o*-phenylphenol in citrus fruits using gas chromatography with atomic emission or mass spectrometric detection. *J. Agric. Food Chem.* 54, 5736, 2006.

142. Tanaka, A. et al., Application of an improved steam distillation procedure in residue analysis. Part II. Determination of 2-phenylphenol in citrus fruits by a spectrophotomet-ric method. *Analyst* 106, 94, 1981.

143. Dellacassa, E., Martinez, R., and Moyna, R., Simultaneous determination of *o*-phenyl-phenol, imazalil, and thiabendazole residues in citrus fruits by TLC-densitometry. *J. Planar Chromatogr. Mod. TLC* 6, 326, 1993.

144. Paseiro-Losada, P., Simal-Lozana, G., and Simal-Gandara, J., Determination of biphe-nyl residues in citrus fruit by derivative infrared spectrophotometry. *J. Assoc. Off. Anal. Chem.* 73, 632, 1990.

145. Noda, K. et al., Estimation and determination of nisin in processed cheese by the agar well method using *Micrococcus flavus* as a test organism. *J. Antibact. Antifung. Agents - Nippon* 17, 205, 1989.

146. Leung, P.P., Khadre, M., Shellhammer, T.H., and Yousef, A.E., Immunoassay method for quantitative determination of nisin in solution and on polymeric film. *Lett. Appl. Microbiol.* 34, 199, 2002.

147. Rossano, R. et al., New procedure for the determination of nisin in milk. *Biotechnol. Tech.* 12, 783, 1998.

148. Nandakumar, R., Nandakumar, M.P., and Mattiasson, B., Quantification of nisin in flow-injection immunoassay systems. *Biosens. Bioelectron.* 15, 241, 2000.

149. Pfeiffer, P., and Orben, C., Bestimmung von Nisin in Wein nach Festphasenextraktion und Hochleistungsflüssigkeitschromatographie (Determination of nisin in wine by high-performance liquid chromatography after solid-phase extraction). *Dtsch. Lebensm. Rundsch.* 93, 47, 1997.

150. Ough, C.S., Dimethyl dicarbonate and diethyl dicarbonate. *Food Sci. Technol.* 57, 343, 1993.

151. Ogawa, S. et al., Colorimetric determination of boric acid in prawns, shrimp, and salted jelly fish by chelate extraction with 2-ethyl-1,3-hexanediol. *J. Assoc. Off. Anal. Chem.* 62, 610, 1979.

152. Oshima, M., Motomizu, S., and Jun, Z., Fluorimetric determination of boric acid by high-performance liquid chromatography after derivatization with chromotropic acid. *Anal. Sci.* 6, 627, 1990.

153. Perez, E.L., Rioz, A., and Valcárcel, M., Mechanized sample work-up interfaced with flow system in flow reversal mode for the determination of boric acid in adulterated shellfish. *Anal. Chem.* 69, 91, 1997.

154. Scotter, M.J. et al., A rapid capillary gas chromatographic method for the determination of propionic acid and propionates in bread and bread products. *Food Addit. Contam.* 11, 295, 1994.

155. Beljaars, P.R., Dijk, R. van, Verheijen, P.J.J., and Anderegg, M.J.P.T., Gas chromatographic determination of propionic acid and sorbic acid contents of rye bread: interlaboratory B study. *J. AOAC Int.* 79, 889, 1996.

156. Suortti, T., and Sirviö, H., Determination of fungistatic quaternary ammonium compounds in beverages and water samples by high-performance liquid chromatography. *J. Chromatogr.* 507, 421, 1990.

157. Taylor, R.B., Toasaksiri, S., and Reid, R.G., Determination of antibacterial quaternary ammonium compounds in lozenges by capillary electrophoresis. *J. Chromatogr. A* 798, 335, 1998.

158. Ford, M.J., Tetler, L.W., White, J., and Rimmer, D., Determination of alkyl benzyl and dialkyl dimethyl quaternary ammonium biocides in occupational hygiene and environmental media by liquid chromatography with electrospray ionisation mass spectrometry and tandem mass spectrometry. *J. Chromatogr. A* 952, 165, 2002.

159. Retho, R. et al., Dosage de l=EDTA dans les conserves de crustacés (Determination of EDTA in preserved crustaceans). *Ann. Fals. Exp. Chim.* 83, 145, 1990.

160. Kvasnička, F., and Miková, K., Determination of EDTA in mayonnaise by on-line coupled capillary isotachophoresis - capillary zone electrophoresis with UV detection. *J. Food Comp. Anal.* 9, 231, 1996.

161. Robards, K., and Dill, S., Analytical chemistry of synthetic food antioxidants: a review. *Analyst* 112, 933, 1987.

162. Page, B.D., High performance liquid chromatographic determination of seven antioxidants in oil and lard: collaborative study. *J. Assoc. Off. Anal. Chem.* 66, 727, 1983.

163. Page, B.D., Liquid chromatographic method for the determination of nine phenolic antioxidants in butter oil: collaborative study. *J. Assoc. Off. Anal. Chem.* 76, 765, 1993.

164. Boussenadji, R., Porthault, M., and Berthod, A., Microbore liquid chromatography with electrochemical detection for the control of phenolic antioxidants in drugs and foods. *J. Pharm. Biomed. Anal.* 11, 71, 1993.

165. Viñas, P., Hernández Córdoba, M., and Sánchez-Pedreño, C., Determination of ethoxyquin in paprika by high-performance liquid chromatography. *Food Chem.* 42, 241, 1991.

166. Pinho, O., Ferreira, I.M.P.L.V.O., Oliveira, M.B.P.P., and Ferreira, M. A., Quantification of synthetic phenolic antioxidants in liver pâtés. *Food Chem.* 68, 353, 2000.

167. Hall III, C.A., Zhu, A., and Zeece, M.G., Comparison between capillary electrophoresis and high-performance liquid chromatography separation of food grade antioxidants. *J. Agric. Food Chem.* 42, 919, 1994.

168. Schüep, W., and Keck, E., Measurement of ascorbic acid and erythorbic acid in processed meat by HPLC. *Z. Lebensm. Unters. Forsch.* 191, 290, 1990.

169. Kutnink, M.A., and Omaye, S.T., Determination of ascorbic acid, erythorbic acid, and uric acid in cured meats by high performance liquid chromatography. *J. Food Sci.* 52, 53, 1987.

170. Arakawa, N. et al., Separative determination of ascorbic and erythorbic acid by high-performance liquid chromatography. *J. Nutr. Sci. Vitaminol.* 27, 1, 1981.

171. Speek, A.J., Schrijver, J., and Schreurs, W.H.P., Fluorimetric determination of total vitamin C and total isovitamin C in foodstuffs and beverages by high-performance liquid chromatography with pre-column derivatization. *J. Agric. Food Chem.* 32, 352, 1984.

172. Tuan, S.H., and Wyatt, J., Separation of ertyhorbic (isoascorbic) acid from ascorbic acid by HPLC. *J. Micronutr. Anal.* 3, 119, 1987.

173. Thompson, C.O., and Trenerry, V.C., A rapid method for the determination of total L-ascorbic acid in fruits and vegetables by micellar electrokinetic capillary chromatography. *Food Chem.* 54, 43, 1995.

174. Marshall, P.A., Trenerry, V.C., and Thompson, C.O., The determination of total ascorbic acid in beers, wines and fruit drinks by micellar electrokinetic capillary chromatography. *J. Chrom. Sci.* 33, 426, 1995.

17 Measuring Radioactive Contaminants in Foods

Andras S. Szabo and Sandor Tarjan

CONTENTS

17.1 INTRODUCTION

The problem of radioactive contamination of the biosphere and the need for protection against such radioactive effects belong to the problem of pollution of our environment. Because of the existence of contaminating sources (e.g., radioactive emission of nuclear power plants, from normal operation or accidents), the regular checking of the radioactivity levels in the environment is an ongoing and important task. Assessment of any release of radioactivity into the environment is important for the protection of public health, especially if the released radioactive contamination can enter the food chain.

Assessment demands rapid, reliable, and practical techniques and methods for analysis of various radionuclides. After the Chernobyl disaster (1986), there is a serious need for laboratories capable of performing rapid and reliable determination of radionuclides in a variety of sample materials (e.g., water, air, soil, fodder) and capable of handling large numbers of samples.

17.2 PATHWAYS

Figure 17.1 shows the major pathways of radionuclide contamination in humans. The source of release of radionuclides and the local conditions at the site at which contamination occurs determine one or more critical pathways—for example, the season of the year determines the magnitude of contamination of different foods and agricultural products to a great extent. The main purpose of analysis should be fast identification of critical samples and the most important radioactive isotopes so that necessary actions can be carried out rapidly.

17.3 SAMPLES OF MAIN INTEREST

Only those foods and agricultural products and only those radionuclides whose consumption contributes significantly to population exposure should be sampled and

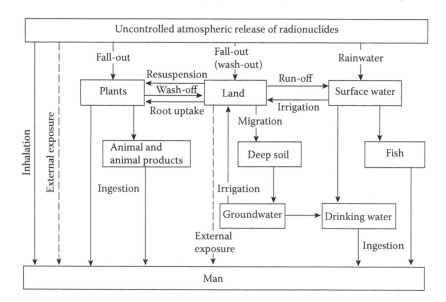

FIGURE 17.1   Major pathways of radionuclide contamination in humans.

analyzed, respectively. If, for example,[131]I is being released to cow or sheep pastures, its concentration in milk will provide far more meaningful information than its concentration in air or its deposition. Radionuclide sampling should be based on the understanding of agricultural practices and food-processing technologies and knowledge of food-consumption statistics.

It is recommended that food analyses be based on the measurement of radionuclides in individual food items rather than in a mixed diet sample. The analysis of individual foods only can correctly indicate whether, and which, countermeasures should be taken to reduce radiation doses. Analyses of individual foods should preferably be performed after preparation, taking into account the effect of food-processing and kitchen activities, such as washing, peeling, cleaning, and cooking.

17.3.1 Milk and Milk Products

In many countries, milk and milk products are important components of the diet. Milk can be collected on a daily basis, and it can be analyzed in liquid or dried form. The composition of milk from cows is practically identical all over the world, but there is a significant difference in chemical composition between milk samples from cows and samples from goats and sheep.

Milk is likely to be contaminated mainly by radioactive cesium and iodine within a few days after release of the contaminants. The radioactive contamination of milk is mainly based on the content of radionuclides in drinking water and contaminated feed, but may occur, in part, via inhalation by the animals, as well.

17.3.2 Grains, Rices, and Cereals

Bread and rice are basic foodstuffs. If the contamination occurs during the vegetation period, the radionuclides will be transported and accumulated through the plant growth process. After harvesting, grain and rice are subject to contamination only during storage, and only the outer layers would be contaminated.

17.3.3 Meat and Meat Products

In nonvegetarian diets, meat is one of the main sources of radiocesium. Contamination of meat is mainly the result of grazing by animals, but contaminated drinking water might also be an important factor.

17.3.4 Aquatic Organisms

Contamination of fish in rivers, lakes, and seas may constitute a significant pathway for the uptake of radiocesium by humans. In general, ocean fish will not take up as much radiocesium as freshwater fish because of the salinity of the water and the depth of the ocean, as well as other reasons. It should be mentioned that some aquatic organisms—mussels, shellfish, and some species of macroalgae—take up huge amounts of contaminants from the seawater (enrichment factors can be even more than 100) and can also be used as biological indicators.

17.3.5 Vegetables and Fruits

Green leafy vegetables can be a very significant pathway for the uptake of short-lived radionuclides. Of course, fruits and root vegetables may also become contaminated. Although mushrooms and berries can be contaminated markedly, only in very rare cases would they contribute significantly to the ingestion dose.

17.3.6 Water and Drinks

Rainwater and snow are early indicators of radioactive contamination, and in some places, drinking water can be a significant pathway for the uptake of short-lived nuclides. Drinking water is a potentially important pathway, directly or through its use in the preparation and processing of foods and drinks. Time delay and water treatment can reduce the contamination levels markedly.

17.4 RADIONUCLIDES OF INTEREST

Although several hundred radionuclides are produced by nuclear weapon tests or are present in irradiated reactor fuel, only a very limited number contribute significantly to human exposure. Radioactive noble gases (e.g., ^{85}Kr and ^{133}Xe) are not considered because they do not contribute significantly to internal exposure via the food chain, although their effect via inhalation—as external exposure—is not negligible in every case. Table 17.1 shows fission and activation products that may contribute significantly to human exposure.

17.4.1 Natural Radioactivity

In nature, including those in the food chain—independent of the man-made, artificial radionuclides caused by nuclear explosions, nuclear power plants, or nuclear installations—there are a number of radioactive isotopes of natural origin. These isotopes—e.g., ^{40}K, ^{87}Rb, the series of ^{232}Th, ^{235}U, and ^{238}U decays (e.g., ^{226}Ra and ^{210}Pb)—play an important role in the radioactivity of foods and also in the radiation burden on human beings, but these isotopes are not considered contaminants and therefore are not discussed in this chapter.

Some radionuclides (e.g., ^{14}C) are of natural origin, but they can be produced as activation products as well.

17.4.2 Fission Products

The source of radionuclides produced in the fission process is the testing of nuclear weapons and the release of radionuclides from nuclear reactors (e.g., the Chernobyl nuclear power plant accident in 1986). The mass number of the isotopes produced in a nuclear fission is between 70 and 170. The distribution of the fission products of ^{235}U as a function of the mass number is shown in Figure 17.2.

17.4.3 Activation Products

Radionuclides classified as activation products are produced in nuclear reactors or other nuclear facilities mainly by the reactions of neutrons with fuel and construction materials, including transuranic elements.

TABLE 17.1
Fission and Activation Products That May Be of Concern in Human Exposure

Products	Nuclides	Half-Life[a]	Major Decay
Fission products	^{89}Sr	50.5 d	beta
	^{90}Sr, ^{90}Y	28.7 a, 64.1 h	beta, beta
	^{95}Zr, ^{95}Nb	64.09 d, 35.0 d	beta, gamma, beta, gamma
	^{99}Mo, ^{99m}Tc	2.747 d, 6.006 h	beta, gamma, beta, gamma
	^{103}Ru, ^{103m}Rh	39.272 d, 56.116 min	beta, gamma, beta, gamma
	^{106}Ru, ^{106m}Rh	372.6 d, 29.92 s	beta, beta, gamma
	^{129m}Te	33.6 d	beta, gamma
	^{131}I	8.021 d	beta, gamma
	^{132}Te, ^{132}I	76.856 h, 2.3 h	beta, gamma, beta, gamma
	^{137}Cs, ^{137m}Ba	30.0 a, 2.55 min	beta, gamma
	^{140}Ba, ^{140}La	12.7561 d, 1.6779 d	beta, gamma, beta, gamma
	^{144}Ce, ^{144}Pr	284.45 d, 17.28 d	beta, gamma, beta, gamma
Activation products	^{3}H	12.35 a	beta
	^{14}C	5730 a	beta
	^{55}Fe	2.75 a	EC
	^{59}Fe	44.53 d	beta, gamma
	^{54}Mn	312.5 d	EC, gamma
	^{60}Co	5,27 a	beta, gamma
	^{65}Zn	243.9 d	EC, gamma
	^{134}Cs	754.2 d	beta, gamma
	^{239}Np	2.355 d	beta, gamma
	^{241}Pu, ^{241}Am	14.35 a, 432.0 a	beta, alpha, gamma
	^{242}Cm	162.94 d	alpha
	^{238}Pu	87.7 a	alpha
	^{239}Pu	24110 a	alpha
	^{240}Pu	6563 a	alpha
	^{242}Pu	373500 a	alpha

[a] Half-life is given in minutes (min), hours (h), days (d), or years (a).

17.4.4 RADIONUCLIDES OF SPECIFIC INTEREST IN FOODS

In regard to internal exposure to radionuclides from ingestion of foods and drinks, the most important radionuclides are the following:

Gamma emitters: ^{134}Cs, ^{137}Cs (^{137m}Ba), ^{131}I
Beta emitters: ^{89}Sr, ^{90}Sr, ^{3}H
Alpha emitters: ^{238}Pu, ^{239,240}Pu, ^{241}Am, ^{242}Cm

In special cases, food samples can also contain some other radionuclides (e.g., ^{54}Mn, ^{55}Fe, ^{59}Fe, ^{60}Co, ^{65}Zn, ^{95}Zr, ^{95}Nb, ^{103}Ru, ^{106}Ru, ^{110m}Ag, ^{125}Sb, ^{140}Ba, ^{140}La, ^{141}Ce, or ^{144}Ce).

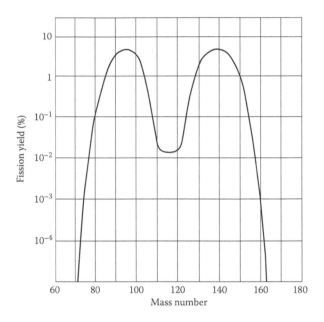

FIGURE 17.2   Distribution of the fission products of ^{235}U.

17.5 COLLECTION AND PREPARATION OF SAMPLES

Care should be taken that the sample is representative of and suitable for the specific purpose of the measurement. Samples may not be in the proper physical form for analysis: they may require reduction in size, homogenization, drying, filtering, evaporation, refrigeration, chemical treatment, ashing, and so on.

After collection, the samples must be properly stored to avoid degradation, decomposition, spoiling, and additional contamination. When long periods of storage are needed, it may be preferable to convert the samples to a more stable form. Drying and/or ashing reduce the weight and volume of the samples and allow extended storage of the samples. However, the temperature must be carefully determined in these operations to avoid the loss of radionuclides.

Samples may be dried in an oven at 105°C without a significant loss of radionuclides except radioiodines. (It is strongly dependent on the chemical form of I.) For ashing, a temperature of 450°C is recommended, because above 400°C–500°C significant loss of radiocesium will occur. If only radiostrontium and transuranic isotopes are to be measured, temperatures up to 600°C may be used.

17.6 ANALYTICAL METHODS AND MEASUREMENT TECHNIQUES

Of the approximately 2000 isotopes known today, the number of stable ones is only about 300. Radioactivity is a physical phenomenon based on radioactive decay. The cause of the decay—which is a statistical process—is the fact that the nuclides try to attain the minimum energy state and to establish a proton:neutron ratio corresponding to the minimum energy. If, for example, the nucleus has neutrons in excess, the solution is beta radiation, because then the mass number remains unchanged; however, the atomic number (the number of the protons) increases.

The application of the analytical processes and chemical methods for the measurement (enrichment and quantitative determination) of the different radioactive isotopes depends to a decisive extent on the measuring equipment available, level and type of radioactive contamination of the samples, and the required detection level (e.g., 0.1 Bq/kg) as well. The nuclear-measuring equipment consists of several units (e.g., lead shield, semiconductor detector, multichannel analyzer), depending on the nature of the measurement task. For determination of gamma emitters, direct instrumental analysis can be performed without chemical separation, and in the case of a not-very-low contamination level, even nondestructive testing can be performed.

17.6.1 GROSS BETA ACTIVITY DETERMINATION

A considerable proportion of the radioactive isotopes (e.g., ^{40}K, ^{90}Sr, and ^{137}Cs) are beta-radiating. For the detection of beta radiation, GM-tube detector or beta-sensitive scintillation measuring heads are mainly used, but recently the liquid scintillation technique has become widespread in use. Measuring the gross beta activity provides useful information regarding the total beta radioactivity of the investigated samples, including the activity of natural origin and of artificial origin. Of course, this gross beta activity is not nuclide-specific and does not give information about the contamination level.

17.6.2 GAMMA SPECTROMETRY

An outstanding advantage of gamma spectrometry is its ability to measure the gamma-ray-emitting isotopes without the need for chemical separation. In special cases, it can be performed directly even in the original sample without pretreatment, as a nondestructive measurement. Gamma ray spectroscopy allows quantitative, nuclide-specific determination of radionuclides. Measuring time varies according to sample type, contamination level, required detection limits, detection geometry and efficiency, and radionuclides of interest.

To achieve a reliable measurement, the following items are recommended:

1. High-purity germanium or lithium-drifted germanium [Ge(Li)] detectors, with good energy resolution (approximately 2 keV for the 1.33 MeV gamma peak of ^{60}Co).
2. Effective detector shield for low-level background (e.g., 5- to 10-cm-thick lead walls).
3. Preamplifier and amplifier.
4. Multichannel analyzer (MCA) with a minimum of 4096 channels. The MCA should be connected to a keyboard and display screen for input and output data and interaction with a computer. Software packages are available for MCAs based on PCs.
5. Data evaluation software.

17.6.3 ALPHA SPECTROMETRY

In some cases—for example, measurement of Pu isotopes—alpha spectrometry is necessary to perform the analysis. Since the specific ionizing capacity of the alpha particles is very large, and therefore the penetration range of the radiation is unusually short, a special measurement technique (e.g., in a vacuum chamber with surface barrier Si detector, or alpha-radiography method) is required to carry out the measurements. If the contamination level concerning the alpha particles emitting isotopes is rather high, it is possible to perform the measurements even directly, without chemical separation. But in routine analysis, chemical separation should be carried out beforehand.

17.6.4 RADIOCHEMICAL MEASUREMENTS

Many reliable methods are known for the determination of beta and alpha radiation-emitting radionuclides, based on the chemical separation of the components before the activity measurements. The methods mentioned here have been selected on the basis of experience.

17.6.4.1 Analysis of Strontium

The most commonly used method for separation of strontium involves chemical precipitation, and it can be applied to all kinds of food samples. Although this method is rather time-consuming, it is reliable and safe. Different methods for calculation of ^{89}Sr and ^{90}Sr activities are known; in the case of ^{90}Sr activity, the measurement can be carried out based on the determination of the activity of daughter ^{90}Y also, without waiting for equilibrium. If the activity of the Sr-fraction is made up of the activity of ^{89}Sr and ^{90}Sr (^{90}Y), then the ^{90}Sr content, due to the very different half-lives of ^{89}Sr and ^{90}Sr, can also be calculated on the basis of the measurements of the two activities taken at different times.

A special application of liquid scintillation technique is the measurement of Cerenkov radiation, produced by beta emitters.

17.6.4.2 Analysis of Tritium

Tritium is measured by liquid scintillation counting of a portion of a distilled sample. To prevent interference with radioiodine, several reagents can be added. The method is normally adequate for routine determination; however, in case of low tritium concentration in water, electrolytic enrichment can be applied.

17.6.4.3 Analysis of Transuranic Elements

The method for determination of transuranic elements (plutonium, americium, and curium) is based on alpha spectrometry, after sequential radiochemical separation. The chemistry of curium is very similar to that of americium, so there is no need for a curium tracer; the americium tracer serves as a yield monitor for both elements.

17.7 GROSS ALPHA ACTIVITY DETERMINATION

Chemical treatment of samples requires too much time, chemicals, and human resources as well. If we need to know only the activity of alpha emitters—below or above the limit—then gross alpha activity determination is enough for fast screening. A typical monitoring device is the LB gas flow proportional detector system.

17.8 EVALUATION OF CONTAMINATION LEVEL OF FOODS

The level of radioactivity in foods is due to radioactive substances of natural and artificial origin. Natural radioactivity is mostly a function of geological circumstances. Among the isotopes of artificial origin—man-made radioactivity—that are emitted into the biosphere as a consequence of nuclear weapon tests or from nuclear power plants or other nuclear facilities, ^{90}Sr, ^{131}I, ^{134}Cs, and ^{137}Cs are of paramount importance. The contamination level of foods depends on many factors; the most important ones are as follows:

- Geographical location and climatic, meteorological features
- Mechanical structure, chemical composition of the soil, the applied agrotechnical and meliorative methods
- Species and vegetation period of the food plants
- Species, age, and feeding of animals
- Methods of food processing

To evaluate the level of contamination, we should take into account the level of natural radioactivity, mainly that of ^{40}K. If the contamination level is low, it means that the natural radioactivity represents a significantly higher activity than does artificial contamination.

The maximum contamination level due to atmospheric nuclear weapon tests occurred between 1961 and 1964, when the radioactive contamination of foods was higher by one order of magnitude than the values measured at the beginning of the 1980s. Unfortunately, as a consequence of the nuclear power plant disaster in Chernobyl (1986), the contamination level increased significantly in most European countries—the effect was negligible in America, for example, that is very far from the Ukraine—far surpassing the values measured at the beginning of the 1960s. After 1986, the contamination level decreased significantly, and since 1990, the contamination level is as low as it was before the disaster. Figure 17.3 shows the average yearly contamination level of milk and fodder in Hungary between 1975 and 2001.

17.9 CONCLUSIONS

For measurements of the radiocontamination level of foods, different techniques are applied; the methods are sensitive, reliable, and safe and are based mainly on the counting of beta and gamma radiation of the radionuclides but in special cases of alpha radiation as well.

The contamination levels (e.g., ^{137}Cs) of foods at the beginning of the twenty-first century in European countries—not including the northern part of the Ukraine and the southern part of Belorussia—are rather low; the total radioactivity is dominantly based on the activity of natural isotopes (e.g., ^{40}K).

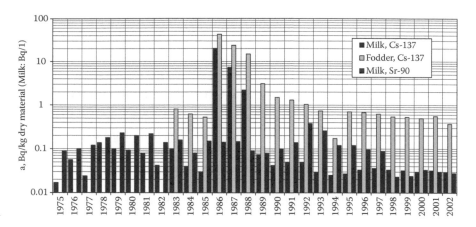

FIGURE 17.3 Specific activity of ^{137}Cs and ^{90}Sr in milk and fodder (countrywide average) in Hungary from 1975 to 2002.

Although the contamination level is rather low and there is no need to apply any special decontamination method, it is worth mentioning that the different procedures of food preparation and technology (e.g., washing, peeling, cooking) can further decrease the level of contamination.

REFERENCES

International Basic Safety Standards for Protection against Ionizing Radiation and for the Safety of Radiation Sources. Safety Series No. 115, IAEA, Vienna, 1996.

B. Kiss, A.S. Szabo. Radioactive contamination of the foodstuffs in the years following the Chernobyl accident. *J. Food Phys.,* 71–78, 1990.

Radiological Monitoring Network of Ministry of Agriculture and Regional Development. *Radioisotopes in Food and Environment,* Radiological Monitoring Network of Ministry of Agriculture and Regional Development, Budapest, Hungary, 2002.

T. Scott, R.F.W. Schelenz, eds. Measurement of radionuclides in food and the environment. In *IAEA Guidebook,* Tech. Rep. Series No. 295, IAEA, Vienna, 1989.

A.S. Szabo. Calculation method of 89-Sr and 90-Sr in food samples. *J. Food Phys.,* 63–67, 1988.

A.S. Szabo. Radioactive contamination of the foodstuffs before Chernobyl. *J. Food Phys.,* 93–100, 1988.

A.S. Szabo, B. Kiss. Radioactive pollution of foodstuffs in Hungary in consequence of the Chernobyl accident. *J. Food Phys.,* 71–84, 1989.

A.S. Szabo. Estimation of the radioactive contamination of the milk on the basis of the fodder contamination. *J. Food Phys.,* 43–48, 1990.

A.S. Szabo. *Radioecology and Environmental Protection.* Ellis Horwood, New York-London, 1993.

A.S. Szabo. Some thoughts concerning the radioactive contamination of foodstuffs 10 years after the Chernobyl disaster. *J. Food Phys.,* 101–109, 1996.

A.S. Szabo. 15 years after the Chernobyl disaster. *J. Food Phys.,* 105-109, 2000/2003.

S. Tarjan, Gy. Kis-Benedek. Radioactive contamination of foodstuffs, effective equivalent dose from food in Hungary, between 1986 and 1990. *J. Food Phys.,* 51–82, 1991.

United Nations Scientific Committee on the Effects of Atomic Radiation. *Sources, Effects and Risks of Ionizing Radiation,* United Nations Scientific Committee on the Effects of Atomic Radiation, New York, 1988.

18 Rapid Analysis Techniques in Food Microbiology

Francisco Diez-Gonzalez and
Yildiz Karaibrahimoglu

CONTENTS

18.1 INTRODUCTION

The term "rapid" as applied to a microbiological method was first used in the early 1970s,[1] but the idea of increasing the speed of analysis had been recognized by Weaver in 1954.[2] The definition of "rapid method" is somewhat subjective, and it is prone to different interpretations according to advances in detection technology.[3] The simplest definition of a rapid analysis technique in food microbiology is a method that obtains a similar result in a shorter period of time than an existing standard protocol. The ultimate rapid microbiological method is the one that obtains an instantaneous real-time measurement from the presence and level of a particular microorganism in a given food. This ideal rapid method, however, has yet to be developed. While such a technique could only be imagined, recent advances in molecular biology, immunology, nanotechnology, microarray technology, and bioelectronics suggest that one day it will become a reality.[4]

The first rapid microbiological methods developed were those intended to facilitate the identification of bacteria by biochemical tests.[2] The individual reactions used to be very laborious and time-consuming, but the development of miniaturized biochemical assays became very popular since the 1950s.[5] Techniques that could accelerate the determination of viable cell count were also among the early rapid methods. The spiral plating method was developed to automate the plate counting process, and the Petrifilm® system (3M Co., St. Paul, MN) provided a convenient alternative to standard plate agar methods.[6] The measurement of adenosine triphosphate using bioluminescence also provided a rapid estimation of bacterial numbers.[7]

The development of new and faster microbiological methods is driven by the need to prevent food-borne illnesses and to meet regulatory and quality standards such as Hazard Analysis Critical Control Point programs. A faster identification of strains implicated in outbreaks has proved to be invaluable to stop more infections, and detection in finished products before consumption has even averted potential outbreaks.[8] Rapid microbial detection methods have not only been useful in the recall of contaminated foods, but many of them are now standard methods that facilitate food trade. Under current market conditions, food companies must have in place a quick and reliable testing system that can be used in field applications and that allows the minimum preshipping storage time, which in turn leads to lower operational costs. In future, with real-time online microbiological testing, it can be envisioned that microbiological testing will not be a cause of delay for the release of foods into the market.

18.2   CLASSIFICATION OF TECHNIQUES

Since the first rapid techniques were developed, the number and variety of methods continues to increase every year. Despite the wide diversity of the rapid methods, they can be classified based on the underlying biological principles. The four major principles that dictate the categorization of methods are (1) growth characteristics, (2) biochemical reactions, (3) nucleic acid sequences, and (4) antibody recognition. Based on the first two principles, specialized growth media have been developed and arrays of reactants have been built to identify specific microorganisms. The last two principles have allowed the use of specific nucleic acid probes and immunoassays to detect microorganisms with particularly high specificity.[3]

Rapid microbiological methods can also be classified on the basis of their purpose or outcome. Rapid techniques can be aimed to enumerate a certain population, determine the presence or absence of a specific group, and identify an isolated strain.[9] Rapid quantification methods have been largely based on cultivation techniques that shorten the time of traditional most probable number (MPN) and plating protocols, but recent reports have shown that real-time polymerase chain reaction (PCR) could also be used for bacterial enumeration.[10] Among the presence/absence techniques, immunoassays and DNA-based formats have proved to be the most effective.[3,11] The rapid identification or typification of isolates has been conducted using biochemical, immunological, and molecular techniques. Because of space limitations, this chapter covers some of the most current rapid analysis techniques in food microbiology. For an exhaustive list of commercial rapid methods, visit the Association of Official Analytical Chemists International and Food and Drug Administration Bacteriological Analytical Manual Web sites.[12,13]

18.3 CULTURE METHODS

Rapid culture techniques are a very diverse group of methodologies, and many different formats have been commercialized. Methods that reduce the analysis time; facilitate the cultivation, isolation, and identification of microorganisms by automation; eliminate steps; and use specialized selective and chromogenic substrates are considered among the culture methods. Automated diluters speed sample preparation. Automated diluters allow the user to place an amount of food product sample to be tested into either a stomacher bag or a sterile blending jar that has been preweighed. Manninen et al.[14] determined that the accuracy of delivery for most samples by gravimetric diluters was in the range of 90%–100%.

Automated-plating systems distribute liquid samples onto the surface of a rotating prepoured agar plate, eliminating the need to perform manual serial dilutions. Using a laser counter, an analyst can obtain an accurate count in a few seconds, compared with several minutes in the procedure of counting colonies by the naked eye. Manninen et al.[15] and Garcia-Armesto et al.[16] evaluated the spiral plating system against the conventional pour plate method using both manual count and laser count and found that the results were essentially the same for bacteria and yeast. The latest versions of the spiral platers have been introduced as Autoplater® (Spiral Biotech, Inc., Norwood, MA) and Whitley Automatic Spiral Platter® (Microbiology International, Inc., Frederick, MD). Colony-counting systems use light scanners and special software to count and analyze colonies on a plate.

The hydrophobic grid membrane filter (HGMF) technique was an important modification of conventional plating methods.[17] The HGMF is essentially a membrane filter that has been divided into growth compartments by the application of a hydrophobic grid. These compartments are termed growth units rather than "colony-forming units." The count on the HGMF is determined by an MPN calculation, and it is considered more accurate than that in traditional pour or spread plates and conventional MPN.[18] An example of this system is the ISOGRID® system (Neogen, Inc., Lansing, MI) that is an automatic instrument capable of counting in seconds. Stressed or injured cells can be resuscitated on a repair medium, and the filters can be transferred to a selective medium, thus increasing recovery. HGMF can detect very low numbers of microorganisms by the filtration of large volumes. The HGMF technique can be used for detection of coliforms, *Escherichia coli*, *E. coli* O157:H7, and *Salmonella*.

The Petrifilm system eliminates pour plates and replaces them with rehydratable nutrients that are fixed into a series of films along with a cold water-soluble gelling agent. Currently, Petrifilm is offered for eight different types of microorganisms, and it facilitates analysis by reducing preparation time. Petrifilm is very easy to use and requires very little laboratory expertise and equipment. In the Petrifilm series 2000 rapid coliform count plates, presumptive coliform colonies may begin to appear at the 6th hour of incubation. Redigel® system (3M Co., St. Paul, MN) uses methoxy pectin as the gelling agent, which is present in the liquid medium supplied. The advantage of this method is minimized media preparation. Sample amounts can be varied from 0.1 to 5 mL on a plate. Colonies can be counted exactly as in conventional standard plates. Chain and Fung[19] conducted an extensive evaluation of the above rapid methods against the conventional standard plate count and found that the alternative systems were comparable to the conventional method and at a lower cost.

The SimPlate method (BioControl, Inc., Bellevue, WA) uses biochemical activity, not growth, and the key factor is the short incubation period. A smaller number of bacteria are needed to produce a detectable signal. The SimPlate method is designed for obtaining coliform, *E. coli*, *Campylobacter*, yeast, and mold counts. Beuchat et al.[20] compared SimPlate to Petrifilm, Redigel, and pour plate methods and they found no significant differences. Colilert (Idexx Co., Westbrook, ME) is a system specific for coliforms and *E. coli* based on chromogenic substrates of the β-D-galactosidase and β-D-glucuronidase enzymes.

18.4   BIOCHEMICAL TECHNIQUES

Microbiological assays based on specific biochemical reactions were some of the first rapid methods used in food microbiology.[5] In a strict sense, selection or enrichment culture methods that use colorimetric reagents as part of the identification of presumptive positives can also be considered biochemical techniques, but for the purpose of this section, we concentrate on those assays intended for the identification of microbial isolates. In general, biochemical assays are arrays of individual reactions that test phenotypic characteristics of microorganisms. The types of reactions in biochemical identification kits include (1) ability to grow on a specific substrate, (2) degradation of a specific compound, (3) production of specific enzymes, (4) cell staining, and (5) ability to grow anaerobically, among others.

The first rapid biochemical techniques for food analysis were commercialized in the early 1970s.[5] Since then more than 20 different formats have been marketed, and many of them have been validated by AOAC International.[3,12] A wide variety of microorganisms can be identified using biochemical kits, and specific reaction combinations are available to identify different groups of microorganisms. For food applications, there are kits to detect *Enterobacteriaceae*, *Vibrionaceae*, *Listeria*, *Enterococcus*, *Campylobacter*, *Bacillus*, anaerobes, and nonfermenters. Typically, most of the biochemical kits provide results within 24 hours, but some recently developed products offer results just after 4 hours. One of the first multiple biochemical reaction kits and probably the most popular format is the API® system developed and commercialized by bioMerieux, Inc. (Hazelwood, MO). One of the advantages of the API systems and other kits developed earlier is the availability of a very large microbial database.

18.5 IMMUNOASSAYS

18.5.1 IMMUNOASSAY PRINCIPLES

An immunoassay can be defined as an analytical technique that utilizes antibodies as part of its protocol. Antibodies are immunoglobulin proteins (Ig) produced by B-lymphocytes of the animal immune system in response to an antigen. Immunoassays exploit the unique ability of antibodies to specifically recognize part of the antigen's molecule (epitope) and tightly bind to it.[21] Antibodies are typically synthesized to bind large molecules such as proteins. This unique characteristic has allowed the application of immunoassays to almost any imaginable field of analytical chemistry and biology. Immunoassays have been used in microbiology for detection and identification of a wide variety of microorganisms.

The first immunoassays in food microbiology were antibody-based methods that allowed the serotyping of food-borne pathogens with the purpose of distinguishing among different strains. Serotyping of bacterial pathogens has been conducted since the 1920s, and it is still a very important component of analytical microbiology. Serotyping is not normally considered a rapid method, but its wide use set the basis for the first rapid immunoassays. The chemical modification of antibodies with fluorescent molecule markers in 1942[22] and the development of enzyme-linked antibodies in 1971[23] set the basis for many of the current immunoassay formats. The development of the technology to obtain monoclonal antibodies by Kohler and Milstein[24] was another major foundation of the modern immunoassays.

The diversity of currently available immunoassays is so enormous that their classification is sometimes difficult. In general, immunoassays can be classified on the basis of the precipitation of the antibody–antigen complex, the excess or limiting concentration of reagents, the inclusion of a separation step (heterogeneous), and the type of detection method. In this section, the four most relevant immunoassays in food microbial analysis are discussed.

18.5.2 ANTIBODY–ANTIGEN PRECIPITATION

The agglutination of bacterial cell–antibody complexes was the characteristic that allowed the widespread use of serotyping, and this trait has been further exploited and perfected by commercial rapid agglutination assays. The ability of some antibodies to bind more than one antigen particle and the fact that an antigen molecule could have multiple sites for polyclonal antibody binding are the foundation of the immunoassays based on the detection of the antibody–antigen precipitate (Figure 18.1). Antibody–antigen precipitation assays could be considered as one of the major types of light-scattering immunoassays.[25] Two variants of antibody–antigen precipitation assays have proven to be quite effective to speed up the detection of food-borne pathogens: latex agglutination (LA) and immunodiffusion.[26]

LA immunoassays are basically suspensions of colored latex beads that have been coated with antibodies to specifically bind bacterial cells or toxins and form a clump or aggregate that can be visually observed. Most of the existing commercial kits are

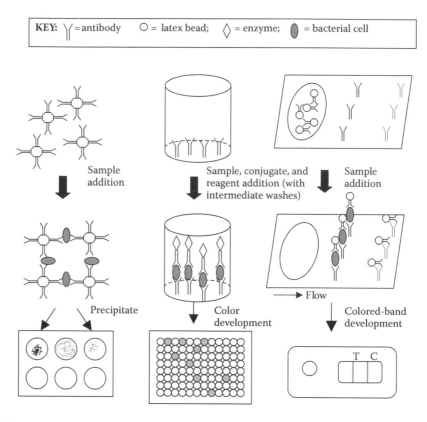

KEY: γ = antibody   O = latex bead;   ◇ = enzyme;   ● = bacterial cell

Sample addition

Sample, conjugate, and reagent addition (with intermediate washes)

Sample addition

Precipitate

Color development

Flow

Colored-band development

T   C

FIGURE 18.1   Latex agglutination, ELISA, and lateral-flow device immunoassays.

designed to bind whole cells, and they can typically detect more than 10^7 bacterial cells.[26] When the antigen is a toxin, the assay is called reverse passive latex agglutination and there is no visual cluster; instead, development of opacity is the positive proof. The major advantages of LA assays are simplicity and speed, but the ability of LA assays to detect target bacteria in enrichment media is limited and it could be affected by solid particles. Other advantages of LA kits are their relatively low cost and robustness. LA kits have been preferably used for quick confirmation of isolates rather than for detection purposes. More than 10 companies currently manufacture LA kits.[12,13]

One of the most successful immunoassays based on the principle of antibody–antigen precipitation is an immunodiffusion method for rapid detection of *Salmonella* in foods, which is marketed as the *Salmonella* 1–2 Test® (BioControl, Inc., Bellevue, WA). This method combines the motility of this bacterium in an enrichment media with the specificity of flagellar antibodies to form an easily visualized precipitate.[27] The *Salmonella* 1–2 Test has been validated as an AOAC Official Method, and it has been effective in detecting this pathogen in a variety of foods after approximately 14 hours of enrichment.[28] Unfortunately, this rapid method only detects motile strains, and it has not been adapted to the detection of other food-borne pathogens.

18.5.3 Enzyme-Linked Immunosorbent Assay

The conjugation of antibodies with enzymes has proved to be one of the most powerful tools for immunoassays, and enzyme-linked immunosorbent assay (ELISA) kits have been the most popular format in food detection. ELISA is based on the principle that antibodies can still bind to antigens and enzymes can still catalyze reactions while being covalently attached. In general, the major components of a typical ELISA kit are primary antibody, antigen, antibody–enzyme conjugate, species-specific antiserum, solid phase, substrate, and detection method.[29] The antibody–enzyme conjugate can directly bind the antigen or can be used as part of the species-specific serum that binds an antibody (primary antibody) that binds the antigen. The ELISA methods can be classified into three major types: simple, sandwich (or capture), and competitive. When these methods use only primary antibodies, they are referred to as direct methods, and when a species-specific serum (secondary antibody) is used, they are referred to as indirect methods.

Simple ELISA protocols consist of an antibody–enzyme conjugate that specifically binds an antigen that has been previously immobilized onto a surface. The main difference between the simple and the sandwich ELISA is that the latter first immobilizes the antigen by using antibodies attached to a solid phase, and another set of antibodies also binds to the immobilized antigen. The simple and sandwich formats are designed to detect the antigen by the increase of signal caused by the reaction catalyzed by the antigen-bound immobilized antibody–enzyme conjugate. In contrast, competitive ELISA relies on the disappearance of signal in the presence of the antigen by using competitive antibodies or antigens.

Most commercially available ELISA kits for detection of microorganisms and toxins in foods are sandwich formats. The capture antibodies can be attached to different solid phases such as dipsticks, membranes, and pipet tips,[3] but microtiter plate wells have been the preferred matrix because of the potential to spectrophotometrically quantify the response. In traditional ELISA, the two most common enzymes linked to antibodies have been horseradish peroxidase (HRP) and alkaline phosphatase. HRP has many substrates that can yield products with absorbance at different wavelengths or that can emit light when degraded.

The typical sandwich ELISA test involves the following steps: (1) addition of sample and reagents, (2) initial incubation, (3) blocking, (4) washing, (5) addition of enzyme–antibody conjugate, (6) second washing, (7) addition of substrate, (8) color development, and (9) reading.[29] During the initial incubation, pathogen cells or toxin molecules are captured by immobilized antibodies. The blocking step prevents nonspecific attachment of antibody conjugates. After the substrate is converted and color is developed, the reading can be done at specific wavelengths in a microtiter plate reader.

The detection limit of commercially available ELISA kits is typically 10,000 CFU/g or higher for the detection of bacterial cells, and those kits that detect proteins require concentrations in the range of 10 μg/L.[30] ELISA techniques that detect the presence of shiga toxins in enrichment media are capable of yielding a positive reaction if at least 10^8 shiga toxin-producing cells per milliliter are present in the enrichment media.[31] ELISA formats have been very successful because of their speed, sensitivity, ease of use, and cost-effectiveness. Some of the limitations of ELISA, however, are the dependence on liquid media, the interference of contaminants, the excessive specificity, and the fact that sensitivity is sometimes below the regulatory level.[29] There are more than 10 commercial ELISA manufacturers, but TECRA® (Tecra, Inc., French Forest, Australia) is the only brand that has kits for as many as six food-borne pathogens.[13]

18.5.4 IMMUNOCHROMATOGRAPHY

A chromatographic technique is typically a separation procedure that employs a stationary and a mobile phase. When antibodies are added either in the mobile or stationary phase this can be referred to as immunochromatography. Among the immunochromatographic methods, lateral-flow devices have been very popular for identifying *Salmonella, E. coli* O157:H7, *Listeria,* and *Campylobacter.*[32] Lateral flow devices are disposable plastic cards that hold a membrane strip containing two types of antibodies immobilized at two separate cross-section areas of the strip (Figure 18.1). Enrichment samples are typically added to lateral flow cards through an inoculation chamber that is saturated with latex- or colloidal gold-labeled antibodies specific to the pathogen or toxin molecule. The labeled antibodies will migrate along the membrane, and the labeled antibody–antigen complex will be captured at the first set of immobilized antibodies specific to the antigen. When sufficient labeled complexes are captured, a colored line or spot can be visualized. As control, most lateral flow devices include a second set of immobilized antibodies that will capture the labeled antibodies in the absence of the antigen as the sample continues to migrate through the membrane, and this binding will also yield a colored signal.

The popularity of lateral flow devices has been a result of their simplicity and ease of use. Because there is only one step involved in using these devices, the chances for error are largely eliminated, as compared to ELISA. In addition, the results can be obtained in as little as 10 minutes. Similar to ELISA, the sensitivity of lateral flow devices is dependent on the enrichment step in which the devices can detect 10^4 or more cells per milliliter. A number of lateral flow devices have been reviewed by AOAC International, and they can have less than 1% and 5% false negatives and positives, respectively.[32] There are at least five commercial brands of lateral flow devices, and VIP® (Neogen Corp., Lansing, MI) and Reveal® (BioControl, Inc., Bellevue, WA) brands have cards for *E. coli* O157:H7, *L. monocytogenes,* and *Salmonella.*

18.5.5 AUTOMATED IMMUNOASSAYS AND ELECTROIMMUNOASSAYS

Traditional ELISA is in itself a rapid method for identification of food-borne bacteria, but several automated ELISA technologies have been developed to eliminate the potential for operator error and to reduce the various steps needed to complete the assay. The VIDAS® system commercialized by bioMerieux, Inc. (Hazelwood, MO) was the first automated ELISA in the early 1990s for food-borne pathogen detection.[33] The VIDAS system is classified as an enzyme-linked fluorescence assay and utilizes antibodies attached to a solid-phase receptacle. At least five other automated ELISA formats have been commercialized. Among them is the Unique® assay recently introduced by Tecra, Inc. (French Forest, Australia).[34] Automated ELISA techniques have certainly increased the speed of analysis, but they are no more sensitive than traditional ELISA kits.

More recently, a number of electroimmunoassay techniques have been developed with the purpose of improving the sensitivity of ELISA or lateral-flow devices. An electroimmunoassay is based on the generation of electronic responses as a result of formation of antibody–antigen complexes.[34] Detex® and PATH*IGEN*® are two automated electroimmunoassay systems currently commercialized by Molecular Circuitry, Inc. (King of Prussia, PA) and Igen International, Inc. (Gaithersburg, MD) for the detection of food-borne pathogens that have been approved by AOAC International.[12] The Detex system consists of capture antibodies attached to an electrode gap surface and a second set of sandwich antibodies linked to colloidal gold. After addition of antigen-containing sample and formation of sandwich complexes, an electrical signal is generated after deposition of silver ions bridging the electrode poles. The PATH*IGEN* system is an electrochemiluminescence-based immunoassay that also uses a sandwich antibody format, but the capture antibodies are attached to magnetic beads and the detection signal is light emission. The PATH*IGEN* antibodies are labeled with $Ru(bpy)_3^{21}$, which is oxidized after antibody complex formation and emits light when reduced with tripropylamine. Both of these systems, however, still require enrichment or preenrichment steps of at least 6 hours.

18.6 NUCLEIC ACID–BASED TECHNIQUES

18.6.1 Types

All microorganisms possess distinct sequences of DNA and RNA, and because of this feature common to all life forms, the application of nucleic acid–based assays for detection and identification of food-borne bacteria and viruses has been extremely useful. The complementary nature of single-stranded DNA and RNA has provided the basis for all rapid formats and has allowed the development of a wide array of nucleic acid-based assays. In food microbiology, the detection and identification of pathogens has typically targeted specific genes that encode for specific virulence factors such as shiga toxins for *E. coli* O157:H7,[35,36] lysteriolysin for *Listeria*,[37] and *invA* for *Salmonella*[38] or ribosomal RNA (rRNA).[39] Methods that use either probes or the PCR principle have been by far the most effective, and several of those formats have been successfully commercialized. Before the rapid adoption of PCR, a number of nucleic acid-based protocols that exploited the ability of bacteriophages to specifically infect bacteria were also developed, but very few of them remain on the market. This section discusses the most relevant molecular probes (MPs) and PCR assays used in food microbiology.

18.6.2 Molecular Probes

A MP is a single-stranded oligonucleotide or small nucleic acid molecule that has been designed to be capable of specifically binding to a complementary DNA or RNA sequence and to trigger a detection signal as a result of this hybridization.

Typical MPs have between 15 and 30 base pairs, but in DNA hybridization, probes as large as 1.8 kb have been used.[40] In 1983, Fitts et al.[41] designed one of the first DNA hybridization methods to detect *Salmonella* in foods by using a radiolabeled probe obtained from a random chromosomal fragment with a sensitivity of approximately 10^6 cells per milliliter. Radiolabeled probes were limited for widespread application, but the application of alternative labels and a variety of sensing systems allowed the active development and marketing of MP-based techniques.

In general, there are two major types of MP protocols: those that detect target nucleic acid bound to a membrane and those that use immobilized probes.[11] The first group of MP assays originated from the pioneering development of Southern blotting by E.M. Southern,[42] who first employed membranes to immobilize DNA. In this type of method, bacterial colonies or liquid culture is initially transferred to nylon or nitrocellulose membranes where they are lyzed and their DNA is released. The free DNA can be fixed onto the membranes by high temperature or UV light[36] and then hybridized with specific probes that had been previously labeled with typically biotin, fluorescine, or digoxigenin. The hybrid immobilized DNA can be visualized using avidin or antibody conjugates for the probe label that might be linked to enzymes, fluorescine, gold, or rhodamine. The DNA hybridization assays that use immobilized probes are typically based on microtiter plates and also require the release of nucleic acids from a colony or a pure culture. The microbial nucleic acid hybridizes with the capture probe, and these assays have frequently been designed to include a reporter probe that binds to the captured DNA and can be detected based on its specific label.

Among the many MP assays that have been reported in the literature, there are only two formats that have been successfully commercialized since the early 1990s. The first of these systems is the GENE-TRAK® assay currently marketed by Neogen Corp. (Lansing, MI). The first GENE-TRAK technique used random chromosomal radiolabeled DNA probes, and the second generation used a combination of three probes that captured rRNA fragments detected by an antifluorescine antibody–enzyme conjugate.[39–41] The current GENE-TRAK format uses three types of DNA probes: a polydeoxythimidylic acid fragment attached to a dipstick or to a microtiter well, a probe containing polydexydenylic acid at one extreme and a sequence specific for a region of rRNA, and another rRNA-specific probe covalently bound to fluorescine or directly to HRP.[39,43] The detection of this "sandwich" complex can be achieved by light emission in a luminol-mediated reaction catalyzed by an antibody–HRP conjugate. Specific GENE-TRAK kits exist for the analysis of *E. coli*, *Salmonella*, *Listeria*, *Campylobacter*, and *Staphylococcus aureus*.[3]

The second commercially available nucleic acid hybridization kit is the AccuProbe® distributed by Gene-Probe, Inc. (Salem, MA). Similar to GENE-TRAK, it also targets rRNA sequences, but it does not involve immobilization. In the AccuProbe assay, chemioluminescently labeled probes are subjected to differential hydrolysis after mixing with target DNA, and the protected double-stranded hybrids can be visualized by light emission.[3] Accuprobe protocols have been developed for the detection of *Listeria*, *Campylobacter*, and *Staphylococcus aureus*.[3,43] A number of researchers have compared the specificity and sensitivity of Accuprobe and

GENE-TRAK assays for *Campylobacter* and *Listeria*, and for the most part the performance of both protocols is quite similar.[44,45] Both assays have demonstrated very high sensitivity, but it is limited to approximately 10^6 CFU per assay.[3,39] Because of this limitation, these commercial probe kits are mostly well used for a rapid identification of isolates, rather than for rapid detection.

Another nucleic acid method that is not normally classified among the MP-based methods is ribotyping.[33] Ribotyping is the first rapid automated DNA-based fingerprinting technique developed by Qualicon, Inc. (Wilmington, DE) and commercialized as the RiboPrinter® microbial characterization system. Ribotyping has been quite useful in the food industry to quickly identify isolates. The RiboPrinter is capable of performing DNA restriction digestion, electrophoresis, and membrane hybridization automatically. The core of ribotyping is the use of specific probes designed against ribosomal RNA genes, which hybridize to the DNA digestion fragments after these have been separated in a gel. The DNA fragment bands can be visualized by means of a chemiluminescent agent. Despite the automation capabilities, ribotyping has not been capable of replacing other more laborious DNA fingerprinting methods used in epidemiology, such as pulsed field gel electrophoresis because of the superior strain discriminatory power of those methods.

18.6.3 POLYMERASE CHAIN REACTION

18.6.3.1 PCR Principles

The PCR is an *in vitro* DNA amplification method that revolutionized every life sciences field and has become a routine procedure in most biology laboratories. PCR was invented by Kary Mullins in 1983 by taking advantage of the unique properties of DNA replication catalyzed by DNA polymerase.[46] The development of PCR was due to the ability of DNA probes to bind target DNA serving as primers for the enzymatic replication of DNA. In contrast to the MP technology described above, PCR requires the use of pairs of primers specific for flanking regions of each of the two DNA strands in the target DNA.

The steps involved in PCR are the following: (1) denaturation, (2) primer hybridization (annealing), and (3) extension (replication).[47] The separation of DNA strands (denaturation) is achieved by heating at temperatures greater than 94°C for a few minutes, and the annealing of primers to complementary sequences in the target DNA occurs after cooling the reaction to less than 65°C. Once the primers bind, DNA replication catalyzed by a DNA polymerase proceeds in the 5′ to 3′ direction for a few minutes. After the extension step, the cycle is repeated as many times as needed (typically from 15 to 40 cycles). Theoretically, each cycle duplicates the target DNA once, and the number of molecules increases exponentially with additional cycles.

The first PCR protocols utilized DNA polymerase that had been purified from *E. coli*, but fresh enzyme had to be added to the reaction after each cycle because of its inactivation at denaturation temperatures.[46] The discovery that allowed the development of current PCR formats was the use of thermostable polymerases. The first of this type of enzymes was obtained from *Thermus aquaticus* bacteria and is known as *Taq* polymerase. A wide variety of other thermostable polymerases have been successfully marketed. The development of thermocycler devices also allowed the rapid application of PCR to almost any imaginable field.[47]

18.6.3.2 Traditional PCR

PCR was first enthusiastically adopted for research, and an extensive number of articles have reported the utilization of PCR for detection of food-borne pathogens.[35,48] Once the reaction conditions are optimized, PCR protocols can easily be adapted as routine tests. Traditional PCR is typically divided into three major stages: (1) DNA amplification stage, (2) separation of PCR products, and (3) detection of products. The amplification stage involves all the aforementioned reaction steps performed in a thermocycler. Separation of produced DNA fragments is typically conducted by agarose gel electrophoresis, and the detection of PCR products is achieved by visualization of bands onto gels under UV light after staining with ethidium bromide.

Although traditional PCR has been a very powerful technique for research, its adoption for large-scale industrial analysis of foods has been hampered by its relatively high sophistication and its labor-intensiveness. Suppliers of equipment and PCR reagents designed ready-to-use reaction mixtures such as those commercialized with the TaqMan® brand by Applied Biosystems, Inc. (Foster City, CA). Commercial formats have also been able to simplify traditional PCR by automated electrophoresis and the use of hybridization probes.[43] The application of PCR for the rapid microbiological analysis of foods has also been limited because of interferences caused by food components, which can lead to weak responses and false negatives.[3] A wide variety of substances can affect annealing; proteins and fats may inhibit polymerization; and proteinases that can inactivate polymerases are frequently found in food samples.[49] In addition to these shortcomings, traditional PCR cannot discriminate between dead and live bacteria.

Despite the usefulness of PCR in microbiological analysis in research, commercial assays did not become available until the late 1990s.[50] The BAX® System from Qualicon, Inc. (Wilmington, DE), the Probelia® system from Sanofi Diagnostics Pasteur (Paris, France), and the TaqMan were the first PCR assays marketed for the microbiological analysis of foods.[43] The first BAX assays were designed for the detection of *Salmonella, Listeria, L. monocytogenes,* and *E. coli* O157:H7 and were based on a semiautomated system that included a thermocycler and prestained gels. Probelia formats have been available for the detection of *Salmonella, Campylobacter, L. monocytogenes,* and *E. coli* O157:H7, and they relied on a semi-automated thermocycler in which the detection was done by spectrophotometric detection of hybridization of PCR products with reporter probes. Initial TaqMan assays included protocols for *Salmonella* and *E. coli* O157:H7, but the detection of PCR products was achieved by a fluorescence system. TaqMan took advantage of the $5' \rightarrow 3'$ exonuclease activity of *Taq* polymerase by including specific DNA probes that bind the target DNA downstream of the replication direction. These probes have a set of fluorescent reporter and quencher dyes attached, and the fluorescent reporter dye is released as a result of the probe hydrolysis. As the PCR reaction proceeds, the amount of free reporter dye increases and the fluorescence increases.

18.6.3.3   Real-Time PCR

The term "real-time PCR" has been applied to those PCR protocols in which the detection of the amplification products is achieved as the products are synthesized. The first online measurements of PCR products were conducted using ethidium bromide and UV light with the purpose of measuring PCR reaction kinetics in the early 1990s.[43] Since that time several technologies have been developed, and now real-time PCR has become a very powerful and popular technique for the detection and identification of bacteria in foods. The core of real-time PCR resides in the use of a fluorescent dye system included in the reaction and a thermocycler equipped with fluorescence-detection capability. Among the six existing fluorescent systems, the SYBR Green I dye is the only format that uses a nonspecific DNA labeling compound. The other five detection protocols use DNA probes and are based on the principle of fluorescence resonance energy transfer (FRET).[43]

The SYBR Green I dye is capable of binding the minor groove of double-stranded DNA molecules and only fluoresces when attached to DNA. As DNA is synthesized, the fluorescent signal increases with SYBR Green I. The FRET phenomenon is defined as the ability of certain chemical compounds or moieties to absorb and/ or emit the energy produced by an excited fluorophore when the distance between these compounds is approximately 70–100 Å.[43] In most FRET-based real-time PCR systems, the quencher and the fluorophore are typically incorporated into one primer or probe that does not fluoresce alone, but when the reaction takes place and the target gene is amplified, both moieties are separated and fluorescence is observed.

To date, there are at least five different FRET systems: TaqMan, molecular beacons, Scorpions probes, Amplifluor® probes (Talron Scientific & Medical Products Ltd, Rehovot, Israel), and LightCycler® hybridization probes (Róche Diagnostics, Corp., Basel, Switzerland) (Figure 18.2).[51] As described above, the TaqMan system

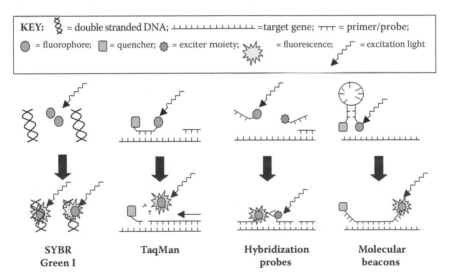

SYBR Green I   **TaqMan**   **Hybridization probes**   **Molecular beacons**

FIGURE 18.2   Comparison of four real-time PCR reporter systems: SYBR Green I, TaqMan, hybridization probes, and molecular beacons.

separates the fluorescent dye from the quencher by nuclease hydrolysis. Molecular beacons are DNA oligonucleotides designed with complementary sequences at their ends such that they form a hairpin structure with the quencher and the fluorophore in close proximity. When the noncomplementary part of the probe binds to the target DNA, the FRET pair is separated and signal is emitted. Similar to the molecular beacons, Scorpions and Amplifluor probes employ hairpin-forming nucleotides. The LightCycler hybridization probes are the only FRET format in which fluorescence is produced when two dyes present in separate probes get in close proximity after specifically binding to the target DNA (Figure 18.2).

The commercially available real-time PCR systems for the detection of food-borne pathogens are still only TaqMan, BAX, and Probelia brands.[43] TaqMan utilizes its own proprietary probe-hydrolysis technology while BAX assay now utilizes the SYBR Green I dye to obtain real-time measurements. The Probelia format licensed to Bio-Rad Laboratories (Hercules, CA) has now been modified to include molecular beacons for detection. All of these real-time PCR systems are marketed as a semiautomated system in which enrichment samples are first lyzed and then loaded onto a fluorescence-detecting thermocycler that is connected to a personal computer recording fluorescence after every cycle. By using capillary PCR reactors, such as the LightCycler, the reaction time can be reduced by half. Despite the fact that total PCR reaction times can be as low as 20 minutes, all available formats still depend on a previous enrichment culture step to detect food-borne pathogens in foods.

The major advantages of PCR are its extreme sensitivity and specificity, and real-time PCR dramatically reduces analysis time. Several researchers have demonstrated that PCR can detect as little as one cell in the reaction mixture, but the limitation for most PCR protocols resides in the ability of the pre-PCR steps to concentrate enough cells. To achieve bacterial concentration, proprietary enrichment media and DNA extraction with resins are often needed.[43] Since lyzed culture enrichment extracts of approximately 10 mL are often added to PCR reactions, the theoretical detection limit of PCR is roughly 100 cells per mL. Future PCR formats would likely involve automated concentration steps that would be able to reduce this detection limit.

18.7   FUTURE OF RAPID FOOD MICROBIAL TECHNIQUES

The development of rapid microbial methods is one of the most active areas in the field of food microbiology. The quick determination of the presence of pathogenic bacteria that have been ruled as adulterants, such as *E. coli* O157:H7 and *Listeria monocytogenes*, continues to be the major driver in the development of detection methods. Because of the need for faster detection methods, the potential demand for new rapid techniques will likely continue to be very strong. New assay formats will continue to be developed, thanks to the advancement in fields such as molecular biology, immunology, nanotechnology, microarray technology, and bioelectronics. Some of the technologies that will soon be commercialized for the detection and identification of food-borne pathogens are biosensors, DNA microchips, microarrays, incorporation of effective separation and concentration steps, and in-packages alert systems.[3,4] While the "ultimate" rapid microbiological techniques (as defined in the section "Introduction") are being developed, the food industry will continue to use traditional and current rapid microbial detection methods in the near future.

ACKNOWLEDGMENTS

The authors would like to thank Drs. James Smith (USDA/ARS) and Purnendu C. Vasavada (University of Wisconsin-River Falls) for critically reviewing this chapter.

REFERENCES

1. Johnson, H. H. and Newsom, S. W. B., *Rapid Methods and Automation in Microbiology,* Learned Information, Ltd., Oxford, 1976.
2. Weaver, R. H., Quicker bacteriological results, *Am. J. Med. Technol.* 20, 14–26, 1954.
3. Feng, P., Development and impact of rapid methods for detection of foodborne pathogens, in *Food Microbiology: Fundamentals and Frontiers*, Doyle, M. P., Beuchat, L. R., and Montville, T. J. ASM Press, Washington, D.C., 2001.
4. Fung, D. Y. C., Predictions for rapid methods and automation in food microbiology, *J. AOAC Intl.* 85, 1000–1002, 2002.
5. Fung, D. Y. C., Cox, N. A., and Bailey, J. S., Rapid methods and automation in the microbiological examination of foods, *Dairy Food Sanit.* 6, 292–296, 1988.
6. Chain, V. S. and Fung, D. Y. C., Comparison of redigel, petrifilm, spiral plate system, isogrid, and aerobic plate count for determining the numbers of aerobic bacteria in selected foods, *J. Food Protect.* 54 (3), 208–211, 1991.
7. Ward, D. R., LaRocco, K. A., and Hopson, D. J., Adenosine triphosphate bioluminescent assay to enumerate bacterial numbers on fresh fish, *J. Food Protect.* 49, 647–650, 1986.
8. Swaminathan, B., Beebe, J., and Besser, J., Investigation of foodborne and waterborne disease outbreaks, in *Manual of Clinical Microbiology*, Murray, P. R. ASM Press, Washington, D.C., 1999.
9. Mossel, D. A. A., Marengo, C. M. L., and Struijk, C. B., History of and prospects for rapid and instrumental methodology for the microbiological examination of foods, in *Rapid Analysis Techniques in Food Microbiology*, Patel, P. Chapman and Hall, London, 1994.
10. Ibekwe, A. M., Watt, P. M., Grieve, C. M., Sharma, V. K., and Lyons, S. R., Multiplex fluorogenic real-time PCR for detection and quantification of *Escherichia coli* O157:H7 in dairy wastewater wetlands, *Appl. Environ. Microbiol.* 68, 4853–4862, 2002.
11. Smith, T. J., O'Connor, L., Glennon, M., and Maher, M., Molecular diagnostics in food safety: rapid detection of food-borne pathogens, *Irish J. Agric. Food Res.* 39, 309–319, 2000.
12. Coates, S., Rapid Test Kits, AOAC International, 2002.
13. Feng, P., Rapid methods for detecting food-borne pathogens. Appendix 1, 8th edn. Center for Food Safety and Applied Nutrition, FDA, Silver Spring, MD, 2001.
14. Manninen, M. T. and D. Y. C., F., Estimation of microbial numbers, from pure bacterial cultures and from minced beef samples by reflectance colorimetry with Omnispec 4000, *J. Rapid Methods. Autom. Microbiol.* 1, 41–44, 1992.
15. Manninen, M. T., Fung, D. Y. C., and Hart, R. A., Special system and laser colony scanner of enumeration of microorganisms, *J. Food Safety* 11, 177–181, 1991.
16. Garcia-Armesto, M. R., Otero, A., Rua, J., Moreno, B., and Garcia-Lopez, M. L., Evaluation of the spiral plating system for the routine assessment of indicator microorganisms in raw ewe's milk, *J. Food Prot.* 65, 1281–1286, 2002.

17. Sharpe, A. N. and Michaud, G. L., Enumeration of high numbers of bacteria using hydrophobic grid-membrane filters, *Appl. Microbiol.* 30, 519–524, 1975.
18. Entis, P., Brodsky, M. H., Sharpe, A. N., and Jarvis, G. A., Rapid detection of Salmonella spp. in food by use of the ISO-GRID hydrophobic grid membrane filter, *Appl. Environ. Microbiol.* 43, 261–268, 1982.
19. Chain, V. S. and Fung, D. Y. C., Comparison of Redigel, Petrifilm, Spiral plate system, ISOGRID and standard plate count for the aerobic count on selected foods, *J. Food Prot.* 54, 208–212, 1991.
20. Beuchat, L. R., Copland, E., Curlu-W., N. L. S., Danisavich, T., Gangap, V., King, B. W., Lawlis, T. L., Lekin, R. O., Okwusoa, J., Snutw, C. F., and Townsend, D. M., Comparison of the SimPlate total plate count method with Petrifilm, Redigel, and conventional pour-plate methods for enumerating aerobic microorganisms in foods, *J. Food Prot.* 61, 14–18, 1998.
21. Stanker, L. H. and Beier, R. C., Introduction to immunoassays for residue analysis, in *Immunoassays for Residue Analysis—Food Safety*, Stanker, L. H. and Beier, R. C. American Chemical Society, Washington, D.C., 1996.
22. Coons, A. H., Creech, H. J., Jones, R. N., and Berliner, E., The demonstration of pneumococcal antigens in tissues by the use of fluorescent antibody, *J. Immunol.* 45, 159, 1942.
23. Van Weemen, B. K. and Schuurs, A. H. W. M., Immunoassay using antigen–enzyme conjugates, *FEBS Lett.* 15, 232–236, 1971.
24. Kohler, G. and Milstein, C., Continuous culture of fused cells secreting antibody of predefined specificity, *Nature* 256, 495–497, 1975.
25. Price, C. P. and Newman, D. J., Light-scattering immunoassays, in *Principles and Practice of Immunoassays*, Price, C. P. and Newman, D. J. MacMillan Reference, Ltd., London, 1997.
26. Feng, P., Impact of molecular biology on the detection of foodborne pathogens, *Mol. Biotech.* 7, 267–278, 1997.
27. Oggel, J. J., Nundy, D. V., and Randall, C. J., Modified 1–2 Test™ system as a rapid screening method for detection of *Salmonella* in foods and feeds, *J. Food Prot.* 53, 656–658, 1990.
28. Warburton, D. W., Feldsine, P. T., and Falbo-Nelson, M. T., Modified immunodiffusion method for detection of *Salmonella* in raw flesh and highly contaminated foods: collaborative study, *J. AOAC Intl.* 78, 59–68, 1995.
29. Crowther, J. R., *ELISA Theory and Practice* Humana Press, Totowa, NJ, 1995.
30. Candisch, A. A. G., Immunological methods in food microbiology, *Food Microbiol.* 8, 1–14, 1991.
31. MacKenzie, A. M. R., Lebel, P., Orrbine, E., Rowe, P. C., Hyde, L., Chan, F., Johnson, W., and McLaine, P. N., Sensitivities and specificities of premier *E. coli* O157 and premier EHEC enzyme immunoassays for diagnosis of infection with verotoxin (Shiga-like toxin)-producing *Escherichia coli*, *J. Clin. Microbiol.* 36, 1608–1611, 1998.
32. Bird, C. B., Miller, R. L., and Miller, B. M., Reveal for *Salmonella* test system, *J. AOAC Intl.*, 625–633, 1999.
33. Fung, D. Y. C., Rapid methods and automation in microbiology: A review, *Irish J. Agric. Food Res.* 39, 301–307, 2000.
34. Brunelle, S., Electroimmunoassay technology for food-borne-pathogen detection, in *IVD Technology Online* 2001.
35. Strockbine, N. and Olsvik, O., PCR detection of heat-stable, heat labile, and shiga-like genes in *Escherichia coli*, in *Diagnostic Molecular Microbiology Principles and Applications*, Persing, D. H., Smith, T. F., Tenover, F. C., and White, T. J. ASM Press, Washington, D.C., 1993.

36. Todd, E. C. D., Szabo, R. A., MacKenzie, J. M., Martin, A., Rahn, K., Gyles, C., Gao, A., Alves, D., and Yee, A. J., Application of a DNA hybridization-hydrophobic-grid membrane filter method for detection and isolation of verotoxigenic *Escherichia coli*, *Appl. Environ. Microbiol.* 65, 4775–4780, 1999.

37. Koo, K. and Jaykus, L. A., Detection of *Listeria monocytogenes* from a model food by fluorescence resonance energy transfer-based PCR with an asymmetric fluorogenic probe set, *Appl. Environ. Microbiol.* 69, 1082–1088, 2003.

38. Manzano, M., Cocolin, L., Astori, G., Pipan, C., Botta, G. A., Cantoni, C., and Comi, G., Development of a PRC microplate-capture hybridization method for simple, fast and sensitive detection of Salmonella serovars in food, *Mol. Cell. Probes* 12, 227–234, 1998.

39. Mazola, M. A., Detection of microorganisms in foods using DNA probes targeted to ribosomal RNA sequences, *Food Biotechnol.* 14, 173–194, 2000.

40. Olsen, J. E., Aabo, S., Hill, W., Notermans, S., Wernars, K., Granum, P. E., Popovic, T., Rasmussen, H. N., and Olsvik, Ø. Probes and polymerase chain reaction for detection of food-borne bacterial pathogens, *Intl. J. Food Microbiol.* 28, 1–78, 1995.

41. Fitts, R., Diamond, M., Hamilton, C., and Neri, M., DNA-DNA hybridization assay for detection of *Salmonella* spp. in foods., *Appl. Environ. Microbiol.* 46, 1146–1151, 1983.

42. Sambrook, J., Fritsch, E. F., and Maniatis, T., *Molecular Cloning: A Laboratory Manual*, 2nd. ed. Cold Spring Harbor Laboratory Press, Cold Spring Harbor Laboratory, 1989.

43. Rijpens, N. P. and Herman Lieve, M. F., Molecular methods for identification and detection of bacterial food pathogens, *J. AOAC Intl.* 85, 984–995, 2002.

44. Ransom, G. M., Dreesen, D. W., Rose, B. E., and Lattuada, C. P., Assessment of three nucleic acid hybridization systems for detection of *Campylobacter spp.* in poultry products, *J. Food Prot.* 57, 703–709, 1994.

45. Beumer, R. R., Te Giffel, M. C., Kok, M. T. C., and Rombouts, F. M., Confirmation and identification of *Listeria spp*, *Lett. Appl. Microbiol.* 22, 448–452, 1996.

46. Mullis, K. B. and Faloona, F. A., Specific synthesis of DNA in vitro via a polymerase-catalyzed chain reaction, *Meth. in Enzymol.* 155 (21), 335–350, 1987.

47. Newton, C. R. and Graham, A., *PCR*, 2nd. ed. Springer, New York, 1997.

48. Bassler, H. A., Flood, S. J. A., Livak, K. J., Marmaro, J., Knorr, R., and Batt, C. A., Use of a fluorogenic probe in a PCR-based assay for the detection of *Listeria monocytogenes*, *Appl. Environ. Microbiol.* 61 (10), 3724–3728, 1995.

49. Rossen, L., Norskov, P., Holmstrom, K., and Rasmussen, O. F., Inhibition of PCR by components of food samples, and DNA-extraction solutions, *Intl. J. Food Microbiol.* 17, 37–45, 1992.

50. Fach, P., Dilasser, F., Grout, J., and Tache, J., Evaluation of a polymerase chain reaction-based test for detecting Salmonella spp. in food samples: *Probelia Salmonella spp*, *J. Food Prot.* 62, 1387–1393, 1999.

51. Pfaffl, M. W., Fluorescence detection chemistry in kinetic RT-PCR, Munchen Technical University, 2003.

19 Analysis of Natural Toxins in Foods

Tõnu Püssa

CONTENTS

19.1 INTRODUCTION

Food is a very complex system consisting of an endless number of high and low molecular substances, mostly of natural origin. A majority of these substances are indispensable for the normal functioning of the human body. On the other hand, food also contains substances, both natural and synthetic (anthropogenic), capable of evoking smaller or bigger health disorders; that is, food can contain toxic substances or toxins. In principle, according to the famous Paracelsus, every substance, even a normal constituent of food may become toxic, if its concentration exceeds the definite lowest observed adverse effect level (LOAEL) during short time (acute toxicity) or is consumed for a longer period on a somewhat lower level (chronic toxicity). To guarantee the safety of food, it is necessary to control it. Since a majority of poisonous substances are present in food in very low concentrations (usually nanograms or micrograms per gram), this control mostly needs very sensitive, selective, and precise methods of analysis. Usually, it is like searching for a needle in the haystack. During last decades, fast development of those analysis methods has enabled to discover and quantify ultralow concentrations of toxic substances that earlier nobody even supposed to be present in food. Determination of tiny amounts of toxins has little sense without comparing the results with the toxicological characteristics of these substances such as LOAEL or acute lethal doses. Since often we have to do with human acute intoxications, the possible short analysis time is also an important parameter in selecting the most appropriate method(s).

Since the number of the analysis methods of various food toxins is enormous, yet growing, it is impossible to give direct references to all of them in such a short review. In this chapter, we try to refer mainly to the recent review articles and to the most important specific papers on the analysis methods. Although attention is focused on the recent publications describing contemporary methods, the simpler and somewhat "old-fashioned methods" are also mentioned to provide a more complete picture. Moreover, the somewhat older methods are often used as reference methods.

A natural food toxin can be defined as a substance synthesized either by a plant species, an animal, or microorganisms and that is harmful to the organism consuming this food, taking into account the real content of this substance in that particular food and its everyday consumption. The ability to detect the toxins is imperative to ensure the safety and quality of our food supplies.

Chemically, natural toxins are a very broad and versatile group of substances. These include compounds, both low and high molecular compounds, that are natural products of the life of plants, animals, or microorganisms and that can be included in our diet in amounts harmful for health and life. A majority of the natural toxins are low molecular substances, although the most dangerous toxins such as the botulinum toxin are bacterial proteins. Undoubtedly, microorganisms are the primary cause of food spoilage and food-borne illnesses.

Natural food toxins can be classified based on their chemical structure (hence, physical, chemical, and biochemical properties), the natural source, the mechanisms of their toxic effect, and so forth. This chapter shortly describes the most important groups of analysis methods and presents the respective review articles related to analysis of food. This is followed by reviews of the methods of analysis of toxins based on their natural source and most essential structural feature.

Based on their natural source, the toxins can be classified as follows:

1. Plant toxins
2. Toxins of microorganisms
3. Animal toxins including seafood toxins

In fact, the situation is more complicated. For example, groups 2 and 3 are often mixed and the division is artificial. Many poisonous compounds that historically belong to the group of seafood toxins are actually produced by unicellular microorganisms, such as algae or bacteria, ingested from sea by higher organisms, such as fish or shellfish, that thereafter are consumed as food by humans or other higher animals. Officially, these compounds have so far been considered as fish or shellfish toxins.

ABBREVIATIONS

APCI—atmospheric pressure chemical ionization
C(Z)E—capillary (zone) electrophoresis
ELISA—enzyme-linked immunosorbent assay
ESI—electrospray ionization
GC-MS—gas chromatography-mass spectrometry
HPLC-DAD—high-performance liquid chromatography with diode array ultraviolet detection
HPLC-FLD—high-performance liquid chromatography with fluorimetric detection
HPLC-PDA—high-performance liquid chromatography with photodiode array detection
HPTLC—high-performance thin layer chromatography
IAC—immunoaffinity clean-up columns
LC-MS—liquid chromatography-mass spectrometry
LC-MS/MS—liquid chromatography-tandem mass spectrometry
LC-TOF-MS—liquid chromatography-time of flight mass spectrometry
LOD—limit of detection
MALDI-TOF-MS—matrix-assisted laser desorption/ionization time of flight mass spectrometry
RIA—radioimmunoassay
UHPLC—ultrahigh-performance liquid chromatography
SDS-PAGE—sodium dodecyl sulfate-polyacrylamide gel electrophoresis
SPE—solid-phase extraction
TLC—thin layer chromatography

19.2   SAMPLE PREPARATION

Sample preparation is the first step of any analysis scheme and depends on the analyte(s) of interest and on the analytical method to be used. Sample preparation starts from taking a sample or sampling. The sampling scheme should guarantee the representativity of the sample. It is easy to obtain the representative sample of a homogenous liquid such as milk or water if there is no sediment or solid material or the analyte is distributed more or less evenly as in a definite part of a plant. In this case, the amount of the initial sample is, first of all, determined by the sensitivity of the analytical method. The situation is quite different, for example, in case of fungal growth and mycotoxin production on crop, where the so-called "spot processes" are affected by many conditions. As a result, the distribution of mycotoxins in a lot of a crop can be extremely heterogenous. A single peanut kernel contaminated by aflatoxin in a 10-kg sample is sufficient to cause the aflatoxin content of the whole lot to be above the European regulation limits allowed for peanuts (Blanc 2006). Thus, correct sampling is the most crucial step for obtaining reliable results of mycotoxin content. In these cases, it is therefore necessary to take a large number of small incremental samples at various places distributed throughout the batch to obtain a representative sample. It is recommended to take at least eight subsamples, picked at random from the same batch. If this number of packs is not available, then as many packs as that will make up 1 kg must be taken. For foods such as spices and crisps, it may not be possible to obtain the 500 g necessary for sample preparation. In these cases, at least the minimum amount required by the analytical laboratory must be taken. This minimum depends on the nature of the sample. In general, at least 50 g of sample is required for each replicate, that is, at least 150 g for three samples (Food Standards Agency 2007). As a general outcome, the best way to reduce the overall variability is to increase the laboratory sample size, the degree of sample comminution, the subsample size, and the number of aliquots quantified (Köppen et al. 2010; Whitaker 2006).

Sample preparation should also guarantee the preservation of the initial amount of a toxin (no loss of the toxin by sorption or chemical reactions and no addition of the toxin by contamination) and should usually achieve a significant purification and concentration of the target substance(s). If the sample is not to be analyzed immediately, then deep refrigeration of the sample until analysis, without exposure to oxygen and light, and appropriate stabilization are necessary.

The toxins are determined always from a solution called the extract of the matrix. It means that the analytes must be dissolved in an appropriate solvent. The extraction method is very important—it is possible to select solvent extraction for dissolving compounds of interest, pressurized fluid extraction, supercritical fluid extraction with carbon dioxide, and microwave- or ultrasound-assisted extraction. Elevated temperatures should be avoided. SPE is performed on suitable sorbent (silica or polymer) cartridges in which target analytes are held, part of contaminants are washed out, and the analytes are solubilized and eluted usually in significantly higher concentrations and in a highly purified state (due to the concentrating effect of this procedure). Deproteinization and delipidation of a sample is necessary for chromatography, but it should be performed taking into consideration the fact that various low molecular compounds, including toxins, may form complexes with proteins or lipids.

A special, efficient type of the SPE is IAC. A polyclonal or monoclonal antibody produced against the analyte is immobilized on a gel packed into a small plastic column. The column is initially conditioned with phosphate-buffered saline (PBS) and then a crude sample extract, usually in an aqueous solution, is applied to the column. The use of organic solvents should be avoided. During the application of the sample extract, the analyte binds to the antibody and thereby to the IAC gel. After loading the extract onto the IAC, the gel is washed with PBS to completely remove any coextractives. Finally, the analyte is eluted from the IAC by breaking the antibody–antigen bond. For small molecules, this can be achieved with a small volume of methanol or acetonitrile. Alternatively, the analyte might be eluted with buffers that are less damaging to the antibody. Acidic conditions are required for the elution of macromolecules. IAC clean-up techniques in food analysis have been reviewed by Şenyuva and Gilbert (2010). The quantitative analysis of biomarkers, drugs, and toxins in biological samples using IAC coupled to MS or tandem MS has been reviewed by Tsikas (2010).

19.3 ANALYTICAL METHODS

19.3.1 Spectrophotometric (Colorimetric) Methods

Spectrophotometric methods (UV–Vis, fluorescence) are relatively simple and do not need any expensive equipment. They can be used either when high sensitivity and selectivity are not critical or when the spectral method itself is selective and sensitive. However, in the case of food toxins, this group of methods is used mainly in conjunction with immunochemical methods (ELISA, etc.). Colorimetric methods are also used, for example, for quantification of cyanogenic glycosides (CG) in plants (Brito et al. 2009).

19.3.2 Chromatographic Methods: Detectors

In most cases, column chromatographic methods with different modes of detection of the analytes that have passed the column are the methods of choice in qualitative and quantitative analysis of food toxins for food safety testing and for confirmation. GC-MS is widely used for detecting various toxins in food, but LC-MS of different modes is developing especially in a fast pace. LC-MS is particularly suited for the analysis of food contaminants, since it provides a large amount of information about a complex mixture, enabling the screening, confirmation, and quantitation of hundreds of components in a single analysis. Triple quadrupole (QqQ) mass spectrometry has been the cornerstone technique for screening and confirmation of food contaminants and residues. The majority of current LC-MS/MS-based contaminants and residue analysis relies on the high sensitivity and selectivity of the selected reaction monitoring mode of QqQ-MS/MS. LC-TOF-MS has also been established as a valuable technique for the routine control of the wholesomeness of food. In this sense, TOF techniques can record an accurate full-scan spectrum throughout the acquisition range and have resulted in an excellent tool for the unequivocal target and nontarget identification and confirmation of food microcomponents. Recently introduced tandem mass spectrometers, having both target and nontarget identification features, such as quadrupole linear ion trap (QqLIT, LTQ, or Q-trap), quadrupole time-of-flight (QqTOF), LTQ-Fourier transform ion cyclotron resonance mass spectrometry (FTICR-MS), LTQ-Orbitrap, and so forth, have allowed the development of several new methods for detection of toxins.

A very recent review of the use of various techniques of LC-MS for the purpose of food safety, including analysis of various natural toxins, has been published by Malik et al. (2010).

19.3.3 Electrophoretic Methods

There is an increasingly growing interest in the application of capillary electrophoresis (CE) to food research, reflected by the high number of articles and reviews published on this topic. CE can be coupled to MS, to electrochromatography, to chemiluminescence detection, to microemulsion EKC (MEEKC), and to sample treatment prior to CE-MS. An exhaustive and critical review up to 2007 of the analysis of natural toxins by capillary electromigration techniques was provided by Garcia-Cañas and Cifuentes (2007). Subsequently, the analysis of natural toxins by CE techniques was updated in a second review by Garcia-Cañas and Cifuentes (2008). A recent review focused on the different applications of particular CE techniques in food analysis has been published by Vallejo-Cordoba and Gonzalez-Cordova (2010).

SDS-PAGE is an appropriate method for determination and semiquantitation of proteinous toxins in food.

19.3.4 Immunochemical Methods: Immunosensors

A review dedicated to the recent developments in immunochemical methods for the detection of mycotoxins, with a particular emphasis on simultaneous multiple analyte determination, has been published by Goryacheva et al. (2007). This includes a high-throughput instrumental analysis for the laboratory environment (microtiter plate ELISA, different types of immunosensors, fluorescence polarization immunoassay, and CE immunoassay), as well as rapid visual tests for on-site testing (lateral-flow, dipstick, flow-through, and column tests). For each type of immunoassay, perspectives for multiple analyte application are discussed. Various immunoassays and immunosensors of mycotoxins have been reviewed by Maragos and Busman (2010) also. Most ELISA techniques for the detection and measurement of pyrrolizidine alkaloids (PA) based on antibodies have been applied against retrorsine. They tend to have mixed cross-reactivity to other PAs and can be varied to provide total PA screening or more specific targeted methods (Crews et al. 2010). Immunology-based methods for detection of pathogens including *Clostridium botulinum* have been recently reviewed by Velusamy et al. (2010).

Immunoassays along with recombinant technologies have provided new insights into the detection of toxins. A review summarizing the methods used to produce recombinant antibodies and to tailor their properties, together with the examples focused on food quality and safety analysis, has been very recently published by Vehniäinen et al. (2011).

19.3.5   Biological Methods: Biosensors

The need for quick and cheap detection of the natural toxin levels in the food and environment has led to the search for systems alternative to the currently employed analytical methods such as spectrophotometry, GC-MS, thin layer chromatography (TLC), or LC-MS. Biosensors used as sensitive, simple, rapid, cost-effective, and portable detectors are ideal for food monitoring. They are defined by the International Union of Pure and Applied Chemistry as integrated receptor–transducer devices, which are able to provide selective quantitative or semiquantitative analytical information using a biological recognition element (RE). Biosensors combine a RE with a suitable signal transduction method so that the binding or reaction between the target and the RE is translated into a meaningful signal. However, most biosensors still have a few drawbacks. Mostly, they allow the detection of just one analyte, although recently some multianalyte sensors have been developed (Taitt et al. 2008).

Biosensors can be classified according to the type of RE as enzymatic, whole cell, and affinity-based devices. Enzymatic biosensors measure the selective inhibition of the activity of enzymes by a specific target. These biosensors for the detection of contaminants in food have been described in several reviews (Amine et al. 2006; Manco et al. 2009). Another frequently used RE, especially for the monitoring of environmental pollutants, is whole cells such as bacteria, fungi, yeast, and animal or plant cells. These whole cell biosensors detect the responses of cells after exposure to a sample, which are related to its toxicity. The third group of RE is the affinity-based RE that specifically binds to individual targets or groups of structurally related targets. Affinity-based sensors are very sensitive, selective, and versatile since affinity-based RE can be generated for a wide range of targets. Antibodies have long been the most popular affinity-based REs. An overview of affinity biosensors in food analysis has been published by Patel (2006). A wide variety of antibody biosensors reported for different food and environmental applications exists, which are summarized and discussed by Van Dorst et al. (2010).

Various biosensors for detection of botulinum toxin and other bacterial toxins have been reviewed by Velusamy et al. (2010) and for marine toxins by Campas et al. (2007).

19.3.6   Mouse Bioassay: Functional Analysis

The mouse bioassay (MBA) that was developed in 1937 to check toxicity in the extracts of mussels (Sommer and Meyer 1937) is still in use for the analysis of seafood toxins, botulinum toxins, etc. The assay provides an indication about the overall toxicity of the sample, but it is not able to differentiate among individual toxins. In most countries, MBA is the reference method for marine toxins, with the exception of domoic acid for which the reference method is HPLC. If two of three mice die within a defined time frame, the test is positive, but if only one dies, the test is negative. The methods are fast and sensitive but nonspecific and also demand the sacrificing of a large number of animals. Depending on the solubility of the compounds, there are two types of bioassays: with hydrophilic solvents, for paralytic toxins (saxitoxin and analogs) and with lipophilic solvents, for most of the other toxins. Various versions of the MBA have been described by Botana et al. (2009). MBA is considered to be an old-fashioned, cruel method and is prone to errors. Therefore, alternative *in vitro* methods classified as functional, immunological, and analytical assays have been proposed (Campbell 2010; Botana et al. 2009).

Functional assay is a detection method that, with the purpose of quantification, uses the mechanism of action of the toxin molecule. Either receptor of the toxin group (functional assay) or other biochemical binders (e.g., antibodies, nucleic acids, or amino-acid sequences) (biochemical assay) can be used as a basis for the assay.

The main advantage of using receptors is that they show a response that is proportional to the toxic potency of the group, hence mimicking the nature of the bioassay; therefore, a functional assay would be synonymous with an activity assay. Biochemical binders do not show a correlation between toxicity and binding. Functional and biochemical assays provide similar results only if the cross-reactivity matches exactly the toxic equivalent factor (TEF), but, generally, this is not common. Overall, receptor-based functional assays are the best option. If the receptor is difficult to isolate, as for a sodium channel, then it is possible to obtain similar results with cell lines rich in the receptor. In general, functional assays have a LOD in the microgram range, which is several times lower than the LOD of bioassays (160 μg/kg), although a functional assay can work in the nanogram range in some cases (Botana et al. 2009).

TEF or relative toxicity factor expresses the toxicity of the compounds with the same mechanism of action in comparison to a reference, usually the most potent toxic congener. The toxicity of the individual congeners may vary by orders of magnitude. TEFs of toxins are calculated either by intraperitoneal injection of mice or by comparing direct effects on their receptors if the amount of a toxin is not enough to use animals. TEFs are very important for quantifying the total toxicity of a group of substances, thus providing an amount of a single reference toxin based on the toxicity of each analog relative to that reference toxin. For example, for the paralytic shellfish toxins, the reference compound is saxitoxin, and all other toxins with a similar effect should be expressed as the quantity of saxitoxin calculated from their amount and relative toxicity compared with saxitoxin. For instance, if a sample that has toxicity 1 contains 1 mg of saxitoxin, and another analog is analyzed and the calculated amount is 2 mg, but its toxicity is half that of saxitoxin, then the amount of this analog can be expressed as 1 mg of saxitoxin. Okadaic acid is the reference compound for diarrhetic shellfish toxins (Botana et al. 2009).

19.4   PLANT TOXINS

19.4.1   ALKALOIDS

Alkaloids have been isolated from the roots, seeds, leaves, or bark of members of at least 40% of plant families. The families particularly rich in alkaloids include Amaryllidaceae, Compositae, Leguminosae, Liliaceae, Papaveraceae, and Solanaceae. A recent review of the methods of analysis of various medicinally valuable alkaloids including alkaloids in food has been written by Steinmann and Ganzera (2010).

19.4.1.1 Pyrrolizidine Alkaloids

PAs are found in more than 250 plant species, and more than half of them are toxic. For humans, the most important PA-containing plants are ragwort (*Senecio*), crotalaria (*Crotalaria*), heliotrope (*Heliotropium*), coltsfoot (*Tussilago farfara*), and comfrey (*Symphytum officinale*); most toxic PAs are retronecine, petasitenine, and senecionine. Humans can come into direct contact with plants such as coltsfoot or comfrey via vegetable salads and herbal teas and with viola (*Viola* x *wittrockiana*) or bugloss (*Echium*) via honey, milk, or birds' eggs. Extremely high concentrations of PAs (30–70 µg/kg) have been detected in the honey collected in the foothills of the Swiss Alps in Europe. Comfrey is a traditional remedy for the treatment of arthritis and sprains, and coltsfoot decoctions are used against coughs. Besides having desired bioactive constituents, both herbs contain liver-toxic PA. For humans, the interval of the toxic daily doses of PAs is 0.1–10 mg/kg body weight (bw), according to the World Health Organization (WHO); 0.015 mg/kg bw is the lowest total daily dose of PAs (on the basis of comfrey alkaloids) that has produced adverse effects. Since the content of PAs in the leaves and roots of comfrey varies in the interval 450–8300 mg/kg, the actual degree of dose the comfrey tea drinkers get varies quite much (Rietjens et al. 2005).

Analysis—Comprehensive reviews of analysis methods of PAs, including sample preparation, have been published by Colegate and Gardner (2007) and, recently, by Crews et al. (2010).

Extraction—Since PAs, both the free bases and the N-oxides, are basic polar compounds, they are soluble in hydrophilic solvents such as diluted aqueous acids or methanol and their mixtures, and these solvents have been used most frequently. The extraction yield of PAs from plants with the use of dilute (0.05–1 M) aqueous sulfuric or hydrochloric acids is very good. In a comparison of 21 extraction methods applied to comfrey, Mroczek et al. (2006) reported extraction with methanol, using percolation with steeping in methanol at different temperatures. Quantitative spectrophotometric determination of the product of reaction with Erlich's reagent (4-dimethylaminobenzaldehyde) showed that a large-scale extraction with a relatively high volume of methanol (500 ml per 5 g comfrey) produced the highest yields. Altamirano et al. (2005) compared the methods of extracting PA free bases and N-oxides from comfrey powder. Simple solvent extraction with less polar solvents has been used to extract PAs from plants and other matrices, including honey, milk, and eggs. A disadvantage of this approach is that the PA N-oxides are less soluble in relatively nonpolar solvents such as dichloromethane and may be left in the aqueous phase. This means that two analyses have to be made to determine the full profile of PA free bases and N-oxides: the first of the extracted PA free bases alone and the second of the total free bases obtained by reduction of the N-oxides (Crews et al. 2010).

Water or methanol extracts usually require additional clean-up prior to chromatography to reduce the concentration of coextracted materials. SPE techniques have been based on three methods. The first method involves using a column of inert diatomaceous earth, such as ChemElut or Extrelut. This material adsorbs aqueous extracts in a solid matrix while a water-immiscible organic solvent is passed through the column to extract the PAs. The second approach is using columns filled with nonpolar phases such as octylsilane or octadecylsilane (C_8 or C_{18}); C_{18} phases gave better recoveries than C_8 phases. The third and most popular method uses polymeric strong cation exchange (SCX) SPE to isolate both free bases and N-oxides (Cao et al. 2008). SPE based on SCX resins has great advantages in that free PAs and their N-oxides can be isolated simultaneously with a high yield and free from many coextracted interferents.

Chromatography—There is a wide variety of chromatographic methods used to study PAs.

TLC with both silica and aluminum oxide adsorbents has been used to separate PAs. Unsaturated PA bases are first oxidized to the N-oxides with hydrogen peroxide, and these are reduced to dehydropyrrolizidine by heating with acetic anhydride or *o*-chloranil whereupon the pyrrole gives a violet-blue dye. Stelljes and coauthors used TLC to separate PA fractions before identification of a number of PAs by GC–MS (see reference 62 in Crews et al. 2010).

GC–MS has been used to identify many PAs, whereby soft chemical ionization techniques are preferred. As plants frequently contain several PAs of the same molecular mass, the fragmentation produced by electron-impact ionization can usually distinguish between similar compounds. GC-MS data from about 100 PAs using five different stationary phases have been tabulated by Witte et al. (1993). Unsaturated monoester PAs esterified with common acids (senecioic, angelic, and tiglic) have a protonated molecular ion [M]+ of *m/z* 237, and unsaturated monoester PAs where the acid is senecioic or angelic have a protonated molecular ion of *m/z* 237. Where the bases are saturated, the protonated molecular ion has *m/z* of 239. Ions characteristic of unsaturated PAs are those of *m/z* 93, 120, 136, 137, and 138. Macrocyclic PAs with a retronecine base typically have intense ions of *m/z* 136, 120, 119, and 93 with relatively weak molecular ions (Crews et al. 2010 and reference 71 therein).

HPLC methods have the advantage of allowing the determination of free bases and N-oxides in a single analytical run without prior reduction of the oxides. Under reversed-phase conditions, the PA free bases elute before their N-oxides under acidic mobile-phase conditions. HPLC has been used to separate the PA N-oxides independently from their free bases using a C_{18} column with methanol-aqueous buffer mobile phases. Detection is often by UV at 220 nm for unsaturated PAs. Derivatization has been used to increase the specificity and LOD of UV analysis of PAs separated by HPLC. The products can be measured by UV, nuclear magnetic resonance (NMR), or MS detection (Zhang et al. 2007).

LC-MS is the method of choice for analysis of PAs, and a large number of papers have been published in this field where positive ionization modes are usually preferred. Good recent examples of LC-MS determination of PAs have been published by Altamirano et al. (2005), Liu et al. (2009), Jiang et al. (2009), Joosten et al. (2010), Zhou et al. (2010), and Steinmann and Ganzera (2010).

Liu et al. (2009) published a LC-MS$_n$ method for the determination of several alkaloids in comfrey in the positive ESI mode. MS^2 and MS^3 spectra were found to be especially helpful in differentiating symviridine N-oxide and echimidine, two structurally closely related minor alkaloids. Additionally, the effect of different extraction procedures (pressurized hot water extraction and refluxing) on the content of lycopsamine was studied. This number was markedly decreased if the samples were extracted under pressure at 60°C compared with a higher temperature, without applying pressure. This observation is in contrast to a similar study of the same group on the determination of senkirkine and senecionine in ragwort (Jiang et al. 2009). These compounds are also PAs, but the results were comparable with the extraction procedures of both senkirkine and senecionine. This discrepancy was explained by the unstable nature of lycopsamine. Joosten and colleagues studied PAs in *Jacobaea vulgaris*. This plant is responsible for livestock losses and the contamination of milk and honey (Joosten et al. 2010).

ELISA—Most ELISA techniques for the detection and quantitation of PAs are based on antibodies produced against retrorsine. They tend to have mixed cross-reactivity to other PAs and can be varied to provide general PA screening or more specific targeted methods. Competitive inhibition-ELISA was used to estimate total PA content of plants and feed (Lee et al. 2001). The cross-reactivity of 16 PAs was investigated, and it was found that there was cross-reactivity between the N-oxide and free base forms. Antibodies to retronecine and monocrotaline have been produced and detected using an avidin-biotin antibody ELISA. Retronecine was obtained from the hydrolysis of monocrotaline, succinylated and coupled to bovine serum albumin or ovalbumin. Roseman et al. developed sensitive quantitative ELISA techniques for detecting retrorsine and monocrotaline using immunogens based on quaternary pyrrolizidinium salts. Three procedures studied had varied cross-reactivity to other PAs but none for unrelated alkaloids (Crews et al. 2010).

Other methods—A comprehensive review of the use of CE for analysis of miscellaneous alkaloids, including PAs, has been written by Gotti (2010). Many PAs and N-oxides have been shown to exhibit chemiluminescence on reaction with tris(2,2′-bipyridyl) ruthenium(II) or acidic potassium permanganate. The products are determined using flow injection analysis (Gorman et al. 2005).

19.4.1.2 Solanine-Group Glycoalkaloids

Toxic steroidal alkaloids α-solanine, α-chaconine, and tomatine, belonging also to the group of terpenoids, are found in high concentrations in plants of the genus *Solanum* such as potato (*S. tuberosum*), eggplant (*S. melongena*), and *S. lycopersicon* and in tomato (*Lycopersicon esculentum*) and some other species. Solanine has both fungicidal and pesticidal properties, being one of the plant's natural defenses. The human body converts solanines into a poison called solanidine. Human solanine poisoning is primarily displayed by gastrointestinal and neurological disorders. Symptoms include nausea, diarrhea, burning of the throat, heart arrhythmia, headache, and dizziness. Hallucinations, loss of sensation, paralysis, jaundice, dilated pupils, and hypothermia have been reported in more severe cases. The most problematic has been, of course, potato, especially the tubers that are sprouting, green, infected by mildew, injured, or tainted. The normal content of the solanine-group glycoalkaloids (SGAs) in the tubers is 20–100 mg/kg; during sprouting as well as exposure to sun, this number can rise up to 5000 mg/kg. Most of the alkaloids are located in the periderm of the tubers. The total daily toxic dose of solanine and chaconine (LOAEL) for humans is 2–5 mg/kg, the lethal dose being 3–6 mg/kg. These two doses are extremely close to each other and are partly overlapping. The highest legislatively allowed content is 200 mg/kg of tubers; this number is planned to reduce to 100 mg/kg (Püssa 2008). The principal toxins of potatoes and tomatoes and their toxicity mechanisms have been exhaustively reviewed by Barceloux (2009).

Analysis—Analysis methods of SGAs have been reviewed by Friedman (2004), Friedman and Levin (2009), and Barceloux (2009). The glycoalkaloids (GA) content in potato tubers is usually estimated by HPLC. Ieri et al. (2011) have elaborated a rapid LC-DAD-MS method for simultaneous determination of phenolic acids, glycoalkaloids, and anthocyanins in pigmented potatoes. Other methods used for the analysis of potato extracts containing glycoalkaloids include spectrophotometry, TLC, isotachophoresis, GC, GC-MS, CE-MS, immunoassays, and biosensing methods. Enzymatic biosensors for SGAs are reviewed by Amine et al. (2006). Although these methods usually generate similar results, advantages of the widely used HPLC procedures include analysis at room temperature and simultaneous analysis of individual structurally related glycoalkaloids and hydrolysis products without derivatization. The extraction-clean-up steps are of high importance in all analyses. The usefulness of ELISA, based on monoclonal antibodies, was evaluated to measure potato glycoalkaloids extracted from potatoes, potato leaves, sprouts, and processed potato products (French fries, potato chips, and potato skins). The results from the ELISA for total GAs were then compared with those obtained by HPLC analysis of the same extracts for the sum of α-chaconine and α-solanine. The agreement between the results of these two methods reinforces the validity and utility of both HPLC and ELISA.

19.4.2 Cyanogenic Glycosides

These glucosides containing cyanic ($-C\equiv N$) group and releasing highly toxic hydrogen cyanide (HCN) at their enzymatic hydrolysis are found in many dietary plants including cassava (particularly important), sweet potato, yam, corn, sugar cane, peas, beans, almonds, cherries, apricots, apples, pears, and so forth. Hydrolysis of these glucosides is promoted by physical destruction as well as by stress (drying, freezing, and cooking) and is catalyzed by the enzymes β-glucosidase and hydroxynitril lyase. Among the 20 CGs identified so far, the most important are four—amygdalin, dhurrin, linamarin, and lotaustralin. The most well-known and widely spread CG is amygdalin, which is present in the bones of the fruits of drupaceous plants. The amygdalin content in various fruits is different—the highest, about 2.5%–3.5%, is in bitter almonds. Serious health injuries can occur when bitter almonds are consumed; the sweet almond contains little amygdalin and is harmless. Eating of some stones of apricot, plum, or cherry does not cause any intoxication.

Serious food toxicological problems are connected with the roots of cassava (*Manihot esculenta* Crantz), and the leaves of linseed contain linamarin that is hydrolyzable under catalysis by an intrinsic β-glucosidase called linamarase when the tissues are broken down. The use of cassava products for food or feed is strongly influenced by the presence of potentially toxic CG that may have residual levels in cassava food after processing. Under drought conditions, the total cyanide content of cassava flour becomes more than double, which leads to cyanide poisoning and outbreaks of konzo. Tropical ataxic neuropathy occurring in Africa, West Indies, and South India probably results from a continuous year-long intake of cyanogens from a monotonous diet of cassava (Bradbury and Denton, 2010). The cassava roots are classified into three categories according to their potential toxicity to humans and animals: (1) nontoxic (less than 50 ppm HCN in fresh roots), (2) moderately toxic (50–100 ppm HCN in the fresh pulp), and (3) dangerously toxic (above 100 ppm of fresh pulp). The maximum cyanide level recommended by the WHO for fresh cassava intake is 10 mg of HCN/kg bw (Cumbana et al. 2007). Fermentation pretreatment of cassava in West Africa to get the roasted product called gari removes much more CGs than the various methods used in other parts of Africa. Thus, the average total cyanide content of cassava flour in a good season in Mozambique is about 45 ppm and of gari in West Africa is about 20 ppm. The WHO safe level for cyanogens in cassava flour is still 10 ppm (Bradbury and Denton 2010).

Analysis—There are many methods for CG analysis, especially from the cassava roots and products, described in the literature and reviewed by Brito et al. (2009). Only some methods use direct analysis of linamarin; see Campa et al. (2000) and Sornyotha et al. (2007) for examples. A majority of the methods require a three-step procedure: extraction of CG from plants, hydrolysis of glycosides into free cyanide, and determination of cyanide. All the methods available in the literature are based on some general principles: since linamarase is inactive at low pH, the extraction of CG from the plant is performed by diluted acid. Linamarin is then removed and submitted to a series of hydrolyses until cyanide becomes free cyanide. Enzymatic hydrolysis using linamarase is a simple method, but the enzyme is expensive and often difficult to obtain in developing countries. An alternative that avoids the use of linamarase is hydrolysis in 2 M H_2SO_4 at 100°C, which is not simple but gives good results. Among the methods cited in the literature, one can find titration of cyanide with $AgNO_3$ and reaction with alkaline picrate. Three colorimetric methods have been evaluated. Cumbana et al. (2007) have described a picrate paper method. Free cyanide was determined by adapting the method of Smith, based on colorimetric reaction of cyanide with picrate. Despite its very good sensitivity, this method has the disadvantage that the color reading has a great variability due to the noninhibited action of linamarase. Brito et al. (2009) have improved this method. Bradbury et al. (1999) have used picrate paper kits immobilizing linamarase in a small filter paper disc. The most widely used coloring method is based on the König reaction, in which free cyanide is oxidized into cyanogen halide by chloramine-T or N-chlorosuccinimide. Another method uses a specific electrode for cyanide and a voltmeter to measure the potential difference. Another method described by Cooke (1978) was one of the first enzymatic ones, requiring the addition of linamarase and the use of pyridine and pyrazolone as indicators. This method also has a good sensitivity and very good specificity to cyanide and good accuracy. Esser et al. (1993) developed an enzymatic method adapting the Cooke's method by retaining the addition of linamarase while replacing the indicators by isonicotinate and dimethyl barbiturate.

Among the methods described in the literature, the most internationally recognized is the Esser's method in which the linamarase extracted from cassava tissues is used for linamarin hydrolysis, and the released free cyanide is used to establish the standard curve. The linamarase is also used in excess for linamarin hydrolysis in extracts of cassava releasing free cyanide (Bradbury 2009).

19.4.3 SAPONINS

Saponins are glycosidic surface active substances (from Latin *sapo* = soap) found in soya beans, sugar beet, peanuts, spinach, broccoli, alfalfa, potato, apples, and in other plants and fruits. On the basis of their chemical structure, saponins are divided into two large groups—steroidal and triterpenic saponins.

Saponins are very toxic substances; because of their surfactant activity, they are able to decompose the cellular membranes. They can elicit blood hemolysis, especially in the case of cold-blooded animals. For warm-blooded animals, the small perorally administered doses of saponins are generally harmless, since saponins are decomposed by enteral microflora. In addition, they are poorly absorbed and their effect is inhibited by blood plasma. Nevertheless, in case of large doses of saponins nausea, vomiting, diarrhea, and dizziness can occur.

Glycyrrhizin (Gl) or glycyrrhizinic acid is an interesting triterpenoid saponin that is present in the roots and rhizomes of the licorice (liquorice) (*Glycyrrhiza glabra*, *G. uralensis*) in the form of ammonium and potassium salts. Mixtures containing licorice extract are used for preparation of drugs and candies. The enzyme glucuronidase catalyzes hydrolysis of Gl into glycyrrhetic acid and diglucuronic acid. Gl possesses a number of useful effects on the organism—protects the liver against the toxic effect of tetrachloromethane, has antivirus effect (HIV, influenza A and B, herpes simplex 1 and 2, hepatitis B and C) both *in vitro* and *in vivo*, and helps regenerating the inflammatory tissue. The medicinal use of licorice extracts is rather broad: treatment of cough and inflammations of the upper respiratory tract, gastritis, and gastric ulcer. Licorice extract containing Gl, which is 200 times sweeter than saccharose, is used for making various licorice sweets.

At the same time, Gl or, more correctly, its enteral hydrolysis product glycyrrhetic acid has toxic effects also. A long-term consumption of high doses of Gl increases blood pressure and causes water retention in the body through increase in the sodium content in blood (hypernatremia). Hypokalemia is also formed in the blood. The European Commision recommends 100 mg for the upper daily limit of this compound. Studies performed in Finland showed that eating of licorice can cause early birth. Licorice cannot be consumed by persons with type 2 diabetes, high arterial blood pressure, and low K^+ concentration in blood; licorice is not recommendable in the case of hepatitis and during pregnancy. Consumption of licorice together with drugs increasing blood pressure and with corticoids and heart stimulaters is not allowed. In many cases, we have to deal with intoxication by licorice, which is caused by unawareness. The symptoms of the Gl poisoning are elevated blood pressure, edema, feeling of burning, weakness, dark urine, etc. In case of the intoxication, the consumption of licorice should be stopped. Medical treatment is usually unnecessary. However, a fatal case has been described with a 34-year-old woman who was suspected to have suffered lethal acute intoxication from eating nothing but liquorice over a period of several months. The liquorice ingredient Gl and its metabolite glycyrrhetic acid, which elicits a mineralocorticoid effect, were determined in the type of liquorice the woman had consumed by using LC-MS/MS (Albermann et al. 2010). In addition, a fast and sensitive procedure for the quantification of glycyrrhetic acid, including a simple sample preparation procedure, was developed. The method was proven to be accurate and precise. In a liquorice ingestion experiment, 200 g of liquorice had to be eaten. Afterward, the concentrations of glycyrrhetic acid in the blood up to 434 ng/ml were measured. Excluding other causes of death, the woman is believed to have died from a lethal hyperglycemic coma.

Analysis—Recently, determination and the selective extraction of Gl from liquorice roots has gained great interest. Several papers report analytical methods for the determination of Gl with different techniques such as HPLC, CE, GC, and high performance thin layer chromatography (HPTLC) (Cirillo et al. 2011). In all these techniques, different kinds of sample preparation methods are required: plant extracts are, indeed, complex mixtures with a large variety of chemical compounds. Among the extraction techniques, SPE-based Molecularly Imprinted Polymers was used by Cirillo et al. (2011) as an efficient approach for purification of analytes from complex matrices and for the preconcentration of samples.

A number of techniques, including CZE and micellar kinetic CE, have been applied to the determination and quantification of Gl in fresh or dried liquorice roots, in root extracts, and in formulations of Glycyrrhiza species (Lauren et al. 2001; Wang et al. 1998). However, all these methods depend on the analysis of isolated Gl or of the corresponding aglycone after hydrolysis. In a very recent study, a qualitative and quantitative evaluation of the secondary metabolites present in liquorice root has been carried out using LC coupled with ESI/MS and tandem ESI-MS/MS to obtain a full metabolite profile of the drug (Montoro et al. 2011).

19.4.4 Grayanotoxins

Grayanotoxins (GT) are present in rhododendron (*Rhododendron*) and other plants of the family Ericacca, for example, azalea *(Rhododendron* subgenus *Azaleastrum*). The other names of the substance are andromedotoxin, acetylandromedol, and rhodotoxin. From plant nectar, GT passes to honey and may cause honey or rhododendron intoxication. Because of the growing areas of rhododendrons, the intoxications are mainly caused by highland or coastal honey. According to the Old-Roman writer Pliny the Elder, the nations inhabiting the coasts of the Black Sea had used the toxic rhododendron honey as an unconventional weapon against armies of Kyros the Elder (401 BC) and Pompeius (69 BC). In recent times, documented poisoning has occurred in Turkey and Austria. Grayanotoxins are also present in western azalea (*R. occidentalis*), Californian rosebay (*R. macrophyllum*), and laurels (*Kalmia latifolia*; *K. angustifolia*) (Püssa 2008).

From the chemical point of view, GTs are polyhydroxylated cyclic diterpenes. They bind to the specific sodium channels of the cell membranes, hindering their inactivation and leaving cells excited. The symptoms of the poisoning are ample salivation, sweatening, dizziness, weakness and paraesthesia of limbs and face, low blood pressure, and bradycardia. In the case of higher doses, coordination disorders, severe progressve muscular weakness, bradycardia, and paradoxically ventricular tachycardia as well as the Wolff–Parkinson–White syndrome can appear. The pathophysiology, signs, symptoms, clinical course, and treatment of grayanotoxin/mad honey poisoning have been discussed by Gunduz et al. (2008). There are 18 forms of GTs that comprise a number of closely related chemical structures. GTs I–IV is a unique class of toxic diterpenoids that are polyhydroxylated cyclic hydrocarbons. The principle toxic isomer in rhododendron is GT III, although GT I and GT II are present in lower amounts. GT I is also toxic and GT II is less toxic (Koca and Koca 2007).

Analysis—Grayanotoxins can be isolated from the so-called "mad honey" or any other commodity by typical extraction procedures for naturally occurring terpenes. The toxins are identified by paper electrophoresis or TLC. Grayanotoxins have been detected in samples of rhododendron material by GC also (Koca and Koca 2007). Since grayanotoxins are unstable on heating and have low vapor pressure; they require derivatization before the GC analysis (Tasdemir et al. 2003; Terai and Tanaka 1993). Also, the toxins in biological samples can be determined by LC-MS/MS (Holstege et al. 2001). Rhododendron samples were analyzed for levels of toxins by LC-TOF/MS (Hough et al. 2010), providing accurately determined molecular masses of the compounds that are compared with the theoretical masses. Isomers of grayanotoxin were present in all rhododendron samples at concentrations of up to 9283 mg/kg.

19.4.5 LECTINS

Lectins are thermolabile proteins of nonimmune origin, which reversibly bind carbohydrate groups of glycoproteins, glycolipids, or polysaccharides with high affinity. They have been found in more than 800 edible plants but are especially widely distributed in seeds of leguminous plants such as beans (including soy and peanuts) and the products made from them, peas, and so forth. They are generally named according to the plant they are extracted from. Since many, but not all, lectins are capable of agglutinating erythrocytes and other cells, they are often called hemagglutinins.

Some plant lectins can exert adverse effects when one eats a raw plant. Large amounts of lectins can damage the heart, kidneys, and liver; lower blood-clotting ability; destroy the lining of the intestines; and inhibit cell division. Cooking neutralizes lectins to some extent, and digestive juices further destroy them. If the thermal treatment is inefficient, residual lectin activity can remain, which poses moderate risks for consumers' health, as shown by several reports of outbreaks.

The common antinutritional effect of lectins is caused by binding of the lectin molecules to the membranes of intestinal cells. This binding is followed by a nonspecific inhibition of both active and passive transport of important nutrients such as amino acids, fats, vitamins, and minerals through the cell wall and necrosis of the cells of intestinal epithelium. Long-term consumption of raw legumes may lead to growth retardation and even to goiter. An acute systemic exposure to lectins may cause fatal injury to the liver and other organs (Vasconcelos and Oliveira 2004).

Phytohemagglutinin—One of the most potent and more studied lectins is phytohemagglutinin (PHA) from the seeds of common bean (*Phaseolus vulgaris*). Available assays to reveal PHA contamination in food are based on immunological tests or on the measurement of hemoagglutinating activity on hemolysate sample. These methods are sensitive, but very time-consuming. Furthermore, because of the not-rare occurrence of false positive results arising from nonspecific response, a nontoxicological method to confirm results is desirable. MS is the only alternative method presently available to detect PHA and related lectins in legumes, flours, and food samples, with high specificity. The direct observation by MALDI-TOF-MS can be used for semiquantitative measurement of lectins in food after appropriate isolation. Because of the complexity of composition of the flour samples, which prevents direct MS analysis, a procedure for lectin purification has to be developed, based on extraction of seed proteins followed by affinity column purification using immobilized standard glycoprotein, α_1-acid glycoprotein. This procedure isolates the undenatured lectin molecules still present in the samples as distinct from the molecules denatured by industrial heat treatment. With this assumption and with the efficient step of column washing to eliminate aspecifically bound proteins, this analytical method can isolate lectins in native form, and therefore it isolates potentially toxic lectins still present in the samples. A step forward will be the scaling-down of this procedure to an automated miniaturized scale through immobilization of the glycoprotein on a solid support to obtain an affinity sensor capable of detecting low-ppm amounts of native lectins in legumes and in derived foods (Nasi et al. 2009).

Ricin—A lethally toxic lectin ricin is prepared from castor beans, which contain large quantities of this protein consisting of two subunits. The US Centers for Disease Control has set a possible minimum amount of 500 μg as the lethal dose of ricin for humans if exposure is from injection or inhalation. One molecule of ricin is capable of killing one cell. Therefore, it is important to detect ricin in food and water. Conventional methods such as RIA, fluorescence-based fiber optic immunoassay, aptamer microarrays, mass-sensitive biosensor, microelectrochemical biosensors, and electrochemiluminescent assay are effective for ricin detection (Zhuang et al. 2010). Zhuang et al. (2010) have developed a novel silica coating magnetic nanoparticle-based silver enhancement immunoassay for rapid electrical detection of ricin toxin using interdigitated array microelectrodes as electrodes.

19.4.6 ENZYME INHIBITORS

Although many foods of plant and animal origin contain inhibitors of proteases, amylases, and lipases, it is reasonable to discuss only about the proteases here. These extremely proteolysis-resistant proteins are present in all plants, but in large quantities in soya bean (*Glycine max*—the Kunitz inhibitor) and in other beans, peas, beet, cereals (corn, wheat, rye, rice), and in clover, potato, etc. Mostly, they are trypsin inhibitors, but inhibitors of chymotrypsin and carboxypeptidase B can also be found. It is unlikely that anybody eats a raw source of the inhibitor so much that it causes a serious poisoning. But it has been shown that eating of various protease inhibitors can also increase the risk of growth retardation, pancreatic hypertrophy, and cancer. During binding of the inhibitor molecule to the enzyme molecule, the pancreas receive the signal for synthesis of new enzyme molecules, which finally leads to the hypertrophy of pancreatic tissue.

Heating usually destroys the proteinous inhibitor molecules by denaturation. In the case of potatoes, microwave processing and boiling are more effective than cooking. Since many vegetables are eaten without processing or after short-term cooking, they can still have, at least partly, inhibition ability of the digestive enzymes (Püssa 2008).

Analysis—The standard methods of measuring protease inhibitors in foods by enzyme assays often give inaccurate results, especially with processed samples having low residual activity.

Traditionally, SDS-PAGE is widely employed to detect and characterize proteins. However, this method is time-consuming (more than 2 hours for 10–20 samples) and with low accuracy (10%–50%) for quantitation. Size exclusion chromatography in which molecules are separated on the basis of their hydrodynamic volumes is another method for separating proteins, which provides significantly better quantitation with high accuracy but requires about 30 minutes for a sample. On the other hand, microfluidic or Lab-on-a-Chip (LoaC) devices and their application to sensitive chemical and biological analyses have been recently reported. The accuracy of the method is higher (5%–20%) than that of the SDS-PAGE (Goetz et al. 2004). The LoaC technique that has the potential for a fast, reliable, and automatable analysis in the clinical diagnosis, separation, and quantitation of proteins has been reported to be a high-throughput, automated alternative to traditional SDS-PAGE. The chip-based protein analysis is comparable in sensitivity, accuracy, and reproducibility to SDS-PAGE stained with standard Coomassie Blue. To achieve this, the LoaC method uses noncovalently bound fluorescent dyes that bind to the SDS-protein complexes on the chip. The resolution and linear dynamic range are improved. Absolute quantitation accuracy and reproducibility is improved in comparison to SDS-PAGE and is comparable to batch-based quantitation methods, such as Lowry and Bradford's method. The LoaC system has several additional advantages over conventional SDS-PAGE. However, more challenges still exist in sample preparation, such as the extraction of a target fraction from solid samples or from fragile and complex biological samples. Torbica and coauthors have recently compared three methods to detect enzyme activity (trypsin inhibitor assay) and quantify the amount of soybean Kunitz trypsin inhibitor (SDS-PAGE and LoaC methods) in soybean experimental lines (Torbica et al. 2010).

19.4.7 MUSHROOM TOXINS

Among thousands of mushrooms growing in forests and meadows, only some hundreds are intrinsically toxic. By far not all of these hundreds are harmful for people, since nobody eats them. However, quite many poisonings have occurred with mushrooms picked in forests that have mistakenly been considered as edible.

Really fatally toxic are two mushrooms of genus *Amanita*, green death cap (*Amanita phalloides*) and white death cap (*A. virosa*), which contain amatoxins (α-, β-, and γ-amanitins) and phallotoxins such as phalloidin, phalloin, phallacidin, and phallolysin. In addition, *A. virosa* contains virotoxins. Amatoxins are also constituents of the other species of *Amanita*. Amatoxins have been found, although in lower concentrations, in the genera *Galderina* and *Lepiota* also. They are thermostable and insoluble in water; they cannot be destroyed either by boiling or cooking as well as by drying of the mushrooms.

Phallotoxins induce severe gastroenteritis, appearing 4–8 hours after the ingestion of the white death cap, but due to negligible absorptivity, this group of toxins has no essential role in mushroom poisonings.

Analysis—The analysis of amanitin(s) and/or phalloidin has been performed using HPLC (Muraoka et al. 1999), CE and CE coupled to MS (Rittgen et al. 2008), and LC coupled to MS or to tandem MS/MS (Chung et al. 2007). In the study by Chung and coauthors, hydrophilic interaction liquid chromatography (HILIC) on an amide-based stationary phase in combination with ESI-MS/MS in positive ionization mode was successfully employed to separate and identify polar mushroom toxins, including amanitins and phallotoxins that are cyclic oligopeptides and muscarine that is a quaternary ammonium compound, in mushrooms. Valley-to-valley separation achieved upon the optimization of the mobile phase and the solvent gradient by HILIC for confirmatory analysis of the concerned polar toxins could not be achieved through the use of RP-LC. Ahmed and coworkers (2010) used LC-TOF-MS with ESI in establishing a detailed procedure for simultaneous analysis of α-and β-amanitins and phalloidin in mushrooms using microcystin (MC) RR as the internal standard. Since no data were available for TOF mass spectra for the *Amanita* toxins, the authors recorded mass spectra also both in the single-stage and tandem MS modes. A combined extraction procedure with acidified methanol (0.1% TFA) plus SPE was followed. For LC separation, again initial reversed-phase column was substituted by a new normal-phase column for better separation of toxins.

19.5 TOXINS OF MICROORGANISMS

19.5.1 MYCOTOXINS

Mycotoxins include more than 250 various toxins that are in favorable conditions being produced by at least 120 various micro- or mold fungi (molds). The most important and most studied mycotoxins originate from the species of genera *Aspergillus*, *Fusarium*, *Penicillium*, and *Claviceps*. According to their origin and chemical structure, they are divided into aflatoxins, ochratoxins, trichothecenes, fumonisins, zearalenone, patulin, citreoviridin, ergot toxins, and so forth.

Aflatoxins (AT) are structurally related coumarine derivatives, altogether at least 13 types, whose molecules contain a domain of condensed dihydrofuran. They are produced by microfungi, mostly belonging to the species *Aspergillus flavus* that has given the name to the whole group of toxins. The four main aflatoxins B_1, B_2, G_1, and G_2 are produced by the microfungi in plant feed or food raw material that has not been sufficiently dried after harvest but stored as half-dry at relatively high temperatures. Letters B—blue and G—green indicate the color of the respective aflatoxin band at the TLC plate irradiated by UV. AT are potent mutagens, teratogens, and carcinogens in the animal tests. The most widespread aflatoxin B_1 (ATB_1) is the most potent carcinogen of all; its carcinogenicity is observed already in animal tests at a daily dose of 10 μg/kg bw. ATB_1 is also acutely toxic ($LD_{50} = 0.3–18$ mg/kg bw) to all studied animal, bird, and fish species. When a domestic animal eats contaminated feed, aflatoxins move into its tissues and milk; in the mammal, aflatoxins of B-group are converted by hydroxylation to aflatoxin M_1 (ATM_1) belonging to the M-group of aflatoxins.

Ochratoxins are a group of seven stable derivatives of isocoumarine linked via amide bond to amino acid phenylalanine, which are produced by the microfungi

Aspergillus ochraceus and *Penicillium verrucosum.* These fungi contaminate barley, corn, wheat, oats, rye, green coffee beans, peanuts, grape juice and wine, cocoa, dried fruits, and spices. Concentrations determined are in the intervals 0.3–1.6 μg/kg of ppb in cereals, 0.8 μg/kg in coffee, and 0.01–0.1 μg/kg in wine. Ochratoxin A (OTA) has caused proximal tubular lesion of kidneys and hepatic degeneration in animal tests. The toxicity of OTA substantially depends on animal species—its acute per-oral LD_{50} is 0.2 mg/kg in case of dogs and 59 mg/kg in case of mice. Existence of a direct link between the ingestion of high doses of OTA and endemic nephropathy has been proved for humans and pigs in the Balkan countries.

Trichothecenes are a group of 12,13-epoxytrichothecenes produced on various cereal grains mainly by microfungi of the genus *Fusarium.* Here belong T-2 toxin, neosolaniole, diacetylnivalenol, deoxyvalenol (alias DON or vomitoxin), HT-2 toxin, fusarenone and others, altogether over 20 compounds. Trichothecenes inhibit protein synthesis in the cell and induce apoptosis. In addition, at least a part of them is considered to influence the serotonergic pathways of the brain as well as induce the expression of a number of cytokines. The role of the last-mentioned effects in the intoxications is not clear yet.

Patulin (PAT; 4-hydroxy-4H-furo[3,2-c]pyran-2(6H)-one is produced as a secondary metabolite by various species of genera *Penicillium, Aspergillus*, and *Byssochlamys. P. expansum* is the main patulin-producing fungus causing rotting of apples and pathogenesis of many fruits and vegetables. The mechanism of the adverse effect of patulin is covalent binding to cellular nucleophiles, particularly proteins and SH-groups of glutathione. As a result of covalent cross-linking with thiol and aminogroups, essentially denatured proteins such as inhibited protein tyrosin phosphatase are formed.

Neurotoxic citreoviridin is produced in yellow rice by *Penicillium citreoviride.* Ingestion of this substance causes vomiting, convulsions, and paralysis of hind legs and sides of the animals followed by disorders in the functioning of the respiratory and cardiovascular systems, general paralysis, fever, gasp, coma, and arrest of respiration.

From history, serious acute ergot intoxications or ergotisms are known, appearing after eating of grain (particularly rye) contaminated with parasitic microfungus *Claviceps purpurea* (more rarely *C. paspali*) that forms nodules in the spikelets of gramineous plants. Respective toxic substances are ergot alkaloids, derivatives of lysergic acid—ergotoxin, ergotamine, ergomethrine, and so forth, altogether about 50 different substances. Depending on the proportion of various alkaloids in the food, either gangrenous or convulsive (CNS) form of the ergotism will develop. Nowadays ergotism is rare, but it still occurs and if at all then in large numbers.

Analysis—A number of recent reviews (Shephard 2008; Cigić and Prosen 2009; Turner et al. 2009; Fernández-Cruz et al. 2010; Maragos and Busman 2010; Köppen et al. 2010) have been published on the methods of analysis of mycotoxins.

Sample preparation—Mycotoxin analysis in food and feed is a multistep process consisting of sampling, sample preparation, toxin extraction from the matrix (usually with mixtures of water and polar organic solvents), extract clean-up, and finally detection and quantitative determination. Proper sampling, milling, and homogenization procedures are a prerequisite for obtaining reliable analytical results because of the heterogeneous distribution in the so-called "hot spots" of mycotoxins in food commodities (Whitaker 2006). For example, a single peanut kernel heavily contaminated by aflatoxin in a 10 kg sample (20,000 seeds) is sufficient to cause an aflatoxin content of the whole lot being above the European regulation limits allowed for peanuts (Blanc 2006). For sampling of material for mycotoxin analysis, see the section "Sample Preparation." Liquid–liquid partitioning, SPE, supercritical fluid extraction, gel permeation chromatography, IAC, and multifunctional clean-up columns are used for the purification of extracts (Turner et al. 2009).

Chromatography—Chromatographic methods for analysis of mycotoxins include TLC, HPLC-DAD, HPLC-FLD, LC-MS or LC-MS/MS, and GC coupled with electron capture, or flame ionization detector, or MS. From this wide range of techniques, in recent years, HPLC combined with tandem MS (LC-MS/MS) or sequential MS (MS_n) detection using APCI or ESI interfaces has become the method of choice for mycotoxin and food analysis, due to ease of handling, high sensitivity, and accuracy and due to the possibility of simultaneous quantitative determination of the most important mycotoxins belonging to different chemical classes without the need for derivatization (Maragos and Busman 2010; Sulyok et al. 2010; Spanjer et al. 2008; Di Mavungu et al. 2009; Berthiller et al. 2007; Songsermsakul and Razzazi-Fazeli 2008). $LC-MS_n$ and LC-MS allow the enhancement of sensitivity using the single reaction monitoring mode for the determination and the identification of the selected mycotoxin by both characteristic sequential losses and enhanced product ion scans to confirm the fragmentation pattern. For some mycotoxins, better results were obtained using APCI, although ESI shows higher sensitivity for the majority of mycotoxins and is, therefore, used most frequently. To avoid negative effects of coeluting matrix components on the ionization efficiency of the selected mycotoxins, several HPLC-MS/MS-based multimycotoxin analysis methods include a clean-up step. In contrast, Spanjer et al. (2008) reported direct analysis of 33 mycotoxins extracted from peanut, pistachio, wheat, maize, cornflakes, raisins, and figs after simple dilution of the raw extract without further purification. Determination of 87 analytes in wheat and maize in two consecutive chromatographic runs (positive and negative ESI mode) on a triple quadrupole linear ion trap instrument was reported (Sulyok et al. 2007). Recently, a variety of multimycotoxin analysis methods have been reported, showing the special interest of researchers in high-throughput multimycotoxin routine analysis and therefore, the number of methods will continue to rise in the next years (Beltran et al. 2009, 2011; Di Mavungu et al. 2009; Monbaliu et al. 2010; Sulyok et al. 2010; and others).

GC-based methods were used for quantitative determination of mycotoxins as well (Turner et al. 2009). However, some mycotoxin-specific problems, such as nonlinearity of calibration curves, poor repeatability, matrix-induced overestimation, and memory effects from previous sample injections, have led to the increasing use of HPLC-MS(/MS).

Immunoassays and TLC methods are widely used for both quantitative and semiquantitative monitoring of mycotoxins (Maragos and Busman 2010; Turner et al. 2009; Cigić and Prosen 2009). TLC enables the screening of large numbers of samples at a low operating cost as well as the identification of target compounds using UV–Vis spectral analysis. ELISAs are widely used for routine mycotoxin measurement due to the availability of test kits for practically all relevant mycotoxins. In recent years, rapid methods including flow through membrane-based immunoassays, immunochromatographic assays, fluorimetric assays with IAC or SPE clean-up, as well as fluorescence polarization methods have become increasingly important because of their ease of use and cost-effectiveness. Most of these rapid methods provide qualitative or semiquantitative results and are recommendable as screening methods. A review by Goryacheva and coauthors (Goryacheva et al. 2007) is focused on recent developments in immunochemical methods for detection of mycotoxins, with a particular emphasis on simultaneous multiple analyte determination. This includes high-throughput instrumental analysis for the laboratory environment (microtiter plate ELISA, different kinds of immunosensors, fluorescence polarization immunoassay, and CE immunoassay), as well as rapid visual tests for on-site testing [lateral-flow, dipstick, flow-through, and column tests (Köppen et al. 2010)].

Capillary electrophoresis—In spite of the great separation power, short analysis time, and versatility of CE techniques, they have not gained as much popularity as HPLC for the analysis of mycotoxins. Achieving low enough LOD might present a serious problem in CE; thus the methods developed were for those mycotoxins that could be detected by fluorescence (Cigić and Prosen. 2009). The combination of CE with sensitive fluorescence-based detection methods has been described for AT (references 47 and 165 in Turner et al. 2009). Fast separations can be accomplished by CE in aqueous buffer solutions, excluding the need for organic solvents.

OTA was determined in wine by CE-LIF after combined extraction with SPE and supported liquid membrane (SLM). SLM combined with off-line SPE was a good alternative to the use of immunoaffinity columns before CE analysis. The SLM was formed by impregnating a polytetrafluorethylene (PTFE) membrane with a water-immiscible organic solvent. In the off-line SPE step, the nonpolar sorbent C-18 selectively retained the target OTA (Almeda et al. 2008).

Novel methods—There is a need to develop and validate novel analytical methods for rapid, sensitive, and accurate determination of mycotoxins in cereals and cereal-based products to properly and quickly assess the relevant risk of exposure and to ensure that regulatory levels fixed by the European Union (EU) or other international organizations are met. Advances in the analysis of *Fusarium* mycotoxins in cereals and its quality assurance have been reviewed by several authors (Visconti 2007; van Osenbruggen and Petterson 2005 in: Visconti et al., 2007). A variety of emerging methods have been reported for the rapid analysis of mycotoxins (including *Fusarium* toxins). They are based on novel technologies including lateral flow devices, membrane-based flow-through enzyme immunoassay, near-infrared spectroscopy (NIR), molecularly imprinted polymers, and surface plasmon resonance biosensors (Goryacheva et al. 2007; Krska et al. 2008; Maragos 2004; Zheng et al. 2006).

A paper by Visconti et al. (2007) summarizes the recent results obtained in the author's laboratory in the analysis of *Fusarium* mycotoxins applied to cereals and cereal-based products. In particular, novel methods have been developed based on fluorescence polarization and Fourier transform-near infrared spectroscopy (FT-NIR) for rapid quantification of DON in wheat and wheat-derived products;

HPLC-FLD for simultaneous determination of T-2 and HT-2 in cereal grains; and LC-MS/MS for simultaneous determination of DON, NIV, T-2, and HT-2 in cereals and cereal-based food products and for simultaneous determination of DON, T-2, HT-2, ZEA, and FBs, together with aflatoxins and ochratoxin A (OTA), in maize after multimycotoxin IAC.

Ergot alkaloids—A comprehensive review of the analysis including TLC, immunoassays, CZE, LC-FLD, and LC–MS/MS of ergot alkaloids has been published by Krska and Crews (2008). As the authors say, the major focus of this paper was to report on the latest developments for the determination of major ergot alkaloids and their epimers in cereals and cereal-derived products. Information about the stability and the availability of calibrants, sampling issues, extraction and clean-up strategies, and a variety of final separation and detection techniques is also provided.

19.5.2 BACTERIAL TOXINS

Microorganisms are actually the primary cause of food spoilage and food-borne illnesses. The ability to detect pathogenic organisms and their toxins is very important for ensuring the safety and quality of food supplies. Numerous technologies have been developed to enumerate the total and groups of microorganisms as well as to detect and identify specific pathogens and toxins present in foods. Advanced technologies, including convenience-based, antibody-based, and nucleic acid-based assays, have revolutionized the detection methodology for microbial pathogens and toxins in foods. In addition, the unique approach of the rapidly evolving field of nanotechnology has contributed to pathogen and toxin detection in foods (Ge and Meng. 2009).

The most dangerous food pathogens are anaerobic bacteria producing botulinum toxin.

19.5.2.1 Botulinum Toxin

Botulinum neurotoxin (BoNT) produced by the Gram-positive bacteria species *Clostridium botulinum, C. barati,* and *C. butyrium* in seven serotypes is the most toxic of the bacterial exotoxins and among the most poisonous substances known. This antigenous protein is estimated to be 1000, 10,000, or 100,000 times more acutely toxic than ricin, tetrodotoxin (TTX), or nicotin, respectively, by weight. The LD_{50} of BoNT is approximately 1 ng/kg in humans. Alternatively, by extrapolation from primate studies, the lethal amount of crystalline BoNT serotype A has been estimated to be 70 ng orally for a 70-kg human. The high toxicity of BoNTs poses a great challenge for detection of BoNTs and diagnosis of botulism.

Natural cases of BoNT intoxication can occur through three primary mechanisms: consumption of contaminated food (food-borne botulism), infection of an open wound (wound botulism), and infections in the intestines of infants (infant botulism). In addition, inhalational botulism is a fourth intoxication mechanism that can result from intentional release of aerosols in acts of warfare or terrorism.
Analysis:

Mouse bioassay—The gold-standard method for BoNT detection is MBA, which can detect BoNT to <10 pg/mL. The quantity that is lethal is one mouse lethal dose

(MLD), corresponding approximately to 10 pg of BoNT. In the case of a 0.5-ml injection, the LOD of BoNT would be 2 MLD/ml (Cai et al. 2007; Grate et al. 2010). This assay requires intraperitoneal injection of samples into mice followed by observation for 2–4 days. The MBA comprises a three-part protocol: in a screening assay, a limited number of sample dilutions are made and injected into mice, and the mice are observed for 48 h; if the mice die of botulism symptoms, the sample is diluted further and injected into mice until a dilution that does not kill the mice is found; after the end point is reached, the toxin type is determined by using mice protected with monovalent, serotype-specific antitoxin administrated 30 minutes before injection of the toxin samples. At each stage of analysis, the mice are observed for 48 hours. Thus, although sensitive, the assay is laborious and time-consuming, and it poses potential hazards to the personnel during injection. *In vitro* assays to replace the MBA are highly desirable. Nevertheless, the MBA is currently the only accepted method to confirm BoNTs (Cai et al. 2007).

Continuous efforts are being made to develop sensitive *in vitro* assays to replace the MBA, together with an emphasis on rapidity of the assays. Reviews on the *in vitro* methods have been published by Cai et al. (2007) and very recently by Grate et al. (2010). A wide variety of analysis methods has been developed for BoNTs (see Table 19.1). Historically, two primary approaches—structure-based and functional approach, have been used to develop *in vitro* assays. Immunoassays use antibodies to recognize and to bind to epitopes on the surface of the three-dimensional structure of the toxin. Hence, these are structure-based assays and are generally carried out as ELISAs where the enzyme provides amplification. The colorimetric detection is being constantly replaced by more sensitive luminescence detection. Novel approaches are immuno-PCR and microarrays (Grate et al. 2010, Cai et al. 2007). Alternatively, functional assays have been developed to detect the proteolytic activity of the toxin; some of them also use structural recognition to capture and to

TABLE 19.1
Main Techniques Used for the Detection of *Clostridium botulinum* Neurotoxins

Assay	LOD	Total Analysis Time
Mouse bioassay	20 pg/ml	4–6 days
ELISA	100 pg to 1 ng/ml	4–6 hours
ELISA–ELCA	8 pg/ml	5–6 hours
Biosensors	150 pg or 5 ng/ml	10 minutes
Strips	10 ng/ml	15–20 minutes
Electroluminescence sensor	5 pg/ml	1 hour
Endopeptidase immunosorbent	5 pg/ml	5–6 hours
Endopeptidase-MS	20 pg/ml	3–4 hours
Immuno-PCR	5 pg/ml	5–6 hours
Liposome-PCR	0.02 fg/ml	3 hours
Cellular-based assay	pg-ng/ml	2–3 days

Source: Cai et al., *Crit. Rev. Microbiol.*, 33, 109–125, 2007.

concentrate the toxin before assaying the catalytic activity. High-sensitive fluorescence functional assays and methods using MS or biosensors have also been elaborated (Grate et al. 2010). A liposome-PCR assay that is several orders more sensitive than other methods of detection of biological toxins including botulinum toxin has been elaborated by Mason et al. (2006).

Wei et al. (2009) have developed a functional aptamer-based electrochemical voltammetric sensor for detection of BoNT, where steric hindrance is applied to achieve specific signal amplification via conformational change of the aptamer. Under optimized experimental conditions, a high signal-to-noise ratio was obtained within 24 hours with a LOD of 40 pg/ml. Typically, either type of *in vitro* assay requires hours, rather than days required by MBA.

Although typical ELISA requires 4–6 hours for completion, there is a need for completing assays in 1 hour to enable the rapid screening of potentially contaminated samples, diagnosis of potentially exposed humans, and determination of serotype for serotype-specific therapies. A number of fast immunostrip tests for BoNT have been developed. The sensitivity of these tests is usually lower than that of the ELISA. For example, Sharma and Whiting (2005) have elaborated a strip test for detection of BoNT in food samples. Gold nanoparticles were used to label the antibody against BoNT. The LOD is reported as 10 ng/ml for BoNT/A and BoNT/B, and 20 ng/ml for BoNT/E.

19.6   SEAFOOD TOXINS

Animal seafood, both fish and shellfish, can be an unexpected source of various intoxications that are actually caused by various phycotoxins accumulated by the higher marine organisms from the unicellular alga ingested. Usually, these substances are not poisonous to the fish or shellfish but are poisonous to the humans consuming the seafood. Among these toxins, there are dangerous neurotoxic substances such as TTX, saxitoxin and derivatives that induce paralytic shellfish poisoning (PSP), domoic acid and derivatives that cause amnesic shellfish poisoning (ASP), ciguatoxins responsible for ciguatera poisoning, and the palytoxins. Less toxic substances are causative agents of diarrhetic shellfish poisoning (DSP), such as the okadaic acid, or yessotoxins, neurotoxic brevetoxins, etc. A comprehensive review of all aspects of seafood toxins, including analysis, has been published by Botana (2008).

Most seafood or marine toxins are detected using animal (mostly mouse) bioassays as the official method. The need to have an alternative to MBA methods enabling identification and quantitation of every toxicant is of highest importance. Vilariño and coauthors have classified the other methods that are currently in use for detection of marine toxins as nonanalytical methods that, as that of MBA, do not enable identification of the particular compounds of a toxin group present in the sample but yield an overall estimate of toxin content and analytical methods that enable identification and quantification of the toxins as long as there are standards, usually many, available. The first group comprises antibody-based methods such as ELISA that are available for detection of most toxin groups, the receptor- or target-based methods, cell-based assays, and tissue-based assays. The analytical methods are HPLC-FLD, HPLC-UV, LC-MS, and LC-MS/MS that have been elaborated for most groups of seafood toxins. Nevertheless, the information obtained by the analytical methods does not offer any insight into the toxicology of a sample (Vilariño et al. 2010).

Physicochemical methods are more prone to generating larger errors. In many cases, immunomethods exhibit the same difficulties, while receptor-based methods have been developed to measure the concentrations of toxin present in samples based on the relative toxicity of all analogs (Botana et al. 2010). Nevertheless, the biochemical methods identify only the compounds for which the binder was developed, unrelated to toxic potential, while receptor-based or functional assays recognize the potential toxicity of the toxins, hence mimicking bioassay, but they cannot identify single analytes in the sample. The best approach is a combination of a functional assay, which provides information about sample toxicity quickly and cheaply, and a confirmatory method, which identifies the profile of toxins in the sample. An example can be the method worked out by Caillaud et al. (2010), in which cell-based assay of seafood toxins is coupled to HPLC. A review of current developments in functional assays for marine toxins has been published by Botana et al. (2009) and a review considering wide range of methods of analysis of versatile seafood toxins by Campas et al. (2007).

19.6.1 Fish Toxins

19.6.1.1 Tetrodotoxin

This extremely poisonous substance, acutely 10,000 times more toxic than cyanide ion is found in various marine animals such as porcupine fish, ocean sunfish, parrot fish, in some terrestrial animals, and so forth. But the most famous organism containing TTX is delicious puffer fish or fugu (*Takifugu niphobles*), inhabiting in Chinese and Japanese waters. This fish and the first two of the above-mentioned species are the most toxic sea organisms for humans. Puffer fish is consumed mainly in Japan, but also in the United States as a delicacy provided that the dish has been prepared by a specialist and does not contain toxic dose of TTX. For that, very toxin-rich liver, intestines, skin, and sexual glands of puffer fish must be extremely skillfully and completely removed. Because of noncompetent preparation of puffer fish, about 200 deaths occur yearly throughout the world, about 50 of them in Japan. According to other data, 646 cases of TTX intoxication were registered in Japan in 1974–1983, 179 of which had a fatal end (Püssa 2008).

TTX is an extremely potent and heat-stable nerve toxicant, a specific blocker of sodium channels of the skeletal muscles, with human lethal dose about 10 μg/kg bw that means about 1 mg per an adult individual. Except some bacterial toxins such as botulism toxin, only palytoxin from some corals and maitotoxin are more toxic than TTX and saxitoxin. TTX is accompanied in nature by its analogs such as 6-epiTTX, 11-deoxyTTX, 4-epiTTX, and others. There is a widespread hypothesis that TTX and its analogs are actually products of bacteria, living in the host organisms.

Analysis—MBA elaborated for PSP is the best method for control of the fish. The toxicity of an extract is expressed in terms of mouse units (MU), where 1 MU is defined as the amount of TTX required for killing of a 20-g male of ddY or

ICR strain in 30 minutes. LOD is about 0.2 mg of TTX (1 MU) per assay (Hwang and Noguchi 2007). Use of various biosensors for analysis of TTX is reviewed by Campas et al. (2007).

Chromatography—Products of alkaline degradation of TTX can also be determined as trimethylsilyl derivatives by GC-MS (Matsumura 1995). A HPLC-FLD method with postcolumn alkaline treatment is also worked out for estimation of TTX (Huang et al. 2008). Nakagawa et al. (2006) have elaborated a HILIC-ESI/ MS system in single ion mode for quantification of TTX and its deoxy analogs, 5-deoxyTTX, 11-deoxyTTX, and others, which were not detectable with LC-FLD. Jang and coauthors have recently used this approach to quantify TTXs in various tissues of sea puffer fish, *Fugu niphobles,* and in the whole body of the brackish water puffer fishes, *Tetraodon nigroviridis* and *T. biocellatus.* Identification of these four deoxy analogs in the ovarian tissue of *F. niphobles* was further confirmed by HILIC-LC/MS/MS (Jang et al. 2010).

Matsumoto and coauthors (Matsumoto et al. 2007) have studied TTX in liver slices of the puffer fish after extraction with 0.1% acetic acid by ultrasonication and heating in a boiling water bath. The extract was defatted with dichloromethane and centrifuged, the resulting supernatant was ultrafiltered, and the filtrate was analyzed for TTX by reversed-phase LC/ESI-MS. TTX detection at m/z 320 corresponding to the protonated molecular ion $[M+H]^+$ was achieved using the selected ion recording mode.

Hwang and Noguchi (2007) have compiled a versatile review of all aspects of TTX, including detection methods. Apart from the aforementioned methods CE, capillary isotachophoresis, UV spectroscopy, fast atom bombardment, and ESI-^{1}H NMR spectrometry should also be noted.

19.6.1.2 Ciguatoxins

Poisonings caused by Ciguatera fish toxin or ciguatoxin (CTX) is one of the four most important seafood intoxications caused by consuming fish inhabitating the coral reefs. With approximately 50,000 reported yearly cases, ciguatera poisoning is the most common nonbacterial, fish-borne poisoning in the United States. As ciguatoxin vectors, representatives of over 400 fish species can be defined. Content of harmful amount of CTX in a fish is practically unpredictable; it is not possible to recognize the toxic fish either visually or organoleptically. Fortunately, most of the tropical coral reefs are not ciguatoxic, and the outbreaks of ciguatoxin poisoning are mostly localized. CTXs are secondary metabolites produced by the marine benthic dinoflagellate of the genus *Gambierdiscus.* Numerous congeners of CTXs have been identified according to differences in their molecular structure. The main Pacific ciguatoxin (P-CTX-1), the most toxic ciguatoxin, has intraperitoneal $LD_{50} = 0.25$ µg/ kg bw in case of mice and is able to cause human intoxication at the concentration of 0.1 µg/kg of fish meat. The main Caribbean ciguatoxin (C-CTX-1) is less polar and tenfold less toxic in humans. Symptoms of ciguatoxin intoxication may appear less than 6 hours after eating the toxic fish. The first symptoms—nausea, abdominal cramps, and vomiting, emanate from the digestive tract. Furthermore, neurological disorders will appear. In the case of severe poisoning, cardiac disorders may also occur. The victims will usually recover in few days, but serious repeating disorders of the nervous system may stay for months or even years (Püssa 2008).

Analysis—Methods of CTXs from both microalga and fish have been thoroughly described by Caillaud et al. (2010) and Botana (Editor) (2008). Sample preparation depends on the source, with regard to the elimination of interferences and the expected toxin profiles. These two factors differ significantly between fish and microalga. Methods for CTX determination may also determine the grade of purity of extracts required for the analysis. For example, purification of extracts through various SPE steps after solvent partition is often applied when analyzing CTX content using LC-MS/MS and cell-based analysis (CBA) methods (Dickey et al. 2008). A rapid extraction protocol using LC-MS/MS analysis has recently been developed to improve the usefulness of the screening of CTX-containing fish (Lewis et al. 2009). The high diversity and structural complexity of CTX congeners present at trace levels in different matrices have greatly hampered the development of reliable methods for their determination. The standard MBA that is probably the most reliable *in vivo* bioassay system is often used for the surveillance and determination of CTX-contaminated fish. *In vitro* CBA, competitive receptor binding assay (RBA), immunoassays, HPLC-UV, HPLC-FLD (after derivatization), or MS/MS methods are also described by Caillaud et al. (2010). Lewis et al. (2009) reported HPLC-MS/MS analysis following a ciguatoxin rapid extraction method, where a limit of quantification of 0.1 ng/g was achieved.

19.6.1.3 Palytoxin

Palytoxin (PLTX) is a polyether with a very large and complex molecule that has both lipophilic and hydrophilic areas. It presents the longest continuous carbon atom chain known to exist in a natural product, second to maitotoxin. This toxin was first isolated from *Palythoa toxica* and was subsequently reported in dinoflagellates of the genus *Ostreopsis*. Although PLTX has so far been associated with ciguateric fish poisoning (CFP), recent evidence suggests that PLTXs should be excluded from CFP toxins. NMR and LC–MS/MS techniques have enabled the isolation of 10–15 new analogues from dinoflagellates ever since their first discovery. MBA, hemolytic assays, cytotoxicity and immunoassays, HPCE, HPLC-UV, and HPLC-FLD have been reviewed by Riobó and Franco (2011). Different LC–MS methods, recently reviewed by Riobó and Franco (2011) and Ciminiello et al. (2011), have been used for identifying and quantifying the PLTX-group toxins in seawater and phytoplankton. The presence of the PLTX-group toxins in samples is confirmed by the presence of the m/z 327 fragment ion, the bicharged ion (m/z 1340), and/or the tricharged ion (m/z 912). Another characteristic in the MS spectrum of PLTXs is the cluster generated by multiple losses of water molecules. Further research of the development of PLTX analysis methods is essential for future application in routine testing of shellfish tissues.

19.6.2 Shellfish Toxins

Several species of marine phytoplankton produce toxins that bioaccumulate in bivalve shellfish through their filter-feeding nature. When ingested by humans, shellfish contaminated with these toxins can lead to severe illness. Various organisms belonging to the group shellfish, such as clams, lobsters, mussels, oysters, scallops, etc., that have ingested toxic alga, particularly dinoflagellates, can also be very toxic. Shellfish are especially toxic during intensive blooming of alga, when the seawater contains 200 or more microorganisms per milliliter. The toxicity of shellfish is proportional to the concentration of alga in water, and it disappears 2 weeks after the disappearance of the toxic phytoplankton. The toxins can be classified according to the symptoms they induce when they are consumed, and they can also be grouped according to their water solubility. The poisonings are divided into four groups— paralytic, diarrhetic (DSP), neurotoxic (NSP), and amnesic (ASP) shellfish poisonings (Püssa 2008).

19.6.2.1 Paralytic Shellfish Poisoning

Toxins that cause PSP are 25 structurally related imidazoline-guanidinium alkaloids produced by a number of sea dinoflagellates and bacteria as well as fresh-water cyanobacteria accumulated by filter-feeding bivalve shellfish and specific herbivorous fish and crabs. Most well-known are saxitoxins, named by the shellfish *Saxidomus giganteus* in which this toxin was first identified and produced by cyanobacteria and dinoflagellates *Alexandi(n)umi, Gymnodinium catenatum,* and others; neosaxitoxin; anatoxin (*Anabaena flos-aquae*); and gonyautoxins. Nowadays, PSP has turned into a global toxicological issue.

These toxins specifically block the excitation current in nerve and muscle cells, resulting in paralysis and other symptoms. Ingestion of 4 mg of saxitoxin may have lethal end without immediate medical aid. LD_{50} of saxitoxin in case of both guinea pigs (intramuscularly) and mice (intraperitoneally) is about 5 µg/kg bw.

In fresh water, the same toxin is produced by blue alga *Anabaena circinalis* that can pass the toxin to freshwater shellfish such as *Alathyria condola*. Nevertheless, no information concerning saxitoxin poisonings by fresh-water organisms is available. Because of the high polarity of the molecule, saxitoxin dissolves well in water and is stable at both acidic and neutral pH values even at high temperatures. Boiling in slightly alkaline water that inactivates the toxin and discarding the broth help to escape from PSP with saxitoxins.

Anatoxins is a group of low molecular neurotoxic alkaloids that were first found in fresh-water blue alga *Anabaena flos-aquae*. Anatoxins include secondary amines anatoxin-a and homoanatoxin-a as well as anatoxin-a(S), which is phosphoric acid ester of *N*-hydroxyguanine. Similar to acetyl choline, anatoxin-a and homoanatoxin-a (S) are postsynaptic depolarizing neuromuscular blocking substances that tightly bind to the nicotinic acetylcholine receptor. Unlike acetylcholine, acetylcholinesterase is unable to inactivate anatoxins by hydrolysis, and the muscle will stay in the contracted state. Anatoxins cause fast death of mammals via breath paralysis; their LD_{50} for mice is about 250 µg/kg bw. Anatoxin-a(S) is an inhibitor of cholinesterase; its LD_{50} for mice is 25–50 µg/kg bw. Since anatoxin-a(S) is an organophosphorous compound (the only one natural!), its toxic effect is similar to the effect of synthetic organophosphorous pesticides (Püssa 2008).

Analysis—MBA is the worldwide PSP official method used in monitoring programs. However, several limitations, including the pH-dependent high variability and low sensitivity in addition to the strong opposition to animal sacrifice, have led to the search of alternative methods for PSP detection. These alternatives include immunoassays, RBAs, cell assays, and several chemical methods. The most common chemical method uses a combination of HPLC-FLD with either precolumn or postcolumn oxidation. The ability of PSP toxins to be easily converted into fluorescent derivatives has been the basis for their detection (Rourke et al. 2008). These methods have the advantage of detecting and quantifying individual PSP toxins by the use of standards, even though certified material for each toxin is not available. The two main methodologies used in the analysis of PSP toxins involve an isocratic separation of the toxins followed by a postcolumn oxidation or a precolumn oxidation of toxins, which is followed by a gradient separation of the oxidation products (Lawrence et al. 1995). The Lawrence method has been adopted as an official method to detect PSP toxins, and it has been approved by the EU for monitoring these toxins. However, this method seems to be useful mainly for official PSP control in certain samples since its performance depends on the toxic profile of the sample (Ben-Gigirey et al. 2007). The major impediments to widespread use of this method are the coelution of oxidation products and the amount of time required to process samples containing significant amounts of PSP toxins (Ben-Gigirey et al. 2007). Nowadays, the postcolumn oxidations methods have been subject to continuous modifications to reduce the run-times of analysis (Rourke et al. 2008). The precolumn and postcolumn HPLC methods have been compared recently by Rodríguez et al. (2010).

Enzymatic biosensors for analysis of anatoxins have been described by Amine et al. (2006).

19.6.2.2   Diarrhetic Shellfish Poisoning

The major lipophilic toxin groups causing DSP include okadaic acid (OA) and related dinophysistoxins, pectenotoxins, yessotoxins, and azaspiracids as well as the cyclic imines gymnodimines and spirolides, all belonging to the inducers of DSP. With the exception of the cyclic imines, the maximum levels of lipophilic toxins permitted in shellfish to be sold in the market in Europe have been set out in the 2004 EU regulation (reference 1 in Fux et al. 2007). The impact of the cyclic imine toxin group on human health is yet to be determined, and currently no regulatory limits have been established for this group. The reference method described in the EU regulation for lipophilic toxins is the MBA, but there is a growing need to develop and implement non-animal-based methods of toxin detection. Recent advances in analytical instrumentation have enabled the development of alternative methods such as LC–MS/MS. This technology is becoming the method of choice for the detection and quantification of several marine toxins and has been used by several research and monitoring laboratories following in-house validation (Fux et al. 2007). A rapid method for the detection of marine toxins was developed by Fux et al. (2007) using an UHPLC system coupled to the latest generation MS system. The analysis of 21 lipophilic marine toxins in the flesh of mussels and oysters was achieved on a C_{18} column using a water–acetonitrile gradient with a cycle time of 6.6 minutes, reducing analysis time by more than a factor of 2 compared to HPLC while maintaining peak resolution. Linear ranges, limits of detection, and limits of quantification were established for OA, pectenotoxins-2, azaspiracids-1, yessotoxins, GYM, and 13-desmethylspirolide C. The method was found to be accurate when using a triplicate methanolic extraction. Compared with previous studies, the matrix effects occurring in the method presented are diminished.

19.6.2.3   Neurotoxic Shellfish Poisoning (NSP)

Dinoflagellate *Karenia brevis* (formerly known as *Ptychodiscus brevis* or *Gymnodinium breve*), inhabiting mainly the Gulf of Mexico and southern US coast and also coasts of New Zealand, produces lipophilic and heat-stable polyether nerve poisons brevetoxins A, B, and C that are toxic to fish assimilating these substances via gills but not to molluscan shellfish. *K. brevis* causes mass death of fish in the period from November to March. Brevetoxins cause nausea, diarrhea, and paresthesia some minutes after ingestion of shellfish. The symptoms of the intoxication are paresthesias of lips, tongue, and throat, reversal of hot and cold sensations; fever; dizziness; supraventricular tachycardia; and broadening of pupils. Recovery occurs usually in 24 hours. Although brevetoxins are capable of killing the test animals in case of different routes of administration, including peroral, no human deaths have been described. Brevetoxin poisoning can be mixed up with ciguatera poisoning caused by fish (see the section "Ciguatoxins"). Toxin molecules are unstable both in strong acid (pH < 2) and alkali (pH > 10) (Püssa 2008).

Analysis—Reviews concerning various analysis methods of brevetoxins by MBA and by pharmacology-based and structure-based methods have been published by Watkins et al. (2008) and Plakas and Dickey (2010).

Mouse bioassay—Toxicity of shellfish exposed to *Karenia brevis* blooms has been historically assessed by MBA. Results are expressed in mouse units (MU) per 100 g tissue, as derived from dose-response tables. One mouse unit is the amount of crude toxin that will kill 50% of test animals within 15.5 hours. The MBA is still considered the "gold standard" for shellfish toxicity assessment and continues to be required, for example, for the reopening of shellfish beds in Florida and other areas after regulatory closures. As a monitoring tool, MBA has several disadvantages. It lacks sensitivity, specificity, and precision, is time-consuming, and is ethically questionable. Moreover, MBA as performed does not reflect the full complement of bioactive toxins in the brevetoxin-contaminated shellfish (Plakas and Dickey 2010).

Functional methods are cytotoxicity assays, based on the action of brevetoxins on voltage-gated sodium channels, and RBA, based on brevetoxin affinity for its sodium channel receptor.

The cytotoxicity assay is a highly sensitive and useful screening assay for the presence of brevetoxin-like activity in shellfish extracts. Nevertheless, it has several disadvantages. Sample turnaround is generally slow, because of long incubation times, while its microplate format improves throughput. Specificity for brevetoxins is limited to recognition of sodium channel activity. Matrix components can influence response. A multilaboratory study found the cytotoxicity assay to be highly variable when applied to shellfish extracts and to be unsuitable as a replacement for MBA (Plakas and Dickey 2010).

RBA is simple, rapid, and sufficiently sensitive for use as a monitoring tool for brevetoxins in shellfish and has been considered a potential alternative to MBA. Still, as with the cytotoxicity assay, matrix effects are of concern, as components of tissue extracts can interfere with specific binding of brevetoxins to the receptor and increase nonspecific binding.

The structure-based methods are divided into immunoassays and LC-MS methods. Immunoassays offer high sensitivity and specificity and, depending on format, are suitable for field applications and high-throughput analyses. An antibody of high specificity for B-type brevetoxin backbone structure has been produced, with little or no cross-reactivity with A-type brevetoxins or with any other marine toxins. ELISA and RIA platforms have been developed. The competitive ELISA format is capable of measuring brevetoxins with high sensitivity in a variety of matrices, including shellfish, seawater, and human clinical specimens. The LOD for brevetoxins in spiked oyster tissues was 0.025 mg/kg, with an assay time of 4 hours.

LC-MS methods for quantifying individual brevetoxins and their shellfish metabolites in complex mixtures require analytical standards for each. At present, few metabolite standards are commercially available. Most promising applications of LC-MS are for confirmation of brevetoxins in shellfish and for monitoring specific brevetoxins as markers of composite toxin burden and toxicity. Hua and Cole (2000) described diagnostic ions for the A- and B-type brevetoxin backbone structures, by using LC-MS/MS with ESI. Wang et al. (2004) presented mass spectra for the more abundant brevetoxin metabolites in Eastern oyster. Ishida with coauthors have described an LC-MS/MS selected reaction monitoring method, ESI positive and negative ion modes, for monitoring major brevetoxin metabolites in New Zealand shellfish, with LOD in sub-ppb region (Ishida et al. 2004b and 2006 in: Plakas and Dickey 2010).

19.6.2.4 Amnesic Shellfish Poisoning

ASP is caused by domoic acid (DA), first isolated from macroscopic red alga *Chondria armata* with Japanese name *domoi*. Only recently, it was established that the same substance was the causative agent of ASP first diagnosed on the Island of Prince Edward in Canada. The initial source of the toxin is a diatomic alga of genus *Pseufonitzschia*, such as *P. pungens* f. *multiseries*, *P. australis*, and so forth. The alga is eaten by krill, which in turn are consumed by whales and other sea animals as well as birds. It is interesting that when the content of DA in the alga rises very high, the krill gives up eating alga that in turn favors flourishing of colonies of the toxic alga.

DA that contains three carboxylic groups resembles by its molecular structure to kainic acid, a glutamate receptor agonist. DA is a strong amino acid neuroexcitant that influences ionotrophic glutamate receptors of nervous cells, particularly in brain. These receptors are actually ion channels that open and close the passage of cations through nervous cell membranes by the action of L-glutamate as a ligand. These activities release several intracellular cascades either directly by letting in cations or indirectly by mediation of secondary messengers such as cAMP or reactive organic species. Induction of such a neuronal imbalance by DA as an agonist may cause disorders in the functioning of brain or even formation of permanent brain injuries.

DA is well soluble in water and poorly soluble in organic solvents. It is toxic to mammals with LD_{50} of 3.6 mg/kg bw (mice, intraperitoneally). In Canada, ASP caused death of four humans in 1987, of 200 cormorants in 1991, and of 50 sea lions in 1998. The allowable limit of content of DA in molluscs in the United States, Canada, and EU is 20 mg/kg (Püssa 2008, p. 223).

Analysis—All the following data is from Lefebvre and Robertson (2010) and concentrations are in nanograms of DA per milliliter.

Mouse bioassay—Working range (wr) >200. Provides an estimate of total toxicity, but is insensitive, nonspecific, and nonconfirmatory and exhibits high variability between mice.

Direct ELISA—LOD: 0.01; wr: 0.01–10; highly sensitive, selective for DA with little interference, robust, and suitable for a variety of matrices. Disadvantages: nonconfirmatory, small working ranges.

Radio-ligand binding assay—LOD: 5; wr: 5–200. High reproducibility and sensitivity. Use of a tritiated ligand requires a regulated laboratory with personnel trained in radiation safety; scintillant waste and radioactive waste disposal can be problematic. Confirmatory method required.

LC–(UV-DAD)—LOD: 10; wr: 10–10,000. Accurate, high resolution, accepted regulatory method, highly reproducible. Matrix effects; sample cleanup required.

LC-MS/MS—LOD: 2; wr: 2–10,000. Provides structural confirmation; high sensitivity and resolution; quantitative for all isomers. Enables simultaneous use with analysis of other toxin groups. Ionization suppression by matrix requires clean-up.

UV-DAD-CE—LOD: 150, wr: 150–8,000. No clean-up required; sample dilution is sufficient and matrix is eliminated during the run; high resolving power (10 times lower than LC-UV); reduced buffer/solvent use. Requires coupling to MS for confirmation.

19.6.2.5 Microcystins and Nodularins

These toxins are produced by fresh-water blue alga or cyanobacteria of genera *Anabaena*, *Nodularia*, *Nostoc*, *Oscillatoria*, and *Microcystis*. More than 50 various MC congeners are known. Nodularins (ND) have been found in the brackish water alga *Nodularia spumigena* as well as in the common mussel *Mytilus edulis*.

MCs are cyclic heptapeptides that are both specific hepatic toxicants and hepatic carcinogens. NDs are cyclic pentapeptides. Like ocadaic acid that causes DSP, MCs and NDs are potent inhibitors of serine/threonine protein phosphatases. By binding to the same sites of these enzymes as ocadaic acid, MCs and NDs act as tumor promoters. Unlike many other cell types, since these compounds are able to penetrate hepatocyte membranes, the liver is especially sensitive to the toxic effect of MCs and NDs. These substances spoil the normal balance of phosphate groups in cellular cytoskeleton by inhibiting phosphatases. This action is followed by the collapse of the cytoskeleton and all hepatocytes. The process is terminated by death of the organism caused by serious acute hepatic damage.

In principle, human poisoning by MDs and NDs can be caused by eating of fish or shellfish contaminated with these toxins. But contamination of drinking water with these substances is actually considered to be a more serious toxicological issue that concerns humans as well as wild and domestic animals. There is a well-known case of poisoning with cyanobacterial toxins and deaths in the renal dialysis hospital in

Brazil. MCs were found in the salmon and striped perch (bass) that died from acute hepatic damage and in shrimps. Similar to shellfish, shrimps are considered to be the vector of toxins here (Püssa 2008).

Analysis—SPE on tandem Oasis-HLB and silica gel/plus silica gel tandem cartridges was used for prepurification of MC from fish tissue samples with subsequent analysis by HPLC-PDA (Xie and Park 2007). SPE clean-up is strongly recommended since SPE resulted in a decrease in LOD when there was no clean-up in the sample preparation.

Various techniques such as GC-MS, HPLC-UV, LC-MS, CE, and phosphatase inhibition assay have been used to detect and quantitate these toxins. LC combined with different detectors, such as UV and MS, can identify and quantify MCs in freshwater, cyanobacterial blooms, fish, shellfish, and other biological samples (Mekebri et al. 2009).

LC-MS provides specificity and good sensitivity. The matrix effects and signal response in LC-MS analysis of six MC and nodularin-R were studied in mussels and liver samples from the common eider and rainbow trout by triple quadrupole MS with ESI. The recorded matrix effects were not severe; all the studied toxins could be detected with sufficient LOD in all matrices. LC-MS analysis is also suitable for detoxification studies of MC and ND. Cysteine conjugate was identified as the main detoxification product in a mussel sample that was exposed to toxic cyanobacteria in an aquarium (Karlsson et al. 2005).

A LC-ESI–MS/MS method has been developed and validated to identify and quantify trace levels of MCs (MC-LA, LF, LR, LW, RR, and YR) with a single chromatographic run in water, bivalves, and fish tissue, with enhanced sensitivity and specificity. The method enables both confirmation and quantification. Chromatography also allows determination of certain MC metabolites (Desmethyl-LR and -RR). By using LC-ESI–MS/MS in multiple reaction monitoring mode, the LOD and quantitation for the MC studied were determined to be between 0.2 and 1 pg on a column (5:1 S/N ratio). These values are below the 2 pg LODs found in literature (Mekebri et al. 2009).

REFERENCES

Ahmed, W.H.A., Gonmori, K., Suzuki, M., Watanabe, K., Suzuki, O. 2010. Simultaneous analysis of α-amanitin, β-amanitin, and phalloidin in toxic mushrooms by liquid chromatography coupled to time-of-flight mass spectrometry. *Forensic Toxicol.*, 28:69–76.

Albermann, M.E., Musshoff, F., Hagemeier, L., Madea, B. 2010. Determination of glycyrrhetic acid after consumption of liquorice and application to a fatality. *Forensic Sci. Int.*, 197:35–39.

Almeda, S., Arce, L., Valcárcel, M. 2008. Combined use of supported liquid membrane and solid-phase extraction to enhance selectivity and sensitivity in capillary electrophoresis for the determination of ochratoxin A in wine. *Electrophoresis*, 29:1573–81.

Altamirano, J.C., Gratz, S.R., Wolnik, K.A. 2005. Investigation of pyrrolizidine alkaloids and their N-oxides in commercial comfrey-containing products and botanical materials by liquid chromatography electrospray ionization mass spectrometry. *J. AOAC Int.*, 88:406–412.

Amine, A., Mohammadi, A., Bourais, I., Pallischi, G. 2006. Enzyme inhibitor-based biosensors for food safety and environmental monitoring. *Biosens. Bioelectron.*, 21:1405–1423.

Barceloux, D.G., 2009. Potatoes, Tomatoes, and Solanine Toxicity (*Solanum tuberosum* L., *Solanum lycopersicum* L.). *Disease-a-Month*, 55:391–402.

Beltran, E., Ibañez, M., Sancho, J.V., Cortés, M.I., Yus, V., Hernandez, F., 2011. UHPLC–MS/MS highly sensitive determination of aflatoxins, the aflatoxin metabolite M1 and ochratoxin A in baby food and milk. *Food Chem.*, 126:737–744.

Beltran, E., Ibañez, M., Sancho, J.V., Hernandez, F. 2009. Determination of mycotoxins in different food commodities by ultra-high-pressure liquid chromatography coupled to triple quadrupole mass spectrometry. *Rapid Commun. Mass Spectrom.*, 23(12):1801–1809.

Ben-Gigirey, B., Rodríguez-Velasco, M.L., Villar-Gonzalez, A., Botana, L.M. 2007. Influence of the sample toxic profile on the suitability of a high performance liquid chromatography method for official paralytic shellfish toxins control. *J. Chromatogr., A*, 1140:78–87.

Berthiller, F., Sulyok, M., Krska, R., Schuhmacher, R. 2007. Chromatographic methods for the simultaneous determination of mycotoxins and their conjugates in cereals. *Int. J. Food Microbiol.*, 119:33–37.

Blanc, M. 2006. Sampling: the weak link in the sanitary control system of agricultural products. *Mol. Nutr. Food Res.* 50:473–479.

Botana, L.M. (Ed.) 2008. *Seafood and Freshwater Toxins: Pharmacology, Physiology, and Detection*, Second Edition, CRC Press: Boca Raton, FL.

Botana, L.M., Alfonso, A., Botana, A., Vieytes, M.R., Vale, C., Vilariño, N., Louzao, C. 2009. Functional assays for marine toxins as an alternative, high-throughput-screening solution to animal tests. *Trends Anal. Chem.*, 28:603–611.

Botana, L.M., Vilariño, N., Alfonso, A., Vale, C., Louzao, C., Elliott, C.T., Campbell, K., Botana, A.M. 2010. The problem of toxicity equivalent factors in developing alternative methods to animal bioassays for marine-toxin detection. *Trends Anal.Chem.*, 29:1316–1325.

Bradbury, J.H., Denton, I.C. 2010. Simple method to reduce the cyanogen content of gari made from cassava. *Food Chem.*, 123:840–845.

Bradbury, J. H. 2009. Development of a sensitive picrate method to determine total cyanide and acetone cyanohydrin contents of gari from cassava. *Food Chem.*, 113:1329–1333.

Bradbury, M.G., Egan, S.V., Bradbury, J.H. 1999. Picrate paper kits for determination of total cyanogens in cassava roots and all forms of cyanogens in cassava products. *J. Sci. Food Agric.* 71:595–601.

Brito, V.H.S., Ramalho, R.T., Rabacow, A.P.M., Moreno, S.E., Cereda, M.P. 2009. Colorimetric method for free and potential cyanide analysis of cassava tissue, *Gene Conserve*, 8:841–852. Available at http://www.geneconserve.pro.br/artigo077.pdf.

Cai, S., Singh, B.R., Sharma, S. 2007. Botulism diagnostics: From clinical symptoms to *in vitro* assays. *Crit. Rev. Microbiol.*, 33:109–125.

Caillaud, A., de la Iglesia, P., Darius, H.T., Pauillac, S., Aligizaki, K., Fraga, S., Chinain, M., Diogène, J. 2010. Update on methodologies available for ciguatoxin determination: Perspectives to confront the onset of ciguatera fish poisoning in Europe [1]. *Mar. Drugs*, 8:1838–1907.

Campa, C., Schmitt-Kopplin. Ph., Cataldi, T.R.I., Bufo, S.A., Freitag, D., Kettrup, A. 2000. Analysis of cyanogenic glycosides by micellar capillary electrophoresis. *J. Chromatogr. B*, 739:95–100.

Campas, M., Prieto-Simon, B., Marty, J.-L. 2007. Biosensors to detect marine toxins: Assessing seafood safety. Review, *Talanta*, 72:884–895.

Campbell, K., Valiriño, N., Botana, L.M., Elliot, C.T. 2011. A European perspective on progress in moving away from the mouse bioassay for marine-toxin analysis. *Trends in Anal. Chem.,* 30:239–253.

Cao, Y., Colegate, S.M., Edgar, J.A. 2008. Safety assessment of food and herbal products containing hepatotoxic pyrrolizidine alkaloids: interlaboratory consistency and the importance of N-oxide determination. *Phytochem. Anal.*, 19:526–533.

Cirillo, G., Curcio, M., Parisi, O.I., Puoci, F., Iemma, F., Spizzirri, U.C., Restuccia, D., Picci, N. 2011. Molecularly imprinted polymers for the selective extraction of glycyrrhizic acid from liquorice roots. *Food Chem.*, 125:1058–1063.

Chung, W.-C., Tso, S.-C., Sze, S.-T. 2007. Separation of polar mushroom toxins by mix-mode hydrophilic and ionic interaction liquid chromatography-electrospray ionization-mass spectrometry. *J Chromatogr. Sci.*, 45:104–111.

Cigić, I.K., Prosen, H. 2009. An overview of conventional and emerging analytical methods for the determination of mycotoxins. *Int. J. Mol. Sci.*, 10:62–115.

Ciminiello, P., Dell'Aversano, C., Dello Iacovo, E., Fattorusso, E., Forino, M., Tartaglione, L. 2011. LC-MS of palytoxin and its analogues: State of the art and future perspectives. Toxicon 57:376–389.

Colegate, S.M., Gardner, R., 2007. LC–MS of alkaloids: qualitative profiling, quantitative analysis, and structural identification, in: *Modern Alkaloids: Structure, Isolation, Synthesis and Biology*, E. Fattorusso, O. Taglialatela-Scafati (Eds.), Wiley-VCH, Weinheim, 369–409.

Cooke, R.D. 1978. An enzymatic assay for the total cyanide content of cassava. *J. Sci. Food Agric.*, 29:345–352.

Crews, C., Berthiller, F., Krska, R. 2010. Update on analytical methods for toxic pyrrolizidine alkaloids. *Anal. Bioanal. Chem.*, 396:327–338.

Cumbana, A., Mirione, E., Cliff, J., Bradbury, J.H. 2007. Reduction of cyanide content of cassava flour in Mozambique by the wetting method. *Food Chem.*, 101:894–897.

Dickey, R.W. 2008. Ciguatera Toxins: Chemistry, Toxicology and Detection. In *Seafood and Freshwater Toxins: Pharmacology, Physiology, and Detection*, 2nd ed.; Botana, L.M., (Ed.); CRC Press: Boca Raton, FL; pp. 479–500.

Di Mavungu, J.D., Monbaliu, S., Scippo, M.L., Maghuin-Rogister, G., Schneider, Y.J., Larondelle, Y., Callebaut, A., Robbens, J., Van Peteghem, C., De Saeger, S. 2009. LC-MS/MS multi-analyte method for mycotoxin determination in food supplements. *Food Add. Contam. A*, 26:885–895.

Esser, A.J.A., Bosveld, M., Van der Crift III, R.M., Voragen, A.G.J. 1993. Studies on the quantification of specific cyanogens in cassava products and introduction of a new chromogen. *J. Sci. Food and Agric.*, 63:287–296.

Fernández-Cruz, M.L., Mansilla, M.L., Tadeo H.L. 2010. Mycotoxins in fruits and their processed products: Analysis, occurrence and health implications. *J. Adv. Research*, 1:113–122.

Food Standards Agency. 2007. *Sampling advice: Mycotoxins in foodstuffs*. Food Standards Agency, UK. Available at http://www.food.gov.uk/multimedia/pdfs/mycotoxinsguidance.pdf.

Friedman, M. 2004. Analysis of biologically active compounds in potatoes (*Solanum tuberosum*), tomatoes (*Lycopersicon esculentum*), and jimson weed (*Datura stramonium*) seeds. *J. Chromatogr. A*, 1054:143–155.

Friedman, M., Levin, C.E. 2009. Analysis and Biological Activities of Potato Glycoalkaloids, Calystegine Alkaloids, Phenolic Compounds, and Anthocyanins. In: *Advances in Potato Chemistry and Technology*, Elsevier, Amsterdam, 127–161.

Fux, E., McMillan, D., Bire, R., Hess, P. 2007. Development of an ultra-performance liquid chromatography—mass spectrometry method for the detection of lipophilic marine toxins. *J. Chromatogr. A*, 1157:273–280.

Garcia-Cañas, V., Cifuentes, A. 2007. Detection of microbial food contaminants and their products by capillary electromigration techniques. *Electrophoresis*, 28:4013–4030.

Garcia-Cañas, V., Cifuentes, A. 2008. Recent advances in the application of capillary electromigration methods for food analysis. *Electrophoresis*, 29:294–309.

Ge, B., Meng, J. 2009. Advanced Technologies for Pathogen and Toxin Detection in Foods: Current Applications and Future Directions. *JALA*, 14:235–241.

Goetz, A., Kuschel, M., Wulff, T., Sauber, C., Miller, C., Fisher, S., Woodward, C. 2004. Comparison of selected analytical techniques for protein sizing, quantitation and molecular weight determination. *J. Biochem. Biophys. Methods*, 60:281–293.

Gorman, B.A., Barnett, N.W., Bos, R. 2005. Detection of pyrrolizidine alkaloids using flow analysis with both acidic potassium permanganate and tris(2,2′-bipyridyl)ruthenium(II) chemiluminescence. *Anal. Chim. Acta*, 541:119–124.

Goryacheva, I.Y., De Saeger, S., Eremin, S.A., Van Peteghem, C. 2007. Immunochemical methods for rapid mycotoxin detection: Evolution from single to multiple analyte screening: A review. *Food Add. Contam.*, 24:1169–1183.

Gotti, R. 2010. Capillary electrophoresis of phytochemical substances in herbal drugs and medicinal plants. *J. Pharm. Biomed. Anal.*, in press; doi:10.1016/j.jpba.2010.11.041.

Grate, J.W., Ozanich Jr., R.M., Warner, M.G., Marks, J.D., Bruckner-Lea, C.J. 2010. Advances in assays and analytical approaches for botulinum-toxin detection. *Trends Anal. Chem.*, 29:1137–1156.

Gunduz, A., Turedi, S., Russell, R.M., Ayaz, F.A. 2008. Clinical review of grayanotoxin/mad honey poisoning past and present. *Clin. Toxicol.*, 46, 437–442.

Holstege, D.M., Puschner, B., Le, T. 2001. Determination of Grayanotoxins in biological Samples by LC-MS/MS. *J. Agric. Food Chem.*, 49:1648–1651.

Hough, R.L., Crews, C., White, D., Driffield, M., Campbell, C.D. Maltin, C. 2010. Degradation of yew, ragwort and rhododendron toxins during composting. *Sci. Total Environ.*, 408:4128–4137.

Hua, Y., Cole, R.B. 2000. Electrospray ionization tandem mass spectrometry for structural elucidation of protonated brevetoxin in red tide algae. *Anal. Chem.*, 72:376–383.

Huang, H.-N., Jie Lin, J., Lin, H.-L. 2008. Identification and quantification of tetrodotoxin in the marine gastropod *Nassarius* by LC–MS. *Toxicon*, 51:5774–779.

Hwang, D.-F., Noguchi, T. 2007. Tetrodotoxin Poisoning. *Adv. Food Nutr. Res.*, 52:141–236.

Ieri, F., Innocenti, M., Andrenelli, L., Vecchio, V., Mulinacci, N. 2011. Rapid HPLC/DAD/MS method to determine phenolic acids, glycoalkaloids and anthocyanins in pigmented potatoes (*Solanum tuberosum* L.) and correlations with variety and geographical origin. *Food Chem.*, 125:750–759.

Jang, J.-H., Lee, J.-S., Yotsu-Yamashita, M. 2010. LC/MS Analysis of tetrodotoxin and its deoxy analogs in the marine puffer fish *Fugu niphobles* from the southern coast of Korea, and in the brackishwater puffer fishes *Tetraodon. nigroviridis* and *Tetraodon biocellatus* from Southeast Asia. *Mar. Drugs*, 8:1049–1058.

Jiang, Z.J., Liu, F., Goh, J.J.L., Yu, L.J., Li, F.Y., Ong, E.S., Ong, C.N. 2009. Determination of senkirkine and senecionine in *Tussilago farfara* using microwave-assisted extraction and pressurized hot water extraction with liquid chromatography tandem massspectrometry. *Talanta*, 79:539–546.

Joosten, L., Mulder, P.P.J., Vrieling, K., vanVeen, J.A., Klinkhamer, P.G.L. 2010. The analysis of pyrrolizidine alkaloids in *Jacobaea vulgaris*: a comparison of extraction and detection methods. *Phytochem. Anal.*, 21:197–204.

Karlsson, K.M., Spoof, L.E.M., Meriluoto, J.A.M. 2005. Quantitative LC-ESI-MS analyses of microcystins and nodularin-R in animal tissue—Matrix effects and method validation. *Environ. Toxicol.*, 20:381–389.

Koca, I., Koca, A.F. 2007. Poisoning by mad honey: A brief review. *Food Chem.Toxicol.*, 45:1315–1318.

Köppen, R., Koch, M., Siegel, D., Merkel, S., Maul, R., Nehls, I. 2010. Determination of mycotoxins in foods: current state of analytical methods and limitations. *Appl. Microbiol. Biotechnol.*, 86:1595–1612.

Krska, R., Crews, C. 2008. Significance, chemistry and determination of ergot alkaloids: A review. *Food Add. Contamin.*, 25:722–731.

Lauren, D.R., Jensen, D.J., Douglas, J.A., Follet, J.M. 2001. Efficient method for determining the glycyrrhizin content of fresh and dried roots, and root extracts, of Glycyrrhiza species. *Phytochem. Anal.,* 12:332–335.

Lawrence, J.F., Menard, C., Cleroux, C., 1995. Evaluation of prechromatographic oxidation for liquid chromatographic determination of paralytic shellfish poisons in shellfish. *J. AOAC. Int.,* 78:514–520.

Lee, S.T., Schoch, T.K., Stegelmeier, B.L., Gardner, D.R., Than, K.A., Molyneux, R.J. 2001. Development of enzyme-linked immunosorbent assays for the hepatotoxic alkaloids riddelliine and riddelliine N-oxide. *J. Agric. Food Chem.,* 49:4144–4151.

Lefebvre, K.A., Robertson, A. 2010. Domoic acid and human exposure risks: A review. *Toxicon,* 56:218–230.

Lewis, R.J., Yang, A.J., Jones, A. 2009. Rapid extraction combined with LC-tandem mass spectrometry (CREM-LC/MS/MS) for the determination of ciguatoxins in ciguateric fish flesh. *Toxicon,* 54:62–66.

Liu, F., Wan, S.Y., Jiang, Z.J., Li, S.F.Y., Ong, E.S., Osorio, J.C.C. 2009. Determination of pyrrolizidine alkaloids in comfrey by liquid chromatography–electrospray ionization mass spectrometry. *Talanta,* 80:916–923.

Malik, A.K., Blasco, C., Pico, Y. 2010. Liquid-chromatography-mass spectrometry in food safety. *J. Chromatogr. A.* 1217:4018–4040.

Manco, G., Nucci, R., Febbraio, F. 2009. Use of esterase activities for the detection of chemical neurotoxic agents. *Protein Pept. Lett.,* 16:1225–1234.

Maragos, C.M., Busman, M. 2010. Rapid and advanced tools for mycotoxin analysis: a review. *Food Addit. Contam.,* 27:688–700.

Mason, J.T., Xu, L., Sheng, Z.M., O'Leary, T.J. 2006. A liposome-PCR assay for the ultrasensitive detection of biological toxins. *Nat. Biotechnol.* 24:555–557.

Matsumoto, T., Nagashima, Y., Kusuhara, H., Yuichi Sugiyama, Y., Ishizaki, S., Shimakura, K., Shiomi, K. 2007. Involvement of carrier-mediated transport system in uptake of tetrodotoxin into liver tissue slices of puffer fish *Takifugu rubripes. Toxicon,* 50:173–179.

Matsumura, K. 1996. Tetrodotoxin concentration in cultured puffer fish, *Fugu rubripes. J. Agric. Food Chem.,* 44:1–2.

Mekebri, A., Blondina, G.J., Crane, D.B. 2009. Method validation of microcystins in water and tissue by enhanced liquid chromatography tandem mass spectrometry. *J. Chromatogr. A,* 1216:3147–3155.

Monbaliu, S., Van Poucke, C., Detavernier, C., Dumoulin, F., Van De Velde, M., Schoeters, E., Van Dyck, S., Averkieva, O., Van Peteghem, C., De Saeger, S. 2010. Occurrence of mycotoxins in feed as analyzed by a multi-mycotoxin LC-MS/MS method. *J. Agric. Food Chem.,* 58:66–71.

Montoro, P., Maldini, M., Russo, M., Postorino, S., Piacente, S., Pizza, C. 2011. Metabolic profiling of roots of liquorice (Glycyrrhiza glabra) from different geographical areas by ESI/MS/MS and determination of major metabolites by LC-ESI/MS and LC-ESI/MS/MS. *J. Pharm. Biomed. Anal.,* 54:535–544.

Mroczek, T., Widelski, J., Glowniak, K. 2006. Optimization of extraction of pyrrolizidine alkaloids from plant material. *Chem. Anal.,* 51:567–580.

Muraoka, S., Fukamachi, N., Mizumoto, K., Shinozawa, T. 1999. Detection and identification of amanitins in the wood-rotting fungi *Galerina fasciculata* and *Galerina helvoliceps. Appl. Environ. Microbiol.,* 65:4207–4210.

Nakagawa, T., Jang, J., Yotsu-Yamashita, M. 2006. Hydrophilic interaction liquid chromatography electrospray ionization mass spectrometry of tetrodotoxin and its analogs. *Anal. Biochem.,* 352:142–144.

Nasi, A., Picariello, G., Ferranti, P. 2009. Proteomic approaches to study structure, functions and toxicity of legume seeds lectins. Perspectives for the assessment of food quality and safety. *J. Proteomics.,* 72:527–538.

Patel, D. 2006. Overview of affinity biosensors in food analysis. *J. AOAC Int.*, 89: 805–818.

Plakas, S.M., Dickey, R.W. 2010. Advances in monitoring and toxicity assessment of brevetoxins in molluscan shellfish. *Toxicon*, 56:137–149.

Püssa, T. 2008. *Principles of Food Toxicology*, CRC: Boca Raton, FL.

Rietjens, I.M., Martena, M.J., Boersma, M.G., Spiegelenberg, W., Alink, G.M. 2005. Molecular mechanisms of toxicity of important food-borne phytotoxins. *Mol. Nutr. Food Res.*, 49:131–158.

Riobó, P., Franco, J.M. 2011. Palytoxins: biological and chemical determination. *Toxicon*, 57:368–375.

Rittgen, J., Pütz, M., Pyell, U. 2008. Identification of toxic oligopeptides in *Amanita* fungi employing capillary electrophoresis-electrospray ionization-mass spectrometry with positive and negative ion detection. *Electrophoresis*, 29:2094–2100

Rodríguez, P., Alfonso, A., Botana, A.M., Vieytes, M.R. Botana, L.M. 2010. Comparative analysis of pre- and post-column oxidation methods for detection of paralytic shellfish toxins. *Toxicon*, 56:448–457.

Rourke, W.A., Murphy, C.R., Pitcher, G., van de Riet, J.M., Burns, B.G., Thomas, K.M., Quilliam, M.A. 2008. Rapid postcolumn methodology for determination of paralytic shellfish toxins in shellfish tissue. *J. AOAC Intern.*, 91:589–597.

Şenyuva, H., Gilbert, J. 2010. Immunoaffinity column clean-up techniques in food analysis: A review. *J.Chromatogr. B*, 878:115–132.

Sharma, S.K., Whiting, R.C. 2005. Methods for detection of *Clostridium botulinum* toxin in foods. *J. Food Prot.*, 68:1256–1263.

Shephard, G.S. 2008. Determination of mycotoxins in human food. *Chem.Soc.Rev.*, 37:2468–2477.

Sommer, H., Meyer, K.F., 1937. Paralytic shellfish poisoning. *Arch. Pathol.*, 24:560–598.

Songsermsakul, P., Razzazi-Fazeli, E.A. 2008. Review of recent trends in applications of liquid chromatography-mass spectrometry for determination of mycotoxins. *J Liq. Chromatogr. Relat. Technol.*, 31:1641–1686.

Sornyotha, S., Khin Lay Kyu, Ratanakhanokchai, K. 2007. Purification and detection of linamarin from cassava root cortex by high performance liquid chromatography. *Food Chem.*, 104:1750–1754.

Spanjer, M.C., Rensen, P.M., Scholten, J.M. 2008. LC-MS/MS multimethod for mycotoxins after single extraction, with validation data for peanut, pistachio, wheat, maize, cornflakes, raisins and figs. *Food Addit. Contam.*, 25:472–489.

Steinmann, D., Ganzera, M. 2010. Recent advances on HPLC/MS in medicinal plant analysis. *J. Pharm. Biomed.* Anal., 55:744–757.

Sulyok, M., Krska, R., Schuhmacher, R. 2010. Application of an LC–MS/MS based multimycotoxin method for the semi-quantitative determination of mycotoxins occurring in different types of food infected by moulds. *Food Chem.*, 119:408–416.

Sulyok, M., Krska, R., Schumacher, R. 2007. A liquid chromatography/tandem mass spectrometric multi-mycotoxin method for the quantification of 87 analytes and its application to semi-quantitative screening of moldy food samples. *Fresenius J. Anal Chem.*, 389:1505–1523.

Taitt, C.R., Shriver-Lake, L.C., Ngundi, M.M., Ligler, F.S. 2008. Array Biosensor for Toxin Detection: Continued Advances. *Sensors*, 8:8361–8377.

Tasdemir, D., Demirci, B., Demirci, F., Dönmez, A.A., Baser, K.H.C., Rüedi, P., 2003. Analysis of the volatile components of five Turkish Rhododendron species by headspace solid-phase microextraction and GC–MS (HS-SPME–GC–MS). *Z. Naturforsch.*, 58c:797–803.

Terai, T., Tanaka S. 1993. Gas chromatographic determination of grayanotoxins. *Chem. Express*, 8:381–384.

Tsikas, D. 2010. Quantitative analysis of biomarkers, drugs and toxins in biological samples by immunoaffinity chromatography coupled to mass spectrometry or tandem mass spectrometry: a focused review of recent applications. *J. Chromatogr. B*, 878:133–148.

Torbica, A.M., Živančev, D.R., Nikolić, Z.T., Đorđević, V.B., Nikolovski, B.G. 2010. Advantages of the Lab-on-a-Chip Method in the Determination of the Kunitz Trypsin Inhibitor in Soybean Varieties. *J. Agric. Food Chem.*, 58:7980–7985.

Turner, N.W., Subrahmanyam, S., Piletsky, S.A. 2009. Analytical methods for determination of mycotoxins: A review. *Anal. Chim. Acta*, 632:168–180.

Vallejo-Cordoba, B. and Gonzalez-Cordova, A.F. 2010. Capillary electrophoresis for the analysis of contaminants in emerging food safety issues and food traceability. *Electrophoresis*, 31:2154–2164.

Van Dorst, B., Mehta, J., Bekaert, K., Rouah-Martin, E., De Coen, W., Dubruel, P., Blust, R., Robbens, J. 2010. Recent advances in recognition elements of food and environmental biosensors: A review. *Biosens. Bioelectron.*, 26:1178–1194.

Vasconcelos, I.M. and Oliveira, J.T.A. 2004. Antinutritional properties of plant lectins. *Toxicon*, 44:385–403.

Vehniäinen, M., Leivo, J., Lamminmäki, U. 2011. Directed antibody-engineering techniques and their applications in food immunoassays. *Trends Anal. Chem.*, 30:219–226.

Velusamy, V., Arshak, K., Korostynska, O., Oliwa, K., Adley, K. 2010. An overview of foodborne pathogen detection: In the perspective of biosensors. *Biotechnol. Adv.*, 28:232–254.

Vilariño, N., Louzao, M.C., Vieytes, M.R., Botana, L.M. 2010. Biological methods for marine toxin detection. *Anal. Bioanal. Chem.*, 397:1673–1681.

Visconti, A., De Girolamo, A., Lattanzio, V.M.T., Lippolis, V., Pascale, M., Solfrizzo, M. 2007. Novel analytical methods for Fusarium toxins in the cereal food chain. In: Proceedings of the Conference Advances and prospects of research of *Fusarium* mycotoxin contamination of cereals, 11–13 September 2007, 2010 [online] at www.symposcience.fr/exl-doc/colloque/ART-00002145.pdf

Wang, P., Li, S.F.Y., Lee, H.K. 1998. Determination of glycyrrhizic acid and 18-beta-glycyrrhetinic acid in biological fluids by micellar electrokinetic chromatography. *J. Chromatogr. A*, 811:219–224.

Wang, Z., Plakas, S.M., El Said, K.R., Jester, E.L., Granade, H.R., Dickey, R.W. 2004. LC/MS analysis of brevetoxin metabolites in the Eastern oyster *(Crassostrea virginica)*. *Toxicon*, 43:455–465.

Watkins, S.M., Reich, A., Fleming, L.E., Hammond, R. 2008. Neurotoxic shellfish poisoning. *Mar. Drugs*, 6:431–455.

Wei, F., Ho, C.-M. 2009. Aptamer-based electrochemical biosensor for Botulinum neurotoxin. *Anal. Bioanal. Chem.*, 393:1943–1948.

Whitaker, T.B. 2006. Sampling for food mycotoxins. *Food Add. Contam.*, 23:50–61.

Witte, L., Rubiolo, P., Bicchi, C., Hartmann, T. 1993. Comparative analysis of pyrrolizidine alkaloids from natural sources by gas chromatography–mass spectrometry. *Phytochemistry*, 32:187–196.

Xie, L., Park, H.-D. 2007. Determination of microcystins in fish tissues using HPLC with a rapid and efficient solid phase extraction. *Aquaculture*, 271:530–536.

Zhang, F., Wang, C., Xiong, A., Wang, W., Yang, L., Wang, C., Bligh, S.W.A., Branford-White C.J. 2007. Quantitative analysis of total retronecine esters-type pyrrolizidine alkaloids in plant by high performance liquid chromatography. *Anal. Chim. Acta*, 605:94–101.

Zheng, M.Z., Richard, J.L., Binder, J. 2006. A review of rapid methods for the analysis of mycotoxins. *Mycopathologia*, 161:261–273.

Zhou, Y., Li, N., Choi, F.F., Qiao, C.F., Song, J.Z., Li, S.L., Liu, X., Cai, Z.W., Fu, P.P., Lin, G., Xu, H.X. 2010. A new approach for simultaneous screening and quantification of toxic pyrrolizidine alkaloids in some potential pyrrolizidine alkaloid-containing plants by using ultra performance liquid chromatography-tandem quadrupole mass spectrometry. *Anal Chim. Acta*, 681:33–40.

Zhuang, J., Cheng, T., Gao, L., Luo, Y., Ren, Q., Lu, D., Tang, F., Ren, X., Yang, D., Feng, J., Zhu, J., Yan, X. 2010. Silica coating magnetic nanoparticle-based silver enhancement immunoassay for rapid electrical detection of ricin toxin. *Toxicon*, 55:145–152.

Index